AF443456

Advances in Fusion and Processing of Glass II

Related titles published by the American Ceramic Society:

Corrosion of Materials by Molten Glass (Ceramic Transactions Volume 78)
Edited by George Pecoraro, James C. Marra, and John T. Wenzel
© 1996, ISBN 1-57498-019-X

Ceramic Joining (Ceramic Transactions Volume 77)
Edited by Ivar E. Reimanis, Charles H. Henager, and Antoni P. Tomsia
© 1997, ISBN 1-57498-022-X

Fractography of Glasses and Ceramics III (Ceramics Transactions Volume 64)
Edited by J.P. Varner, V.D. Fréchette, and G.D. Quinn
© 1996, ISBN 1-57498-007-6

Experimental Techniques of Glass Science
Edited by Catherine J. Simmons and Osama El-Bayoumi
© 1993, ISBN 0-944904-58-0

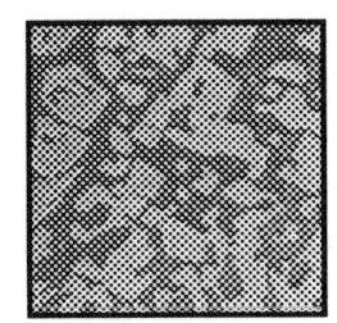

Ceramic Transactions
Volume 82

Advances in Fusion and Processing of Glass II

Edited by

Alexis G. Clare
Alfred University

Linda E. Jones
Alfred University

Published by
The American Ceramic Society
735 Ceramic Place
Westerville, Ohio 43081

Proceedings of the Fifth International Symposium on the Advances in Fusion and Processing of Glass held in Toronto, Canada, July 17-31, 1997.

For information on ordering titles published by the American Ceramic Society, or to request a publications catalog, please call 614-794-5890.

Printed in the United States of America.

1 2 3 4–01 00 99 98

ISSN 1042-1122
ISBN 1-57498-045-9

Dedication

These proceedings are dedicated to Oleg Mazurin in honor of his 70th birthday.

Photo by Margaret Rasmussen, New York State College of Ceramics at Alfred University.

Prof. Oleg Mazurin (left) with Prof. Phil Bray at Alfred University, 1990

Contents

ix

Advances in Materials for Glass Making

Surface Engineering/Coating

Advances in Glass Finishing

Preface

These are the Proceedings of the Fifth International Symposium on the Advances in Fusion and Processing of Glass held in Toronto Canada from 27th to 31st of July 1997. These conferences were convened by the Institute of Glass Science and Engineering and the NSF Industry-University Center for Glass Research (CGR) at Alfred University, Alfred, NY, USA. The first was at Alfred University in 1988, chaired by Professor L. David Pye, the then-Director of the CGR. Since then this conference has alternated between America and Germany as a result of the close association with the German Glass Industry (HGV). The second in the series was in Dusseldorf in 1990, the third in New Orleans in 1992 and the fourth in Wurzburg in 1995. The next conference will again return to Germany under the Chairmanship of our Keynote Speaker Professor Dr. Helmut A. Schaeffer. The fifth in the series was dedicated to Professor Oleg Mazurin's 70th Birthday, and was celebrated in great style at the Conference. Professor Mazurin's colleague Oleg Prokhorenko has provided a history of Professor Mazurin's achievements.

The editors would like to thank the conference Chairs without whom this Conference would not have been possible; Manoj Choudhary, Bob Condrate, Kathleen Richardson, George Pecoraro, Harry Russell, Ron Schroeder, Jim Shell, Jim Varner, Arun Varshneya, and Frank Woolley. We would also like to thank our sponsors; Corning Incorporated, The New York State College of Ceramics at Alfred University, Owens Corning and P.P.G.

The editors greatly appreciate our Conference Coordinator Mrs. Marlene Wightman without whom **nothing** would have happened. The key to a successful conference is to have good talks and good management, Marlene is the latter, the contributors were the former. Last but not least, the editors appreciate the skills of Mrs. Barbara Timbrook who's organization of the manuscripts enabled us to get a timely publication.

Alix Clare and
Linda E. Jones
November 1997

THE SCIENTIST WHO ENSURED THE EASY ACCESS TO INFORMATION
ON GLASS AND MELT PROPERTIES
On the occasion of 70th birthday of O.V. Mazurin

Oleg A. Prokhorenko
Thermex Company
24/2 Odoevskogo
St.Petersburg, 199155, Russia

SHORT BIOGRAPHY OF OLEG V. MAZURIN

Oleg Vsevolodovich Mazurin was born in city Nizhnii Novgorod (which is located on Volga river in the center of the European part of Russia) on July 21, 1927. In 1937 was moved to Leningrad - cultural and scientific center of the former Soviet Union (now St.Petersburg) where lives and actively works until now (except 1941-1944).

In 1950 Oleg V. Mazurin was graduated from Lensoviet Technological Institute one of the leading school and research center on apply chemistry and technology in former Soviet Union (where he studied in the Department of Glass Technology). In 1953 Oleg V. Mazurin received Ph.D. (Chemical Sciences) degree and began to work on the Department of Glass Technology as an Assistant Professor. During 10 years systematic research of electrical conductivity of glasses was performed by Oleg V. Mazurin and in 1963 he received the second scientific grade, Dr. of Science (Technical Sciences). In 1964 Oleg V. Mazurin was invited to the Institute of Silicate Chemistry of the Academy of Sciences of the USSR and was elected by the Scientific Counsel of the Institute on the position of the head of the Laboratory of Glass Properties which he occupies until now.

SCIENTIFIC INTERESTS OF OLEG V. MAZURIN

Oleg V. Mazurin is a good teacher. Working very much and fruitful Oleg Mazurin teaches his students and young colleagues to find most effective ways for the best result. He was a supervisor of more than 20 post-graduate students and due to that he received in 1973 the title of professor. In 1991 when Russian science was in extremely bad conditions he started to attract the best specialists of all of ages from different scientific organization of Saint-Petersburg to work in the cooperation with the Laboratory of Glass Properties for commercial projects. At present the laboratory headed by Oleg V. Mazurin having minimal support from the federal government has become a good training center for scientists who wish to find new ways to organize researches on glass properties in Russia.

The main fields of scientific interest of Oleg V. Mazurin are as follows: electrical and rheological properties of glasses and melts, thermal expansion, chemical durability, and absorption spectra. The field of his special interest is the phenomenon of structural relaxation in glass-forming melts and correspondingly changes in glass properties in glass transition region. In 70-s and 80-s Oleg V. Mazurin took active part in the development of programs for computer simulation of such changes and also of development and relaxation of stresses arising in the course of annealing and tempering of glass articles as well as in the metal-to-glass and glass-to-glass seals. Oleg V. Mazurin also paid great attention to the study of properties of phase separated glasses and to the use of these data for determination of some structural characteristics of phase separated glasses.

Oleg V. Mazurin is the author or co-author of more than 400 papers (more than 60 of them were published outside Soviet Union or Russia) and more than 15 books (monographs, textbooks, and handbooks). Two of the monographs were published in English ("Electrical Properties of Glass", published by Consultants Bureau in 1970 and "Phase Separation of Glass", published by North-Holland in 1984).

Most well known however are two handbooks prepared for publications by him, M.V.Streltsina, and T.P.Shvaiko-Shvaikovskaya. First handbook, Svoistva Stekol i Stekloobrazuyushchikh Rasplavov (Properties of Glasses and Glass-Forming Melts) was published by Publishing House "Nauka" in 1972-1997. It contains 9 books with nearly 5000 pages. The second one, Handbook of Glass Data was published by Elsevier in 1982-1993. It contains 5 volumes with more than 4000 pages. Both these books are popular enough.

The work of handbooks of glass properties was practically finished several years ago (the publications of the last two books for the Russian edition was postponed only due to the difficulties with funding). However, from the beginning of 1995 this work obtained quite modern prolongation. Oleg V. Mazurin offered the idea and developed the main principals of "electronic information system" (EIS) on glass properties. Generally it is next level in building of database because EIS is the program containing several databases working under common easy-to-use shell. American company SciVision (Lexington, MA) had accepted the proposal of Oleg V. Mazurin and his two co-authors of handbooks and started to help them in the development of commercial version of electronic information system on glass properties which was called SciGlass. The first version of this product was released in 1996. Among the customers who wished to order and buy the first version were a number of famous Universities and Glass-Manufacturing Companies worldwide. The second version of SciGlass will be released in September 1997. The databases of this Information System will contain data for various glass and melt properties of more than 85.000 oxide and halide compositions. A great amount of additional information about measurements of corresponding properties, the possibility of the use of methods of calculations, to develop graphs, to print and copy tables and graphs, and many other options make SciGlass an invaluable source of very useful information practically for any glass specialist. Thus SciGlass created under the leadership of Oleg V. Mazurin is the first flexible information system which makes access to data on glass properties collected by scientists during the long time as well as to data collected by a user (and loaded by himself into the SciGlass) easy. SciGlass combines the best features of good classical handbook and good computer database.

Oleg V. Mazurin was a member of the editorial board of the journal Fizika I Khimiya Stekla (the journal is published simultaneously in English under the title Glass Physics and Chemistry) since 1975 when the journal was founded. In 1991 he was appointed as the editor of this Journal and works in this capacity until now.

From 1972 Oleg V. Mazurin was a member of the Editorial Board of Journal of Non-Crystalline Solids. During several years he was a Regional Editor of this Journal.

Oleg V. Mazurin was a member of organizing committees (or advisory committees of many scientific conferences and meetings both in Soviet Union (or Russia) and in many other countries. In the course of organization of XV International Congress on Glass he was a chairman of the Program Committee of the Congress.

In 1994 he received a Morey Award from American Ceramic Society for "outstanding contribution into glass science."

CONCLUSION

In 1991 R&D department in THERMEX Company (which was organized by the initiative of the Institute of Silicate Chemistry) was established by Oleg V. Mazurin. As the director of the Research and Development Department he began to organize the work on high-quality measurements of glass and melts properties for various companies and research institutions. From the middle of 1993 the main customer for THERMEX Company was the Center of Glass Research at Alfred University (Alfred, State NY). Oleg V. Mazurin had invited excellent specialists from the Institute of Silicate Chemistry (as part-time employees) and some leading specialists from several other scientific institutions to work for THERMEX Company. By the by the company becomes to be one of the leading centers of measurements of glass and melt properties in Russia.

Fining Issues in Glass Processing

Simulation of Fining in Specialty Glass

Fritz W. Krämer
Schott Glaswerke
D-55014 Mainz

Abstract

Fining is one of the most important processes in the melting of specialty glass. To understand and improve this process simulations of fining outside the melting tanks must be done. Fining behavior of glasses can be studied in the laboratory with fining tests, including analysis of gases evolved from a batch, and determination of partial pressure of oxygen, an important analysis when the glass is fined with multivalent ions. These tests can determine the temperature at which fining will occur, what fining agent to use, and what raw material at what grain size distribution should be used.

In addition to this experimental data, mathematical simulation of processes in melting tanks is available. Temperature distribution, flow current velocities, and redox reactions and the fining process can be calculated if the thermodynamic data of redox reactions are known. The behavior of bubbles in the melt during fining can be simulated by combining tank models with a bubble change model. The influence of water on fining in an oxy-fuel fired tank is described.

Introduction

Glassmakers melted glasses empirically for thousands of years before Otto Schott and Ernst Abbe introduced science in glass manufacturing in Germany at the end of the 19th century. After World War I, major steps were taken to study the physics and chemistry of glassmaking. With the recent development of mathematical modeling, glass manufacturers have begun to gain better understanding of melting, fining and processing of glass. However, refining experts quickly realized the need for data on the physical and chemical properties of glass melts. Mathematical models can be used more and more successfully to improve product quality as more reliable input data becomes available.

Nevertheless, glass melting tanks are often still a black box with little known about what is happening. Simulations of the glass melt fining process, analysis of evolved gases (EGA) techniques, and oxygen fugacity measurements are essential. Large glass producers today are carrying out mathematical simulations of the fining process including calculations of the behavior of bubbles.

Fining/Refining

We understand by fining/refining the removal of gas bubbles from the melt and reduction of over-saturation of gases dissolved in the melt. Fining is understood as the gas evolving reactions which occur during temperature increase and refining are the gas dissolving reactions which occur during temperature decrease of the melt.

The basic mechanism of fining involves the rise of bubbles in the melt and diffusion of dissolved gases into bubbles according to the equations:

$$u \propto r^2 / \eta \text{ and } dr/dt \propto \Delta C$$

where u is the bubble rise velocity, r is the bubble radius, η is the viscosity of the melt, and ΔC is the concentration gradient of dissolved gases between bubble melt/interface and the concentration in the melt.

The basic equation of refining is:

$$dr/dt \propto \Delta C$$

which indicates that the gas exchange between bubbles and melt is the important factor.

Fining Prediction by Experimental Simulation

Because the fining behavior of a new glass should be known before it goes into the melting tank, the fining tests are among the most important tests.[1] To perform these tests, the batch with cullet and fining agent (total of about 500g) is melted in platinum crucibles in a gas or electric furnace, depending on what conditions are to be simulated, with a definite temperature program that simulates the melt at fining and refining. The melt is then cooled to ambient temperatures, a core is drilled out and slices are cut. The bubbles, or particles, in the sample slices are counted and classified by an image analyzing system as a measure of fining efficiency, or melting behavior. Figure 1 shows the result of a fining test on a TV glass melt at different concentrations of antimony oxide as the fining agent with 40 wt percent cullet and a fining time of five minutes at 1520° C. The bubble and stone counts decrease as the amount of antimony oxide increased, suggesting that fining and melting behavior improve with addition of a fining agent. Simultaneously, foam (due to release of fining gases) on top of the melt increased with increasing concentration of fining agents.

Another example is the influence of silica particle size distribution on the fining and melting behavior of a borosilicate glass melt. Sieving of the silica increases the fraction of the finer particle sizes. With the proper temperature program, decreasing silica particle size improves fining/refining and melting behavior. However, beyond an optimum size, further decreases lead to worsened fining/refining behavior and increased silica agglomeration.

Advances in Fusion and Processing of Glass II

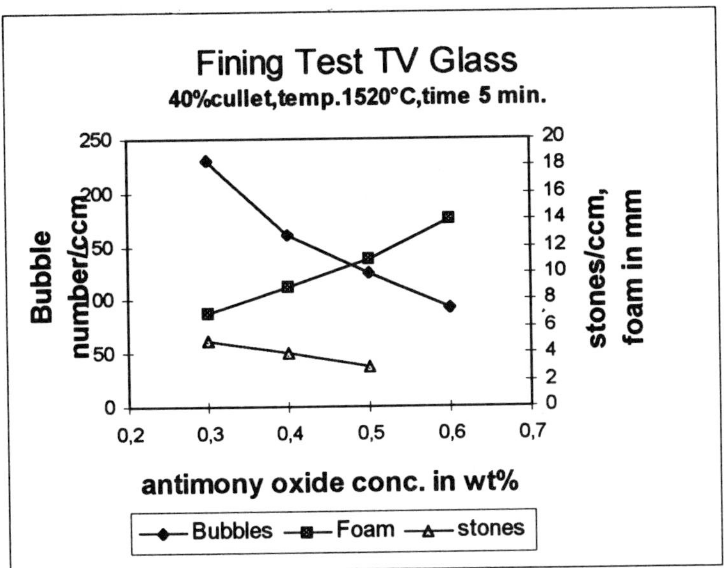

Figure 1 Fining Test with TV Glass

Analysis of Evolved Gases (EGA)

In evolved gas analytical techniques,[2] up to two grams of batch are heated in a carrier gas stream. Gases that are released during melting and fining are analyzed qualitatively and quantitatively, indicating at what temperatures the nitrates decompose or which fining agent works best in a certain temperature range. The amount of gas released can be used for mathematical modeling of fining.

Granulometry	Bubbles/cm^3	particles/cm^3	agglomerations/cm^3
coarest	2100	18	0
	1800	19	0.01
increasing	1500	4	0
finer	1470	6	0
flour	1650	4	0.03
	1950	8	0.1
finest	3000	18	0.8
	4750		2

Table 1
<u>Fining Test-Dependence on Sand Particle Size</u>
Borosilicate Glass made with silica sand with particle size distribution altered by sieving
Fining Temperature: 1570ºC; fining time: 3 h
Refining Temperature 1400ºC; refining time: 1 h

Oxygen Partial Pressure Measurements

Oxygen partial pressure is an important parameter in redox fined melts. The released oxygen diffuses into the existing bubbles, enlarging them and causing them to rise faster to the melt surface. Oxygen partial pressures near 1 bar are helpful for fining. Partial pressure can be measured in glass melts[3] using a ZrO_2 reference electrode and a platinum sensor. Figure 2 shows the oxygen partial pressure in TV glass melts fined with antimony oxide; the upper curve is for 100% batch, while the lower curve is reheated cullet melted previously at 1550° C. The oxygen partial pressure of the melt from batch is near 1 bar at 1200° C while oxygen partial pressure in the melt from cullet is some orders of magnitude lower. Thus fining is more efficient in melting batch than in melting cullet, a factor that must be considered in modeling a tank, which is normally fed with mixed batch and cullet.

Fining Prediction by Mathematical Simulation

Development of a tank model that describes all processes that occur in a glass melting tank would be helpful. A main model would deal with the glass flow, the temperature distribution and the geometry of the tank. Sub-models would describe combustion, energy balance, refractory attacks, batch blanket/melting kinetics, redox fining and refining, homogenization, particle tracing, and residence time of particles and bubbles, bubble behavior, fining and quality indices.

                    Advances in Fusion and Processing of Glass II

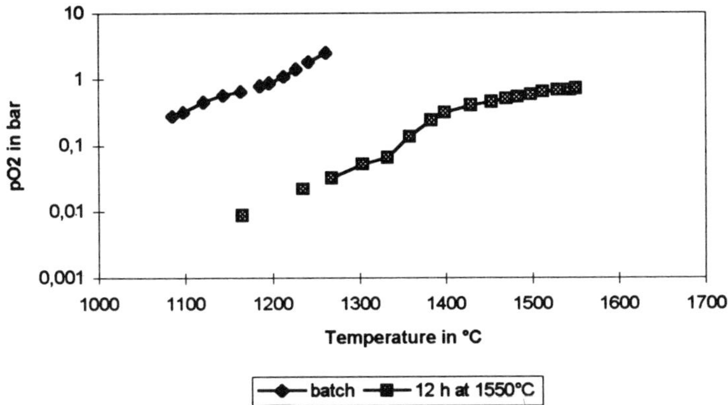

Figure 2 Oxygen partial pressure of TV glass melt fined with 0.65wt% Sb203

Schott Glaswerke uses the TNO model developed by the Institute of Applied Physics, Eindhoven, The Netherlands. The TNO model simulates the glass melting process by means of a 3D mathematical model,[4] which is based on Patankar's finite difference approach. Equations for energy conservation, continuity, and the Navier-Stokes equations of hydrodynamics are solved numerically with an implementation of boundary conditions and interactions of the molten glass with other phases. This allows calculation of temperatures and flow velocities in all three directions at every position in the melt.

Assuming that fining takes place at locations with high temperature and low velocities, this calculation shows the areas in the tank where fining and refining are most effective. But this can be more precisely calculated by a particle tracing sub-model where the trajectories of melting particles and bubbles can be traced. For instance, the bubble size can be calculated under certain conditions. Predictions can be made for when it will rise to the melt surface of the tank and whether it will reach the end product. Particles with different residence times in the tank can be traced. Most important are the particles with minimum residence time, the ones that often determine the output quality of a tank.

TNO has defined the fining index, a characteristic quantity that simulates the fining behavior of bubble trajectories, can be determined. Time t, the square of the difference between the melt temperature T and the starting time for fining (here 1623 K), and the dynamic viscosity η of the melt are integrated over each trajectory to obtain this index:

$$\text{Fining index} = \int_{o}^{t} t * (T - 1623)^2 / \eta$$

Figure 3 shows the fining index and the schematic temperature profile in the TV tank for a bubble on a minimum residence time trajectory. This indicates that effective fining occurs only in the refiner after the hot spot.

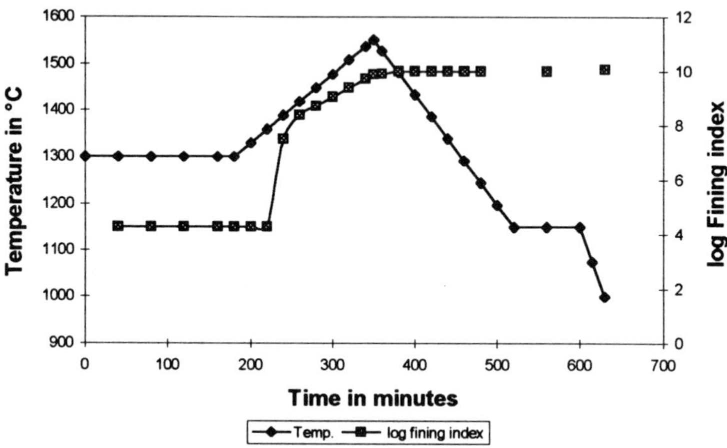

Figure 3 Fining Index of a Tank

Redox Simulation

In specialty glasses, which are often fined with multivalent oxides, redox reactions play an important role. The equilibrium constant K' of the redox reaction

$$M^{(m+n)+} + \frac{n}{2}O^{2-} \Leftrightarrow M^{m+} + \frac{n}{4}O_2$$

$$K' = \exp\left\{\frac{\Delta S}{R} - \frac{\Delta H}{RT}\right\} = \left(\frac{M^{m+}}{M^{(m+n)+}}\right)\left(\frac{p_{O_2}}{p_{O_2}^o}\right)^{\frac{n}{4}}$$

(where ΔS, ΔH, R and p°_{O2} is entropy, enthalpy, gas constant and standard oxygen partial pressure respectively) can be determined by square wave voltammetry[5] (Table 2)

The chemically dissolved oxygen due to redox reaction is

$$C_{O_2} = \frac{80\ n\ d\ c_M}{G_M}\left(\frac{M^{(m+n)+}}{M^{m+}}\right)\left(1 + \frac{M^{(m+n)+}}{M^{m+}}\right)$$

where c_M is the amount of fining agent and G_M is its molecular weight. The units of C_{O_2} are $kg/m^3 \cdot$ bar.

The physical solubility of oxygen (Table 3) can be estimated by comparison with inert gas solubilities.

The redox ratio of the multivalent ions (Sb, Ce, Fe, S), the total amount of dissolved oxygen, and the amount of fining oxygen evolved can be calculated if an open oxygen system during melting and fining is assumed, based on p_{O2} measurements and on the assumption $p_{O2}=1$ bar in the fining region.

During melt cooling, a closed oxygen system with interactions between the redox pairs can be assumed in the refining region:
$$\Sigma c_{O2\ chem} + c_{phys} = const$$
The redox ratio and its associated changes, i.e., color of the glass, can be calculated on the basis of p_{O2} measurements.

Reaction	Constant K´
$Sb^{5+} \Leftrightarrow Sb^{3+}$	$11.25 * 10^6 \exp(-23933/T)$ [6]
$Ce^{+} \Leftrightarrow Ce^{3+}$	$96 * \exp(-6130/T)$ [6]
$Fe^{3+} \Leftrightarrow Fe^{2+}$	$840 * \exp(-16353/T)$ [6]
$S^{6+} \Leftrightarrow S^{4+}$	$7 * 10^{10} \exp(-43640/T)$ [7]

Table 2 Redox equilibrium constants of multivalent ions in TV glass

Gas Concentration in a Melting Tank

Results of the modeling, which have been confirmed for oxygen in pot melts, can be transferred to real tank melts. The local gas concentration in each section of the melting tank can be calculated by solving numerically the mass balance equations for all gases that are dissolved in the melt, using known gas solubilities, gas diffusivities, actual gas concentrations, as well as the source terms. Bubble rise rate and the change in bubble composition with bubble growth or shrinkage can then be calculated for a gas bubble on any trajectory.

Gas	gas kinetic diameter in A	soda lime silica melt at 1400°C $mol/m^3.bar$
H_2O	2.27	0.076
Ar	2.99	0.026
O_2	3.02	0.025
NO	3.09	0.022
H_2S	3.18	0.020
N_2	3.22	0.019
CO	3.23	0.018
CO_2	3.45	0.013
SO_2	3.71	0.009
COS	3.73	0.009

Table 3 Physical gas solubilities estimated from inert gas solubilities[8]

Bubble Behavior

The behavior of bubbles in glass melts is governed by two processes: bubble removal by rise to the melt surface and change of bubble sizes by in-diffusing and out-diffusing gases. These processes are described by Hadamard/Rybczynski's law, Stoke's law, or Fick's law and by the relevant diffusion equations. Several computer models have been used to calculate bubble growth and dissolution.[9] Here the quasi stationary solution was used to calculate bubble behavior with the constants listed in Table 4.

Diffusion constants in m^2 / sec
O_2 : $0.52 * 10^{-2}$ exp(-26864/T)
N_2 : $0.94 * 10^{-4}$ exp(-20321/T)
CO_2 $0.26 * 10^{-4}$ exp(-20838/T)
H_2O : $0.59 * 10^{-7}$ exp(- 9850/ T)

Solubility in kg/ m^3
O_2 :   0.0288   exp (5422 / T)
N_2 : $0.74 * 10^{-2}$ exp (- 5372/ T)
CO_2 : $0.57 * 10^{-8}$ exp (22412 / T)
H_2O : 17.5   exp (-2250 / T)

Gas partial pressure in bar
O_2 : $0.29 * 10^{9}$ exp (-35584 / T)
N_2 :   0.8
CO_2 :   170   exp (-10170 / T)
H_2O : $0.33 * 10^{-2}$ exp (4500 / T)   (conv.)
H_2O : $0.74 * 10^{-2}$ exp (4500 / T)   (oxy-fuel)

Surface tension :   0.36 N / m
Dynamic viscosity: $1.787 * 10^{-4} *$ exp (23047 / T)

Table 4 Properties for Bubble Change Calculations

Figure 4 shows the fining efficiency of a bubble on a minimum residence time trajectory in a TV glass tank for different conditions. The rise height relative to the melt is shown for a CO_2 bubble initially 0.2mm diameter. The lowest curve with the lowest fining efficiency shows the bubble rise with no fining agent and no diffusion to or from the bubble, i.e. only physical fining. The second curve shows the fining efficiency (bubble rise) with antimony fining and diffusion of gases (except water) to and from the initial CO_2 bubble. The third and fourth curves show fining efficiency with chemical antimony fining in the presence of water and diffusion of gases. In an oxy-fuel fired tank (upper curve) fining efficiency is increased by 63 percent, compared to a conventional air-fuel tank (curve 3), due to the larger amount of water which diffuses into the bubble. The bubble contents for these cases are given in Table 5. With this or similar calculations, furnace design and melting and fining process conditions can be optimized to manufacture glass products without blisters and bubbles.

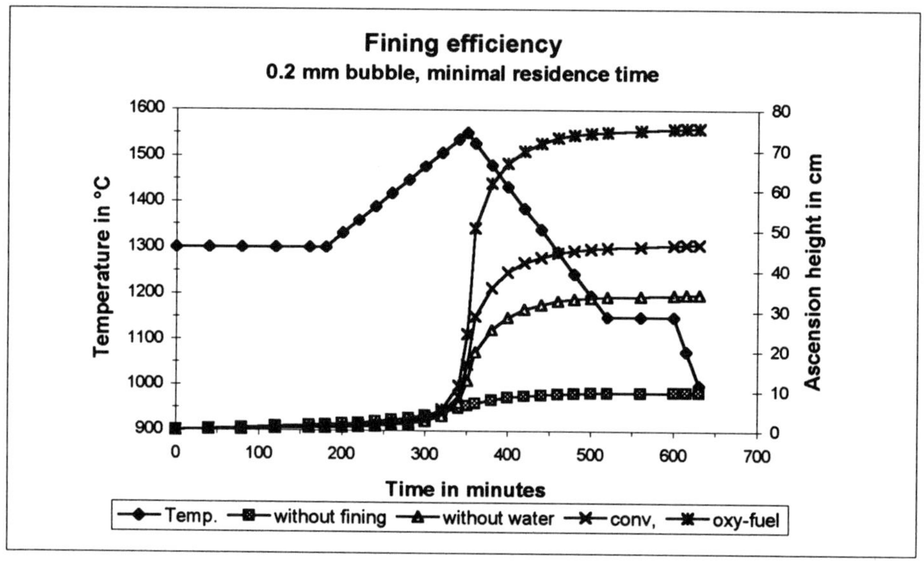

Figure 4 Fining efficiency of TV glass with and without water implementation.

condition	end diameter of the bubble in mm	N_2 in vol%	CO_2 in vol%	H_2O in vol%
without fining agent	0.2	0	100	0
with Sb_2O_3, without H_2O	0.38	78.7	21.3	0
with Sb_2O_3, with 384 ppm H_2O	0.43	68.5	20.7	10.8
with Sb_2O_3, with 541 ppm H_2O, oxy-fuel	0.51	56.5	19.2	24.3

Table 5 Bubble content and bubble size of initial 0.2 mm CO_2-bubble with minimum residence time in a TV tank

                                   Advances in Fusion and Processing of Glass II

References

[l] Mulfinger, H.O. and F. W. Krämer, "Analyses von Blasen und Gasen im Glas," in "Glastechnische Fabrikationsfehler," pp 162-179, 3rd ed. Edited by H. Jebsen-Marwedel and R. Brückner, "Springer Verlag, Berlin, Heidelberg, New York (1980).

[2] F. W. Krämer, "Gasprofilmessungen zur Bestimmung der Gasabgabe beim Glasschmelzprozess," Glastechn. Ber. 53 (1980) pp 177-188.

[3] F. G. K. Baucke, "Development of electrochemical cells employing oxide ceramics for measuring oxygen partial pressures in laboratory and technical glass melts," Glastechn. Ber. 56 K(1983) pp 307-312.

[4] F. Simonis; H. DeWaal; R.C.G. Beerkens, "Influence of furnace design and operation parameters on the residence time distribution of glass tanks, predicted by 3-D computer simulation," Coll. Papers XIV Int. Congr. Glass (1986) New Delhi, Vol. III, pp 118-127.

[5] C. Rüssel and E. Freude, "Voltammetric studies of the redox behaviour of various multivalent ions in soda-lime-silica glass melts," Physics and Chemistry of Glasses 30 (1989) pp 62-68.

[6] R. G. C. Beerkeens and M. van Kersbergen, TNO-Report: "Redox reactions and properties of gases in glass melts," Final Report NCNG-Novem, July 1996.

[7] Estimated from K. Papadopoulos, "The solubility of SO_3 in soda-lime silica melt," Physics and Chemistry of Glasses 14 (1973), pp 60-65 and M. H. Chopinet, I. J. Massol, J. L. Barton, "Factors determining the residual sulfate content of glass," Glastechn. Ber. 56K (1983) pp 596-601.

[8] F. W. Krämer, "Solubility of gases in glass melts," Ber. Bunsen Ges. Phys. Chem. 100 (1996), pp 1512-1514.

[9] F. W. Krämer, "Mathematical models of bubble growth and dissolution in glassmelts," in "Gas Bubbles in Glass" TC 14 Report, Int. Com. Glass, Ed. by TC1/1CG (1985).

THE REFINING OF FLOAT GLASS

Dr. Ian H. Smith
Pilkington Technology Management Ltd.

INTRODUCTION

The refining of Flat glasses has predominantly been carried out using either calcium or sodium sulphate and although the levels of sulphate added has changed significantly, the basic principles of refining flat glasses have been established for many years. It could easily be said therefore that this process was fully understood and little new could be added to the knowledge base. However the very competitive and demanding float glass market particularly for the automotive sector requires that float glass manufacturers face increasingly stringent bubble quality standards. We do not have the luxury therefore of sitting back on our laurels. Bubble quality improvement remains a high priority.

This paper is intended to review the challenges float glass manufacturers face in achieving the very high bubble quality standards demanded by the current flat glass market. Areas of progress will be highlighted along with areas where further detailed scientific work is required. It is hoped to demonstrate that a number of important questions remain unanswered to the level of confidence required by the float glass operator.

It is worthwhile spending a short time reviewing why sulphate?

The first reason would have to be the important role sulphate plays in the melting process. Along with carbon present in the raw materials or added to the batch, it considerably speeds up the melting process, particularly silica dissolution.

The second could be the lack of suitable alternative chemistry. It can be demonstrated that cerium oxide will refine float glass to the required quality, however it has a high cost (an important factor in a high volume market) and its use would leave the important question of silica digestion unresolved.

Arsenic and antimony are not compatible with the float bath as the reducing atmosphere results in a metallic bloom on the glass surface.

Halides also could not be used on the float glass scale due to the potential for volatilisation and pollution problems and the contamination of the float bath.

Recently physical methods of refining float glass (such as the use of vacuum) have been used with some success but economics remain a critical factor.

REFINING MECHANISMS

Having accepted the need for sulphate, let us examine the refining process in a little more detail. Refining requires bubble rise to the surface of the melt as the major bubble removal mechanism and although glass flows in a large float furnace are complex, rise to the surface is governed by the well known Stokes Law:

$$V_s = \frac{g\, d^2\, \rho}{180\, \eta}$$

Where

V_s	=	Velocity of bubble rise (cm s^{-1})
g	=	Acceleration due to gravity (cm s^{-2})
d	=	Bubble diameter (cm)
ρ	=	Density (g cm^{-3})
η	=	Glass viscosity (dyne s cm^{-2} , where dyne = g cm s^{-2}), (or poise)

The dependence of velocity of bubble rise on bubble diameter (d^2) at 1450°C can be illustrated in the following table :-

Bubble Diameter (mm)	Velocity (cm min^{-1})	Time to rise 25.4 cm
0.05	0.00152	280 hrs
0.1	0.00607	70 hrs
0.2	0.0243	18 hrs
0.3	0.0546	8 hrs
0.4	0.0970	4 hrs
0.5	0.152	3 hrs
0.6	0.218	2 hrs

The diffusion of SO_2 and O_2 into a bubble as the temperature of the glass is increased in the refining zone to approximately 1450°C dramatically increases bubble size and speeds rise to the surface.

The role of water in this process also cannot be underestimated and assumes increasing importance with the advent of oxyfuel firing.

Refining cannot be simply be said to occur in the refining stage of a furnace; a significant proportion of the released gas is removed prior to this stage.

Gases requiring refining:

Carbon Dioxide	-	from the carbonates in the raw materials
Nitrogen	-	from entrapped air
Water	-	from the raw materials

Sulphate has a high solubility in the alkali rich eutectic melts formed in the early stages of melting. As silica is incorporated into these, melts this solubility is reduced and SO_2 gas is released. This gas, along with the large volume of gases released during the melting process, can sweep up through the "melt" and clear significant numbers of bubbles to the surface. This can easily be demonstrated by the well known Guinness experiment, in which the mass of bubbles generated in the beer when poured can be seen to clear behind a moving boundary.

If this was an efficient process, float manufacturers would have a much easier life. Unfortunately this not the case as the process is inhibited by the presence of silica and "other measures" have to be taken.

Flows in a furnace are encouraged to rise to the surface in the spring zone or hot spot and in doing so pass through the temperature window for the decomposition of sulphate. This encourages diffusion of the refining gas into existing bubbles and their rapid rise to the surface.

$$SO_3 \iff SO_2 + 0.5\ O_2$$

Beyond this point, temperatures on a float furnace are decreasing and the decrease of bubble size by contraction of the gas and diffusion of soluble gases such as SO_2 and O_2 into the melt become the predominate refining mechanism.

Soluble Gases	Insoluble Gases *
Sulphur Dioxide	Nitrogen
Oxygen	Carbon Dioxide
Hydrogen	Argon
Water	Carbon Oxysulphide

* or gases with low solubility

Again in an ideal world, bubbles will contract to a critical bubble size when surface tension allows complete absorption into the melt. However gas diffusion across a bubble boundary is a two way process since insoluble gases also diffuse into the bubble. A small residual bubble is sometimes left which should be small enough not to be commercially significant, but is not completely removed.

BUBBLE SOURCES

Bubble sources in float glass manufacture can be broadly divided into the following categories:

Melting origin   -   Incomplete refining.

Contamination   -   Raw materials, normally well controlled but the potential for contamination of raw materials always exists particularly when long transport routes are involved.

                     -     Cullet, traditionally "in house" and almost assumed to be clean but facing increasing pressure to recycle more and more down stream cullet with the consequent need to implement raw material supply quality controls.

                     -     Furnace contamination, occasional contamination of glass during melting by refractory pieces particular during repairs.

Refractory sources	-	Corrosion of furnace side wall and bottom refractories is always a potential low level source of bubbles. However as bubble quality improves, the significance of this source is increasing.
		Transpiration through refractory pores and joints, potentially a significant source of bubble.
Devitrification	-	Under stable operation not normally a significant problem but can give rise to bubbles during changing temperature conditions particularly during tint manufacture.
Leaks	-	Leaking of water cooled equipment minimised by correct maintenance.
Reboil	-	Localised heating or temperature imbalance in the furnace provides another potential source of bubbles addressed by standard furnace control measures.

The identification of "outbreak" bubble sources is essential in maintaining high product yields on any float furnace. Pilkington has used a multifaceted approach to address this problem and I would like to review just two of them.

REFINING MODELS

The continuing development of 3D computer models to predict glass flows in furnaces and track bubbles back to their origin in the final glass ribbon is an important tool in identifying possible bubble sources.

The Pilkington model has been developed in house but like all computer models it is only as good as the input data. Examples of these are discussed below:-

Diffusivity and Solubility Data

Diffusivity and solubility data for the most important gases i.e. N_2, CO_2, O_2, SO_2 and H_2O still pose a problem for computer model's in the glass field. Although the situation is improving, only limited data is available from literature and often these values have to be estimated. This can perhaps be illustrated by the recent experience of TC 14 of the ICG who conducted a round robin on the solubility of gases in glass and could not find any reasonable agreement between the results from the participating laboratories although internal consistency was good. I think we can only look forward to taking advantage of the advances being made in this area.

Heat transfer across furnace side walls

Heat transfer across furnace refractory side walls is another important parameter which, if better characterised on a larger scale, would also improve the accuracy of these models.

Measurements of bubble growth/shrinkage rates

Bubble growth/shrinkage at high temperatures is being studied by Pilkington as an input to its 3D model. A number of workers have developed methods to observe bubbles in molten glass. However few appeared to meet the requirements of allowing a bubble of known composition to rise freely in a glass melt at high temperature. A method which allowed these criteria to be met was thus developed building on existing Pilkington methods. The apparatus designed to do this is shown in Figure 1 and consists of a Pentax LX camera, bellows, extension tubes and a 300mm lens mounted on an optical bench.

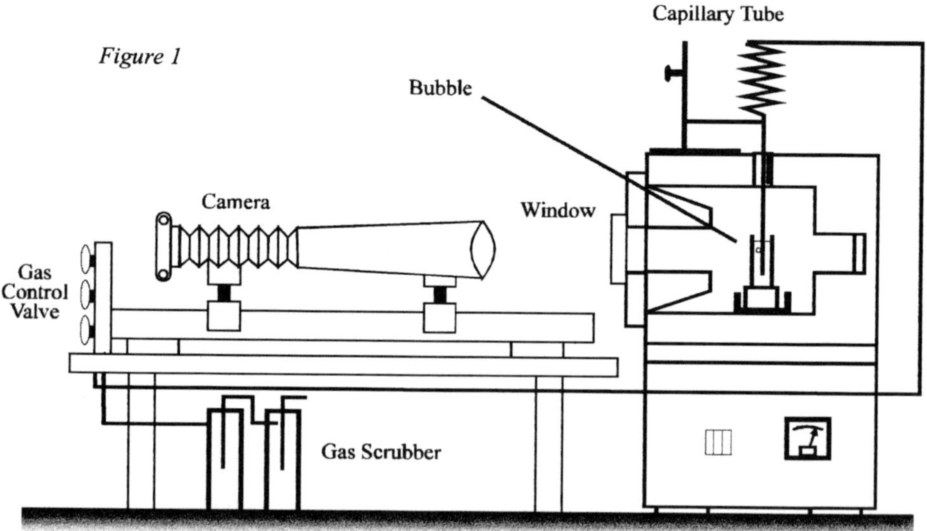

Advances in Fusion and Processing of Glass II

The bubble to be studied is introduced into a vitreous silica cell containing molten glass using a platinum capillary after carefully purging the line to ensure that the gas to be studied is of known composition. The cell has been specially designed with a flat window at the front to enable direct observation of the bubble without distortion and is maintained at the desired temperature using a furnace heated by silicon carbide elements.

Bubbles are photographed using a time lapse photographic system for time intervals between 10 and 240 seconds. From the photographic negative, bubble diameters can be measured and a growth or shrinkage rate determined.

A new vitreous cell is used every day and refilled with standard UK float glass. The cell is held at 1150°C overnight to remove unwanted bubbles before introducing the bubble of known composition.

The results of these measurements are shown in Figures 2, 3, 4, 5 and 6. Figures 2 and 4 show examples of plots of diameter vs. time for an individual nitrogen and carbon dioxide bubble respectively taken from the time lapse photography. A least squares linear regression is plotted for each line.

Figure 2

Linear Regression For Nitrogen Bubble

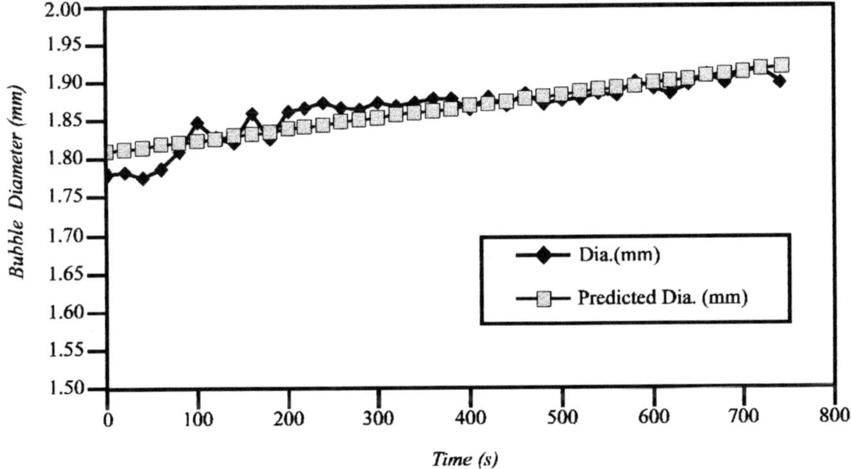

Growth Rates For Nitrogen Bubbles
1150°C Float Glass

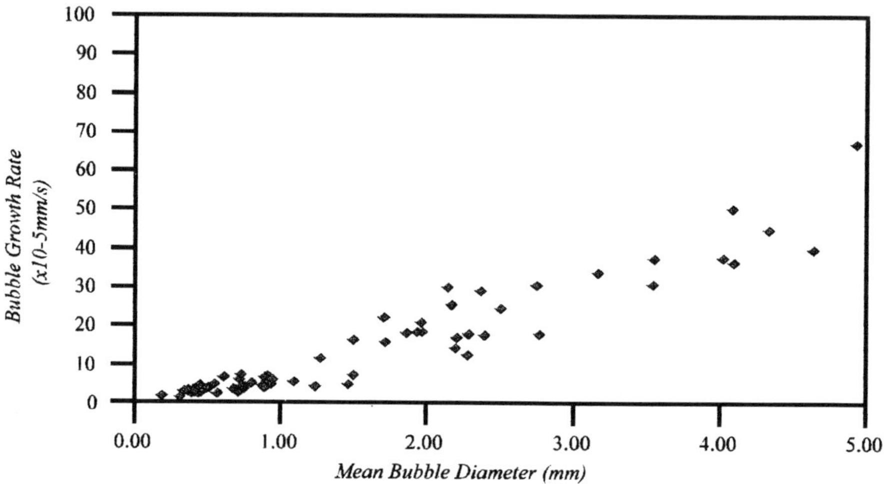

Linear Regression For
Carbon Dioxide Bubble

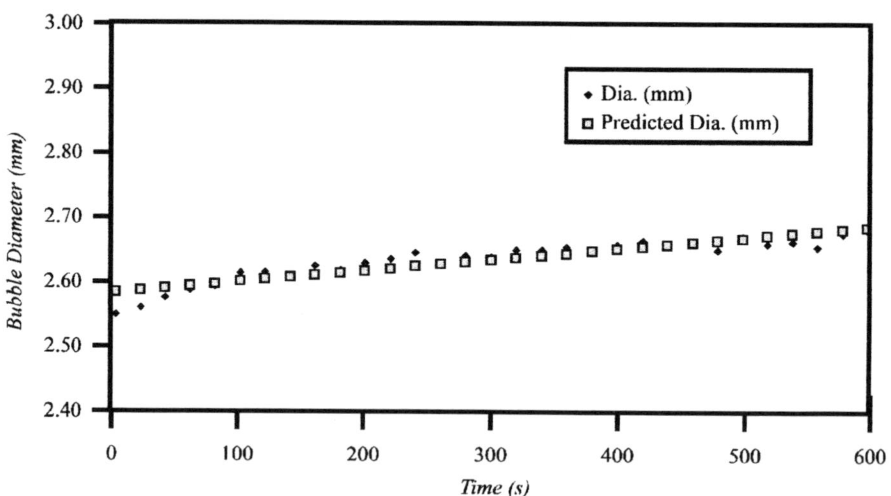

 Advances in Fusion and Processing of Glass II

Growth Rates For Carbon Dioxide Bubbles 1150°C Float Glass

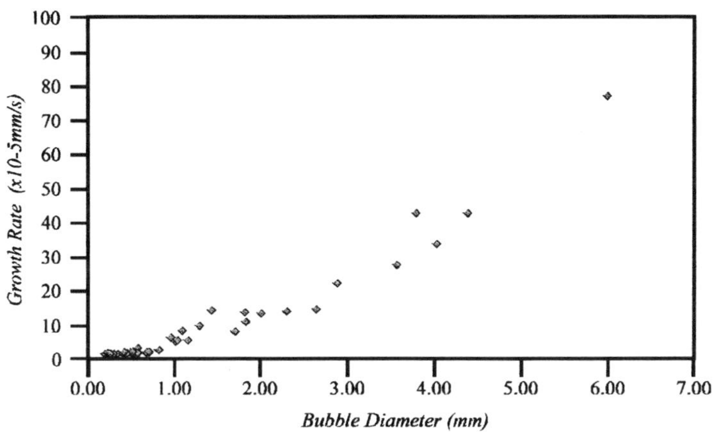

SO2 Bubble Shrinkage Rate at 1200°C

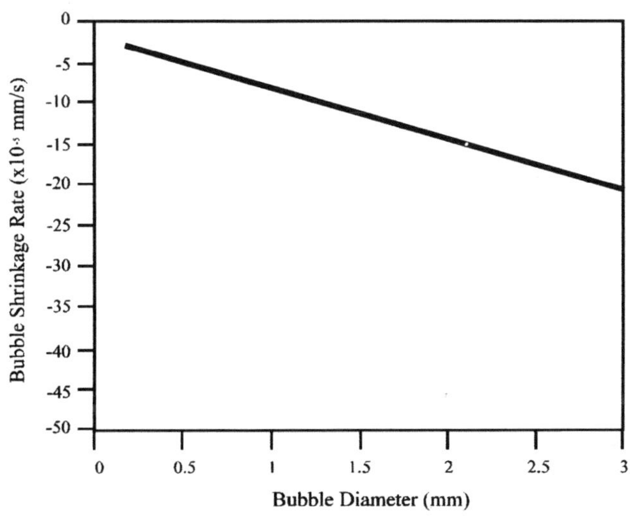

This linear regression enabled a bubble growth rate to be calculated from the slope of the line and a subsequent plot of bubble growth rate vs. bubble diameter for each of the gases to be made. The graphs (Figures 3 and 5) showed that for both gases, bubbles grow at a constant rate and that the bubble growth rate is dependant on bubble diameter.

Figure 6 shows the shrinkage rate of sulphur dioxide bubbles at 1200°C plotted against diameter.

This information on bubble growth/ shrinkage rates has been used to update our 3D computer model.

BUBBLE ANALYSIS

The analysis of the gas content of fault bubbles also provides valuable information in determining bubble source. Pilkington has used mass spectrometry for many years and has recently obtained a Balzers GIA 707 instrument.

Modern mass spectrometers are extremely sensitive and even for small bubbles (0.2mm-0.3mm in diameter) are opening up a new window for bubble sourcing. In addition to the bulk gas analysis, it is now possible to analyse trace quantities of other gases. These are found in ppm levels. We need to understand more thoroughly what these trace gases are indicating about bubble source.

An important parameter Pilkington has used is the estimation of the gas pressure inside a bubble. It can be demonstrated that the maximum theoretical pressure inside a bubble is approximately one third of one atmosphere. This can be reduced by the condensation of sulphur species on cooling from melting temperature. A simple conclusion therefore could be that bubbles with lower than one third of one atmosphere contain sulphur species gained by the bubble as it passed through the refining stage of the furnace.

Traditional methods of analysis of condensates in bubbles involves the use of hot stage microscopy and occasional analysis of the inside of a bubble surface using SEM/EDAX techniques. The use of SEM/EDAX techniques for bubbles is somewhat restricted by the tedious sample preparation required.

Recent advances in laser raman spectroscopic techniques however offer the possibility of "non destructive" analysis of the forms of sulphur seen inside bubbles, which could impact directly on bubble source diagnosis.

                              Advances in Fusion and Processing of Glass II

CONCLUSION

The understanding and application of sulphate refining to float glass has advanced immeasurably over the last 20 years. Significant work remains to be carried out.

ELECTRON EXCHANGE REACTIONS BETWEEN POLYVALENT ELEMENTS

H. Müller-Simon
Hüttentechnische Vereinigung der Deutschen Glasindustrie
Mendelssohnstrasse 75 - 77
60325 Frankfurt/Main, Germany

INTRODUCTION

The color of glass is mostly determined by polyvalent elements and their particular oxidation state. Also, the refining behavior of glass melts is coupled with the distribution of the different oxidation states of polyvalent elements at varying temperatures. In oxidic glasses a polyvalent element A reacts with physically dissolved oxygen according to

$$A^{x+} + \frac{n_A}{4} O_2 \underset{\leftarrow}{\rightarrow} A^{(x+n_A)+} + \frac{n_A}{2} O^{2-} \tag{1}$$

The redox state of A is determined by the oxygen partial pressure with which the melt is in equilibrium. This relation can be expressed by means of a reaction constant

$$K_A(T) = \frac{A^{x+}}{A^{(x+n_A)+}} p_{O_2}^{n_A/4} \tag{2}$$

If a second polyvalent element B is introduced into the melt, the oxygen partial pressure is the same for both elements and the redox ratios will adjust according to their particular reaction constant. However, during cooling the redox ratio of both elements does not remain constant, but the oxidation state shifts due to an electron exchange reaction according to [1].

$$n_B A^{x+} + n_A B^{(y+n_B)+} \underset{\leftarrow}{\rightarrow} n_B A^{(x+n_A)+} + n_A B^{y+} \tag{3}$$

REDOX SERIES

The electron exchange is enforced by the different temperature dependencies of the respective reaction constants. The temperature dependence of the reaction constant is given in the case of element A by the standard reaction enthalpy ΔH_A^o and the standard reaction entropy ΔS_A^o according to

$$K_A(T) = \exp\left(-\frac{\Delta H_A^o}{RT} + \frac{\Delta S_A^o}{R}\right) \tag{4}$$

Therefore, the reaction constant for the combined reaction 3 is given by

$$K_{AB}(T) = \left(\frac{A^{x+}}{A^{(x+n_A)+}}\right)^{n_B} \left(\frac{B^{(y+n_B)+}}{B^{y+}}\right)^{n_A} = K_A^{n_B} \cdot K_B^{-n_A}$$

$$= \exp\frac{n_A \Delta H_B^o - n_B \Delta H_A^o}{RT} \cdot \exp\frac{n_B \Delta S_A^o - n_A \Delta S_B^o}{R} \tag{5}$$

K_{AB} decreases with decreasing temperature if $n_A \Delta H_B^o - n_B \Delta H_A^o \langle 0$. In this case the concentrations of $A^{(x+n_A)+}$ and B^{y+} increase during cooling, i.e. B oxidizes A. Equation 5 provides a measure to predict the direction of the interaction of two polyvalent elements present in the same glass melt. With the aid of equation 5 and literature data of the standard reaction enthalpy[2,3], the following oxidation/reduction series is derived:

$$Cr^{3+}/Cr^{6+} \rangle Mn^{2+}/Mn^{3+} \rangle As^{3+}/As^{5+} \rangle Sb^{3+}/Sb^{5+} \rangle Ce^{3+}/Ce^{4+} \rangle Fe^{2+}/Fe^{3+}$$

where the element which stands more to the left oxidizes all the elements which stand right of it. The series above is in full agreement with experimental findings[4].

FINAL STATE OF THE ELECTRON EXCHANGE

If the standard reaction enthalpy and the standard reaction entropy is known, the amount of the polyvalent element in the oxidized state can be calculated by means of equation 1. Therefore, for given oxygen partial pressure and given total concentrations of different polyvalent elements, the total

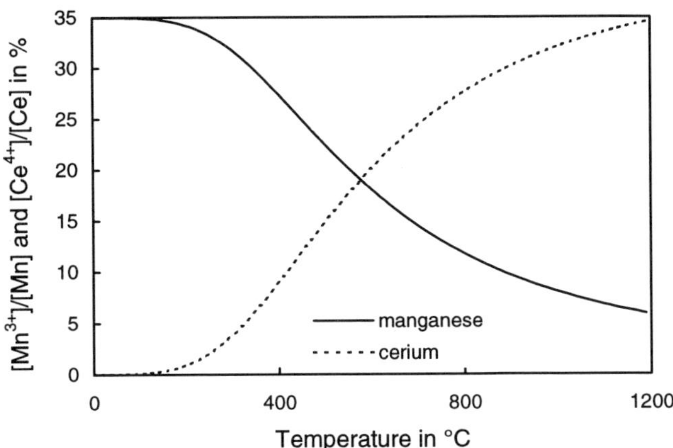

Figure 1: Dependence of the Ce^{4+} and the Mn^{3+} concentration on the temperature if simultaneously present in a sodium borate glass; molar concentration ratio Ce/Mn = 0.84.

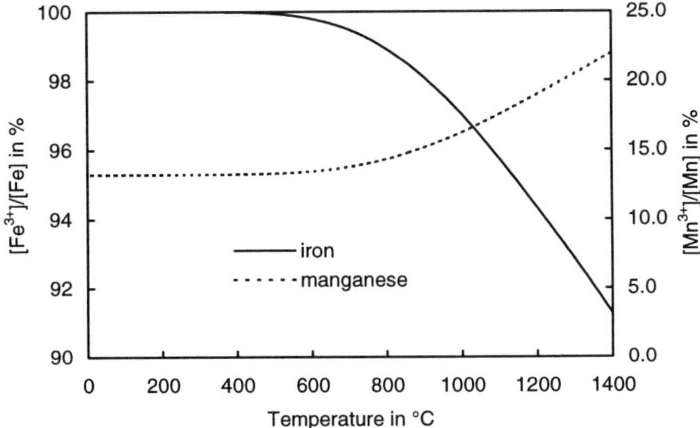

Figure 2: Dependence of the Fe^{3+} and the Mn^{3+} concentration on the temperature if simultaneously present in a soda-lime-silica glass equilibrated with air at 1400°C; molar concentration ratio Fe/Mn = 1.

chemically dissolved oxygen can be calculated. By means of the totally dissolved chemical oxygen the redox ratios of all polyvalent elements can be calculated for any temperature. Such calculations have been carried out for the manganese/cerium pair in a sodium borate glass (fig. 1) and the iron/manganese pair in a soda-lime-silica glass (fig. 2)[5]. The figures show that two different behaviors occur. In principle, the electron exchange can take place until one species is totally consumed. In the case of figure 1 the interaction has not come to an end when the transition temperature (420°C) is reached during cooling. In this case a freezing-in of the reaction can be observed, i.e. the redox ratios in the glass correspond not with the room temperature but with the transition temperature. In the case of iron and manganese in the soda-lime-silica glass the electron exchange reaction has finished above the transformation temperature. For instance, in the case of figure 2, Fe^{2+} has been totally consumed. In this particular case it is possible to calculate the shift of the redox ratios from stoichiometric considerations.

TEMPERATURE DEPENDENCE OF THE OXYGEN PARTIAL PRESSURE

The influence of the electron exchange during cooling appears also in the development of the oxygen partial pressure. In iron containing glasses the oxygen partial pressure is given by

$$p_{O_2} = \left(\frac{[Fe^{3+}]}{[Fe^{2+}]} \right)^4 \exp\left(-\frac{4\Delta H^o_{Fe}}{RT} + \frac{4\Delta S^o_{Fe}}{R} \right) \tag{6}$$

In a glass which contains only iron the development of the oxygen partial pressure is proportional to $\exp(-4\Delta H^o/RT)$. However, if manganese is present as a second polyvalent element, iron is oxidized during cooling (cf. fig. 2). Thus, the Fe^{2+}/Fe^{3+}-ratio increases and a higher oxygen partial pressure is obtained. Indeed, these relations are found in experiments as shown in figure 3[6].

CHROMIUM/IRON INTERACTION IN INDUSTRIAL CONTAINER GLASSES

The redox ratio of chromium has been reported as a function of the iron/chromium ratio equilibrated with air at different temperatures[3]. Figure 4 shows the results, which can be approximated by straight lines. The intercept of the ordinate is determined only by means of the reaction constant of the oxidizing agent according to

$$\frac{[Cr^{6+}]}{[Cr]} = \left(\frac{K_{Cr}(T)}{P_{O_2}^{3/4}} + 1 \right)^{-1} \tag{7}$$

At the intercept of the abscissa the reduced form of the oxidizing species just consumes the oxidized form of the reducing species, i.e. iron and chromium are present in the molar ratio of 1:3. Under these conditions

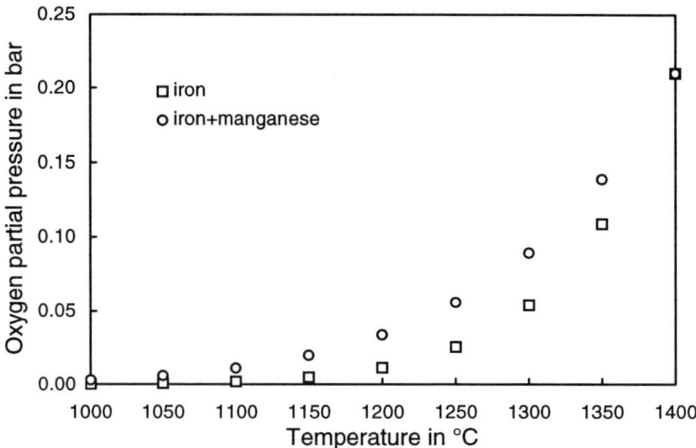

Figure 3: Temperature dependence of the oxygen partial pressure in a soda-lime-silica glass melt[6].

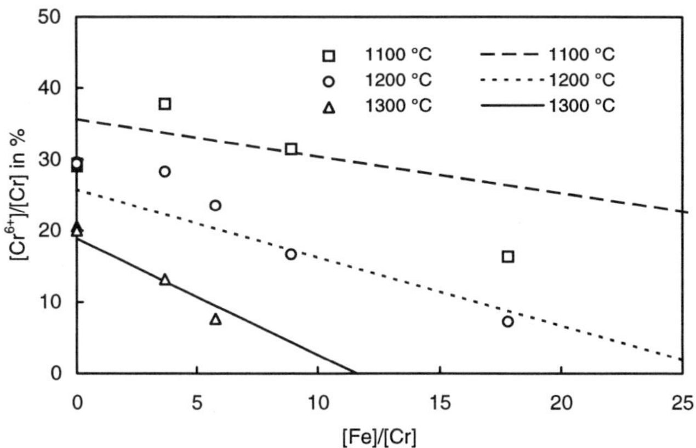

Figure 4: Electron exchange reaction between iron and chromium in a soda-lime-silica glass equilibrated with air at different temperatures.

$$\frac{[\text{Fe}]}{[\text{Cr}]} = 3 \left(\frac{p_{O_2}^{1/4}}{K_{\text{Fe}}} + 1 \right) \left(\frac{K_{\text{Cr}}(T)}{p_{O_2}^{3/4}} + 1 \right)^{-1} \tag{8}$$

proves the iron/chromium interaction. In packaging materials the allowed total amount of the heavy metals such as Pb, Hg, Cd and Cr^{6+} is limited due to governmental regulations. Because of its high toxicity, it is very important to

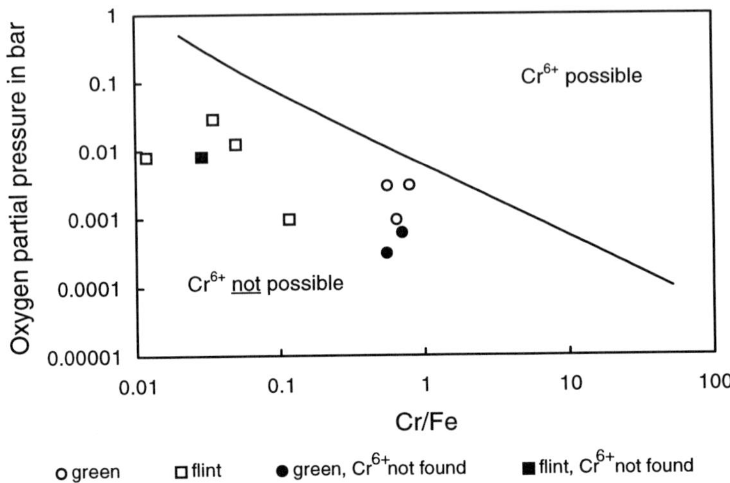

Figure 5:   Region of existence of hexavalent chromium depending on oxygen partial pressure and iron chromium ratio and examples of industrial green and flint container glass.

ensure the absence of Cr^{6+} even in chromium-colored glasses. For a given oxygen partial pressure equation 8 yields the iron/chromium ratio at which the hexavalent chromium vanishes. Figure 5 shows for some examples of industrial container glasses the position in the p_{O_2}-Cr/Fe diagram. Figure 5 also contains the boundary defined by equation 8. Obviously, all the samples lie in the region where no hexavalent chromium exists. The solid symbols mark results for which the absence of hexavalent chromium is checked by chemical analysis.

REFERENCES

[1]A. Paul, „Effect of Thermal Stabilization on Redox Equilibria and Colour of Glass", *J. Non-Cryst. Solids* **71** 269 – 278 (1985).

[2]C. Rüssel, „Online Measurements of Redox Properties in Glass Forming Melts", *GlassResarcher* **3** [1] 4-5 (1993).

[3]H.D. Schreiber; L.J. Peters; J.W. Beckman; C.W. Schreiber, „Redox Chemistry of Iron-Manganese and Iron-Chromium Interactions in Soda Lime Silicate Glass Melts", *Glastech. Ber. Glass Sci. Technol.* **69** [9] 269-277 (1996).

[4]C. Kühl; H. Rudow; W. Weyl, „Oxidations- und Reduktionsgewichte in Farbgläsern (Oxidation and Reduction Equilibria in Coloured Glasses)", *Sprechsaal* **71** [7] 91-93; [8] 104-106;[9] 117-118 (1938).

[5]H. Müller-Simon, „Electron Exchange Reactions between Polyvalent Elements in Soda-Lime-Silica and Sodium Borate Glass", *Glastech. Ber. Glass Sci. Technol.* **69** [12] 387-395 (1996).

[6]R. Brückner, „Properties and Structure of Glasses and Melts versus Preparation", p. 40 in *Glass: Science and Technology* Vol 4A. Edited by D. R. Uhlmann and N. J. Kreidl. Academic Press, Boston, 1990.

IN SITU CHRONOAMPEROMETRIC ANALYSES OF INDIVIDUAL MANGANESE REDOX STATES IN GLASS-FORMING MELTS

Michael W. Medlin, Karl D. Sienerth, and Henry D. Schreiber

Department of Chemistry
Virginia Military Institute
Lexington, VA 24450

ABSTRACT

Chronoamperometry was combined with square wave voltammetry for the *in situ* analyses of a series of manganese-containing glass melts at 1150°C. Glass melts containing 1 wt% Mn were titrated with varying amounts of iron. Analyses monitored the changing manganese redox ion distribution caused by electron exchange reactions with the iron redox ions. This electrochemical technique shows promise for the *in situ* measurement of individual redox states in glass melts.

INTRODUCTION

A target redox is usually specified during commercial glass production so that the resulting glass possesses desirable properties, for example color. Much effort has accordingly been expended in developing quantitative methods for the determination of individual redox state concentrations of multivalent elements in glass. Of crucial importance is the redox of iron, an indication of the percentage of iron as Fe^{2+} and is typically taken as a measure of the redox of the glass. Although there are numerous chemical as well as spectral methods for measuring the redox states in glass, such classical methods are done after the glass has been processed.

It would be advantageous to develop a sensor to monitor the redox of the glass batch during processing, that is *in situ* in a glass-forming melt. Thus, if the target redox is not being met, reductants or oxidants could be added on-line to adjust the redox. Oxygen activity sensors have been used to infer the redox states present in a glass-forming melt [1], while *in situ* square wave voltammetry has been sensitive to multivalent elements' concentrations albeit not specific redox states [2]. Additional information can also be gleaned from *in situ* studies, in particular an understanding of the melt's redox chemistry.

Electrochemistry has been shown to be a promising analytical tool for monitoring *in situ* the redox state of glass-forming melts [3]. The goal of this paper is the illustrate that a technique combining square-wave voltammetry (SWV) with chronoamperometry (CA) is capable of measuring individual redox states of manganese in a borosilicate glass-forming melt.

EXPERIMENTAL METHODS

The glass-forming system used in this study is an alkali borosilicate, previously identified as SRL-131 [4]. Its composition (in wt%) is 57.9 SiO_2, 14.7 B_2O_3, 17.7 Na_2O, 5.7 Li_2O, 1.0 TiO_2, 0.5 ZrO_2, 0.5 La_2O_3, and 2.0 MgO. A series of compositions was prepared containing 1.0 wt% Mn with added amounts of Fe varying from 0.25 to 4.0 wt%. Each composition was equilibrated in air at 1150°C within a Pt crucible for at least 16 hours.

For glasses containing just 1.0 wt% Mn under the experimental conditions, the manganese is present as 0.90 wt% Mn^{2+} and 0.10 wt% Mn^{3+} [4]. In analogous glasses containing just 1.0 wt% Fe, the iron exists as 0.97 wt% Fe^{3+} and 0.03 wt% Fe^{2+} [4]. If both manganese and iron are simultaneously present in the glass melt, a redox reaction occurs as shown by the equation [5]:

$$Fe^{2+} + Mn^{3+} \rightarrow Fe^{3+} + Mn^{2+}.$$

Stoichiometric considerations indicate that for an SRL-131 glass containing 1.0 wt% Mn, Mn^{3+} would be in excess of Fe^{2+} as long as there is less than 3.3 wt% Fe in the system. On the other hand, Fe^{2+} is in excess (meaning that Mn^{3+} is the limiting reactant) when there is greater than 3.3 wt% Fe.

The electrochemical system is sketched in figure 1. Its working electrode is a Pt wire, flame polished to a spherical tip with surface area of 0.25 cm^2. The quasi-reference electrode is simply a Pt wire. The counter electrode is the entire crucible because a heavy gauge Pt post contacts the crucible bottom, also providing a reproducible depth of the entire electrode assembly into the melt. A previous study [6] obtained the reduction potentials of multivalent elements in SRL-131 melts with respect to an oxygen reference electrode. The electrode is connected to an EG&G potentiostat.

SWV scans on each sample provided the positions (reduction potentials) of the Mn ($Mn^{3+} \rightarrow Mn^{2+}$) and Fe ($Fe^{3+} \rightarrow Fe^{2+}$) redox couples with respect to the quasi-reference, as shown on figure 2. Such positions were not greatly shifted from those obtained with the oxygen reference electrode. CA scans were done over a step of 0.6 volt centered on the corresponding SWV reduction potential. For example, the SWV signal for manganese is at 0.33 volt on figure 2; CA scans were then taken from 0.63 to 0.03 volt for the reduction of Mn^{3+} and from 0.03 to 0.63

volt for the oxidation of Mn^{2+}. For the CA scan, the initial voltage was applied for 5 sec, then jumped to the final voltage for 10 sec.

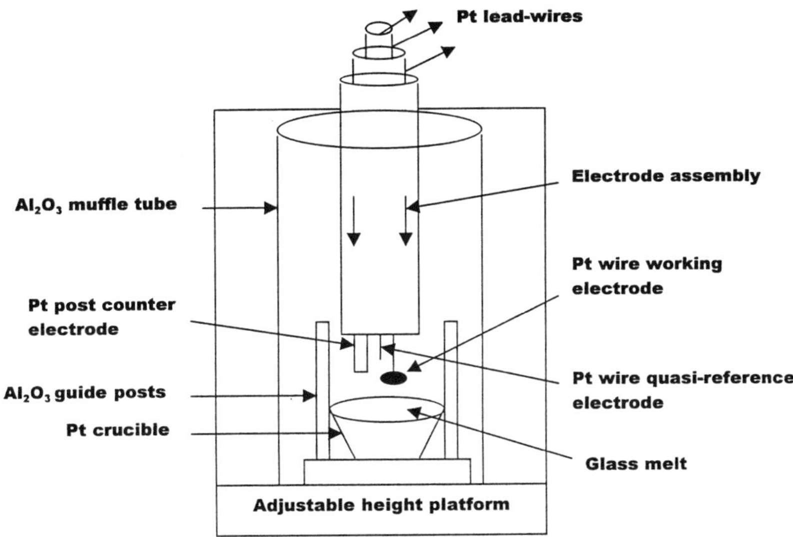

Figure 1. Schematic diagram of electrochemical system.

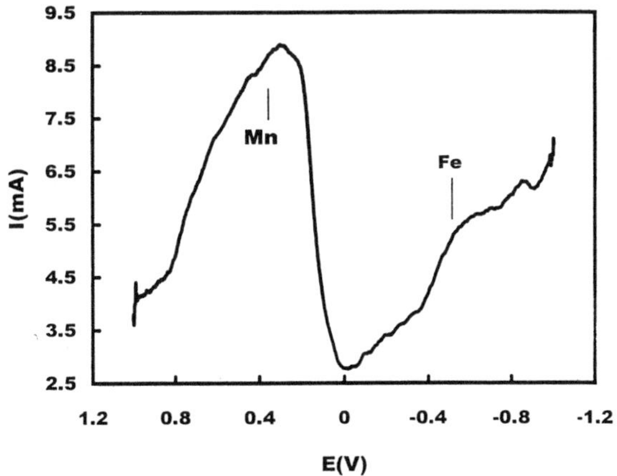

Figure 2. SWV scan at 1150°C of the glass melt containing 1 wt% Mn and 1 wt% Fe. Reduction potentials for each element are indicated.

Typical *in situ* CA scans for manganese are shown in figure 3. The initial spike is the charging of the double layer around the working electrode. The diffusion-controlled region was taken to be the CA signal from 5 to 10 sec. The concentration of the oxidized or reduced ion, depending on the direction of the scan, was calculated from:

$$C^* = [i(t\pi)^{1/2}]/[nFAD_o^{1/2}]$$

where $i(t)^{1/2}$ is the modified current-time plot integrated over the 5 to 10 sec time interval, n is one electron, F is Faraday's constant, A is the surface area, and D_o is the diffusion coefficient of the measured ion approximated as 9.8×10^{-6} cm^2/sec for both manganese ions and as 1.8×10^{-8} cm^2/sec for both iron ions [6]. The percentage of the element M in a specific redox state was then calculated from:

$$\%M_{red} = 100\% \times [C^*_{red}]/[C^*_{red} + C^*_{oxid}]$$

and analogously for the oxidized ion. Because this procedure only provides a relative measure of the redox ion percentages, it could then be calibrated to absolute concentration ratios.

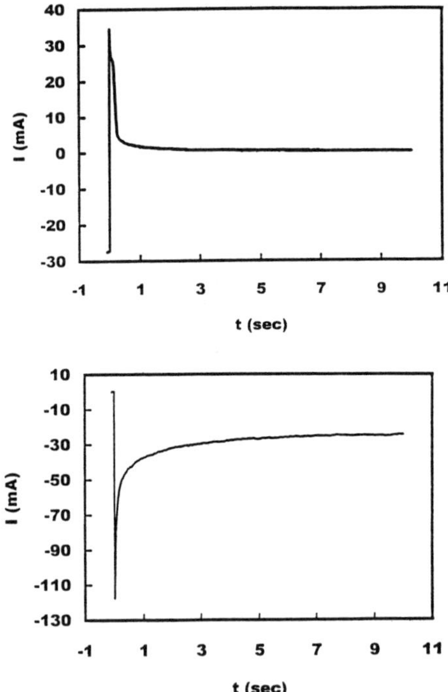

Figure 3. CA scans for manganese in the same system as figure 2. Top = reduction scan, bottom = oxidation scan.

Advances in Fusion and Processing of Glass II

RESULTS

Results obtained from this SWV-CA sensor are shown in table I and figure 4. As expected from a stoichiometric redox titration, the percentage of Mn as Mn^{3+} steadily decreases with increasing amounts of iron in the melt up to the addition of 3 wt% Fe. The CA is still registering appreciable current during the Mn^{3+} reduction scan when 3 wt% Fe is present. However, the current plummets to near zero when 4 wt% Fe is added; and negligible Mn^{3+} is present in the melt. The end-point for the redox titration was stoichiometrically calculated to be 3.3 wt% Fe for glass melts containing 1 wt% Mn.

Quantitative CA results for iron redox state distributions are not as good as for manganese. However, the CA scans for iron are noisier than for manganese, probably due to the weaker iron signal on a sloping background. Interpretation of the iron signal is also complicated because of clustering of the iron ions at this concentration level [3].

DISCUSSION

The dual SWV-CA analytical procedure can follow changes in the manganese redox state distribution *in situ* in a glass-forming melt at 1150°C. SWV is used to locate the exact reduction potential of the species being analyzed. CA scans in both the reduction and oxidation directions are then used to determine relative ion concentrations of the multivalent element. Thus, this procedure shows great promise for application as an *in situ* redox sensor.

That the manganese redox states followed stoichiometric reaction with the iron redox states *in situ* is evidence that the electron exchange reaction between the multivalent elements occurs at the melt temperature and not as a consequence of the quench [6]. Thus, additives may be included in the molten glass and, once equilibrium is established, their effects on the glass redox state can be measured during processing and before quenching to the final product.

Table 1. *In situ* SWV-CA analyses of glass melts containing Mn and Fe.

total concentration		Mn redox distribution		Fe redox distribution	
wt % Mn	wt% Fe	% Mn^{3+}	% Mn^{2+}	% Fe^{3+}	% Fe^{2+}
1.00	0.00	4.2	95.8	--	--
1.00	0.25	1.8	98.2	95.5	4.5
0.99	1.00	1.5	98.5	96.8	3.2
1.01	2.00	1.4	98.6	96.0	4.0
1.00	3.00	0.96	99.0	97.3	2.7
1.00	4.00	0.02	100.	97.6	2.4

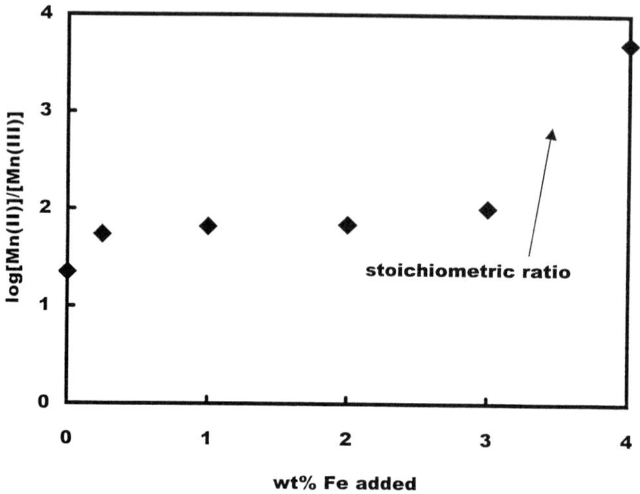

Figure 4. *In situ* redox titration at 1150°C in SRL-131 of 1 wt% Mn by iron.

ACKNOWLEDGMENT

This research has been supported by the NSF University-Industry Center for Glass Research centered at Alfred University.

REFERENCES

[1] F.G.K. Baucke, R.D. Werner, H. Müller-Simon, and K.W. Mergler, "Application of oxygen sensors in industrial glass melting tanks," *Glastech. Ber.* **69**, 57-63 (1996).

[2] O. Claussen and C. Rüssel, "Quantitative *in situ* Determination of Iron in a Soda-Lime-Silica Glass Melt with the Aid of Square-Wave Voltammetry," *Glastech. Ber.* **69**, 283-292 (1996).

[3] C. Rüssel, "On-Line Measurements of Redox Properties in Glass-Forming Melts," *Ceram. Trans.* **29**, 259-266 (1993).

[4] H.D. Schreiber, "An Electrochemical Series of Redox Couples in Silicate Melts -- A Review and Applications to Geochemistry," *Geophys. Res.* **70**, 9225-9232 (1987).

[5] H.D. Schreiber, L.J. Peters, J.W. Beckman, and C.W. Schreiber, "Redox Chemistry of Iron-Manganese and Iron-Chromium Interactions in Soda Lime Silicate Glass Melts," *Glastech. Ber.* **69**, 269-277 (1996).

[6] M..W. Medlin, K.D. Sienerth, and H.D. Schreiber, "Electrochemical Determination of Reduction Potentials in Glass-Forming Melts," *Glastech. Ber.*, submitted.

Advances in Fusion and Processing of Glass II

THE EFFECT OF WATER ON FINING, FOAMING AND REDOX OF SULFATE CONTAINING GLASS MELTS

Ruud Beerkens [1], Paul Laimböck[2], Sho Kobayashi [3]
TNO-Institute of Applied Physics Eindhoven, Netherlands [1]
Heraeus Electro Nite, Houthalen, Belgium [2]
Praxair Inc., Tarrytown NY, USA [3]

INTRODUCTION

For the same batch compositions remarkable differences in the redox state, sulfate retention and water concentrations in glass products from air fired compared to products from oxygen fired furnaces have been found. The redox state of the glass (melt) shifts, depending on the water vapor concentration, present in the furnace atmosphere and the shift depends on the redox number of the batch. Laboratory studies, using different simulated furnace atmospheres and also industrial observations show clearly the impact of water vapor on the colour of the final glass, on the sulfate fining process and on the water concentration in the glass product.

OBSERVATIONS AND PRACTICAL EXPERIENCES OF MELTING IN WATER RICH ATMOSPHERES

- In a water vapor rich (as in oxygen-gas or oxy-fuel oil fired furnaces) atmospheres, the water concentration in the molten glass increases during the melting and fining and reaches a value about proportional to the square root of the water vapor pressure above the melt.
- The water vapor from the furnace atmosphere infiltrates into the melt mainly during the <u>primary melting stages</u> and the higher temperature levels during the <u>course of fining and foaming</u>. An exchange of water with the furnace atmosphere to or from a calm melt (no violent movement at the surface) is very limited.
- At the same time, the redox state (Fe^{3+} /Fe^{2+}) of the glass decreases as the water vapor pressure increases for a soda lime silica <u>batch without reducing components.</u>
- For batch <u>with reducing</u> components, water vapor seems have an <u>oxidizing</u> effect during the melting and/or fining process.
- A higher water vapor pressure in the furnace often will increase the formation of foam due to an increased release of fining gases.
- A water vapor rich atmosphere improves the sulfate fining: the extra amount of dissolving water enhances the removal of bubbles.

In this paper, results of laboratory simulation tests and results of model calculations, in order to understand the impact of dissolved water in the glass melt on the sulfate decomposition, the total quantities of released fining gases and the redox state are presented.

LABORATORY EXPERIMENTS

Soda-lime silica batches have been heated in vitreous silica crucibles up to 1500 °C. During this heating and during the fusion and melting, the furnace atmosphere is controlled, experiments with different water vapor levels and oxygen concentrations have been performed. The water concentration in the glass product after this melting procedure has been analysed by infra-red spectroscopy [1]. The residual sulfur concentrations (usually expressed in wt-% SO_3) in the glass products are determined wet-chemically. The water concentrations and residual sulfur concentrations are determined as a function of the water vapor pressure in the furnace atmosphere, the glass composition and the redox state of the batches used. Also the maximum temperature during the fining process has an important effect on the sulfate (sulfur) retention. Oxygen activity measurements in flint and float glass, molten in dry and wet atmospheres show the reduction the of the melt (oxygen activity decreases by factor 2-3) in the wet atmosphere. For more reducing (green and amber) glasses, the oxidation state shifts in the other direction, then water increases the redox ratio (Fe^{3+}/Fe^{2+}).

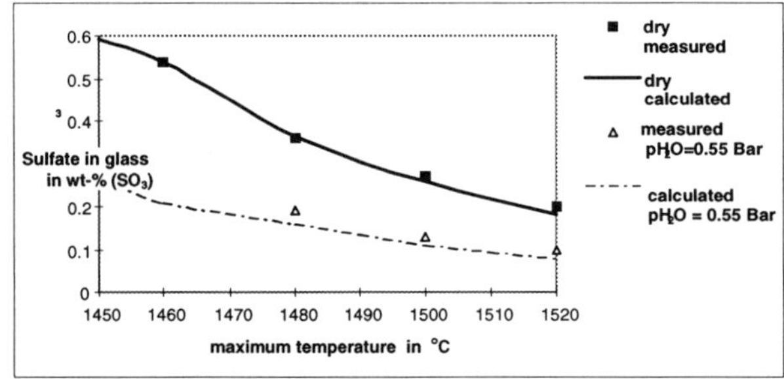

Figure 1 Sulfate concentration in sulfate refined soda-lime glass (no reducing agents in batch) measured and calculated as a function of max. melting temperature in a dry and humid atmosphere

Figure 1 shows the measured sulfate concentration for a soda-lime-silica glass molten at different temperatures without addition of reducing agents, the atmospheres contain no water vapor or 55 vol-% water vapor, similar to oxy-gas fired furnaces.

                                    Advances in Fusion and Processing of Glass II

EFFECT OF GLASS COMPOSITION

Papadopoulos [2] investigated the effects of glass composition and temperature on the sulfate decomposition and sulfate retention in soda-lime glasses without reducing agents. Especially the sodium concentration but also the overall basicity of the melt determine the sulfate retention after melting. For each weight-percent of extra sodium oxide (at a level of about 15 wt-%) the sulfate retention increases by about 25 %. By means of the calculation of the so called J_c factor, determined from the glass composition [2], the sulfate residue in oxidized molten soda lime glasses can be estimated for melting in dry atmospheres:

$$[SO_3] = 1/K(T) \cdot (J_c/ J_{co}) \cdot pSO_2 \cdot \sqrt{pO_2} \qquad [1]$$

$[SO_3]$ = SO_3-concentration in glass (mol.m^{-3} glass melt)
pSO_2 /pO_2 = vapor pressure (bar) of SO_2 or O_2 in the bubbles in the melt
J_c = factor for correcting the equilibrium constant $K(T)$ for a change in glass composition compared to a reference glass.
The reference glass has the composition in mol-% $Na_2O:CaO:SiO_2 =$ 15:10:75 (in mass-% 15.5 : 13.5 : 75.15) and the J_c -value for this standard composition is: $J_{co}= 0.018$;   T is given in Kelvin
$K(T)$ = equilibrium constant $=1.36 \; 10^{10} \exp (-50635/T)$ (mol.m^{-3}. Bar$^{-3/2}$)

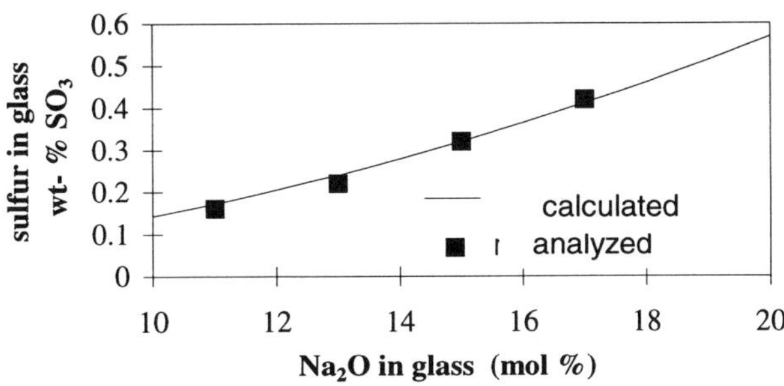

Figure 2   Sulfate concentration in glass, molten in a dry atmosphere at 1480 °C, parameter sodium oxide concentration in the glass : $Na_2O:CaO:SiO_2$ ratio 15-X , 10, 75-X   mol-%

The agreement between the model and the laboratory experiments is well demonstrated in figures 2, for sodium oxide concentrations between 10 and 17 mol-% . Other glass compositions show a good agreement for different alumina concentrations up to about 2-3 wt-%. However, small deviations at larger alumina concentrations have been observed.

The tests show that the values of the vapor pressures of the releasing fining gases in the melt: pO_2 and pSO_2 depend very much on the redox state, the water concentration in the melt and the temperature of the glass melt.

MODELING OF SULFATE BEHAVIOUR

Table 1 shows a survey of estimated thermodynamic values for relevant reactions in the glass melt during the fining process and also important for redox changes during melting. A model [3] has been developed, based on: these data, the roughly estimated SO_2 solubility (0.2 mol/(m^3.bar)) and physical O_2 (about 0.05 mol/(m^3.bar)) solubilities and also based on mass conservation equations in the system glass melt/dissolved gases, for the chemical elements: S, H, C, free oxygen and polyvalent ions.

TABLE I. Redox and Sulphate Reactions in Glass Melts

reaction equilibria (pressure in bar und concentrations in mol/m^3)	estimated enthalpy change ΔH_r^0 (KJ/mol)	estimated entropy change ΔS_r^0 (J/mol/K)	Literature
$SO_4^{2-}(l) \Leftrightarrow SO_3(g) + O^{2-}(l)$	325	103	3
$SO_3(g) \Leftrightarrow SO_2(g) + \frac{1}{2}O_2(g)$	96.3	90.9	4
$SO_2(g) \Leftrightarrow \frac{1}{2}S_2(g) + O_2(g)$	361	72.5	4
$\frac{1}{2}S_2(g) + O^{2-}(l) \Leftrightarrow S^{2-}(l) + \frac{1}{2}O_2(g)$	379	198	5
$Fe^{3+}(l) + \frac{1}{2}O^{2-}(l) \Leftrightarrow Fe^{2+}(l) + \frac{1}{4}O_2(g)$	102	33	6
$M^{(x+k)+}(l) + \frac{n}{2}O^{2-}(l) \Leftrightarrow M^{x+}(l) + \frac{k}{4}O_2(g)$	-	-	6
$\frac{1}{2}H_2O(g) + \frac{1}{2}O^{2-}(l) \Leftrightarrow OH^-$	0	46.3	7
$H_2O(g) \Leftrightarrow H_2(g) + \frac{1}{2}O_2(g)$	248	56.6	6
$H_2S(g) \Leftrightarrow H_2(g) + \frac{1}{2}S_2(g)$	89.2	48.7	4

By this model the composition of released gases during the fining process, the total quantities of released gases from the melt, the redox state (oxygen activity) and the sulfate/sulfide concentration can be calculated, depending on glass composition, furnace atmosphere and temperatures.

   Advances in Fusion and Processing of Glass II

The model assumes that the total pressure of dissolved gases in the melt cannot exceed a value of 1.2 bar, above this value the gases will be released from the melt as fining gases. The water concentrations in the laboratory melts are about in equilibrium with the water vapor in the furnace atmosphere. Under equilibrium conditions, $[H_2O]^e$ in the melt (in mg/kg) is about $[H_2O]^e = 1050 \times (pH_2O)^{1/2}$, (pH_2O in bar) for most commercial soda-lime glass compositions at about 1400-1500 ^{o}C.

However, water concentration analyses of glass products from industrial furnaces show that the water concentration reaches only about 60-70 % of the saturation value. Thus, in practice $[H_2O]^e = \pm 700 (pH_2O)^{1/2}$.

A higher water vapor pressure in the furnace atmosphere will result in a higher water concentration in the melt. A bubble in the glass melt will absorb a certain gas from the melt as the equilibrium pressure of this gas species (CO_2, SO_2, O_2, H_2O, N_2) dissolved in the melt exceeds the vapor pressure of this gas in the bubble. Water, as a relatively fast diffusing gas in melt, rapidly dilutes the bubble gas contents and in this way enhances bubble growth. The dilution also decreases the concentration of the fining gases in the bubble which will increase the driving force for the transfer of the fining gases into the bubbles. During fining, the vapor pressures of nitrogen and carbon dioxide become very small due to the dilution by the fining gases and water:

$$pO_2 + pSO_2 + pN_2 + pCO_2 + p\,H_2O = \pm 1 \text{ bar} \qquad [2]$$
$$pO_2 + pSO_2 \approx 1 \text{ bar} - p\,H_2O \qquad [3]$$

In a water rich glass melt, the pH_2O-value is very high and the total pressure of 1-1.2 bar for all dissolved gases together, is already reached for a relatively moderate decomposition of the sulfates. Then, the required value of $pSO_2 + pO_2$ is much less than 1 bar. These conditions will be reached at much lower temperatures or for lower sulfate concentrations in water rich melts compared to water lean glass melts. This means that dilution of the bubbles by water vapor will decrease the fining onset temperatures and will enhance the fining process. The sulfate addition to the batch can be decreased. Because of the lower vapor pressure of SO_2 and O_2 in the bubbles, due to this water vapor dilution effect, the equilibrium (residual) sulfate concentration in the melt is reduced compared to dryer glass melts. Thus, in oxygen fired furnaces the sulfate retention (at least for glass molten from batches with hardly any reducing agents) will be less compared to the sulfate residue in glass from air fired furnaces.

MODELING RESULTS

Figure 3 shows some more experimental results for sulfate retention in glass compared with the model (parameter is the water vapor pressure).

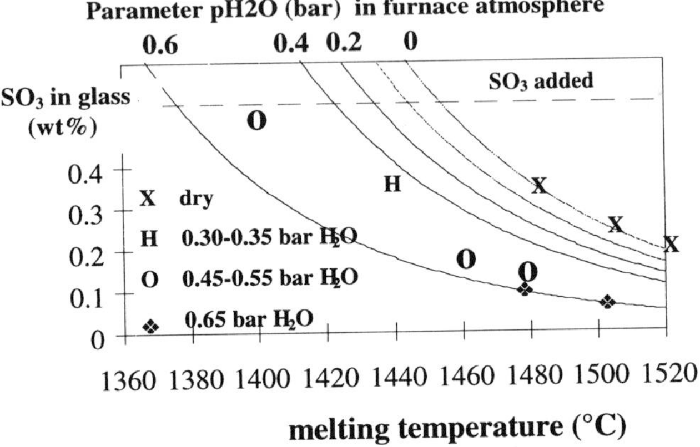

Figure 3. Sulfate retention in soda-lime glass (no reducing agent in the batch), the curves show the results from the modeling studies

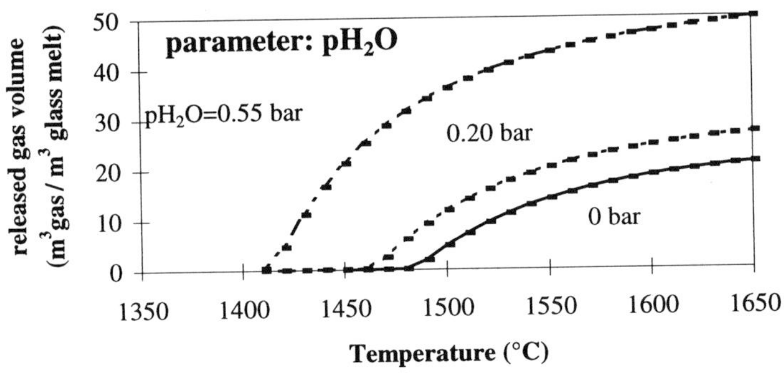

Figure 4.   Calculated quantities of released gases during the sulfate fining of soda-lime glass melt (no reducing agents in the batch)

The agreement between the model and the experiments so far, are relatively good. The sulfate decomposition in a melt <u>without reducing agents</u>, will start at an about 70 °C lower temperature compared to the situation of a dry atmosphere and dry melt. After melting at a maximum temperature of 1500 °C, the sulfate retention in the case of $pH_2O=0.55$ bar, has decreased by about 50 % compared to the value for the air fired ($pH_2O= 0.20$ bar) case. Figure 4 shows the total volume of gases released during the fining process for the three different situations: a dry, air fired and oxygen fired case. For the same amount of added sulfate to the batch, much more fining gases will be released in the 'wet' case.

CONCLUSIONS AND RECOMMENDATIONS

First experimental data show a very good agreement of sulfate retentions determined by a model, based on mass conservation laws and thermodynamics. The model has been applied to predict the effect of changes in batch/glass composition, melting temperature, furnace atmosphere and redox state of the batch on the sulfate retention, redox state of the melt & glass product and the total amount of released fining gases. A water rich atmosphere enhances the sulfate fining but will increase foaming in the fining area. However, often the improved sulfate fining efficiency allows the reduction of the sulfate addition to the batch in oxygen-fired furnaces. This reduction will reduce the foaming caused by the sulfate decomposition and also will lower the SO_2 emissions of the glass furnace. Modifications in the batch, like reduction of the sulfate addition and corrections of the redox seem to be necessary after converting air fired furnaces to oxygen firing.

REFERENCES

[1]Scholze H.; Der Einbau des Wassers in Gläsern. Teil 1: Der Einfluß des in Glas gelösten Wassers auf das Ultrarot-Spektrum und die quantitative ultrarot spektroskopische Bestimmung des Wassers in Gläsern, *Glastech. Ber.* 32 (1959) p. 81-88

[2]Papadopoulos K.; The solubility of SO_3 in soda-lime-silica melts, *Physics. Chem. Glasses* 14 (1973) no. 3 p. 60-65

[3]Laimböck P.R.; Foaming of Molten Glasses, *Ph-D thesis Eindhoven University of Technology* 1997

[4]Barin I.; Thermodynamic data of pure substances, *VerlaggesellschaftmbH*, Weinheim, Germany 1989

[5]Nagashima S., Katsura T.; The solubility of sulfur in Na_2O-SiO_2 melts under various oxygen partial pressures at 1100 °C, 1250 °C, 1300 °C, *Bull. Chem. Soc. Jap.* 46 (1973), p. 3099-3103

[6]Rüssel C., Freude E.; Voltammetric studies of the redox behaviour of multivalent ions in soda-lime-silica glass melts, Physics Chem. Glasses 30, (1989), no. 2, p. 62-68

[7]Franz H., Scholze H.; Die Löslichkeit von H_2O-Dampf in Glasschmelzen verschiedener Basizität, *Glastechn. Ber. 36*, (1963), 9, p. 347-356

keywords: sulphate, water in glass, fining, foaming, redox

CARBON DIOXIDE SOLUBILITY IN SILICATE MELTS

James E. Shelby
New York State College of Ceramics
Alfred University
Alfred, NY 14802

The existing data for the solubility of carbon dioxide in silicate melts is reviewed. The effects of melt composition, temperature, and pressure of carbon dioxide on the solubility are discussed. Measurement of solubility by mass spectroscopy and infrared analysis are also discussed. Models for the dissolution of carbon dioxide are presented and evaluated. In particular, the speciation of carbon dioxide into physically dissolved molecular form and chemically dissolved carbonate ions is considered as a function of the basicity of the melt.

INTRODUCTION

Dissolved gases are found in all viscous melts [1]. Exosolution of these gases can lead to result ranging from small bubbles, which are considered defects in commercial glasses, to violent eruptions in volcanic melts. These gases can enter the melt by entrapment of atmospheric gases during melt formation, by reaction with the atmosphere surrounding the melt, or from the decomposition of batch materials used to form the melt.

Carbon dioxide is particularly common as a dissolved gas in silicate melts. Commercial melts are routinely produced from raw materials such as sodium carbonate, limestone, dolomite, and other carbonates, which decompose to release carbon dioxide during the early stages of the melting process [2]. Furnace atmospheres are rich in carbon dioxide produced by the combustion of natural gas or oil. Geologists are also interested in carbon dioxide, which is one of the two most common volatile components of magmas [3]. Release of carbon dioxide can be explosive during major eruptions and serves as a significant source of carbon dioxide in the earth's atmosphere [1].

MODELS FOR CARBON DIOXIDE SOLUTION

Carbon dioxide can reside in melts as either a physically dissolved molecule or a chemically bonded carbonate ion [1,2]. The ratio of the molecular to the ionic form is termed "speciation", i.e. the division of the gas into different species during the dissolution process [4]. The relationship between melt composition and the speciation of the dissolved gas plays a major role in determining the solubility of carbon dioxide in a given melt [1].

Speciation is believed to be controlled by the basicity of the melt [4,5]. The basicity was originally given as the ratio of the number of moles of basic oxides (primarily alkali and alkaline earth oxides) to the number of moles of acidic oxides (primarily the glassforming oxides such as silica, boric oxide, or phosphoric oxide) in the melt. This concept is useful in analyzing general trends, but does not include more subtle effects due to the differing degree of basicity of different alkali and alkaline earth oxides or acidity of different glassforming oxides. The use of "optical basicity", which is determined by uv-visible spectroscopy of indicator ions such as lead, bismuth, or thallium [6], provides a more quantitative measure of basicity and allows differentiation between the effects of different ions on the overall basicity of a melt.

Equilibrium between molecular CO_2, carbonate ions, and oxygen in the melt can be expressed [1] by the reaction

$$CO_2 + O^{2-} = (CO_3)^{2-},\tag{1}$$

where all of the oxygens in the melt are represented by the doubly negative oxygen ion. Fine and Stolper [4] have suggested that all types of oxygen ions do not react equally with carbon dioxide and that the non-bridging oxygen (NBO) ions will be much more reactive than the bridging oxygen (BO) ions. They then suggest that the appropriate reaction should be expressed as

$$CO_2 + 2O^- = (CO_3)^{2-} + O^0,\tag{2}$$

where O^- and O^0 refer to NBO and BO, respectively. Since the concentration of NBO in silicate melts typically increases with increasing concentration of basic oxides, this reaction indicates that formation of carbonate will become more favorable with increasing basicity of the melt.

The equilibrium described by reaction 2 can be expressed by an equilibrium constant, K, given by the expression

$$K = \frac{\left(a(CO_3)^{2-}\right)\left(aO^0\right)}{\left(aO^-\right)\left(aCO_2\right)}.\tag{3}$$

If the value of K is relatively constant, this expression indicates that the ratio of carbonate to molecular carbon dioxide will be controlled by the ratio of the activities of the BO and NBO. If we then assume that the activities of the oxygen species are proportional to their concentrations, Eq. 3 indicates that an increase in basicity, i.e. a decrease in the ratio of BO to NBO will result in an increase in the carbonate concentration at the expense of the molecular species. In simple terms, this expression predicts that the addition of basic oxides to silica will lead to a shift toward more carbonate and less molecular carbon dioxide in the melt [1].

The bonding of physically dissolved carbon dioxide molecules can be represented by a molecule weakly bonded to the bridging oxygens of the structure, as represented by the structure on the left in Figure 1. Carbonate ions, on the other hand, will be chemically bonded to the oxygen network, as might be represented by the structure on the right in Figure 1. These chemically bonded species will be more tightly bonded into the network and may be considered a bridging species [1,7].

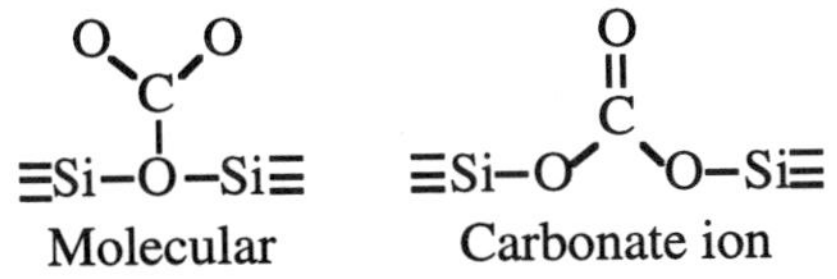

Figure 1: Structural environment of dissolved carbon dioxide in the molecular (left) and carbonate ion (right) forms.

RESULTS OF EXPERIMENTAL STUDIES

Unfortunately, while our model for carbon dioxide solubility in melts relies heavily upon arguments dealing with speciation, very little experimental evidence for changes in speciation with melt composition exists. Most studies of carbon dioxide solubility in simple melts involve measurement of total dissolved concentration and do not differentiate between different species of dissolved carbon dioxide. Verweij, et al. [8], however, did use Raman spectroscopy to demonstrate that carbonate ion exist in potassium silicate glasses containing ≥40 mol% potassium oxide when the carbonate is introduced from batch materials. They found that the carbonate concentration decreased, while the NBO concentration increased, with increasing melting time, which implies that the carbonate acts as a bridging species, as shown in Figure 1.

Results of a number of studies [9-13] of carbon dioxide solubility in sodium silicate melts have been combined in Figure 2. The large degree of scatter in these

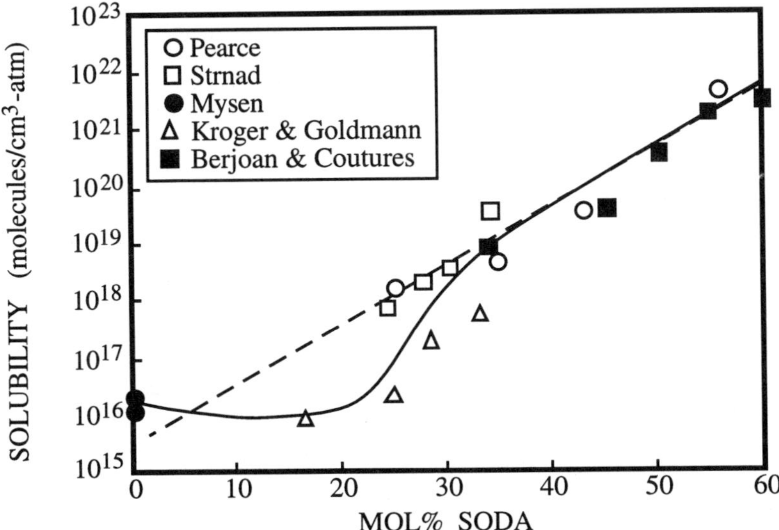

Figure 2: Effect of soda concentration on carbon dioxide solubility in sodium silicate melts. (*Handbook of Gas Diffusion in Solids and Melts*, J. E. Shelby (1996), ASM International, Materials Park, OH 44073-0002, p 186, Fig 8.25)

data are partially due to the difficulties encountered in studies of this type and partially to differences in experimental environments used to produce the data in individual studies. Data from Strnad [10] were obtained by treating the melts at 1000°C, while those of Kroger and Goldmann [12] are for melts equilibrated at 1100°C. Since Pierce [9] obtained data over a range of equilibration temperatures, values shown here were extrapolated to 1000°C. Furthermore, Pierce states that the equilibration times used by Kroger and Goldmann are too short to reach equilibrium, so that their data do not represent true equilibrium values and should lie below the correct values on this curve. Finally, no good values exist for vitreous silica. The values shown here were extrapolated by Fine and Stolper [4] from work by others at 1650°C and high pressure. These values are included since they are the only values available for silica, but they should be considered to be very uncertain.

While there is a large degree of scatter in the data shown in Figure 2, there is clearly a general trend toward increasing solubility with increasing soda concentration, i.e. with increasing basicity of the melt. There is, however, considerable uncertainty regarding the correct shape for the curve of log solubility versus soda concentration for melts containing less than 35 mol% soda. If we ignore the data of Kroger and Goldmann [12], which Pierce [9] believes under estimates the correct solubility due to insufficient equilibration times during the exposure to carbon dioxide, a straight line provides an adequate fit to the data. If we include the data of Kroger and Goldmann, we still can fit the data with a straight line for melts containing more than ≈35 mol% soda. The correct fit to the data in the region between 0 and 35 mol% soda, however, is unclear. The solid line in Figure 2 is intended solely as a guide to the eye, indicating that some significant deviation from a straight line must occur in this region. It is certainly possible that the curve does have some complex shape in this region due to the high acidity of the melts and that the relationship between carbon dioxide solubility and soda concentration changes as the speciation shifts from dissolved molecules to chemically bound carbonate ions.

Observation of a straight line on a log solubility versus composition plot, as definitely occurs for melts containing ≥35 mol% soda, implies that the enthalpy of solution increases linearly with increasing concentration of the compositional variable considered, i.e. soda. An increase in the enthalpy of solution with increasing soda concentration is consistent with the contention that the degree of bonding of carbon dioxide to the network (carbonate formation) increases as the melt becomes more basic via the formation of increasing numbers of NBO.

Unfortunately, most of the studies of carbon dioxide solubility in sodium silicate melts were carried out at only one temperature, so that insufficient data exists to define the relationship between the enthalpy of solution and soda concentration. Results of the studies which were carried out over some temperature range are contradictory. Pierce [9] and Berjoan and Coutures [13] report that carbon dioxide solubility decreases by orders of magnitude with increasing temperature, while Krogen and Goldmann [12] indicate that the solubility increases by a small amount with increasing temperature. Isawe, et al. [14] report that carbon dioxide solubility decreases by a factor of 2 or less with increasing temperature in related rubidium and cesium silicate melts. Swarts [15] reported a similar small decrease in solubility with increasing temperature in soda-lime-silicate melts.

Replacement of soda by lime to form soda-lime-silicate melts reduces the solubility of carbon dioxide significantly [12]. This observation is the reverse from that found for soda-lime-borate melts [16], where replacement of soda by lime increases carbon dioxide solubility. Replacement of silica by boric oxide to form sodium borosilicate melts also reduces carbon dioxide solubility [12].

A large number of studies of carbon dioxide solubility at high pressures have been reported for sodium aluminosilicate melts. Many of these studies directly address the speciation of the carbon dioxide by use of infrared spectroscopy to determine the environment of the dissolved species. Stolper and co-workers [4,5,17-19], for example, determined the location and extinction coefficients for a number of infrared absorption bands due to either molecular CO_2 or carbonate ions in sodium aluminosilicate glasses. Stolper, et al. [17] were able to use infrared spectroscopy to demonstrate that the speciation of carbon dioxide in albitic (sodium aluminosilicate) melts shifts toward the carbonate species with increasing temperature, even though the total carbon dioxide concentration is almost independent of temperature. Increases in pressure also cause a shift in speciation toward the carbonate form.

Study [4] of melts lying on the join $NaAlO_2$-SiO_2 as a function of soda concentration indicate that a decrease in soda content will shift equilibrium toward the molecular form of carbon dioxide. As a result, pure silica melts should contain little, if any, carbonate. These results agree with the contention that the shift toward carbonate ions at the expense of molecular CO_2 is favored by increasing basicity of the melt.

Carbon dioxide primarily dissolves as molecules of CO_2 in natural rhyolitic (alkali aluminosilicate) melts, as would be expected from the low NBO contents of such melts. On the other hand, carbon dioxide dissolves almost entirely in the form of carbonate in very basic, high NBO concentration, basaltic melts [3].

CONCLUSIONS

A model for carbon dioxide solubility based on the speciation of the dissolved gas between molecular and carbonate species can be used to explain trends in solubility as a function of melt composition. Increases in the basicity of the melt, which result from increases in NBO concentration, shift the equilibrium to favor dissolution as carbonate ions, which decreases in NBO concentration shift equilibrium to favor physical dissolution as CO_2 molecules. It is possible that additional effects are due to changes in the identity of the modifier used to produce the NBO in the melt, so that changes in carbon dioxide solubility may also be influenced by chemical differences between various modifier ions. The overall solubility of carbon dioxide is probably due to a combination of the concentration of NBO and the chemical nature of the modifier which causes these NBO to form.

REFERENCES

[1] J. E. Shelby, "Gases in Melts"; pp. 161-200 in *Handbook of Gas Diffusion in Solids and Melts*, by J. E. Shelby, ASM International, Materials Park, OH, 1996.

[2] J. E. Shelby, "Glass Melting"; pp. 25-47 in *Introduction to Glass Science and Technology*, by J. E. Shelby, Royal Society of Chemistry, Cambridge, 1997.

[3] J. E. Dixon, E. M. Stolper, and J. R. Holloway, "Infrared Spectroscopic Measurements of CO_2 and H_2O in Juan de Fuca Ridge Basaltic Glasses," *Earth Planet. Sci. Lett.*, **90** 87-104 (1988).

[4] G. Fine and E. Stolper, "The Speciation of Carbon Dioxide in Sodium Aluminosilicate Glasses," *Contributions to Mineral. & Petrol.*, **91** 105-21 (1985).

[5] G. Fine and E. Stolper, "Dissolved Carbon Dioxide in Basaltic Glasses: Concentrations and Speciation," *Earth Planet. Sci. Lett.*, **76** 263-78 (1985/86).

[6] J. A. Duffy and M. D. Ingram, "Optical Basicity"; pp. 159-84 in *Optical Properties of Glass*, Edited by D. R. Uhlmann and N. J. Kreidl, Am. Ceram. Soc., Westerville, OH, 1991.

[7] J. D. Kubicki and E. Stolper, private communication.

[8] H. Verweij, H. V. D. Boom, and R. E. Breemer, "Raman Scattering of Carbonate Ions Dissolved in Potassium Silicate Glasses," *J. Am. Ceram. Soc.*, **60** [11-12] 529-34 (1977).

[9] M. L. Pierce, "Solubility of Carbon Dioxide and Variation of Oxygen Ion Activity in Soda-Silica Melts," *J. Am. Ceram. Soc.*, **47** [7] 342-47 (1964).

[10] Z.Strnad, "Determination of the Solubility of Carbon Dioxide at Various Partial Pressures in Soda-Silica Melts using Gas Chromatography," *Phys. Chem. Glasses*, **12** [6] 152-55 (1971).

[11] B. O. Mysen, "The Role of Volatiles in Silicate Melts: Solubility of Carbon Dioxide and Water in Feldspar, Pyroxene, and Feldspathoid Melts to 30 kB and 1625°C," *Am. J. Sci.*, **276** 969-96 (1976).

[12] V. C. Kroger and N. Goldmann, "Carbon Dioxide Solubility in Glasses," *Glastechn. Ber.*, **35** [11] 459-66 (1962).

[13] R. Berjoan and J. P. Coutures, "Solubility of CO_2 in Liquids in the Binary System Na_2O-SiO_2," *Rev. Int. Hautes Temper. Refract. Fr.*, **20** 115-27 (1983).

[14] M. Iwase, H. Watanabe, N. Nakayama, H. Hori, J. Ohuchi, and K. Kawamura, "Solubility of CO_2 in Candidate Glasses for Nuclear Waste Immobilization. I. Systems: Cs_2O + SiO_2, Cs_2O + B_2O_3, Rb_2O + SiO_2 and Na_2O + B_2O_3," *Glass Technol.*, **35** [1] 41-47 (1994).

[15] E. L. Swarts, "Gases in Glass," *Ceram. Engr. and Sci. Proceedings*, **7** 390-403 (1986).

[16] C. A. Parker, G. G. Gausman, and J. E. Shelby, "Carbon Dioxide Solubility in Alkali Borate Melts"; pp. 391-96 in *Ceramic Transactions, Vol. 29, Advances in Fusion and Processing of Glass*, Edited by A. K. Varshneya, D. F. Bickford, and P. P. Bihuniak, Am. Ceram. Soc., Westerville, OH, 1992.

[17] E. Stolper, G. Fine, T. Johnson, and S. Newman, "Solubility of Carbon Dioxide in Albitic Melts," *Am. Mineral.*, **72** 1071-85 (1987).

[18] E. Stolper and J. R. Holloway, "Experimental Determination of the Solubility of Carbon Dioxide in Molten Basalt at Low Pressures," *Earth Planet. Sci. Lett.*, **87** 397-408 (1988).

[19] J. G. Blank, E. Stolper, and M. R. Carroll, "Solubilities of Carbon Dioxide and Water in Rhyolitic Melt at 850°C and 750 Bars," *Earth Planet. Sci. Lett.*, **119** 27-36 (1993).

REDUCTION OF SO$_2$ EMISSIONS WITH OXY-FUEL FIRING - "WATER ENHANCED SULFATE FINING"

H. Kobayashi (*) and R.G.C. Beerkens (+)

*Praxair, Inc., Tarrytown , New York, USA
+TNO Institute of Applied Physics, Eindhoven, The Netherlands

INTRODUCTION

SO$_2$ emissions from glass melting furnaces originate from two main sources: sulfur contained in the fuel and sulfates in the feed materials. Heavy oils typically contain 0.5 to 2 %wt sulfur and increase SO$_2$ emissions by about 1.2 to 4.8 kg per ton of container glass produced, or about 250 to 1,000 ppm in the exhaust gas for a typical air fired regenerative furnace. Although some dissolution of SO$_2$ from the furnace atmosphere into the molten glass is expected in the colder region of the furnace, virtually all of sulfur in the fuel leaves the furnace as SO$_2$. Clearly the most obvious step to reduce SO$_2$ emissions from a glass furnace is to reduce the sulfur content of the fuel.

For furnaces fired with natural gas, the only source for SO$_2$ emissions is sulfates contained in the batch materials for fining and redox control. SO2 emissions from the batch sulfate depend on several factors including the total sulfur in the cullet and batch materials, type of glass, redox state of the glassmelt, and firing conditions of the furnace and are typically in a range between 0.4 to 2 kg per ton of container glass produced, or about 100 to 500 ppm in the exhaust gas for a typical air fired regenerative furnace.

The amount of sulfates added in glass batch depends on the type of glass melted. Typical ranges of sodium sulfate used per metric ton of glass product are 6 to 12kg (3.4 to 6.7kg as SO$_3$) for float and oxidized plate glass, 5 to 12kg (2.8 to 6.7kg as SO$_3$) for flint bottle glass, 4 to 7kg (2.2 to 3.9kg as SO$_3$) for green bottle glass, and 2 to 5kg (1.1 to 2.8kg as SO$_3$) for textile fiber glass (E-glass). When other sulfur containing materials, such as cullet, filter dust, slags and calcium sulfate are used in the batch mixture, the amount of sodium sulfate is adjusted to provide the equivalent amount of sulfates and to compensate for the redox conditions. In Figure 1, a typical mass balance of sulfur is shown for flint container glass production. 6.5 kg of sodium sulfate (4.0 kg as SO_3) is used per metric ton of glass product in this example. About 45% (1.8 kg as SO_3) of the sulfate input is retained in glass product and about 55% (2.2 kg as SO_3) evolves mostly as SO_2 gas during batch melting and fining, and exhausted from the furnace. As the flue gas cools down, about 10% (0.2 kg as SO_3)of SO_2 reacts with NaOH, O_2 and H_2O to form condensable sulfate compounds such as Na_2SO_4, $Na_2S_2O_7$, $NaHSO_4$ and H_2SO_4 and released as particulate emissions. SO_2 emissions in this example is 1.6 kg per ton of glass or about 325 ppm in the exhaust gas for a typical natural gas air fired container glass furnace.

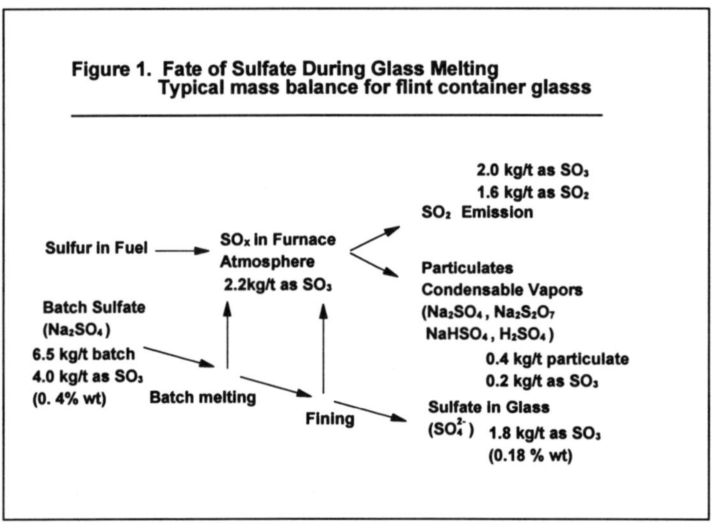

In most commercial glass furnaces, the amount of sulfate in the glass batch has been adjusted to the lowest acceptable level that allows the proper operation of the furnace to achieve good glass quality. So a further reduction in sulfate would presumably result in increased seeds count. Gibbs and Turner[1] suggested the theoretical minimum limit of sulfate requirement for float glass as the amount of sulfate retained in glass plus 0.05 wt% as SO_3 evolved at the fining zone. If we assume 0.25 wt% SO_3 retention in glass, the minimum sulfate requirement is 0.30 wt% SO_3, or equivalent to 5.3 kg of sodium sulfate per metric ton of float glass. The underlying assumption is that 0.05 wt% of sulfate as SO_3 is required for proper fining of gas bubbles in the fining zone of the furnace. The actual amount of sulfate mixed in the batch materials is typically much greater since a significant fraction of SO_2 is released during batch melting by reacting with carbon and other compounds.

In order to reduce SO_2 emissions from a natural gas fired furnace, either (1) the premature release of SO_2 during batch melting has to be reduced, (2) a greater fraction of SO_2 has to be converted to condensable sulfate compounds and removed as particulates, or (3) the fining action of SO_2 has to be replaced with other gases. The first option requires batch and flame adjustments and offers some reduction of SO_2 emissions. It is known that impinging flames and reducing combustion atmospheres tend to accelerate batch sulfate reactions and result in premature release of SO_2 in the batch melting zone. Thus, an adjustment of the burner firing conditions and the furnace atmosphere over the batch area may reduce SO_2 emissions without adversely affecting the glass quality.

The second option is theoretically possible by increasing volatilization of alkali species in the furnace to produce more NaOH and KOH, which react with SO_2 to form more sulfate particulates. However, it is not a desirable option as higher volatilization of alkali species in the furnace causes faster refractory corrosion. Third option requires a new fining gas. Dissolved water has recently been shown to act as an effective fining agent to partially replace sulfate and other fining agents[2].

WATER ENHANCED SULFATE FINING

During the normal sulfate fining process, sulfate in glass melt dissociates at high temperatures to produce a fining gas mixture of $SO_2+1/2O_2$ via the following reaction.

$$SO_4^{2-} \text{ (in melt)} = SO_2 \text{ (gas)} + \tfrac{1}{2}O_2 \text{ (gas)} + O^{2-} \text{ (in melt)} \qquad (1)$$

The above reaction is strongly dependent on temperature and active dissociation of sulfate takes place typically in the temperature range of 1450 to 1500 °C for soda lime glass. When the total pressure of fining gases and other dissolved gases exceeds the pressure in gas bubbles, which is typically 1 to 1.2 bar, the bubbles can grow rapidly in glassmelt by the diffusion of dissolved gases. Since a significant amount of water can dissolve in glass melt as hydroxyls, water can potentially replace the fining reaction of sulfate via the following reaction.

$$2OH^- \text{(in melt)} = H_2O \text{(gas)} + O^{2-} \text{ (in melt)} \qquad (2)$$

However, the equilibrium constant of the above reaction changes very little with temperature. Thus, water can not act alone as the fining agent in the conventional fining process which is based on an increased release of gases at high temperatures. Fortunately, reaction (2) would proceed to the right and produce water vapor when the sulfate fining reaction (1) starts and reduces the partial pressure of H_2O in bubbles. Since the equilibrium partial pressure of water vapor is proportional to the square of the concentration of dissolved water, a high concentration of dissolved water leads to a greatly higher vapor pressure of water vapor and reduces the partial pressure of SO_2 and O_2 required to grow the bubbles. Thus, reactions (1) and (2) promote each other through the mutual dilution effect and the fining action of sulfate is enhanced with dissolved water.

If we assume that a constant volume of fining gas generation is required to achieve the same degree of fining, then 1 1/2 moles of H_2O is required per mole of SO_3. Thus, the theoretical replacement ratio of H_2O to SO_3 is 1.5 on a molar basis or 0.3375 on a weight ratio of H_2O to SO_3. In order to replace 0.05 wt% SO_3, or to reduce about 0.4 kg of SO_2 per ton of glass, about 0.017 wt% H_2O is required, which is relatively small compared with the maximum solubility of water of about 0.11 wt% at 1 bar water vapor. In the conventional air fired furnace, water content of glass is about 0.03 to 0.04 wt%, while that in the oxy-fuel fired furnace is increased by about 0.02 wt%, to 0.05 to 0.06 wt%. Thus, sulfate fining action is expected to be enhanced substantially in oxy-fuel fired furnaces, if the same amount of sulfate is used. Conversely, it should be possible to reduce the amount of sulfate and still achieve good fining results under oxy-fuel firing . It would result in a significant reduction in SO_2 emission. In order to validate the feasibility of the water enhanced sulfate fining process, a mathematical model of a single bubble growth was applied and laboratory experiments were conducted.

MATHEMATICAL SIMULATION WITH A BUBBLE GROWTH MODEL

The bubble growth behavior in the presence of water was modeled using a multi-component diffusion model of a single bubble[3]. Major assumptions of the model are quasi-steady state gas diffusion into a spherical bubble and the bubble ascension rate controlled by the Stokes law. The viscosity of glassmelt is expressed as a function of temperature and glass compositions, including the influence of water content of glassmelt. Five gaseous species, SO_2, O_2, H_2O, N_2 and CO_2 were considered. Figure 2-4 show the results of model calculations for two different fining cases; a relatively low water content, high sulfate melt under air firing and a high water content, low sulfate melt under oxy-fuel firing. An air bubble of 0.2 mm in diameter in soda lime glass melt at 1m depth was modeled at 1500 °C melt temperature. The low water case assumes 0.075 wt% S (0.1875 wt% SO_3) and 50 mol/m^3 (0.039 wt%) H_2O and the high water case assumes 0.05 wt% S (0.125 wt% SO_3)and 90 mol/m^3 (0.070 wt%) H_2O. The redox state of the glassmelt was assumed to be 10 Pascal of O_2 pressure at 1200 °C, which is a normal oxidation state for flint glass.

Figure 2 shows the change of the bubble diameter with time and Figures 3 and 4 show the gas composition in the bubble. The bubble diameter initially increases rapidly due to rapid diffusion of gases to reach a near equilibrium gas composition in the bubble with respect to O_2, H_2O, N_2 and CO_2. The slow steady growth period after about 500 seconds is controlled by diffusion of relatively slow diffusing SO_2. As SO_2 diffuses into the bubble and reduce the partial pressures of other gases below those of the equilibrium values, other gases diffuse into the bubble and the bubble continues to grow. Oxygen is the most important gas composition for bubble growth in this example due to the relatively high oxidation state assumed in the calculation. As shown in Figures 3, water is the second most important fining gas for the high water case and contributes to the bubble growth more that SO_2. By comparison, as shown in Figure 4, water plays only a small part in the bubble growth for the low water case.

Figure 5 shows the time needed to ascend 1 meter in glassmelt for a 0.2 mm air bubble versus water content for various sulfate contents. Three points designated as A, B and C are considered to provide the same fining efficiency. Thus, glassmelt containing 62 mol/m³ (0.048wt%) of water and 0.3% wt sulfate is equivalent to glassmelts containing 77 mol/m³ (0.060 wt%) of water and 0.25% wt sulfate or 93 mol/m³ (0.073 wt%) of water and 0.2 wt% sulfate. This example shows that a decrease in sulfate of 0.05 wt% can be replaced with an increase in water content of about 15 mol/m³ (0.012 wt%).

LABORATORY MELTING/FINING TESTS

A typical flint soda lime glass batch composition was used for melting tests. Calculated compositions of typical test samples are SiO_2 72.3 wt%, Na_2O 14.0%, CaO 10.5%, MgO 1.4%, Al_2O_3 1.5%, SO_3 0.3 to 0.6% (varied), K_2O 0.2%, and Fe_2O_3 0.03%. About 90 to 100 grams of batch materials were first melted in an alumina crucible in an electric furnace at 1250 °C for each test. A gas mixture containing water vapor was introduced into the furnace and also bubbled through the melt for 30 minutes to dissolve water in glass. The water vapor content of the gas mixture was varied by bubbling a nitrogen/oxygen mixture (98% N2 and 2% O2 vol.) through a water bath maintained at a constant temperature. The glassmelt was then heated up to 1450 °C within 20 minutes and kept at the temperature for an additional 20 minutes for fining. These short melting and fining times were chosen so as to allow for observation of sufficient differences in the presence of bubbles. The quality of glass was judged by seven people for three common defects known as seeds (bubbles), grains (stones) and cords, in a scale of 1 (best) to 7 (worst) using 2 mm thick polished glass sections cut from the test samples.

In Table I some of the test results for three different levels of sulfate/water contents are summarized. The water content of glass samples (0.0155, 0.0338 and 0.0617 wt% for Cases A, B and C) are chosen to represent typical commercial glasses produced in an electric furnace, in a gas-air fired furnace and in a gas-oxy fired furnace, respectively. The amount of sulfate added to the batch mixtures were reduced from 0.63 wt% of glass produced for Case A to 0.55 and 0.37 wt% for Cases B and C, since the water content of glass increased. As expected from the equilibrium considerations from reactions (1) and (2), sulfate retention in glass is reduced from 0.40 wt% for Case A to 0.38 and 0.29 wt% for Cases B and C and the sulfate released during fining reactions were also reduced from 0.23 wt% for Case A to 0.17 and 0.08 wt% for Cases B and C. In spite of substantial reduction in SO3 release, the relative seed level was reduced from 4.4 for Case A to 3.1 and 2.6 for Cases B and C.

Table I. Results of laboratory melting/fining tests

Cases	A	B	C
pH_2O (Bar)	0.01	0.2	0.6
SO_3 added (wt%)	0.63	0.55	0.37
SO_3 in glass (wt%)	0.40	0.38	0.29
SO_3 released (wt%)	0.23	0.17	0.08
H_2O in glass (wt%)	0.0155	0.0 338	0.0617
Seeds Level (*)	4.4	3.1	2.6

* Relative scale: 1 = Best, 7 = Worst

CONCLUSIONS

Factors influencing SO_2 emissions from glass melting furnaces were reviewed. An improved fining process that replaces a portion of sulfate with dissolved water was described. Mathematical modeling and laboratory melt studies showed the interaction of dissolved water and sulfate that accelerates the growth of gas bubbles and reduces SO_2 emissions by reducing the amount of required sulfates mixed in the batch materials.

References
1. W.R.Gibbs and W. Turner, "Sulfate Utilization in Float Glass Production", 54th Conference in Glass Problems, The Ohio State University, Nov. 1994.
2. H.Kobayashi and R.G.C. Beerkens, "Water enhanced Sulfate Fining - A method to reduce toxic emissions from glass melting furnaces", Patent pending.
3. R.G.C. Beerkens, "Chemical Equilibrium Reactions As Driving Forces For Growth OF Gas Bubbles During Refining", Glastechnische Berichte 63K (1990) pp222-236.

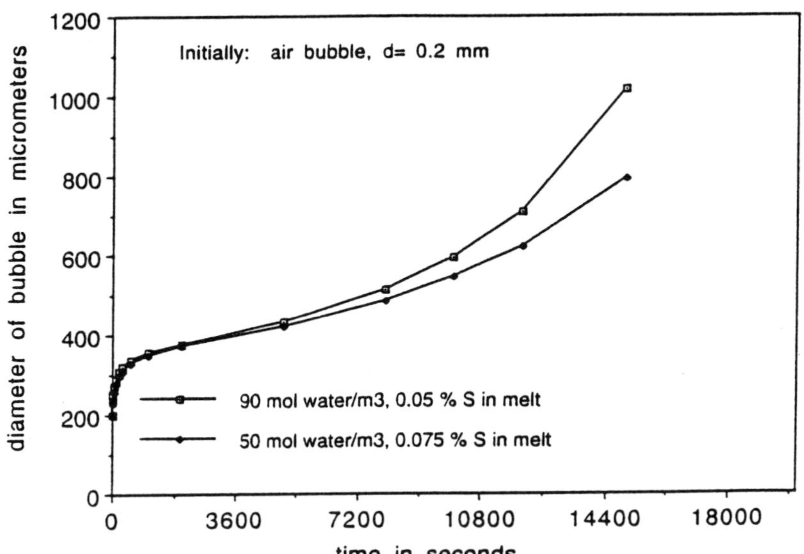

Figure 2   Bubble growth in a molten soda-lime glass at 1500 °C

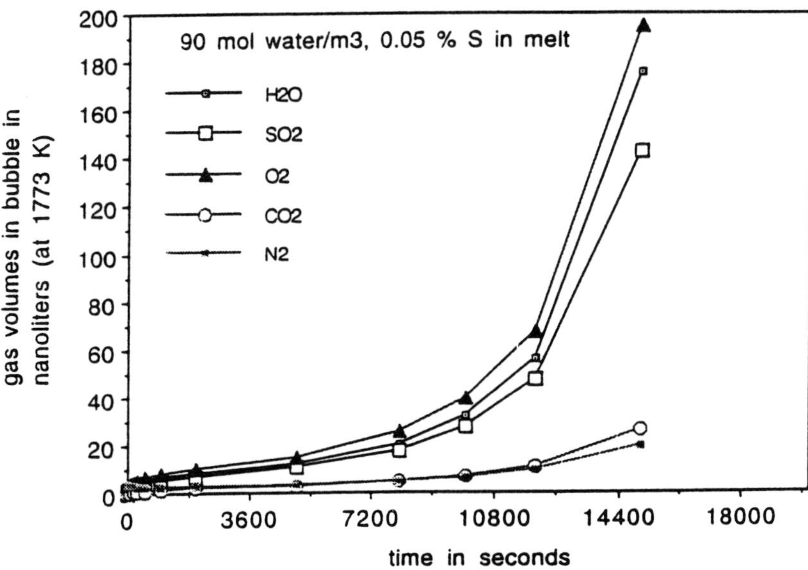

Figure 3   Gas Composition in a growing bubble in a molten soda-lime glass containing 90 mol H_2O/m^3 at 1500 °C

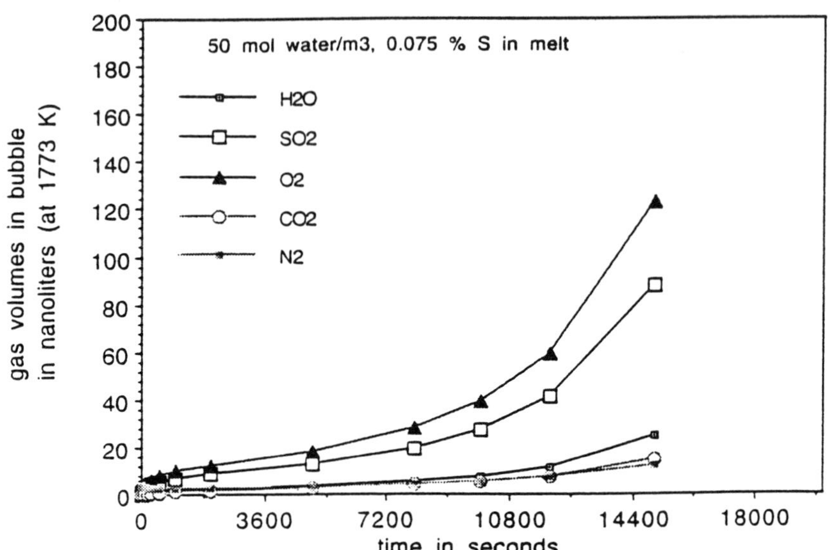

Figure 4   Gas Composition in a growing bubble in a molten soda-lime glass containing 50 mol H2O/m³ at 1500 °C

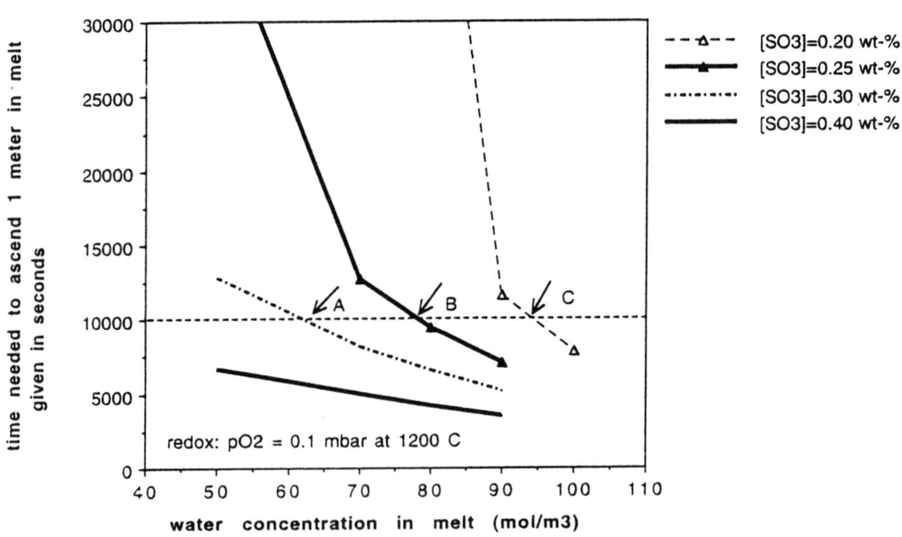

Figure 5   Bubble ascension in 1 meter deep molten soda-lime glass at 1475 °C,   Initial bubble: air, diameter = 0.2 mm

Environmental Issues in Fusion and Processing

ADVANCES IN NUCLEAR WASTE GLASSES - BRINGING THE PAST INTO OUR FUTURE

M. J. Plodinec
Diagnostic Instrumentation and Analysis Laboratory
Mississippi State University
Post Office Drawer MM
Mississippi State MS 39762

C.A. Cicero-Herman*, J.C. Marra*, S.L. Marra^
*Savannah River Technology Center
^Defense Waste Processing Facility
Westinghouse Savannah River Company
Aiken, SC 29802

HIGH-LEVEL WASTE VITRIFICATION

The nation's first and the world's largest facility for the vitrification of high-level nuclear waste (HLW) - the Defense Waste Processing Facility (DWPF) - is now in its second year of radioactive operation. After a ten year construction period and an almost three year non-radioactive test program, DWPF began Radioactive Operations in March 1996. In the development of the melter, the glass compositions, and the vitrification process to immobilize HLW, waste glass technologists applied commercial glass practices and technology to the greatest possible extent.[1] However, there are additional challenges that are faced in the process of immobilizing HLW that are not typically faced by the commercial glass industry:

- The commercial glass process is generally controlled to produce a single glass composition. For HLW, the most important single component of the glass - the waste - varies greatly, resulting in an inherent variability in the final product's composition.
- The waste glass technologist seeks to maximize the amount of waste immobilized in the glass, thus maximizing the variability of the glass produced. Waste loading must be maximized due to the very high cost of disposal of waste packages (at least $100,000 per ton of glass produced).[2]
- In a commercial glass plant, hands-on maintenance is the norm; in a HLW vitrification plant, maintenance can only be performed with tools attached to the end of a crane hook or with manipulators.
- HLW waste glasses contain elements seldom encountered in the production of commercial glasses (e.g., Pu, Tc, Cs).[3]
- Although HLW glasses also contain elements commonly encountered in commercial glass production (e.g., Fe, Ni, Mn), these are usually present in HLW glasses at concentrations seldom encountered in commercial practice.[3]
- Emissions capture and recycle is of much greater importance in HLW vitrification because of the consequences of release of highly radioactive species such as Cs-137.

As a result, the glasses produced in the DWPF are unusual low-melting temperature (1150°C) borosilicate compositions because of the necessity of retaining volatile species such as Cs-137 in the glass. Borosilicate compositions are employed, because the boron enhances the "solvency" of the glass so that large concentrations of iron, nickel, and manganese can be incorporated in the glass.

An average DWPF HLW glass composition is shown in Table I. This represents a hypothetical blend of all the HLW at the Savannah River Site. However, the inherent variability of the waste is likely to cause the concentrations of major waste components to vary by a factor of two, and those of minor waste components to vary by an order of magnitude or more in the glass.

Table I. Composition of a hypothetical glass composition containing a blend of all of the HLW at the Savannah River Site.

Component	Amount (wt%)	Component	Amount (wt%)
Al_2O_3	4.85	MgO	1.36
B_2O_3	7.66	MnO	2.05
$BaSO_4$	0.22	Na_2O	8.58
CaO	1.16	Na_2SO_4	0.10
$CaSO_4$	0.12	NaCl	0.31
Cr_2O_3	0.10	NiO	0.74
Cs_2O	0.08	SiO_2	49.61
CuO	0.40	ThO_2	0.36
Fe_2O_3	12.47	TiO_2	0.65
K_2O	3.47	U_3O_8	0.53
Li_2O	4.40		

Similarly, the melters used for HLW vitrification are not the same as commercial melters. In both the DWPF and at the smaller facility in West Valley, NY, joule-heated melters are used. The ubiquitous combustion melters used in the glass industry are not used for HLW vitrification because of concerns about fires in a remote facility and about volatility of Cs. Electrodes made of Ni-Cr alloys are used instead of the Mo electrodes commonly used in commercial joule-heated melters, because they are more stable than Mo toward the large concentrations of transition metal oxides (e.g., NiO) in the melts. Refractories high in Cr are used instead of the alumina-zirconia-silica (AZS) refractories commonly used in commercial joule-heated melters, again for longer life.

Many years of research and testing led up to the successful radioactive start-up of DWPF. Full scale testing with simulated feed was a major part of the DWPF start-up testing.[4] Much was learned about the operation of this first of a kind, vitrification facility during this test program. This testing provided a large amount of information that assisted in the preparation for actual radioactive operations. However, as with any complex facility difficulties were encountered during the transition to radioactive operations.[5] These difficulties have been overcome and DWPF continues to safely produce waste forms that meet all requirements.

Because of the success of HLW vitrification in the DWPF and in the smaller West Valley facility, vitrification of several other nuclear materials is being actively pursued. These range from Low-Level Waste (LLW) to nuclear materials no longer needed for defense purposes. In order to economically, expeditiously, and safely vitrify these materials, glass compositions and glass melting techniques long known to the commercial glass industry are being utilized in addition to the techniques and methods utilized for the vitrification of HLW.

LOW-LEVEL WASTES

For the vitrification of LLW, more emphasis has been placed on using commercially available melter technologies. This is possible mostly because of the radiation level associated with the wastes, which does not require the shielding and remote operations associated with HLW vitrification. Maintenance of LLW vitrification systems is practical because dose rates are very

Advances in Fusion and Processing of Glass II

low and/or the equipment can be easily decontaminated. One example of the use of commercial technology to treat LLW is the design and construction of the Transportable Vitrification System (TVS). The TVS was a joint effort between the Savannah River Technology Center and EnVitCo, Inc., to fabricate a treatment unit that could be moved to the various sites to treat small LLW streams. The cost of this unit was several orders of magnitude less than for facilities like the DWPF because of the lower radiation levels associated with the LLW. These levels require less radiation shielding and allow for hands-on equipment maintenance. The TVS uses a joule-heated melter with materials of construction similar to commercial glass industry melters.[6]

Vitrification of LLW, however, still presents challenges compared to commercial industry glass making because the waste streams are not completely homogeneous and contain some materials not commonly encountered in the glass industry. Typical LLW that are considered candidates for vitrification include plating line sludges, wastewater treatment sludges, filter aids, soils, ashes, and ion exchange resins. Most of these wastes contain large amounts of silica and lime, which make them excellent candidates for vitrification in the soda-lime-silica system that is widely used in the commercial glass industry. Basic soda-lime-silica glass compositions melt at slightly higher temperatures than borosilicate glasses, but additional additives can be used which lower the melting temperature if volatility of radioactive species is a concern. For some LLW streams, borosilicate glasses, as used in HLW vitrification, are still the preferred glass system due to the ability to use lower melting temperatures. Since most LLW streams are composed of typical glass making constituents, much higher waste loadings can be obtained than with HLW. On average, typical waste loadings are 50 wt%, but waste loadings up to 95 wt% have been obtained for some waste streams.[7]

Like HLW, some LLW streams contain quantities of transition metals and anions not commonly encountered in the commercial glass industry. Transition metals are especially common for LLW which contain hazardous constituents or Low-Level Mixed Wastes. The metals present a challenge when materials of construction, such as Mo electrodes, are used because waste loading must be constrained to avoid the formation of lower melting temperature eutectics. Only minor quantities of anion species can be incorporated in the glass matrix, therefore, waste loadings of the glass are also constrained by these species. Commercial industry experience with limits on the solubility of these materials has been valuable in LLW vitrification.

The vitrification of LLW presents less of a challenge than the vitrification of HLW from an equipment stand-point because commercially available melters and materials of construction can be adopted. However, developing glass formulations and producing acceptable glass wasteforms can be just as difficult a challenge because of the wide range of species that are found in LLW. The waste constituents may differ drastically depending on the generation location or the storage container. Although this is a challenge, it is one that can be met with treatability studies and by using lessons learned from the glass industry on incorporating various species into glass matrices.

EXCESS NUCLEAR MATERIALS

Rare earth borosilicate compositions are being developed to immobilize actinide species no longer needed for defense purposes. The rare earth borosilicate system is currently being applied to the immobilization of weapons usable Pu and the vitrification of Am/Cm solutions.[8,9] The rare earth borosilicate glasses resemble Löffler type glasses once used for optical applications.[10] The unneeded nuclear materials generally do not contain volatile species, but are highly refractory, i.e., difficult to incorporate into glass. Thus, higher melting temperatures can be utilized (and in fact are beneficial) to increase the solvent power of the melt. Boron is also included in the glass to enhance refractory oxide dissolution.

Example compositions for the excess Pu and Am/Cm solution applications are shown in Table II. Achieving a high loading of nuclear material in these glasses is again beneficial, in this case because of the high cost of immobilization and the expense in safeguarding these valuable materials. The rare earth borosilicate family was chosen because its members typically contain high contents of rare earth oxides (typically up to 55 wt%). The chemical similarity of the rare earth oxides and the actinides, indicated that high loadings of actinides in the glass could be achieved.

<u>Table II. Rare earth borosilicate frit compositions for actinide immobilization</u>

Component	Concentration (wt%)	
	Am/Cm Vitrification.	Pu Immobilization.
Al_2O_3	6.1	21.3
B_2O_3	9.2	11.6
BaO	9.0	N/A
Gd_2O_3	N/A	8.7
HfO_2	N/A	6.6
La_2O_3	7.1	10.1
Nd_2O_3	N/A	10.1
PbO	24.4	N/A
SiO_2	44.2	29.1
SrO	N/A	2.5

Bushing-type melters, similar to those used in the fiberglass industry, are being adapted for this application. Both resistively heated and inductively heated melters constructed using Pt/Rh alloys are being developed.[11-12] These melters have been found to be extremely reliable and can be operated at the relatively high temperatures required for actinide dissolution. Additionally, the melters are compact, facilitating installation in radioactive shielded cells or glove boxes. A schematic showing the resistively heated melter being developed for the vitrification of the Am/Cm solutions is shown in Figure 1.

<u>Figure 1. Bushing melter for vitrification of Am/Cm solutions.</u>

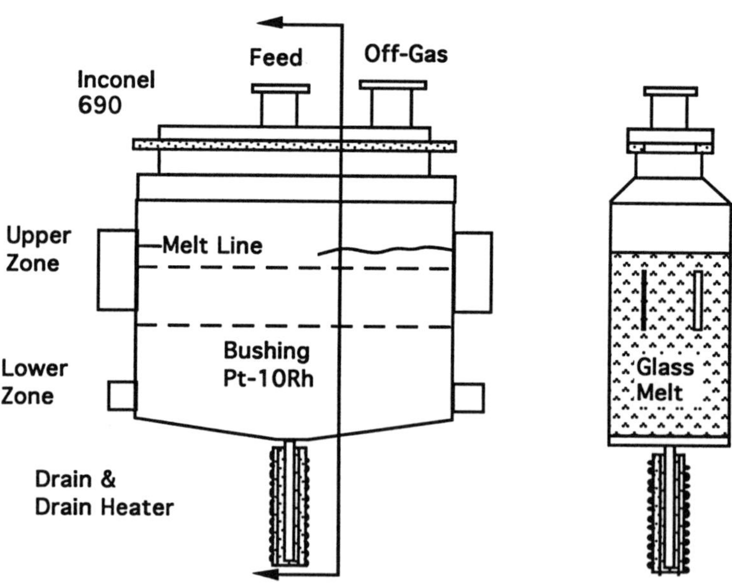

Advances in Fusion and Processing of Glass II

For these excess nuclear materials, the possibility of a nuclear criticality constitutes a special hazard. A nuclear criticality could have devastating consequences, and thus requires especially stringent preventative measures. The rare earth components in the glasses (e.g. Gd and Sm) act as neutron "poisons" helping to prevent a nuclear criticality. Bushing-type melters can be designed in "slab" configurations (basically as vessels, with one horizontal dimension of only three to five inches) or in cylindrical configurations (with diameters less than five inches). Keeping one dimension of the melter less than five inches precludes criticality by ensuring that a mass of fissile material cannot be readily reflected. Therefore, the use of rare earth elements as neutron absorbers in the glass and geometrical limits in the melter design provide two separate means to prevent criticality incidents.

CONCLUSION

Now that successful operation of the DWPF in vitrifying HLW is well underway, attention in the United States is turning toward stabilization of other nuclear materials. In these cases, we can use glass compositions and melters with an established knowledge base in the literature and, more importantly, in commercial practice. Thus, we can take maximum advantage of our past experience in glass technology.

In vitrifying LLW, we can use familiar soda-lime-silica glasses, and the AZS refractories and Mo electrodes commonly used to make them. As discussed above, bushing-type melters are very suitable for the vitrification of nuclear materials no longer needed for defense purposes. Even in this rather specialized case, an existing family of glass compositions can be easily adapted to stabilize these materials. Thus, as we turn our efforts toward these new missions, we are finding that the glass community's past accomplishments are significant signposts toward our future success.

REFERENCES

[1] M.J. Plodinec, "Vitrification Chemistry and Nuclear Waste," *J. Non Cryst. Solids,* **84**, 206-14 (1986).

[2] W.R. McDonell, S.D. Thomas, and C.B. Goodlett, "Costs of Nuclear Waste Glassmaking at Savannah River"; Chapter 21 in *Proceedings - First International Conference on Advances in the Fusion of Glass*, Edited by D. F. Bickford, et al. American Ceramic Society, Westerville, OH, 1988.

[3] J.R. Fowler and M.J. Plodinec, "Projected Compositions and Radiogenic Properties of DWPF Glasses"; pp. 904-910 in *High Level Radioactive Waste Management - Proceedings of the Third International Conference*, 1992.

[4] S.L. Marra, D.E. Snyder, H.H. Elder, J.E. Occhipinti, "The DWPF: Results of Full Scale Qualification Runs Leading to Radioactive Operations"; *Proceedings - Waste Management '96*, Tucson, AZ, 1996.

[5] J.T. Carter, K.J. Rueter, J.W. Ray, O. Hodoh, "Defense Waste Processing Facility Radioactive Operations - Part II Glass Making"; *Proceedings - Waste Management '97*, Tucson, AZ, 1997.

[6] J.C. Whitehouse, P.R. Burket, D.A. Crowley, E.K. Hansen, C.M. Jantzen, R.P. Singer, M.E. Smith, S.R. Young, J.R. Zamecnik, T.J. Overcamp, and I.W. Pence, Jr., "Transportable Vitrification System: Operational Experience Gained During Vitrification of Simulated Mixed Waste"; *Proceedings of Waste Management '97*, Tucson, AZ, 1997.

[7] C.A. Cicero, D.F. Bickford, M.K. Andrews, K.J Hewlett, D.M. Bennert, and T.J. Overcamp, "Vitrification Studies with DOE Low-Level Mixed Waste Wastewater Treatment Sludges", *Proceedings - Waste Management '95*, Tucson, AZ, 1995.

[8]T.F. Meaker, W.G. Ramsey, J.M. Pareizs, D.K. Peeler, and J.C. Marra, "Actinide Solubility in Lanthanide Borosilicate Glass for Possible Immobilization and Disposition"; *Proceedings of the Material Research Society Fall Meeting*, Boston, MA, 1996.

[9]T.F. Meaker, D.K. Peeler, C.M. Conley and J.O. Gibson, "Determination of Solubility Limits of Individual Pu Feeds into a Lanthanide Borosilicate (LaBS) Glass Matrix Without Preprocessing"; p. 167 in *Pu Futures - The Science, Conference Transactions*, Los Alamos National Laboratory and the American Nuclear Society, Sante Fe, NM, 1997.

[10]J.v. Löffler, "Chemical Decolorization," *Glastechnische Berichte*, **10**, 204-11 (1932).

[11]R.F. Schumacher, "Vitrification of Actinides Contained in Platinum Alloy Vessels"; to be published in *Transactions of the American Nuclear Society*, American Nuclear Society, Albuquerque, NM, 1997.

[12]J.C. Marra, K.M. Marshall, R.F. Schumacher, M.P. Speer, J.R. Zamecnik, T.B. Calloway, J. Coughlin, R.P. Singer, J. Farmer, W. Bourcier, D. Riley, B. Hobson and M. Elliott, "Glass Melter Development for Plutonium Immobilization"; p. 167 in *Pu Futures - The Science, Conference Transactions*, Los Alamos National Laboratory and the American Nuclear Society, Sante Fe, NM, 1997.

VITRIFICATION OF FILTER ASHES FROM REFUSE INCINERATION PLANTS WITH TOTAL INCORPORATION OF SULPHUR

Leo Schumacher, Günther H. Frischat
Institut für Nichtmetallische Werkstoffe, Technische Universität Clausthal
Zehntnerstraße 2a
D-38678 Clausthal-Zellerfeld, Germany

EXTENDED ABSTRACT

Since the end of the 70s, great emphasis has been put on the development of the immobilization of radioactive waste and various possibilities have been investigated [1, 2]. Vitrification was preferentially carried out by using borosilicate glasses whose properties can be optimized by various additives. It is reasonable also to develop modified processes of silicate glass formation for the immobilization of nonradioactive, but toxic waste. Refuse incineration produces residues such as fly ash containing a number of potentially hazardous elements, e. g. chromium, lead, cadmium, mercury, arsenic, with filter dust or filter ash being the solid residue arising at dedusting of the raw gases. It is the aim to convert nonrecyclable residual materials into a state allowing a nonhazardous disposal above ground on low grade dumps.

Filter dusts are a mixture of silicates, chlorides, sulphates as well as metal and heavy metal oxides. Due to the addition of calcium oxide during the waste gas purification process the sulphur is incorporated mainly as $CaSO_4$, in some cases up to 15 wt%. Melting of this material under oxidizing conditions delivers a mixture of different phases, a glassy (crystalline) phase, a metallic phase, and a gas phase mainly containing SO_3/SO_2, HCl, etc. The SO_3 content dissolved in the glass phase is as low as 0.3 to 0.03 wt% [3]. One has then again to deal with these phases separately.

The ultimate objective of this work is to develop a vitrification process which yields one phase only containing all compounds. Moreover, the chemical stability of the melting product should be so high that it can be easily disposed of. There is experience that suggests that blast furnace slags display a high sulphide solubility [4], and there is also evidence that natural minerals such as troilite (FeS), wurtzite (ZnS), and alabandite (MnS) belong to the most stable materials known. Therefore, a melting process was developed which guaranteed the total dissolution of the sulphur present in the filter ash in the melt by reducing the sulphate components to sulphide ones [5, 6]. Melting trials were performed with filter ash, SiO_2 and other melting additives, carbon, iron, zinc, titanium, manganese, silicon, and aluminum, added to the batches separately or in combination, with melting temperatures between 1300 and 1400 °C. Several of these batches resulted in melting products with a 100 % incorporation of the sulphur and a hydrolitic resistance falling under the hydrolitic class III at least.

REFERENCES

[1]W. Lutze and C. Ewing, „Radioactive Waste Forms of the Future", North Holland, Amsterdam, 1988.

[2]M. J. Plodinec, „Advances in Processing Nuclear Waste Glasses"; pp. 9.1 - 9.11 in *Advances in the Fusion of Glass*, Edited by D. F. Bickford et al. The American Ceramic Society, Inc., Westerville, Ohio, 1988.

[3]F. W. Krämer, „Influence of Oxygen on the Glass Melt"; pp. 21 - 31 in *Melting History and Workability of Glass*, Fachausschussbericht Nr. 73, Deutsche Glastechnische Gesellschaft, Frankfurt/Main, 1991.

[4]F. D. Richardson and C. J. B. Fincham, „Sulphur in Silicate and Aluminate Slags", *Journal of the Iron and Steel Institute*, **178** 5 - 15 (1954).

[5]L. Schumacher, „Vitrification of Filter Ashes from Incineration Plants with Special Emphasis of Sulphur Incorporation", PhD Dissertation, Technische Universität Clausthal, 1996.

[6]L. Schumacher and G. H. Frischat, „Laboratory Trials to Incorporate Sulphur in the Vitrification of Filter Ashes from Incineration Plants and Hydrolitic Investigation of the Melt Products". *Glastechnische Berichte, Glass Science and Technology* **70** [2] 58 - 63 (1997).

WASTE GAS TREATMENT OF SODA LIME SILICA GLASS FURNACES - INVESTIGATIONS WITH DIFFERENT ABSORPTION AGENTS

U. Kircher
Hüttentechnische Vereinigung der Deutschen Glasindustrie
Mendelssohnstr. 75 – 77
60325 Frankfurt/Main, Germany

INTRODUCTION

Corresponding to the German legislation nearly all glass melting furnaces in Germany are nowadays equipped with waste gas treatment plants. Table 1 shows the typical contents or ranges of waste gas components of soda lime silica glasses, compared with the limits of the TA-Luft[1]. Table 1 shows very clearly, that waste gas treatment plants had to apply also in the container and flat glass industries. The waste gas treatment plants which are used consist typically of an electrostatic precipitator (EP) with a dry absorption stage connected in series. The absorption stage has the task to bind the gaseous waste gas compounds HF, HCl, SO_2, SO_3 and gaseous selenium compounds, so that the threshold limit values can be observed and to protect the electrostatic precipitator (EP) against corrosion by acid waste gas components. As absorption agent mostly powdery calcium hydroxide ($Ca(OH)_2$) is used, which is fed in the flue before a special reaction tower. Sometimes light soda ash is also employed as absorption agent and in the meantime in some special cases also sodium bicarbonate. The different absorption agents have different absorbing properties with respect to the different waste gas components. The absorption agent $Ca(OH)_2$ reacts with the waste gas components HF, HCl, SO_2, SO_3 and gaseous selenium to form CaF_2, $CaCl_2$, $CaSO_3$, $CaSO_4$ and $CaSeO_3$. These compounds can be precipitated in an EP also at high temperature. All these reactions are competitive reactions and partly temperature – dependent. Therefore, the different absorption rates are not constant and depend on the amount of absorption agent, the waste gas composition, the waste gas temperature and probably also on the humidity and the oxygen content. Until now there are sometimes problems with the simultaneous waste gas cleaning of HF, HCl and gaseous selenium. This was the reason for the decision to carry out investigations with different absorption agents.[2]

Table I. Emission concentrations and threshold limit values of glass melting furnaces for soda lime silica glasses

component		concentration	TA-Luft limit
total dust		$150 - 250$ mg/m^3	50 mg/m^3
lead	(Pb)	up to 12 mg/m^3	5 mg/m^3
chromium, total	(Cr)	up to $3,5$ mg/m^3	5 mg/m^3
nickel	(Ni)	up to $0,6$ mg/m^3	1 mg/m^3
vanadium	(V)	up to $2,5$ mg/m^3	5 mg/m^3
selenium, as particulate	(Se)	up to $0,2$ mg/m^3	1 mg/m^3
selenium, gaseous	(Se)	up to $13,0$ mg/m^3	1 mg/m^3
sulfur oxides	(SO$_x$)	$400 - 4000$ mg/m^3	1800 mg/m^3
nitrogen oxides	(NO$_x$)	$400 - 3600$ mg/m^3	$1200 - 3500$ mg/m^3 expect. 800 mg/m^3
HF		$3,5 - 45$ mg/m^3	5 mg/m^3
HCl		$30 - 120$ mg/m^3	30 mg/m^3

MEASUREMENTS

The investigations were carried out in close collaboration with container glass companies operating glass melting furnaces for flint glass where the glass was decolourized by selenium. The waste gas treatment plants of all these furnaces consist of an EP with a dry absorption stage connected in series. The retention rates of these waste gas treatment plants with respect to the emission components HF, HCl and selenium were investigated depending on the type and the amount of the absorption agent. This was done by measuring simultaneously the concentrations and mass flows of the evaporated components in the untreated and treated waste gas. With these data obtained it was possible to calculate the absorption rates.

The measurements were mainly taken on a waste gas treatment plant of three container flint glass furnaces. At this special plant the type and the amount of the

- All concentrations in this paper are referred to dry waste gas in the normal state with 8 % O$_2$ -

Advances in Fusion and Processing of Glass II

absorption agent as well as the waste gas temperature could be varied. The decrease in the waste gas temperature was achieved by quenching either with water or with air. Thus, the influence of humidity on the absorption rates could be investigated, too. As absorption agents calcium hydroxide powder ($Ca(OH)_2$) and light soda ash (Na_2CO_3) were employed.

There is no other waste gas treatment plant to be found in the German glass industry where more parameters can be varied than at this special plant. The total furnace load of the three regeneratively heated flint glass furnaces (end-fired) amounts to 400 t/d. Two of them were heated with heavy fuel oil (sulphur content maximum 1.7 %) and one with natural gas (type L).

The waste gases of the three furnaces are joined together and cleaned in the common waste gas treatment plant. Normally, calcium hydroxide is used as absorption agent. At the bottom of the reaction tower, which is part of the absorption stage, it is possible to spray water into the waste gases to cool them down if necessary.

The filter dust resulting form the treatment is recycled, i. e. remelted with the batch as a batch component.

The investigations were made with three different quantities of each absorption agent (calcium hydroxide and light soda ash) and two different waste gas temperatures of about 390 and 335°C, where the cooling of the waste gases was achieved by quenching either with water or with air. Thus, for each absorption agent nine different operating conditions were studied, i. e. in total 18.

RESULTS OF THE MEASUREMENTS

While using calcium hydroxide as absorption agent, the concentrations in the uncleaned waste gases amounted to 18 mg/m^3 for HF, 45 mg/m^3 for HCl and 6 mg/m^3 for the gaseous selenium compounds. With the measurements using light soda ash as absorption agent, the corresponding initial concentrations differed partly from those obtained during the calcium hydroxide measuring campaign. Whereas the HF contents were quite similar with 18 mg/m^3, the HCl contents were higher and amounted to 42 to 62 mg/m^3, depending on the recycled cullet. The initial concentrations of the gaseous selenium compounds, however, were lower and amounted to 4 mg/m^3. All the measured concentrations of the uncleaned waste gases were higher than the corresponding limits of 5 mg/m^3 for HF, 30 mg/m^3 for HCl and 1 mg/m^3 for selenium.

HF-Measurement Results

Figure 1 shows the obtained absorption rates for HF, depending on the amount of the absorption agents and the waste gas temperature, for the two absorption agents employed. As it can be seen in the figure, the absorption rates for calcium hydroxide range from 82 to 99 % and thus are significantly higher than for light soda ash. Using calcium hydroxide all the measured clean gas concentrations are below the limit of 5 mg/m^3, whereas by using light soda ash all the clean gas concentrations are significantly above the limit, even at higher quantities of the absorption agent. This means in practice that calcium hydroxide has to be applied if there exists a severe fluorine emission problem as it does at the plant investigated.

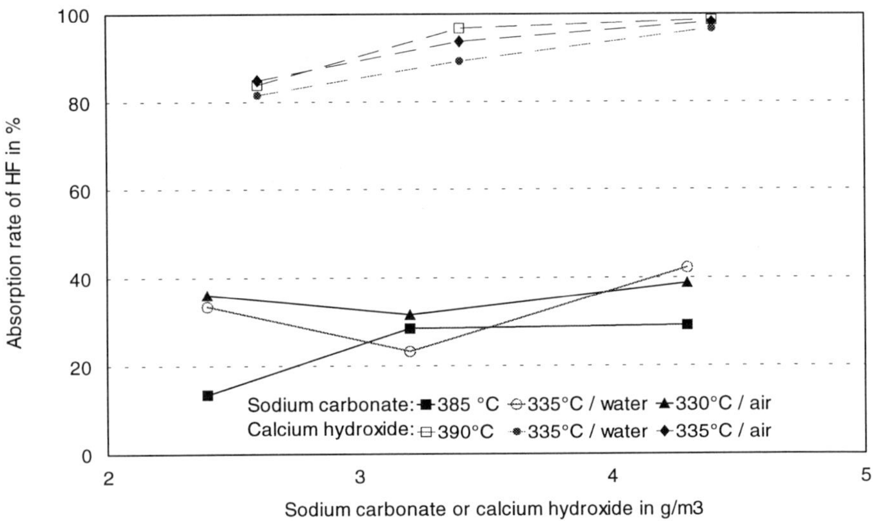

Figure 1. HF absorption rates depending on the amount of light soda ash or calcium hydroxide and on temperature

HCl-Measurement Results

Figure 2 shows the absorption rates for HCl. The absorption rates differ not as much as for HF. By using calcium hydroxide the absorption rates were in the range between 49 and 89 % and by using light soda ash in the range between 49 and 81 %. At the higher temperature of 390 °C calcium hydroxide was the better absorption agent and at the lower temperature, cooling the waste gases by air (330°C), light soda ash is more suitable. In the investigated temperature range between 330 °C and 390 °C, it is possible to observe the limit of 30 mg/m^3 for HCl with both investigated absorption agents. But it is also known, that in a temperature range below 300 °C, the absorption rate for calcium hydroxide is insufficient if the initial HCl content is above 45 mg/m^3.

                    Advances in Fusion and Processing of Glass II

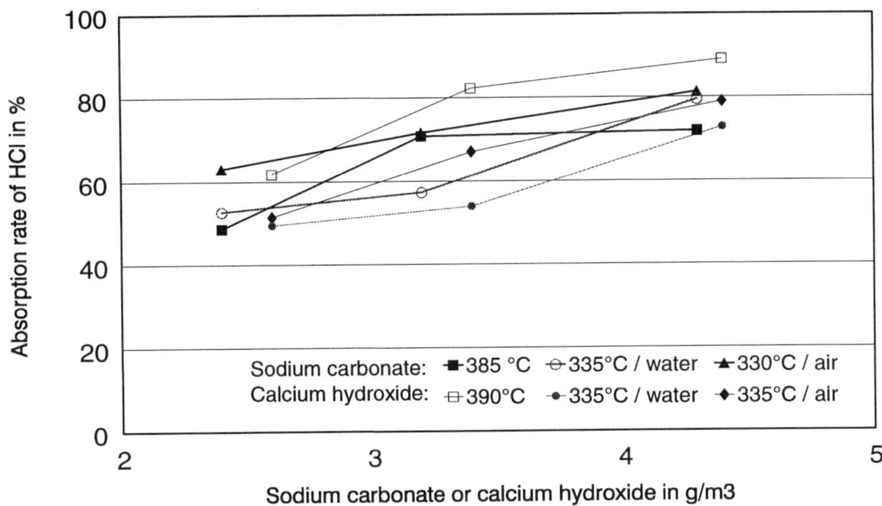

Figure 2. HCl absorption rates depending on the amount of light soda ash or
calcium hydroxide and on temperature

Se Measurement Results

Figure 3 shows the absorption rates obtained for gaseous selenium, depending again on the quantities of the absorption agent used. The absorption rates are increasing with increasing quantities of absorption agents. Further more, it can be seen clearly, that calcium hydroxide, as compared with light soda ash, is by far the better absorption agent for gaseous selenium. This applies at least to the investigated temperature range between 330°C and 390°C. By using calcium hydroxide very high absorption rates can be achieved - up to 97 % -, if the quantity of the absorption agent is high enough and normally, the limit of 1 mg/m^3 can because of that be observed. With the use of light soda ash the absorption rates were in the range between 31 and 59 %. This means that the concentration of gaseous selenium in the uncleaned waste gases should not be higher than 2 mg/m^3 if light soda ash is used as absorption agent.

Finally, it should be pointed out that using calcium hydroxide as absorption agent the highest absorption rates for the emission components HF, HCl and gaseous selenium were found in each case at the higher waste gas temperature of 390 °C. If light soda ash was employed as absorption agent, the higher absorption rates for HF and HCl were achieved at the lower waste gas temperature of 330°C.

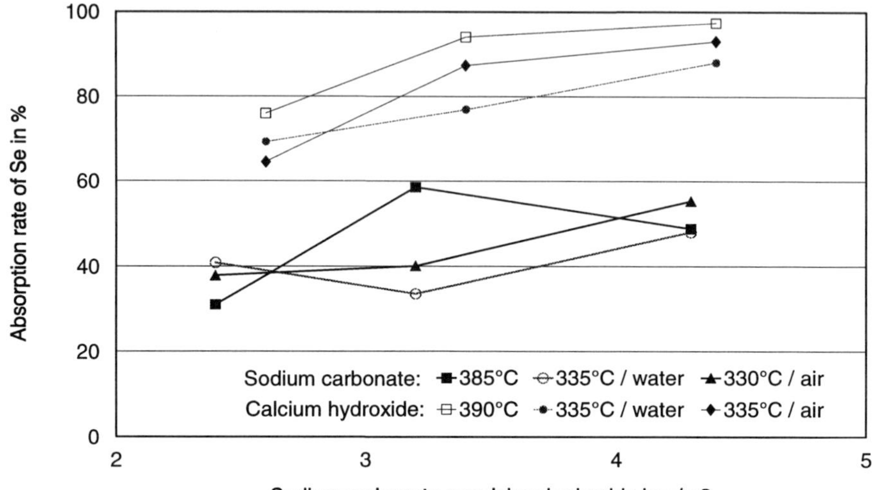

Figure 3.   Selenium absorption rates depending on the amount of light soda ash or calcium hydroxide and on temperature

CONCLUSIONS

In the temperature range about 390°C calcium hydroxide proved to be the better absorption agent with respect to all the waste gas components investigated as compared with light soda ash. Using calcium hydroxide the absorption rates increased with increasing quantities of the absorption agent and with increasing waste gas temperature. For light soda ash, however, the absorption rates for HF and HCl increased with decreasing temperature. But light soda ash can only be used if no significant HF emission problems exists. As a main result of the investigations for industrial practice, it must be pointed out that especially for new glass melting furnaces, and if possible also for the existing ones, waste gas treatment plants, consisting of an electrostatic precipitator with a dry absorption stage connected in series, should be designed for operation in the temperature range between 380 and 400°C. The valid threshold limit value for HF, HCl and gaseous selenium can thus be met in most cases by using calcium hydroxide as absorption agent.

REFERENCES

[1]Technische Anleitung zur Reinhaltung der Luft (TA Luft) v. 27. 2. 1986, *Gem. Ministerialbl.* **A37** [7] 93 (1986).

[2]U. Kircher, „Evapor. of fluorine, chlorine and selenium from glass melts and emiss. reduc. measures", *Glastech. Ber. Glass. Sci. Technol.* **70** [2] 52-57 (1997).

EMISSIONS PRODUCED UPON PROCESSING OF SILICATE GLASS

L.E. Jones, E.D. Larson, T. W. Samadhi and A.G. Clare
New York State College of Ceramics
Alfred University
Alfred, NY 14802

ABSTRACT

Emissions produced during the batching of plate glass have been measured and quantified using high-temperature thermal analysis and mass spectrometry. The emphasis of this paper is the influence of the constituent Na_2SO_4 on the evolution of oxides of sulfur. An on-going study involves the measurement of the rate of thermal decomposition of Na_2SO_4 and SOx release as influenced by the partial pressure of oxygen. Discussed, herein, is the predictive chemistry using thermochemical modeling to evaluate the products of formation from Na_2SO_4 in environments containing air (1mol O_2 and 3.76 mol N_2) and silica.

INTRODUCTION

Emissions production during glass batching and melting is a critical issue for the glass industry. The air toxics produced during glass manufacturing vary as widely as the choices of constituents and processing conditions. The 1990 Clean Air Act Amendments (CAAA) have forced a reevaluation of the influence of glass manufacturing conditions on the evolution of criteria and hazardous air pollutants. Criteria air pollutants include oxides of nitrogen and sulfur (NOx and SOx), ground level ozone (O_3), carbon monoxide (CO), and particulate matter of 10 μm in dimension (PM-10) or less. These criteria air pollutants are emissions for which there exists a minimum or threshold concentrations recognized to produce an impact on health and the environment. The present standards for particulate matter were last revised in 1987. It has been proposed that PM-10 emissions of 50 μg/m^3 annually and 150 μg/m^3 in a 24 hour period are to become stricter. The shift in interpretation is that course particles are those larger than 2.5 μm in diameter, and fine particles are those below 2.5 μm in diameter. The new designation is PM-2.5 and this revised interpretation related to particulates may be widely in place by 1998. A standard of 15μg/m^3 annually and a 24-hour standard of 50μg/m^3 have been proposed for PM-2.5 [1].

In the broad categories of NOx and SOx emissions, nitrogen dioxide (NO_2) and sulfur dioxide (SO_2) are the regulated species. Studies of nitrogen-oxygen chemistry in the context of combustion processes have clearly identified the reaction mechanisms associated with the formation of thermal NOx and the role of the primary product gas NO on the evolution of thermal NOx. Also, many studies have been undertaken on the production of SOx in the context of glass fining. Yet, the influence of combining constituents and the influence of the environment on the rates of evolution of specific species of these oxides are not well understood. Furthermore, the presence of SO_2 has been linked with the formation of PM-2.5 in effluent gas streams. Therefore, the formation of SO_2, the criteria pollutant, is of interest, particularly in the context of glass batching and melting. Standards for SO_2 releases are 0.14 ppm as the 24-hour average and 0.030 ppm, averaged annually [2,3].

PROCEDURE

The approach to this study involved the use of the thermodynamic modeling software HSC Chemistry for Windows produced by Outokumpu Research Oy, Finland. These thermochemical data were used to predict products of glass constituent decomposition in specific environments as well as in the presence of combinations of batch constituents. Thermogravimetric analysis coupled with mass spectrometry was used to study the high temperature evolution of gases. A SETARUM TAG 24 thermal analyzer and a Dycor Mass Spectrometer were the instruments employed in order to study the high-temperature decomposition. A 'typical' plate glass batch composition was evaluated against a commercially available composition. This paper addresses the evolution of SOx from the sodium sulfate (Na_2SO_4) constituent in these batches and the influence of silica and presence of air on SOx release. Work is on-going in which the partial pressure of O_2 on SOx release from Na_2SO_4 is evaluated. From the thermal analysis, regions of steady state mass loss are identified and the gases evolving in these temperature regimes are subsequently identified. Figure 1 is an example of the dramatic zones of mass losses associated with an E-glass batch composition.

From each of these steady state temperature regimes, a temperature is selected and an isothermal thermal analysis is conducted in order to establish the rate of weight loss and concurrent gas evolution. During the isothermal mass loss, gas evolution as well as rates of batch constituent decomposition are measured using mass spectrometry. This decomposition analysis is compared to the thermochemical modeling of the decomposition process.

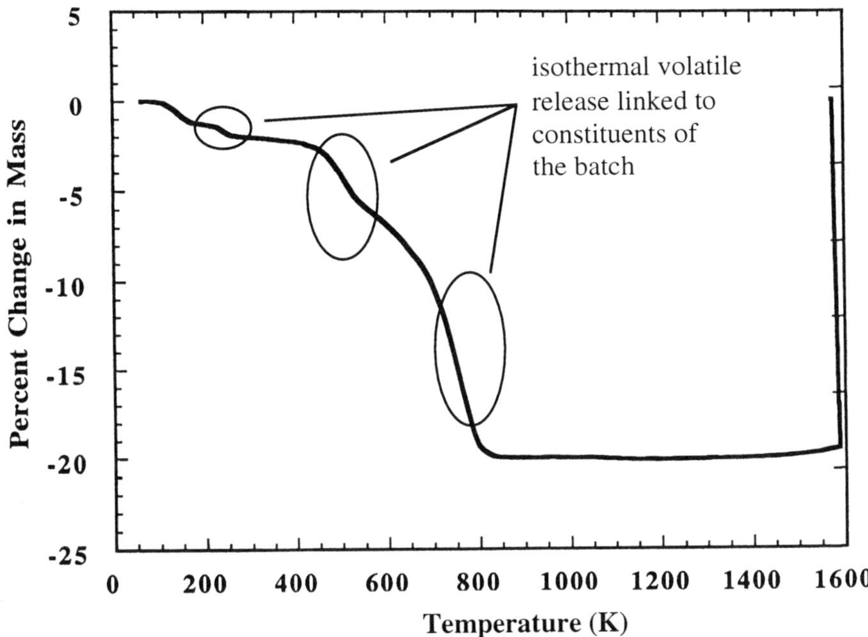

Figure 1. Thermal decomposition of E-glass in which regions of gas evolution are noted.

RESULTS AND DISCUSSION

The plate glass batch composition of interest is listed on a weight % basis in Table 1. The constituent of interest is Na_2SO_4, providing the SO_3 composition in the glass and a small fraction of the Na_2O composition. The total Na_2SO_4 constituent composition in this batch is 1.0 wt. %.

Figures 2 is the thermal decomposition behavior of the silicate glass in air flowing at 10 cc/min. The specimen size is 25.56 mg and the heating rate is 10 °C/min. There are 4 distinct decomposition ranges associated with mass loss and volatile release for this composition. These decomposition ranges for the glass batch are listed in Table 2 along with the primary decomposition species identified by mass spectrometry.

Composition of Glass	Weight %
SiO_2	72.7
Al_2O_3	0.5
SO_3 (from Na_2SO_4)	0.5
CaO (from $Ca(OH)_2$	13.0
Na_2O (from Na_2CO_3)	12.8
Na_2O (from Na_2SO_4)	0.5

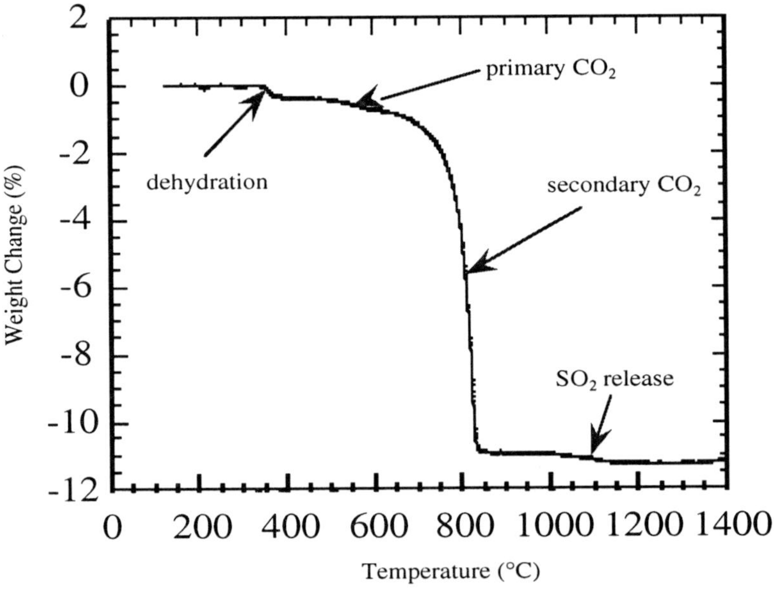

Figure 2. Thermal decomposition of 25.56 mg plate glass batch in air flowing at 10 cc/min.

  Advances in Fusion and Processing of Glass II

Table 2. Decomposition regimes and associated emissions.

Temperature Range, °C	Primary Evolved Gas	Rates of Emissions Release, lbs/hr
300 - 400	H_2O	2.5×10^{-3}
500 - 600	CO_2	-------
700 -850	CO_2	1.0×10^{-2}
1010 - 1130	SO_2	7.7×10^{-4}

The mass spectrum for the decomposition of this plate glass is given in Figure 3.
These data are the effluent gas species produced upon batching 3.7 g of plate glass
at 1400 °C.

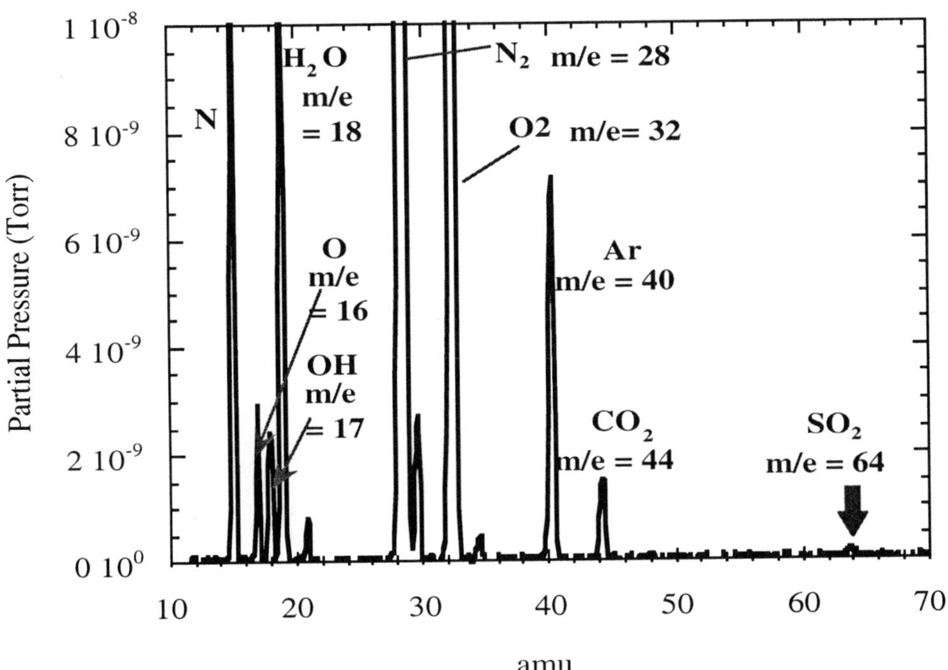

Figure 3. Mass spectrum of emissions produced upon the
decomposition of plate glass in air at 1400 °C.

The partial pressure recorded is the partial pressure of the gas species detected at the quadrapole head of the mass spectrometer. The sampling technique used to acquire these data was to batch all constituents listed in Table 2 simultaneously and to monitor the emission in a continuous mode. By undertaking the analysis in this way, all species previously cited in Table 2 make up the effluent gas. Of particular significance, is the presence of SO_2 at a mass to charge ration (m/e) of 64. The concentration of SO_2 evolved from the batch corresponds to 0.028 vol. % of the total emissions produced.

The following decomposition reactions are considered based upon the constituents that exist in the batch. These reactions and reaction temperatures describe the possible decomposition products and effluent gas species observed in Figure 3.

- Dehydration:

    $Na_2CO_3*H_2O(s)$ — >$Na_2CO_3(s) + H_2O(g)$ (100°C)

    $Ca(OH)_2$ (s) ——-> $CaO + H_2O(g)$ (580°C)

- CO_x Release:

    $Na_2CO_3(s)$ ——-> $Na_2O + CO_2(g)$....$CO(g)$ ...$f(pO_2)$ (850°C)

- SO_x Release:

    $Na_2SO_4(s)$ ——-> $Na_2O + SO_2(g) + 0.5\ O_2(g)$

    $Na_2SO_4(s)$ ——-> SO_3 (g) $+ 0.5O_2$ (g) $+ 2Na$

    SO_3 ------> $SO_2 + SO$

Dehydration can be subdivided into the release of (1) physical water and (2) chemically bound water. The evolution of H_2O observed in our analysis is the result of the decomposition of $Ca(OH)_2$. In the context of the batch, chemically H_2O is evolved at a temperature lower than that predicted in the literature. Oxides of carbon evolve from Na_2CO_3. The evolution of CO or CO_2 is a function of the partial pressure of oxygen and high oxygen concentrations favor CO_2 evolution. There are several potential decomposition reactions linked to the formation of SOx from Na_2SO_4. The decomposition of Na_2SO_4 has been noted to occur at temperatures as high as 1850 °C [4]. The products of this decomposition are described to include Na_2O, and SO_3 or SO_2. Of interest, is the reverse reaction involving SO_2 to form the salt, Na_2SO_3:

    $Na_2O + SO_2$ -----> Na_2SO_3

It has been argued that Na_2O does not exist in the vapor phase, but dissociates further to sodium and oxygen [5, 6].

The decomposition of Na_2SO_4 in air is given in Figure 4. These data were collected for a specimen of sample weight 34.8 mg in 20 cc/min flowing air. The onset of Na_2SO_4 thermal decomposition in air occurred at 1100 °C which is significantly lower than the decomposition temperatures noted previously. The inset in Figure 4 is the mass spectrum of emission collected at 1400 °C in air. These data are given in volume percent having been normalized to the Ar peak concentration which is 1.0 vol. % in these experiments.

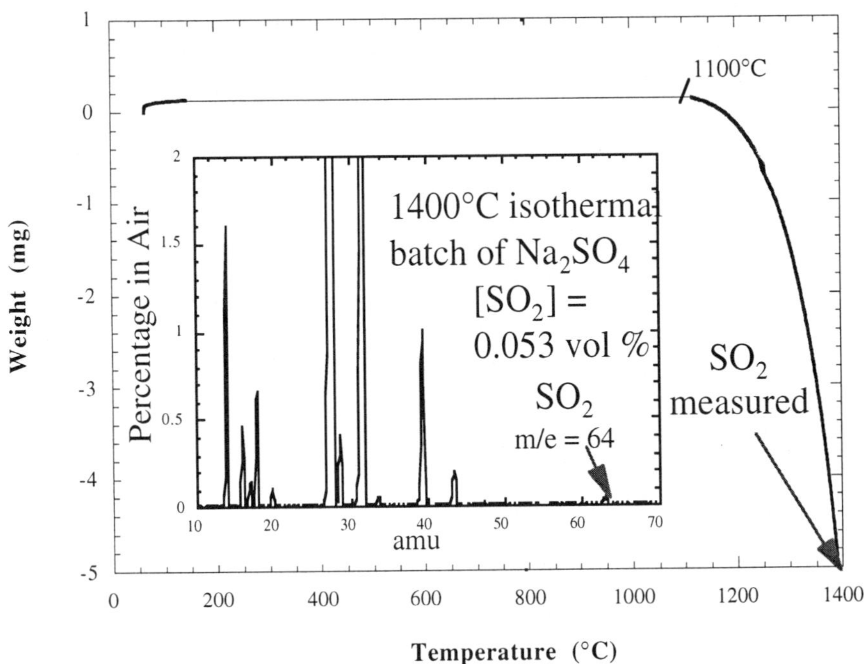

Figure 4. Decomposition of a 34.8 mg sample of Na_2SO_4 in air flowing at 20 cc/min. Inset is the mass spectrum of decomposition recorded at 1400 °C.

The mass loss associated with the decomposition of Na_2SO_4 initiates at 1100 °C and yields SO_2. The concentration of SO_2 evolved is 0.053 vol. %. Sodium sulfate has a molecular weight of 142 g/mole, a density of 2.68 g/cc, and a melting point of 884°C . It decomposition reaction at 1850°C to Na_2O and SO_3 occurs after the phase change according to the following reaction [4]:

$$Na_2SO_4(l) \longrightarrow Na_2O + SO_2(g) + 0.5\ O_2(g)\ ......\ (SO_3).$$

The species $SO_2(g) + 0.5\ O_2(g)$ are often referred to as the composition (SO_3) in the glass literature; however, this volatile fragment is not a primary effluent of SOx in the manufacturing of glass.

In the presence of silica, the sodium sulfate reacts to form a metasilicate at a much lower temperature of 1288°C with the following reaction [7]:

$$Na_2SO_4(s) + SiO_2(s) \rightarrow Na_2SiO_3\ (l) + SO_2(g) + 0.5\ O_2(g).$$

In order to evaluate this chemistry the thermal decomposition of Na_2SO_4 in air and in the presence of SiO_2 is given in Figure 5. It is clear from these data that the presence of SiO_2 accelerates the evolution of SO_2 . The yield of SO_2 is 0.098 vol. %

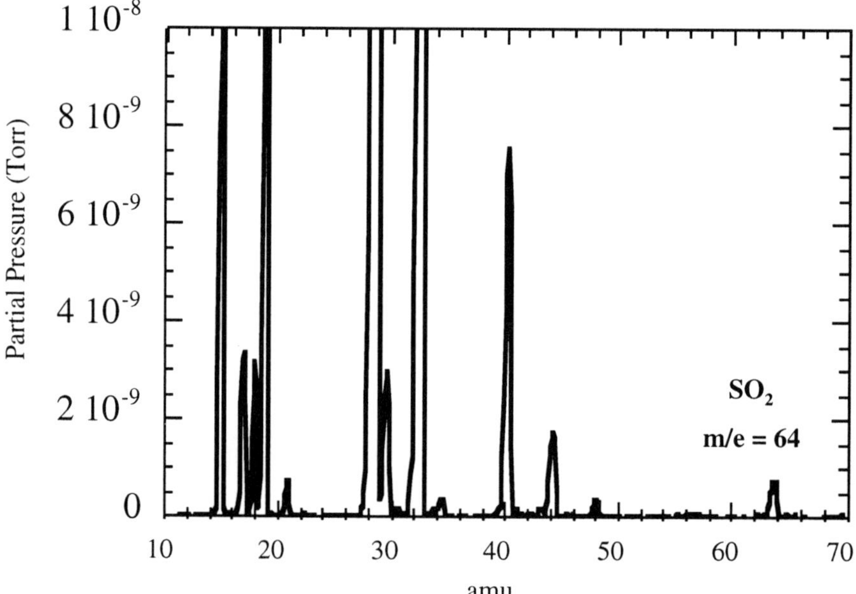

Figure 5. Mass spectrum from the decomposition of Na_2SO_4 and SiO_2 in air at 1400°C. Sample size: 34.8 mg, $[SO_2] = 0.098$ vol. %.

Thermodynamic modeling was employed to assess these decomposition data. Equilibrium calculations for the decomposition of 1 mol Na_2SO_4 in air (1 mol O_2

and 3.76 mol N_2) and the concurrent formation of the oxides of sulfur (SOx); SO, SO_2, SO_3 are given in Figures 6 and 7. These data are plotted as the mol fraction of the species of interest as a function of temperature. In order to calculate these data assumptions regarding the possible product species were made. The following species are considered in these equilibria:

Products: Na_2SO_4, Na_2SO_3, O_2, N_2, SO_3, SO_2, SO, and Na_2O

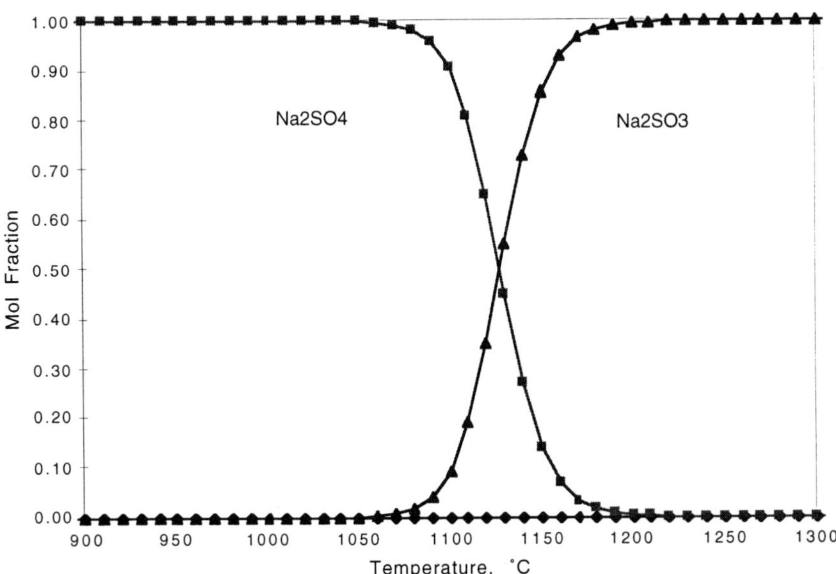

Figure 6. Decomposition of 1 mol Na_2SO_4 in air (1mol O_2 and 3.76 mol N_2) as predicted by thermochemical equilibria calculation. The onset of Na_2SO_4 in air is 1040 ˚C.

In air (1 mol O2 and 3.76 mol N2) the predicted onset of Na_2SO_4 decomposition in is 1040 ˚C and this corresponds to the formation of Na_2SO_3 and the release of SO_2 and SO_3. The primary oxide of sulfur is SO_2 and the maximum concentration evolved is 1.4 x 10^{-6} mol fraction at 1120 ˚C. The concentration of SO_3 produced is extremely small. The maximum concentration of SO_3 evolved in air is 4.8 x 10^{-8} mol fraction at 1110 ˚C.

Plotted in Figure 7 are the SOx concentration evolved from 1mol of Na_2SO_4 alone in the absence of air. The maximum SO_2 and SO_3 concentrations are 2.9 x10^{-7} and 1.7 x 10^{-8} mol fractions, respectively. Both maximum occur at 1140 °C. The presence of air increases the maximum in the SO_2 and SO_3 release and shifts the decomposition temperature lower.

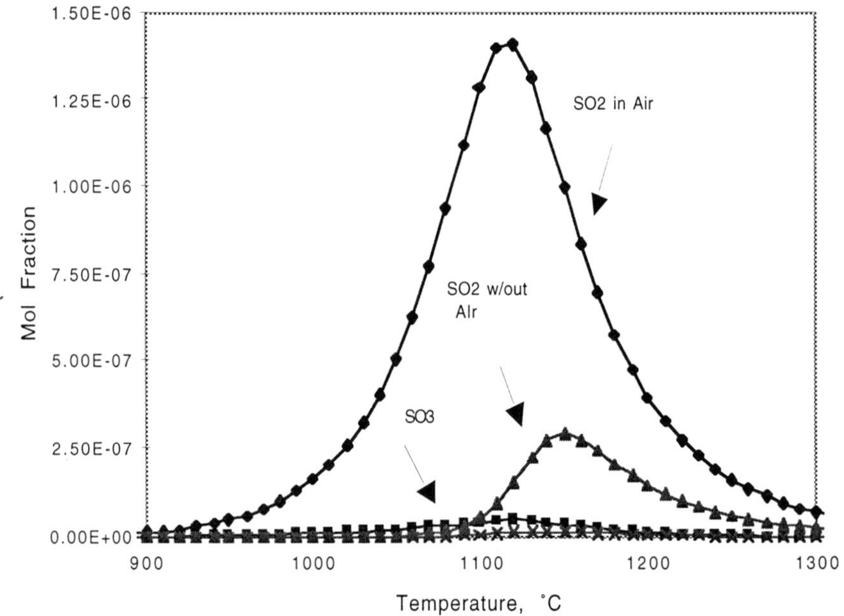

Figure 7. Formation of oxides of sulfur as predicted by thermochemical calculations for the decomposition of 1 mol Na_2SO_4 in air (1mol O_2 and 3.76 mol N_2) and 1 mol Na_2SO_4 alone. The presence of air favors the formation of SO_2.

Of interest in these calculations is the significant concentration of Na_2O produced which is given in Figure 8. No significant concentration of Na_2O is produced via this decomposition in the presence of SiO_2. In the presence of air, the decomposition of Na_2SO_4 yields a maximum concentration of Na_2O of 1.5 x 10^{-6} at 1120 °C. The formation of Na_2O produced mirrors the formation of SO_2. There is a unique shape to the SOx data as predicted via the HSC thermochemistry. It suggests that the evolution of SOx is linked to its presence being tied to the formation of other species. There is a formation and decomposition of these species over a relatively limited temperature range. The concurrent formation of

Advances in Fusion and Processing of Glass II

Na_2SO_3 accounts for the removal of these species in these calculations. These data suggest that in air Na_2SO_3 is the final product and that SOx and Na_2O are the intermediate compositions in a closed system.

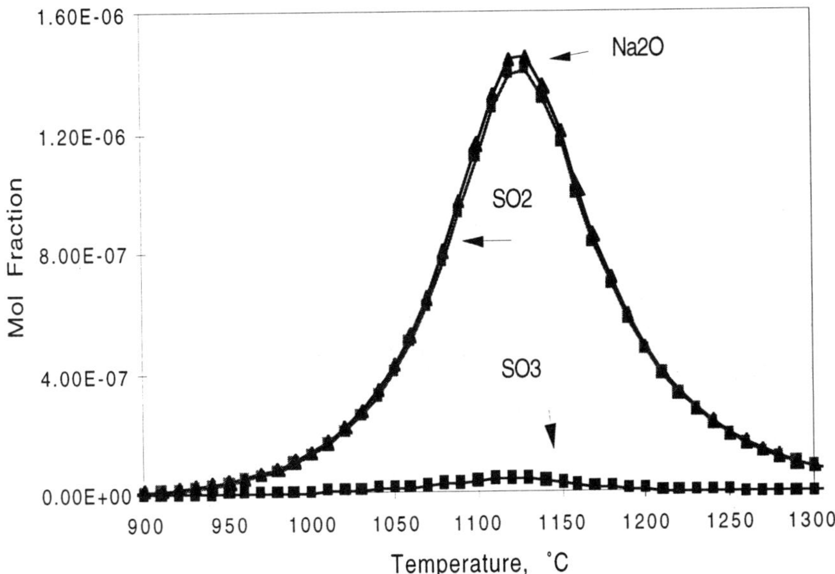

Figure 8. The formation of Na_2O from the decomposition of Na_2SO_4 in the presence of air (1 mol O_2 and 3.76 mol N_2) mirrors the formation of SO_2 as predicted via thermochemistry.

Equilibria calculations were undertaken for the decomposition of 1 mol of Na_2SO_4 with 1 mol of SiO_2 in air. These data appear quite similar to those in Figure 6. However, under these conditions, there is again a shift in the temperature of decomposition.   In the presence of SiO_2 and air the on-set of decomposition of Na_2SO_4 occurs at 850 °C and without SiO2 this decomposition temperature shifts to 1040 °C. The products of decomposition include the metasilicate, Na_2SiO_3. The maximum in the formation of Na_2SiO_3 occurs at 1110 °C to yield 1.8×10^{-2} mol fraction. The decomposition of 1 mol Na_2SO_4 and 1 mol SiO_2 in air produce oxides of sulfur is given in Figures 9.

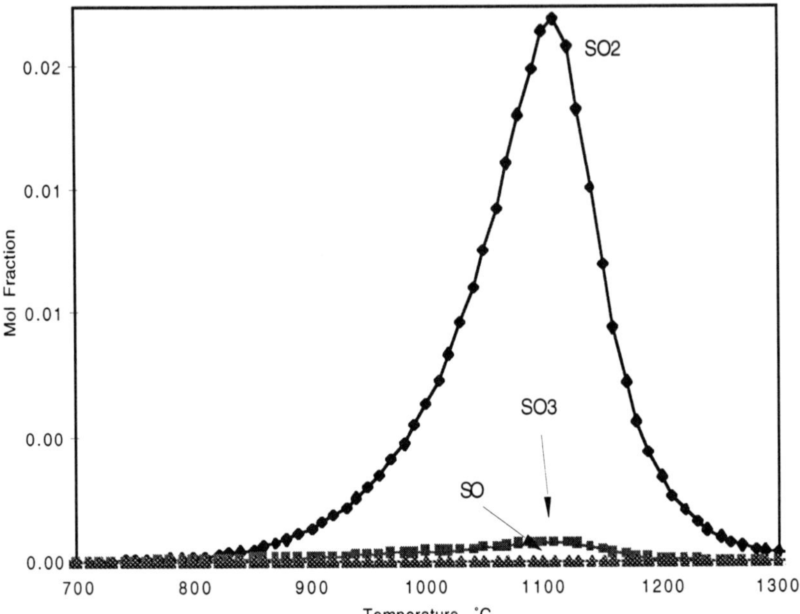

Figure 9. Formation of oxides of sulfur as predicted by thermochemistry for the decomposition of 1mol Na_2SO_4 in air (1 mol O_2 and 3.76 mol N_2) in the presence of 1 mol SiO_2.

Again, SO_2 is the primary oxide of sulfur produced during the decomposition of Na_2SO_4 in air and in the presence of SiO_2. The maximum concentrations of SO_2 and SO3 from this reaction are 1.75 x 10-2 and 6.1 x 10-4 mol fractions, respectively, at 1110 ˚C. These data compare to the decomposition of 1 mol Na_2SO_4 in air in the absence of SiO_2. With SiO_2 present, the concentration of SO_2 produced is 10^{+4} times larger.

The rate of Na_2SO_4 decomposition is of interest simply due to the fact that the control of emissions releases involves an understanding of the rates of release and the controlling parameters associated with the rate of release. Work in this area is on-going. Rates of Na_2SO_4 decomposition in air are given as an initial reference. Figure 10 is the isothermal thermogravimetric analysis of the decomposition of 30 mg samples of sodium sulfate in air at 1100, 1200, 1300, and 1400 ˚C. These data are obtain by heating Na_2SO_4 to the isothermal temperatures of interest . Once the temperature is reached, the oxidizing gas is introduced. Rates in mg - SO_2

                    Advances in Fusion and Processing of Glass II

evolved/ mg- initial wt of Na_2SO_4 per min. are taken from the region of steady state mass loss. Rates taken from these data are plotted in an Arrhenius format in Figure 11. An activation energy of 266 kJ/mol is calculated for the decomposition process in 1 mol O_2 and 3.76 mol N_2.

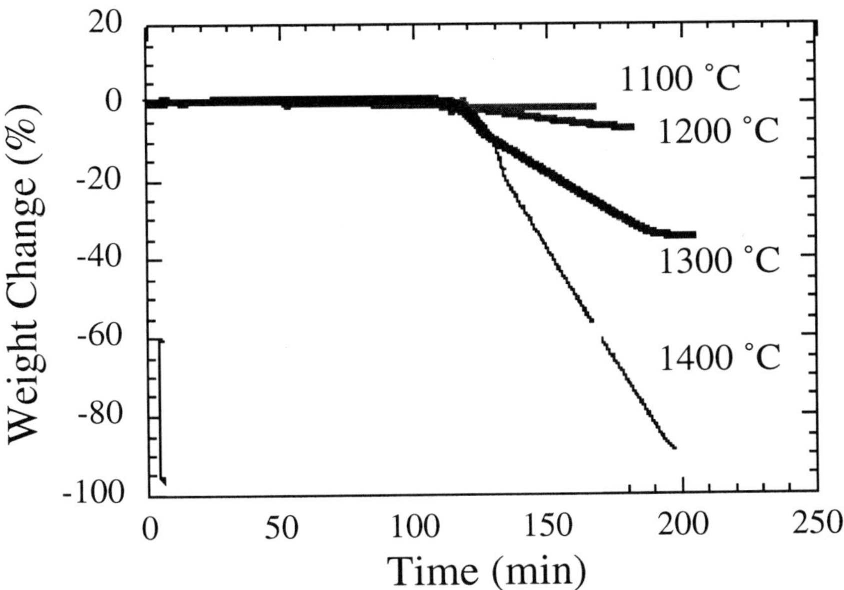

Figure 10. Weight change versus time for Na_2SO_4 decomposition in air (1 mol O_2 and 3.76 mol N_2) at 1100, 1200, 1300, and 1400 °C.

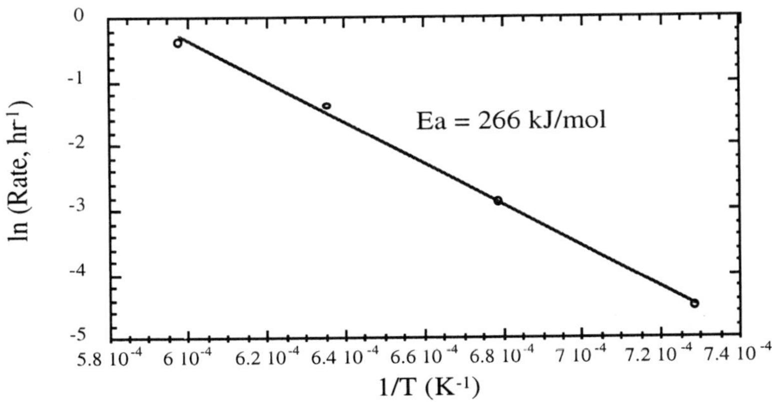

Figure 11. Rates of decomposition of Na_2SO_4 in air (1 mol O_2 and 3.76 mol N_2) as a function of temperature.

SUMMARY

The volatiles evolved from a plate glass composition containing Na_2SO_4 include the release of SO_2 and SO_3 at temperatures in excess of 1100 °C. The environment over Na_2SO_4 does influence the concentration of SO_2 (the primary oxide of sulfur) evolved. In the presence of 1 mol of O_2, the decomposition of 1 mol of Na_2SO_4 yields 1.4 ppmv SO_2 at 1040 °C as predicted via thermochemistries. These data correspond roughly to the temperature and concentration of SO_2 detected via thermal analysis for the decomposition of Na_2SO_4 in air. The decomposition of Na_2SO_4 in air involves the formation of Na_2SO_3 and evolution of SO_2 and SO_3 as well as the formation of Na_2O. The evolution of Na_2O and SO_2 is an intermediate step in the formation of Na_2SO_3. These chemistries are altered in the presence of SiO_2. Na_2O is not a significant volatile product of this reaction. Instead, metasilicates are favored and concentrations of SO_2 increases dramatically. These thermochemical calculations require a reevaluation of earlier assumptions regarding possible products of decomposition in the context of O_2, N_2 and silicate chemistries. A more complete study of the formation of additional metasilicates is on-going. These considerations will evaluate the following species:

$$NO, \ NO_2, \ O, \ O^*, \ S, \ Na_2S, \ and \ Na$$

Rates of decomposition of Na_2SO_4 in air are being evaluated. The Ea for this reaction taken from the steady-steady mass loss in the temperature range from 1100 - 1400 °C is 266 kJ/mol.

REFERENCES

[1]EPA Air Quality Proposed Rules issued November 27, 1996, http://www.epa.gov.

[2] Jones, L. E., Clare, A. G. and Gerling, D."Part II: Emissions Produced by Batching and Melting of Silicate Glass," Glass Researcher, 4 [1] (1994).

[3]Breininger, S. "Guide to Air Pollution Regulations that Affect US Glass Industry," Glass Researcher, 5 (1995).

[4]Jebsen-Marwedel, H. "Glastechnische Fabrikationsfehler: "Pathologische" Ausnahmezustande des Werkstoffes Glas und ihre Behebung; eine Bruckezwischen Wissenschaft, Technologie und Praxis," 3rd Edition, Springer-Verlag; Berlin, New York 1980.

[5]Conroy, A. R., Manring, W. H., and Bauer, W. C., "Part I: The Role of Sulfate in the Melting and Fining of Glass Batch, "THE GLASS INDUSTRY, 84 - 110, 2 (1966).

[6]Brewer, L. and Margrave, J. "Vaporizations of Alkali Metal Oxides," U. of CA Radiation Laboratory Report., UCRL-1864 (1964).

[7]Jebsen-Marwedel, H. and Becker, A., "Der SO3 - Gehalt im Glas," Glastechn. Ber. 8[9] (1930).

THE STATUS OF THE GLASS INDUSTRY IN GERMANY-TECHNOLOGICAL CHALLENGES TO GLASS MELTING AND PROCESSING

Helmut A. Schaeffer
Research Association of the German Glass Industry (HVG)
German Society of Glass Technology (DGG)
Frankfurt/Main, Germany

INTRODUCTION

In the manufacturing of soda-lime-silica based glasses, which represent more than 90% of the global glass production, the major technological challenges are nowadays the improvement of glass quality and thus the improvement of reproducibility of glass product specifications. Modelling of the glass melting process and monitoring of glass melt properties play a decisive role in realizing these objectives. This holds not only for the flat glass industry where it is mandatory to produce almost "optical" glass quality in the bulk and highly reproducible glass surfaces for coating purposes, but also in the container glass industry where stringent customer requirements have to be fulfilled with respect to glass homogeneity and colour and also in view of the production of light weight containers.

OPTIMIZATION OF THE GLASS MELTING PROCESS

The required optimizations of the glass melting process with respect to glass quality have to be adjusted nowadays to energetic, ecological and economic constraints which are enforced upon the glass industry in various degrees depending on the geographical location, but probably most stringently in Germany and Western Europe.

Utilization of Energy

The technological challenges in energy saving are determined by new glass furnace designs, by the choice of fuel, by the exploitation of waste gas heat and by the use of recycled cullet.

Glass Furnace Design: A study on conventionally fired glass furnaces in Germany in the fields of container and flat glass production was carried out recently [1] and compared with the situation in the year 1962 when a similar study was performed. Out of the 100 glass furnaces in Germany roughly 50% are regeneratively end-fired, 40% regeneratively cross-fired, and 10% recuperatively cross-fired. Natural gas has become the main fuel with 56% followed by heavy oil with 44%. Table I shows the glass production, the number of glass melting furnaces and the average glass pull.

Table I. Glass production in Germany for container and flat glass

| year | container glass | | | flat glass | | |
	production [t]	number of units	average pull [t/d]	production [t]	number of units	average pull [t/d]
1962	1.4×10^6	~ 130	~ 30	0.5×10^6	~ 45	~ 33
1993	4.2×10^6	~ 100	~ 115	1.6×10^6	~ 18	~ 237

Typical for the container glass industry in Germany is the high specific load of their furnaces which range between 2.8 and 4.0 $t/(m^2 d)$. More than 50% of the furnaces possess electric boosting. The mean specific energy consumption of all glass furnaces amounted to 4900 kJ/kg in the year 1993. Presently with new furnaces and high rates of recycled glass energy consumptions as low as 3300 kJ/kg can be achieved for the melting of container glass.

Preheating of raw materials and recycled cullet is a convincing measure to reduce energy consumption. It has been reported that up to 20% energy saving can be achieved [2]. A special glass furnace type was introduced for the first time in Germany, namely the so-called $LoNO_x$-melter conbining cullet preheating with recuperative waste heat recovery thus resulting in low specific heat consumption (3300 kJ/kg) and low NO_x emissions (< 500 mg/m^3 at 8% O_2) [3].

Recycling of cullet: The recycling of foreign cullet was started in Germany in the mid-seventies [4]. Ecological constraints, in particular the limited availability of landfill sites, initiated the large scale recycling of glass packaging products such as bottles and jars. The development of glass recycling is shown in Table II for the case of Germany.

Advances in Fusion and Processing of Glass II

Table II. Recycling of packaging glass in Germany

year	domestic sale in 10^6t	recycled glass in 10^6t	recycling rate %
1974	2.3	0.15	6.5
1980	2.5	0.57	23.0
1985	2.4	1.05	43.5
1990	3.3	1.8	54
1992	3.8	2.3	60
1994	3.7	2.8	75
1996	3.6	2.8	75.7

Table III. Container glass production and amount of recycled glass in Germany

year	production of container glass				amount of recycled cullet			
	total in 10^6t	flint in %	amber in %	green in %	total in 10^6t	flint in %	amber in %	green in %
1988	3.3	44	22	34	1.3	23	7	70
1990	3.7	46	21	33	1.8	26	8	66
1992	4.4	50	22	28	2.3	33	12	55
1994	4.4	52	19	29	2.8	47	16	37
1996	4.5	55	19	26	2.8	49	15	36

The recycling rate in Germany has reached a value of close to 76% (in 1996) corresponding to 2.8 x 10^6 tons of recycled cullet. In terms of absolute tonnages of recycled glass the German glass industry is leading, but with respect to national recycling rates it is surpassed by Switzerland (85%), The Netherlands (80%) and Austria (78%). The total recycled glass of the European Container Glass Industry amounts to 7.15 x 10^6 tons in 1995 which corresponds to about 50% of the container glass consumption.

Table III shows the container glass production in Germany with respect to colour and the corresponding amount of recycled cullet.

As compared to the recycling of cullet, the recycling of filter dust was started in Germany only a few years ago. Filter dust is generated as an undesirable by-product as a result of waste gas treatment with the help of an absorption medium and by employing either electrostatic precipitators or bag filters. Utilizing calcium hydroxide as absorption medium the filter dust consists basically of calcium sulphate and unreacted calcium hydroxide. It turned out that the most elegant way to dispose of the filter dust is its addition to the glass batch where it can replace the refining agent. The quantity of filter dust amounts to about 3 kg/t glass, which corresponds e.g. for the German glass industry to about 18×10^3 t of filter dust per year. Thus the recycling of filter dust is an economical relevant step since it relieves the glass industry from the high costs associated with the storage of special waste.

The large scale utilization of recycled cullet requires the removal of impurities in order to ensure stability in glass homogeneity and coloration. The recycled cullet has to be separated from ferrous and non-ferrous metals, ceramic materials (porcelain, stones, slags) and organic components (plastics, paper, food residues). Even though remarkable improvements in the separation of impurities were achieved over a time period of 15 years (see Table IV), yet, no absolute "purification" of the cullet can be realized, especially with respect to differently composed glasses, such as lead oxide - containing glasses (lead crystal glass, TV glass), borosilicate glasses, glass ceramics and fused silica. In view of legal regulations which limit e.g. the level of heavy metals in container glass the container glass industry in Europe is alert to ban the introduction of other glass products into the recycled cullet stream.

Table IV. Average annual data of impurities in processed recycled cullet according to colour

[g/t] = ppm	1981	1987	1994	1996	
cps*)	115	51	31	27	
Al	54	9	2	2	flint glass
Pb	18	2	1	0.1	
Fe	1.6	0.5	1.7	1.4	
cps*)	197	46	23	22	
Al	234	69	17	7	green glass
Pb	98	12	7.5	1.1	
Fe	1.6	0.05	1.4	2.6	
cps*)	234	57	42	74	
Al	173	36	8	4.4	amber glass
Pb	87	8	3	0.4	
Fe	0.4	0.1	1.4	3.2	

*) cps = ceramics, porcelain, stones

     Advances in Fusion and Processing of Glass II

An interesting approach to circumvent the detrimental effect of cps-impurities is grinding and milling of cullet. If the grain size of the ceramic particles does not exceed that of the glass powder, these particles will dissolve completely during glass melting. Larger particles can be removed by sieving the glass powder [5].

The use of recycled cullet has created new challenges for glass melting, but has also improved the state-of-the-art of glass melting.

Waste Gas Cleaning

Further challenges occur due to the fulfillment of legal regulations in the field of waste gas cleaning, in particular in Germany, relating to the precipitation of dust and the absorption of toxic volatiles. The German federal regulation (TA Luft, published in 1986) requires rather stringent threshold limit values for HCl ($30mg/m^3$); HF, Pb, Cr, Sb (5 mg/m^3); Se, Ni, As (1 mg/m^3) and Cd (0.2 mg/m^3). For dust this limit is set at 50 mg/m^3, for SO_x at 1800 mg/m^3, whereas the NO_x-limit is in the process of revision, and values of 800 mg/m^3 have to be attained, while target limits of 500 mg/m^3 are being discussed. All these limits are referred to dry waste gas at standard conditions with an O_2 content of 8%. Regional regulations very often foresee even lower limits. Nearly all glass furnaces in Germany are equipped with waste gas treatment plants using predominately electrostatic precipitators.

In the case of reduction of NO_x emissions no generally accepted technique exists. In Germany the activities in this field are focused to "primary" methods, i.e. preventing the formation of NO_x by lowering flame temperatures and/or minimizing excess air. A radical primary solution for preventing NO_x formation is the employment of oxygen instead of air in furnace combustion. As for "secondary" methods, there are at least two general techniques for the decomposition of NO_x, either reduction by NH_3 or reduction by fuel ("reburning"). Reducing NO_x by NH_3 can be achieved either at high temperatures (900-1100°C) without the help of a catalyser or at lower temperatures (<350°C) utilizing a catalyser. Ample experience has been gathered in German glass plants with all of these NO_x emission reducing techniques [6].

Process Automation

Finally, the permanent challenge of cost-reducing measures, especially in increasing the productivity and the production yields, has accelerated the implementation of process automation, e.g., employment of sensors, data processing systems, furnace control systems and mathematical modeling of the glass melts in the furnace (heat transfer, convective flow, residence time , fining, homogenization, redox state), and thus helps to short-cut engineering developments and decisions [7]. In Germany the emphasis has been placed on the development of sensors for the continuous in-situ characterization of the glass melt with respect to redox control [8].

OXIDATION STATE OF GLASS MELTS

It is well known that the oxidation state or redox state of a glass melt is of great importance with respect to glass melting behavior, glass forming processes and properties of the final glass product [9, 10]. The terms "oxidation state" or "redox state" are descriptive; the measurable quantity is the oxygen partial pressure (in the following: oxygen activity) of the glass melt which affects glass properties predominantly via the ratio of valence states of polyvalent elements, which are added to the glass batch either as coloring (e.g. Fe, Cr, Se, Co, Mn) and/or refining (S, As, Sb) agents or are introduced unintentionally as impurities (Fe, Ti) together with the raw materials.

Properties which are influenced by the redox state of polyvalent elements are e.g. refining behavior, heat absorption, solubility and diffusivity of gases, coloring and decoloring, volatilization of species from the glass melt, surface tension, viscosity, optical transmission. Therefore, the oxygen activity of the glass melt has a pronounced impact on "melt history," "workability," i.e., the forming behavior in the visco-elastic region of the glass melt and finally on the reproducibility of the glass properties of the final product. Industrially melted glasses are never in equilibrium with the oxygen partial pressure of the furnace atmosphere, but are determined by the oxidizing and reducing additions of the glass batch, the convective flows in the glass tank and the temperature history. Especially the varying amounts of organic impurities in recycled glass which give rise to fluctuations of the glass melt properties has stimulated the demand for a continuous in situ measurement and control of the oxidation state.

In Situ Measurement of Oxidation State

Electrochemical methods can be exploited in order to monitor the oxygen activity and the quantities of polyvalent ions in the glass melt. The oxygen activity measurements can be carried out with the help of an electrochemical cell based on an oxygen-ion conducting solid electrolyte (yttrium-stabilized zirconia) as a reference electrode [11]. For some years, experience has been gained in the container glass industry by placing the oxygen activity sensor in contact with the glass melt in the forehearth (1200-1250°C), a location of measurement optimally suited for characterizing the conditioned glass melt prior to the feeding into the glass forming machines [12, 13].

In order to determine the quantities of polyvalent ions in situ voltammetric measurements can be utilized. In particular, square-wave voltammetry [14, 15] has proved to be a powerful tool in conjunction with oxygen activity measurements to determine the concentration and redox state of polyvalent elements. First studies have shown that under industrial conditions polyvalent elements, especially iron, can be detected [16, 17].

Both electrochemical sensors are appropriate for establishing an in situ control system which provides an early warning for deviations of the target oxidation state or the concentration of polyvalent elements. Furthermore, such sensors can be used to control batch additions and furnace process parameters in order to achieve the required redox state and thus to adjust those glass properties which depend on oxygen activity.

Application of Electrochemical Measurements

Electrochemical measurements have proved to be a valuable tool especially in the container glass industry when melting green or amber glass with high portions of recycled cullet. The usefulness of in situ oxygen activity measurements is given e.g. in ensuring colour stability, optimizing colour changes of glass melts, controlling foam formation and maintaining a constant heat transfer into the glass melt. An interesting correlation between oxygen activity and seed formation was found in a furnace melting green glass with 86% cullet, see Figure 1, [18].

Figure 1   Correlation between oxygen activity and number of seeds measured in a green glass melt with 86% cullet

It was observed that a certain decrease of oxygen activity (due to the reducing effect of organic impurities) causes the formation of fine bubbles (so-called "seeds"). The bubble formation (release of SO_3) is triggered by the decomposition of Na_2SO_4 (refining agent) and the fact that the sulfur solubility in the glass melt decreases with lower oxidation state. Therefore, reducing melt conditions give rise to increased SO_3 bubble formation. This example shows the advantage of a continuous monitoring of the oxygen activity to provide an oxidation state which is high enough to prevent the formation of glass defects (seeds).

The usefulness of in situ voltammetric measurements is given e.g. in monitoring the iron, chromium and sulfur content in green or amber glass melts.

As an example, in an industrial amber glass melt the iron content was determined in situ by means of square-wave voltammetry over a period of 9 months. The results are in very good agreement with the data obtained by X-ray fluorescence spectroscopy, see Figure 2, [17]. The changes of iron content are due to fluctuations of the iron concentration of the added recycled cullet to the batch.

Figure 2   In situ determination of iron concentration by means of square-wave voltammetry over a period of 9 months in an industrial amber glass melt

Recently the sensitivity of the voltammetric measurements was improved considerably at the laboratory scale. Iron contents as low as 0.004 mol% Fe_2O_3 can be determined quantitatively [19]. Presently work is carried out in industrial glass melts to determine simultaneously the content of iron, chromium and sulfur via an improved square-wave voltammetry.

CONCLUSIONS AND OUTLOOK

In the manufacturing of soda-lime-silica glasses the major technological challenges are nowadays the improvement of glass quality and the improvement of the reproducibility of product specifications. These objectives can be pursued successfully only by taking into account the present energetic, ecological and economic constraints.

The high energy costs in Germany as well as the stringent regulations for dust and flue gas emissions have forced especially the container glass industry to utilize high portions of recycled glass and to optimize the glass melting process.

The continuous in situ characterization of the redox state of glass melts represents a powerful tool for the detection of deviations from the target oxidation state and thus from the target glass properties. In the near future it will be possible to provide a redox control system combining in situ oxygen activity measurements with simultaneous voltammetric measurements thus resulting in an in situ determination of the concentration and valence state ratio of polyvalent elements in the glass melt.

REFERENCES

[1]B. Fleischmann, "Conventionally fired glass furnaces for the production of container and flat glass in the German speaking region - Results of a HVG study - Part 1. Firing, Furnace Design, Capacity", (in German) *Glastech. Ber. Glass Sci. Technol.* **70** [2] N11-16 (1997).

[2]G. Enninga, K. Dytrych and H. Barklage-Hilgefort, "Practical experience with raw-material preheating on glass melting furnaces", *Glastech. Ber. Glass Sci. Technol.* **65** [7] 186-191 (1992).

[3]R. Ehrig, J. Wiegand and E. Neubauer, "Five years of operation experience with the SORG LoNO$_x$ Melter", *Glastech. Ber. Glass Sci. Technol.* **68** [2] 73-78 (1995).

[4]H.A. Schaeffer, "Recycling of cullet and filter dust in the German glass industry", *Glastech. Ber. Glass Sci. Technol.* **69** [4] 101-106 (1996).

[5]B. Führ, B.-H. Zippe and H. Drescher, "New technology in waste glass recycling: glass grinding", (in German) *Glastech. Ber. Glass Sci. Technol.* **68** [5] N63-69 (1995).

[6]U. Kircher, "Present status of NO_x reduction by primary and secondary measures in the German glass industry"; pp.149-154 in *Proceedings of XVII International Congress on Glass,* Vol. 7. Edited by Chinese Ceramic Society, Beijing, 1995.

[7]R. A. Bauer, G. Peters, E. Muijsenberg and F. Simonis, "Advanced control of glass tanks by use of simulation models"; pp.31-38 in *Proceedings of the 4th International Conference Advances in Fusion and Processing of Glass.* Edited by H. A. Schaeffer and L. D. Pye. Glastech. Ber. Glass Sci. Technol **68** C2 (1995).

[8]H. A. Schaeffer and H. Müller-Simon, "Redox control of glass melts - a tool to improve glass quality"; pp.69-72 in *Glass Production Technology International.* Edited by R. Kent. Sterling Publications Ltd., London, 1995.

[9]H. A. Schaeffer, T. Frey, I. Löh and F. G. K. Baucke, "Oxidation state of equilibrated and non-equilibrated glass melts", *J. Non-Crystalline Solids* **49**, 179-188 (1982).

[10]A. Lenhart and H. A. Schaeffer, "The determination of oxidation state and redox behavior of glass melts using electrochemical sensors"; pp.147-154 in *Proceedings of the XIV. International Congress on Glass*, New Delhi, Vol. I. Edited by Indian Ceramic Society, Calcutta, 1986.

[11]H. Müller-Simon and K. W. Mergler, "Electrochemical measurements of oxygen activity of glass melts in glass melting furnaces", *Glastech. Ber.* **61** [10] 293-299 (1988).

[12]H. Müller-Simon, K.W. Mergler and H.A. Schaeffer, "Monitoring of sulfate refining by means of zirconia-bassed oxygen sensor", *Glastech. Ber.* **63K**, 261-270 (1990).

[13]H. Müller-Simon, K .W. Mergler and H. A. Schaeffer, "Electrochemical monitoring in industrial glass melts"; pp. 39-44 in *Proceedings of the XVII International Congress on Glass*, Vol.6. Edited by Chinese Ceramic Society, Beijing, 1995.

[14]E. Freude and C. Rüssel, "Voltammetric methods for determining polyvalent ions in glass melts", *Glastech. Ber.* **60** [6] 202-204 (1987).

[15]C. Montel, C. Rüssel and E. Freude, "Square-wave voltammetry as a method for the quantitative in-situ determination of polyvalent elements in molten glass", *Glastech. Ber.* **61** [3] 59-63 (1988).

[16]M. Zink, C. Rüssel, H. Müller-Simon and K.W. Mergler, "Voltammetric sensor for glass tanks", *Glastech. Ber.* **65** [2] 25-31 (1992).

[17]H. Müller-Simon and K.W. Mergler, "On-line determination of the iron concentration in industrial amber glass melts", *Glastech. Ber. Glass Sci. Technol.* **68** [9] 273-277 (1995).

[18]M. Beutinger, "Use of recycling glass in the container glass melt" (in German), *Glastech. Ber. Glass Sci. Technol.* **68** [4] N51-58 (1995).

[19]O. Claußen and C. Rüssel, "Quantitative in-situ determination of iron in a soda-lime-silica glass melt with the aid of square-wave voltammetry", *Glastech. Ber. Glass Sci. Technol.* **69** [4] 95-100 (1996).

Modeling of Melting and Forming

MODELING OF GLASS MELTING FURNACES: APPLICATIONS TO CONTROL, DESIGN AND OPERATION OPTIMIZATION

M.G. Carvalho, N. Speranskaia and J. Wang
Instituto Superior Técnico/Technical University of Lisbon - Mechanical Engineering Department
Av. Rovisco Pais, 1096 Lisboa Codex PORTUGAL

M. Nogueira
IrRADIARE, Research & Development in Engineering and Environment, Ltd.
Tagus Park, Núcleo Central 2780 Oeiras PORTUGAL

ABSTRACT

The present paper describes the strategy for implementation of mathematical modeling of glass melting furnaces in control, design operation optimization and training. Prediction codes capable of computing the three-dimensional characteristics of the aerodynamics, mixing, combustion (single or multiphase), pollutants formation and heat transfer in the combustion chamber of glass melting furnaces exist. Furthermore, models for the heat transfer, chemical reactions, phase change and melting down process occurring in the batch region have been developed. Considerable effort has been placed in the development of glass tank models for the fluid flow and heat transfer inside the tank and including sub-models for air bubbling, sand grain dissolution, redox and fining. The present paper describes the state of art of the computer simulation of glass melting furnaces (combustion chamber, batch region and glass tank) and identifies the current drawbacks of such models. Based on the current scenario of research and development, future research priorities are outlined. A mathematical model of combustion chamber, glass tank and batch region is described and its application to an end-port container glass tank furnace is discussed. A measurements campaign was carried out in this furnace. Measurements have been made for in flame mean gas species concentrations of O_2, CO, CO_2, NO_x and mean gas temperature. The results obtained with the model are in good agreement with the measured ones. The paper describes the basic concepts and research strategies involved in an enhanced utilization of modeling technologies for optimized design, operation and control of glass melting furnaces. An innovative procedure for the integration of physically-based and model-based knowledge with advanced optimization algorithms is described. The proposed methodology constitutes a technological contribution towards a new generation of systems able to assist the design, the operation and the control more efficient, cleaner and more intensive glass melting furnaces. Such system may be applicable in operation optimization (if on-line used), design optimization (if off-line used) and operators training (if simulated scenario generation is used). Examples of the implementation of these strategies in industrial environments are presented such as the design optimization of an oxy-fuel furnace, the model-based control and the development of an expert system for operation optimization and training in glass melting furnaces.

INTRODUCTION

Glass is one of the oldest material of the world, but still plays a very important role in the daily life of the humanity. The application of glass may be found in industry and technology, science and art. However, the physicochemical processes of glass formation are still required to be understood.

Glass is normally produced in open-hearth furnaces in which the glass constituents are flame-heated from above to form a melt. The raw materials, sand, limestone, aragonite, dolomite, soda ash, alumna, and recycled glass are fed. The charge reacts and melts, and subsequently exits through the feeders at the opposite end of the furnace.

Energy saving, pollution abatement, increase of furnace lifetime and glass quality improvement are important requirements of the glass industry, which demands new and powerful research methods for better understanding of thermal physical phenomena occurring in a glass furnace, providing information about the furnace performance, improving new furnace design, construction, operation and control, and proposing solutions for glass quality problem. The use of physical models is limited by the difficulty of a confident scale-up criterium of the several properties involved, which is due to the importance of the radiative transfer as well as to the peculiar glass viscosity behavior. Dimensional analysis developed for the fluid dynamics and heat transfer on the beginning of our century gave birth to mathematical modeling approach. Last twenty years with the development of more powerful computers, the mathematical models of glass melting tanks coupled with combustion chamber models have become an important tool for the glass furnace engineers. Now it is possible to solve the complex equations describing the behavior of fluids inside glass furnace and the chemistry of melting, bubble behavior, combustion and to consider the radiative heat transfer.

CURRENT STATE OF COMPUTER SIMULATION OF INDUSTRIAL GLASS FURNACES

Many research works have been published in the literature describing the mathematical models of combustion chamber, batch blanket and glass tank. Several review papers describing the state-of-art in the glass melting furnace modeling have been presented in the literature. Viskanta [1] has presented an overview of the existing glass tank models and Ungan [2] presented the detailed description of the actual trends in glass furnaces modeling including modeling of the combustion space, glass circulation, batch melting models, homogenizing and refining, channels and forehearth, regenerator and refractory models.

Combustion Chamber Models

Gosman et al. [3]presented the prediction procedure for a glass furnace combustion chamber in which they solved the three-dimensional flow field, together with the reaction and heat transfer using the discrete transfer method. The developed prediction procedure was applied to a cross-fired glass furnace burning natural gas. Extending this modeling approach, Carvalho and Lockwood [4] simulated an oil- or gas-fired end-port furnace. This work was aimed to evaluate the operative differences between both firing modes. The use of the discrete transfer method allowed the accurate prediction of the effect of non-uniform fields of radiative properties of the gas in the combustion chamber enclosure.

Advances in Fusion and Processing of Glass II

This work also considered the batch and glass tank through the development of a three-dimensional computer procedure. Carvalho and Nogueira have improved, extended the previously referred model and have used it for improvement of glass furnace design and operation conditions. As an example, Carvalho and Nogueira [5] have described a methodology to evaluate glass quality using three-dimensional mathematical models. Post and Hoogendoorn [6] proposed a three-dimensional modeling procedure for turbulent flow and heat transfer in the combustion chamber of a glass melting furnace. The radiative heat flux was modeled by the Hottel zone method. Lankhorst et al. [7] developed a combustion chamber model which is possible to couple with the glass tank model. Kobayashi et al. [8] presented the three-dimensional combustion chamber space model. This model was used for a comparative study of alkali volatilization in air fired and oxy-fuel furnaces.

Glass Tank Models

Simonis et al. [9] presented a detailed three-dimensional model of glass melts extending the principles of the two-dimensional models of glass-melting tanks. This model allowed the prediction of molten glass residence times in industrial furnaces. This was a significant development since the residence time is one of the critical parameters in the control and operation of glass melting furnaces. Ungan and Viskanta [10] presented a sophisticated model of the circulation and heat transfer in a glass melting tank. Air bubbling and electric boosting were considered as well as a detailed modeling procedure for the batch melting reaction. Glass refining process was simulated by predicting the size distribution and number density of gas bubbles. A zonal model for the radiative heat transfer in the combustion space was applied. Carvalho et al. [11] showed the application of a mathematical model for air bubbling and weir simulation inside glass tank. Schill et al. [12] developed a three-dimensional mathematical model and simulated a large glass furnace.

Batch Melting Models

Mathematical models of batch melting were developed as early as mathematical models of the glass tank. Pugh proposed a simple model for predicting batch melting time [13]. Fuhrmann presented his model for predicting batch melting process [14]. Hrma model simulated the thermodynamics of batch melting [15]. Carvalho developed a model for predicting batch melting process in industrial glass melting furnace [16]. Viskanta and Wu [17] proposed a model for predicting the temperature distribution and heat transfer in the batch. Ungan and Viskanta [18] presents their model for simulating batch melting process in the glass melting tank. Schill [19] presented a batch model based on integrated kinetic, heat transfer and mass transfer approach. Wang [20] developed a quasi-three dimensional batch model, which can simulate batch melting processes for side-fired glass furnace and end-fired glass furnace. The batch model can be coupled together with the combustion chamber model and the glass tank model. Therefore the industrial batch melting process can be simulated [11, 20].

Complete Glass Furnace Models

McConnel and Goodson [21] presented a simplified model of the whole furnace system. Three energy equations were solved for the crown, batch and refractory temperature. The radiation in the combustion chamber was calculated through the Hottel zone model. For the glass-melt streamlines predefined patterns were assumed. The radiative heat transfer was described by a temperature dependent effective thermal conductivity. The results were compared with operating data indicating a good agreement. Mase and Oda [22] solved the two-dimensional flow and temperature pattern for a glass-melt. The batch velocity over the melt was considered constant and an energy balance equation was solved for the batch temperature. The combustion chamber temperature was formulated by using Hottel zone method as in the previously referred work. Carvalho et al. [11, 23] proposed a comprehensive model composed by the combustion chamber sub-model, the glass tank sub-model, and the batch melting sub-model. The sand grain dissolution, redox and fining processes were also considered. The analysis of furnace is therefore possible. Lankhorst et al. [24] developed a complete glass furnace model which includes two separate sub-models for combustion chamber and glass tank. The coupling was performed through the distribution of the heat flux on glass surface. Chmelar et al. [25] presented a complete glass furnace model including three-dimensional glass met flow with bubbling, batch and combustion chamber simulations and is based on numerical solution of transport equations and are enable to simulate flow, temperature and electrical boosting fields.

In recent years with the development of the commercial available software, new glass furnace models were developed. Hoke and Marchiando [26] created their model using the commercial code FLUENT. This model includes both three-dimensional combustion space and three-dimensional melt model and was used for the optimization of burners position and firing rate with the respect to the glass quality. Takamuru et al. [27] presented a model for the combustion chamber which was developed using the commercial available code STAR-CD. This coupled model consists of a three-dimensional combustion chamber model including combustion, heat transfer with radiation exchanges and two-dimensional glass melt model. The obtained data were validated against experimental measurements carried in a test furnace.

DESCRIPTION OF THE IST GLASS MELTING FURNACE MODEL

The model presented in [11, 23] will be shortly described.

The heat transfer, fluid flow, batch melting and combustion phenomena within the furnace are predicted by solving the governing partial differential equations set in its steady state time-averaged form. Mass continuity, transport of momentum, turbulent quantities, conservation energy, and concentration of combustion species are calculated by solving the following general conservation equation:

$$div(\overline{\rho u}\phi) = div(\Gamma_\phi gard\phi) + S_\phi \tag{1}$$

where ρ denotes the density and $\overline{u}$ velocity, Φ represents the transported scalar, Γ_Φ the diffusion coefficient and S_Φ denote the source term. The quantities Φ, Γ_Φ and S_Φ, which are taken for each differential equation, are given in [5]. In the above general formulation the unsteady effects are neglected.

Combustion Chamber Model

The combustion chamber model is able to simulate turbulent fluid flow, heat transfer, combustion and pollutants formation in the combustion chamber of a glass furnace. The thermal fluid behavior of the gas mixture inside the combustion chamber was predicted by solving three-dimensional governing

　　　　　　Advances in Fusion and Processing of Glass II

partial differential equations set in their steady-state time-averaged form using the finite difference/finite volume method.

The combustion model incorporates physical modeling for the turbulent diffusion flame, soot formation and oxidation, NO formation and dissociation and radiative heat transfer. The flow in the combustion chamber is turbulent. A Reynolds decomposition has been followed to obtain a time averaged solution of the considered variables. An ideal gas assumption was considered to predict the thermodynamic behavior of the gaseous mixture present in the combustion chamber. The Bousinesq approximation was considered applicable to gases flow which presents a Newtonian behavior. The combustion chamber model includes several sub-models as following:

<u>Turbulence sub-model</u> The turbulent behavior of the flow inside the combustion space has been modeled using the well established $\kappa-\varepsilon$ model.

<u>Combustion sub-model</u> The combustion sub-model is based on the ideal fast single step reaction between the fuel and the oxidant. A fuel/oxidant mixture fraction has been defined as a passive scalar for which a transport equation is solved, allowing to predict the mass concentration of the combustion related chemical species (CO_2, CO, SO_2, H_2O, N_2 and C_XH_Y). A second mixture fraction transport equation was solved to handle the non-uniform oxidant composition situations. In the present approach, the single step reaction is considered to be controlled by the lean, rich or very rich nature of the local instantaneous mixture. If a rich instantaneous mixture occurs CO is formed. If a very rich instantaneous mixture occurs, part of the fuel is kept unburned. A Gaussian-based PDF curve was used for the time-averaged integration of those instantaneous reaction products concentrations that were assumed to be locally formed.

<u>Soot Formation/Oxidation Model</u> Soot is, for oil flames, a determinant radiative participating specie which strongly influences the radiative flux distribution in the combustion chamber. In the present model a transport equation for the soot mass concentration was solved which includes source terms for the formation/oxidation rates [28,29].

<u>NO_X Formation/Dissociation Sub-model</u> In the present modeling procedure the NO_X model proposed by Carvalho et al. [30] is followed. This model is based on a chemical kinetic approach to predict nitric oxide formation and dissociation. The Zeldovich mechanism, retaining the reverse reactions, is considered. An instantaneous net formation rate from atmospheric nitrogen is calculated.

<u>Radiation Sub-Model</u> The "discrete transfer" radiation prediction procedure [31] was applied to model the radiative heat transfer process inside the combustion chamber enclosure.

<u>Particle Emissions Indicator</u> Most of the particle emissions in a glass melting furnace is due to the volatilization of some chemical substances from glass surface and batch surface, together with significant contribution from the fine particles dragging over the batch region. Particulate mainly comes from volatilization and only 5-15% of the particulate comes from batch, therefore the particulate emissions model considers only the volatilization mechanism [32].

<u>Batch Melting Model</u>
The glass batch, which is fed into tank by batch charger and floats on the surface of the molten glass to form the batch blanket, is heated from top surface by heat radiation from the combustion chamber and heated from bottom surface by heat transfer from hot molten glass underneath. For simulating the batch melting process in an industrial glass tank furnace, a quasi-three dimensional mathematical model of glass batch has been developed. The model does not directly simulate the

complex chemical processes and assumes that the chemical reactions take place when the batch temperature achieves a predefined chemical reaction temperature, and an amount of energy is consumed. The batch will become molten glass and will melt down into glass tank when its temperature arrives at the molten temperature. The thickness of batch is treated as a function of temperature, density, and unmelted batch. Three-dimensional batch model is built as a series of unconnected two-dimensional simulations. The model is able to predict the temperature distribution, the melting down process and the shape of an axially fed or side fed glass batch blankets in industrial glass melting furnaces. The batch melting model is able to be coupled with the combustion chamber model and the glass tank model.

Glass Tank Model

The glass melt is heated by heat flux from the combustion space to the free surface of the molten glass and cooled by heat transfer to the batch blanket. The glass tank model incorporates physical modelling for the flow and heat transfer of the molten glass. The main assumptions for the glass tank model are: the molten glass is a homogeneous, incompressible Newtonian viscous fluid, chemical reactions are neglected, the flow and heat transfer of the molten glass in tank are considered to be at steady-state, the radiation heat transfer inside the glass melt is handled by using an effective thermal conductivity in the energy equation [5]. Mass continuity, transport of momentum and energy equations were solved. Air bubbling sub-model, sand grain dissolution, redox and refining post models are included in the glass tank model [32].

Air Bubbling Model The air bubbling effect was handled using a Lagrangean formulation to predict the air bubble flow, taking in account for the heat and momentum transfer with the glass melt [5].

Sand Grain Dissolution Model Studying the silica sand dissolution process in industrial glass melting furnaces is very helpful for improving the glass melt quality. Due to the high temperatures, and the difficulties associated with measurements, detailed data about the silica sand dissolution in industrial glass melting furnaces is unavailable from publications. Based on theory, mathematical models for studying the silica sand melting process has been developed by researchers. Some results about the numerical simulation of the silica sand dissolution in the glass melt were published [33-40]. Most of these works consists on the simulation of the sand grain dissolution in the glass melt considering the glass melt at steady state and with a constant temperature. With the development of the computer and numerical methods, researchers started to study the sand grain dissolution in industrial glass melting furnaces using the mathematical modelling [23, 39].

A simple mathematical model of sand grain dissolution in glass melt has been developed. The silica sand dissolution in industrial glass melting furnaces is a complex physicochemical process. Principal assumptions and simplifications for the model are made as follows: a sand grain is a spherical particle; the density of the sand grain is constant; equilibrium is maintained at the silica sand - glass melt interface; the concentration in the bulk glass melt is constant.

The dissolution rate of sand grain in glass melt can be expressed as follows [23]:

$$\frac{dR_s}{d\theta} = -\frac{k_s}{\rho_s}(C_i - C_{gl}) \qquad (2)$$

Advances in Fusion and Processing of Glass II

where R_S: radius of sand grain   [m]

θ :   time   [s]

k_s: mass transfer coefficient of SiO_2 in glass melt   [m/s]

ρ_s: density of sand grain   [kg/m^3]

C_i: SiO_2 equilibrium concentration at the sand grain and glass melt   [kg/m^3]
interface

C_{gl}: SiO_2 concentration in bulk glass melt   [kg/m^3]

Redox and Fining Model

The quality of glass product is mainly connected with the fining process occurred in the glass melt. Mathematical models offer a useful tool for this research area. Some studies on numerical simulation of bubble behaviour in the glass melt and fining process have been made [23, 40-44]. Several simple models were set up to describe the bubble growth and ascension in the glass melt. Early models simulated the bubble in the glass melt that was considered in the steady state with a constant temperature [40-43]. During these years, researchers started to study the redox and fining processes in industrial glass melting furnaces [23, 44].

Mathematical models of redox and fining exist. The redox model can predict the concentration of polyvalent ions and the partial equilibrium pressures of fining gases in the glass melt. The fining model can predict the growth and ascension processes of gas bubbles in the glass melt. Both models, as post-models, are able to be coupled with glass tank model and to be used to study the fining process of glass melt in the industrial glass tanks. The optimization of chemical fining agents can be studied and the quality of glass product can be improved.

When the glass batch has been melted down and mutual solution of the liquid phase has taken place, there remains a large amount of gases in the glass melt. The gases may be present in three forms in the glass melt: in bubbles, physically dissolved and chemically dissolved. Bubbles are undesirable in the glass product and must be removed from the glass melt. The fining agents release gases during an increase of the temperature, and these gases will diffuse into bubbles that will grow rapidly, ascend up easily and escape out of the melt from glass surface. These processes are controlled by the redox reactions and gases transfer in the glass melt. The most commonly used chemical fining agents are As_2O_5, Sb_2O_5, Na_2SO_4. The equilibrium constant K(T) of redox reaction in the glass melt can be written as follows [23, 45]

$$K(T) = \frac{\left[M^{x+}\right]_T Po_2(T)^{n/4}}{\left[M^{(x+n)+}\right]_T} = \exp\left(-\frac{\Delta H^*}{R_g T} + \frac{\Delta S^*}{R_g}\right) \tag{3}$$

where $[M^{x+}]$:   concentration of polyvalent ion at x+ valence state   [mol/m^3]
    $[M^{(x+n)+}]$:   concentration of polyvalent ion at (x+n)+ valence state   [mol/m^3]
    $Po_2(T)$:   partial equilibrium pressure of oxygen in glass melt   [N/m^2]
    ΔH^*:   standard enthalpy of redox reaction   J/mol]
    ΔS^*:   standard entropy of redox reaction   [J/(mol.K)]
    R_o: molar gas constant 8.314   [J/(mol.K)]
    n :   number of electrons that is transferred when the polyvalent
                    ion M is converted from one valence state to another

The partial equilibrium pressure of oxygen in the glass melt is expressed [23]:

$$Lo_2(T)Po_2(T) - Lo_2(T1)Po_2(T1)$$

$$= \frac{1}{4}C_M \left(\frac{K(T)}{K(T) + Po_2(T)^{n/4}} - \frac{K(T1)}{K(T1) + Po_2(T1)^{n/4}} \right) \tag{4}$$

where $Lo_2(T)$ and $Lo_2(T1)$ are the physical solubility of oxygen in the glass melt at temperature T and T1 respectively. The concentration of the polyvalent element, C_M, is assumed to remain constant during the melting and fining processes. $Po_2(T1)$ can be measured, $K(T)$ and $K(T1)$ can be calculated using equation 3. $Po_2(T)$ can be calculated using equation 4.

The gas transfer rate into the bubble can be written as [23, 42]:

$$\frac{dn_i}{d\theta} = 4\pi R^2 h_i \left(C_{i,gl} - C_{i,bub} \right) \tag{5}$$

where n_i :   mole of gas i in a bubble   [mol]
    θ :   time   [s]
    R: bubble radius   [m]
    h_i: mass transfer coefficient   m/s]
    $C_{i,gl}$:   the concentration of gas i dissolved in the glass melt   [mol/m^3]
    $C_{i,bub}$:   the concentration of gas i in equilibrium with the partial pressure   [mol/m^3]
                    of gas i in the bubble according to Henry's Law

<u>Coupling Algorithm and Numerical Solution</u>

The comprehensive model is composed of three main sub-models: the combustion chamber, the batch melting, and the glass tank. These three main sub-models are coupled together.

For the bottom surface of batch, the heat flux is calculated by the glass tank model as follows:

$$Q_{G\text{-}B} = k_{eff} \frac{\partial T_{gl}}{\partial y_{gl}} = k_B \frac{\partial T_B}{\partial y_B} \tag{6}$$

                                        Advances in Fusion and Processing of Glass II

For the top surface of batch and the free surface of glass melt, the heat flux is calculated by combustion chamber model as follows:

$$Q_{F\text{-}G} = k_{eff}\,\frac{\partial T_{gl/B}}{\partial y} = H_{gl/B} - J_{gl/B} \qquad (7)$$

where k_{eff} and k_B are thermal conductivity of glass and batch respectively , T_{gl} and T_B the surface temperature of glass melt and batch respectively, $H_{gl/B}$ is the incident energy from combustion chamber, $J_{gl/B}$ the radiosity of molten glass or batch, $T_{gl/B}$ the surface temperature of glass melt or batch.

The simulation of the whole glass furnace was performed in an integrated method. The three main sub-models were coupled by a cyclical iterative way, matched by the relation between the heat flux from the flame to the molten glass surface and the top surface of batch, and the heat flux from the molten glass to the bottom surface. Interface temperatures between flame and batch, flame and molten glass, molten glass and batch were calculated by a cyclical iterative way. The whole procedure was calculated until "convergence" of the coupled process was achieved. A finite difference discretization scheme was used to solve the model equations. As the upstream conditions determine the downstream ones on the batch, a matching downstream technique in its finite difference form is used in the batch model [11].

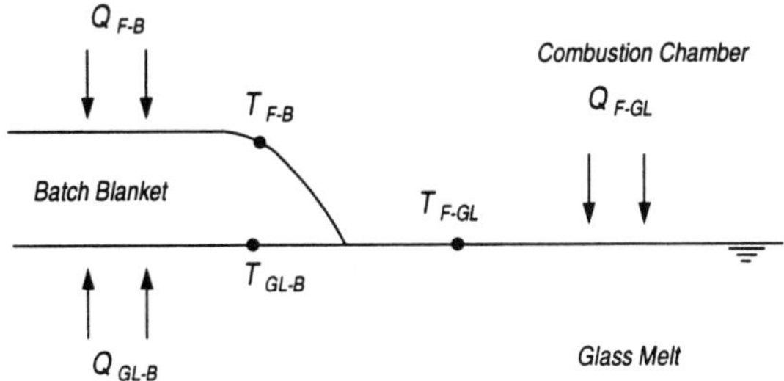

Figure 1. Interface conditions between combustion space, batch blanket and glass melt.

<u>Model Application and Validation</u>

Mathematical models require reliable and detailed data for their validation. Cassiano et al. [46] reported the results for the regenerative, "horseshoe" flow type , oil-fired glass furnace similar to that used in the present study. Barklage-Hilgefort and Sieger [47] studied the effectiveness of flue-gas recirculation and air staging in controlling NO_x emissions. This study involved only flue-gas measurements for the glass melting industrial furnaces. Nakamura et al. [48] investigated NO_x reduction methods and released detailed in-flame data for the scaled-down compartment of the full-scale glass-melting furnace. Costa et al. [49] measured the concentrations of O_2 CO, CO_2 and NO_x and gas temperature for one furnace operating condition. Their results were used for the validation of the numerical simulation models developed by Carvalho et al. [23].

With the objective of validation, the comprehensive model was applied to an end-port container glass tank furnace. The considered glass tank is 10.75 m long, 6.7 m wide and 1.27 m deep. The combustion chamber is 10.95 m long, 7.16 m wide and 2.5 m high. The heat loss from bottom of the tank is considered as 1.5 kW/m^2 and the heat losses from the other tank walls are taken as 3.5 kW/m^2 respectively. The heat loss from crown is considered as 2 kW/m^2 and the heat losses from the other tank walls are taken as 3 kW/m^2 respectively. The flow rate of the combustion oil is 827.4 kg/hr. The pull of the furnace is 166.3 ton/day. For batch blanket, raw materials constitute 75 % by weight and cullet 25 % by weight. The raw materials is not mixed with cullet. Top layer of batch blanket is raw materials and bottom layer of batch blanket is cullet. The inlet height of batch blanket is 0.15 m. The melting mass loss factor R_f is considered equal to 0.15 and the inlet temperature To equal to 50 °C.

Measurements have been made for in flame mean gas species concentrations of O_2, CO, CO_2, NO_X and mean gas temperature.

The sampling of gases for the measurements of O_2, CO, CO_2, and NO_X concentrations was achieved using a water-cooled steel probe. The analytical instrumentation included a magnetic pressure analyzer (Horiba Model CFA-321A) for O_2 measurements and non disperse infrared gas analyzers (Horiba Model CFA-311A) for CO, CO_2, and NO_X measurements. The gas temperature measurements were obtained using uncoated 300 μm diameter platinum/platinu:13 % rhodium thermocouples. Part of measured data are shown in figures 2-5. Details about the furnace measurements can be found from [49].

Above industrial glass melting furnace was simulated using the comprehensive model. Numerical results for the combustion chamber, the batch melting and the glass tank are presented in figure 2 to 14.

The heat flux from the combustion chamber to the surface of the molten glass and the top surface of the batch blanket is shown in figure 2. The heat flux is calculated by combustion model by a cyclical iterative way. The values of the heat flux are very large in the surface of the batch blanket, because the temperature of the top surface of the batch blanket is much lower than the temperature of the free surface of the molten glass.

The batch temperature field and melting rate on the surface of the molten glass are shown in figures 4 and 5. The batch is fed into glass tank by batch charger from the doghouses which located at both side walls. The batch blanket is heated from top surface by heat radiation from the combustion chamber and heated from bottom surface by heat transfer from hot molten glass underneath. When the temperature is higher than the molten temperature, the batch is converted to the molten glass and melting down into the glass tank.

The laminar glass melt flow pattern and temperature field are shown in figures 7 to 8. The temperature of the molten glass is the essential property for the batch melting, glass homogenization and fining which is dependent on heat radiation from the combustion chamber, heat transfer from the glass melt to the batch blanket and inside glass melt, and also acts on the convective effects and the flow. The bubble and the weir force the molten glass mixing strongly and flowing near free surface, has the function of homogenization and fining.

Advances in Fusion and Processing of Glass II

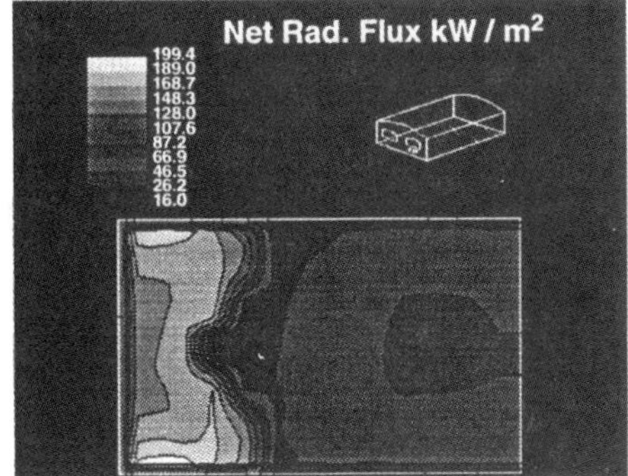

Figure 2. Predicted the heat flux from the combustion space to the surface of the batch blanket and free surface of the glass melt

Figure 3. Predicted flow pattern in the combustion chamber

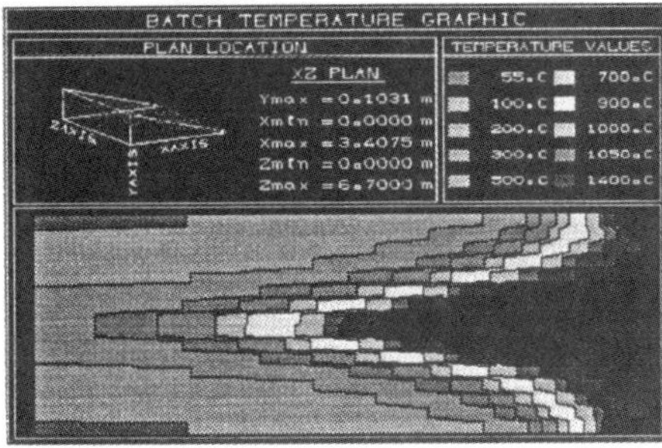

Figure 4. Predicted the temperate with the shape of batch blanket

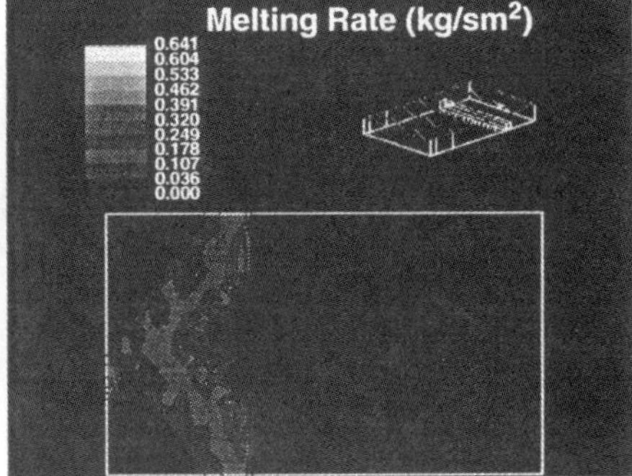

Figure 5. Predicted the batch melting rate on the glass surface

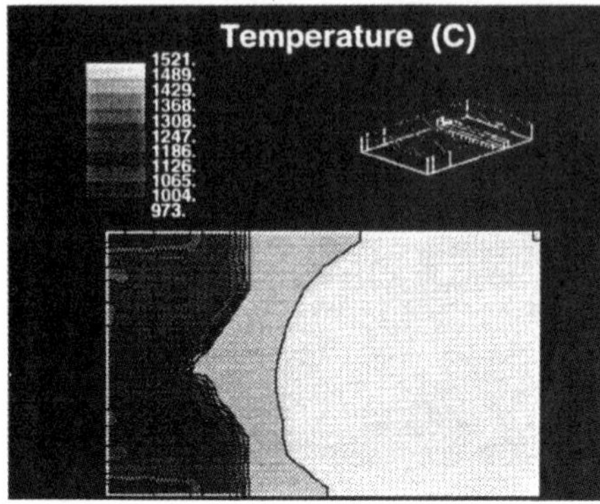

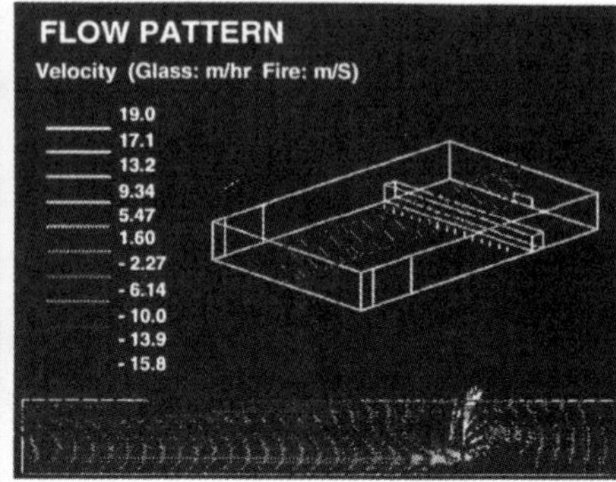

Figure 6. Predicted the surface temperature of the glass melt

Figure 7. Predicted the velocity field of the glass melt

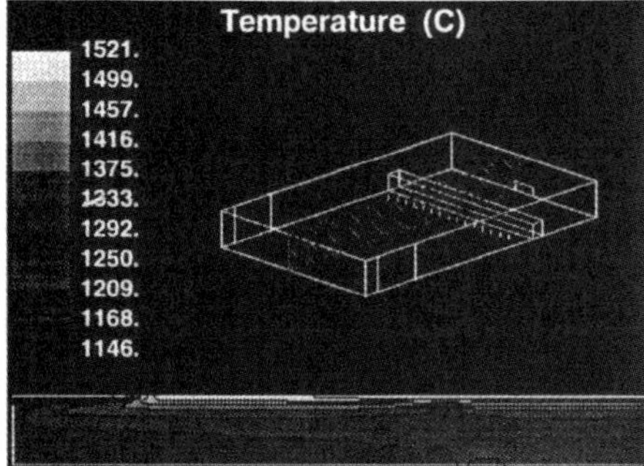

Figure 8. Predicted the temperature field of the glass melt

The results obtained from the model are in good agreement with those obtained from the measurement on an operating furnace (figures 9 to 14 and table I).

　Advances in Fusion and Processing of Glass II

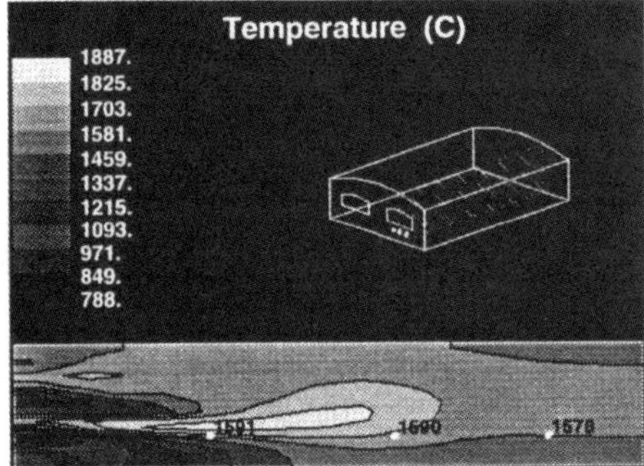

Figure 9. Model predicted temperature field in the combustion chamber and the measurements

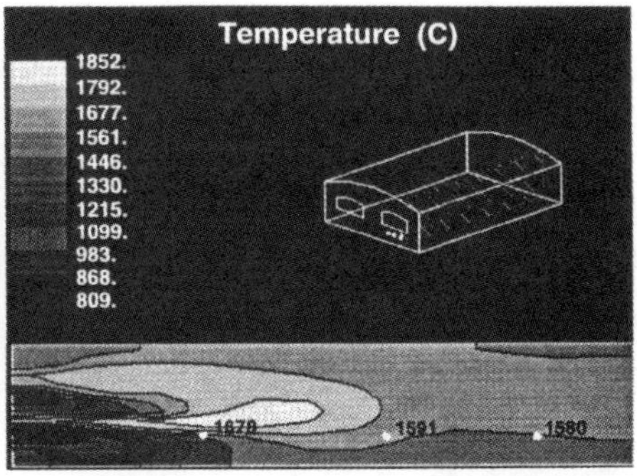

Figure 10. Model predicted temperature field in the combustion chamber and the measurements

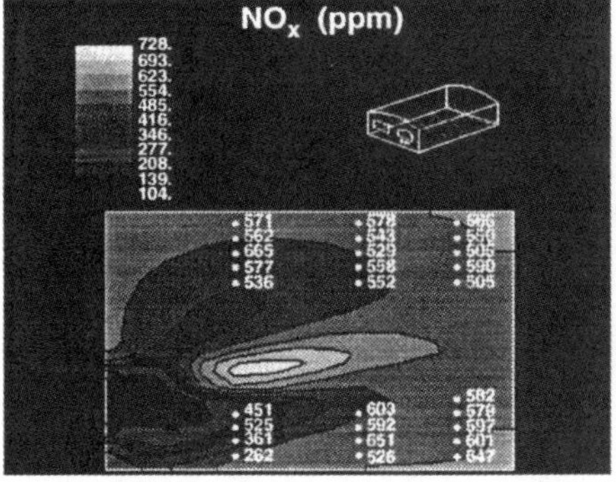

Figure 11. Model predicted NO_x distribution in the combustion chamber and measurements

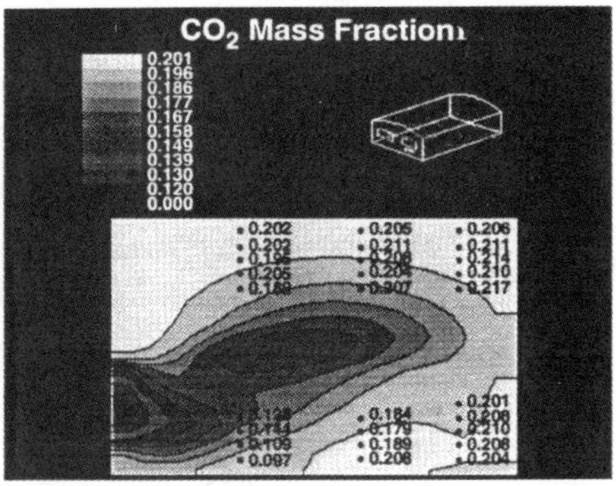

Figure 12. Model predicted CO_2 distribution in the combustion chamber and measurements

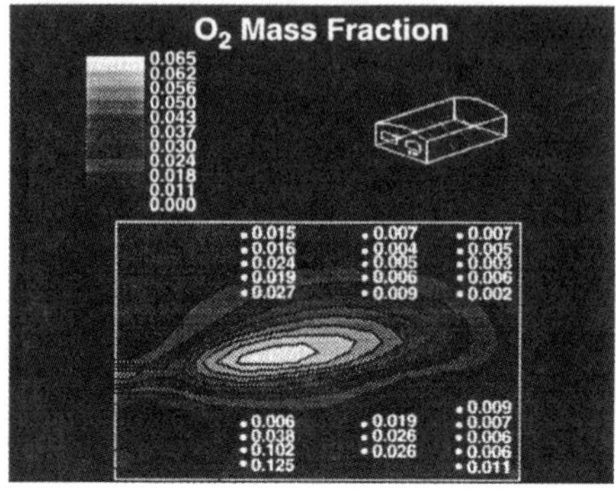

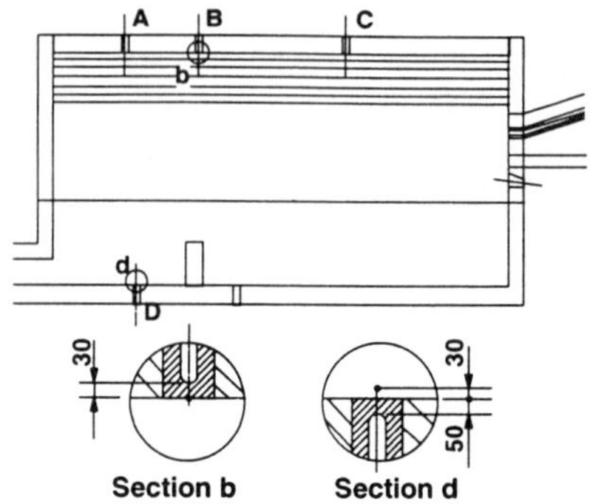

Figure 13. Model predicted O$_2$ distribution in the combustion chamber and measurements

Figure 14. Sketch of end-fired glass melting furnace

Table I. Measured and calculated temperatures (^{o}C)

Point	T_m	T_c	(T_c-T_m)
A	1531	1550	18
B	1502	1548	46*
C	1495	1497	2
D	1249	1301	52*

* A displacement between the location of measurements and calculations exists(see details on Figure 16.)

MATHEMATICAL MODELING APPLICATION TO THE FURNACE DESIGN AND OPERATING CONDITIONS

In the last decade, glass producing companies were faced with an immediate problem of complying with clean air legislation especially in the US and Europe. Mathematical modeling became an useful tool for the design of the oxy-fuel furnace and especially for switching of air-fuel furnaces into the oxy-fuel. Example of such application may be Hoke and Gershtein [26]. Chmelar et al. [50] proposed to use the mathematical model as supporting tool for technical decision for optimal furnace layout. The comparison between air-fuel and oxy-fuel furnace performance was based on the heat balance, crown and superstructure temperature. They also studied the influence of furnace depth on glass melt properties. Bruchet and Lepoutre [51] applied the glass furnace numerical simulator for the parametric study of furnace by changing the furnace height and the fumes exit opening position.

Application Example of an Oxy-Fuel Furnace Optimization

The three-dimensional model was applied to the optimization of the design and operating conditions of industrial oxy-fuel furnace. The model was applied to predict the performance of the furnace for the standard conditions and a parametric study was performed in order to optimize the design and operating conditions of the furnace.

The considered standard geometry is the typical geometry of industrial furnaces used for melting glass or non-ferrous metal. Three different furnace sizes, corresponding to three heating levels: (0.351 MW; 1.053 MW; 1.775 MW), are considered. These three heating levels indicate three possible pull rates since, in the present study the load surface is kept constant and equal to 36.7 kW/m^2. Thus, these three heating levels or melting pull correspond to three values of the area of the heat transfer of the load surface: 9.6 m^2, 28.8 m^2, 48 m^2.

The furnace energy efficiency is given by the ratio of total heat transfer to the load/fuel input energy. In the case of oxy-fuel furnaces, this efficiency is expected to be significantly higher than in the case of conventional furnaces. This is an important requirement for oxy-fuel burning systems, since the saving in energy consumption should overcome the economic cost of the oxygen production.

The predicted emission of NO_x is given accordingly to the ratio of emission of NO_x (kg/s)/heat flux to load (TJ/s). The considered index allows the comparison of the emissions of different furnace sizes and several load pull rates.

The values of specific NO_x emissions are expected to be significantly lower than the produced by conventional air fuel furnaces. This fact, which is due to the limited character of the N_2 concentration in the furnace enclosure constitutes one of the strong motivation for the use of 100% oxy-fuel industrial burning systems.

An extensive parametric study was performed varying several geometry parameters around a reference value for an oxy-fuel geometry arrangement). As an example the influence of the burner arrangement (frontal or staggered) on thermal efficiency is presented here.

The furnace energy efficiency increases with the level of pull, i.e. with the level of the heating power of the considered equipment. From Table II, it may be observed that, in general, the energy efficiency increases with staggered burner arrangement. This effect is due to the more uniform heat flux transferred to the load in the case of the staggered arrangement. This effect may be expected since in the frontal arrangement, the flame region is very limited — large regions without combustion were left between burners. In the present study improvements up to 1% in the thermal efficiency were found changing the burners arrangements from the frontal position to the staggered one.

From the present application, the following main conclusions were withdrawn:

- the NO_x emission of 100% oxy-fuel melting furnaces is found to be 1/50-1/100 times lower than the NO_x specific emission of conventional air-fuel melting furnaces;

-the thermal efficiency of 100% oxy-fuel melting furnaces nearly duplicates the thermal efficiency of conventional melting;

-the furnace thermal efficiency is strongly dependent on several design parameters - variations of 10% may be achieved for different geometrical arrangements;

-the design of 100% oxy-fuel furnaces is a complete multi-dimensional problem - an 100% oxy-fuel furnace presents much larger design freedom degree than a conventional air-fuel furnace;

-in the design of 100% oxy-fuel furnaces the thermal efficiency is particularly important requirement, since the gains in the thermal efficiency should balance the oxygen production cost.

In the case of 100% oxy-fuel furnaces, the development, test and application of advanced design tools is a priority in order to overcome the lack of technological experience in such equipment, the strong requirements for energy efficiency and the urgent character of the adoption of cleaner combustion technologies. From the above considerations, it may be inferred that, three-dimensional modeling is a powerful tool for multidimensional optimization of oxy-fuel furnaces design and operation.

Table II. Influence of the burner arrangement on the furnace energy efficiency.

Heating Power (MW)	Burner Arrangement	Efficiency (W_{fuel}/W_{load})
0.351 1.053 1.775	Staggered	0.56 0.57 0.57
0.351 1.053 1.775	Frontal	0.55 0.56 0.57

<u>Operation Optimization of Glass Melting Furnace</u>

The studied glass furnace is a medium size unit-melter for glass production. The furnace has four staggered natural gas burners at each breast wall burning natural gas.

Twelve burning system architectures were considered for virtual prototyping analysis which was performed through the use of the model. Those burning system architectures combine air/fuel burners, oxy-fuel burners and oxygen lances. The studied cases are listed in table III. The current air-fuel operation condition was used as base case for results comparison and validation. In all the cases the heating power to the glass and batch surface is kept as constant. In the oxygen lances options, the air/natural gas burners geometry is kept constant.

In table IV, an evaluation of the twelve studied cases in terms of pollutant emission and consumption is presented to help the selection of the best burning system architecture. Several gas and oxygen prices were associated to the predicted consumption in order to find the operation price for the considered burning system architectures.

Comparison of the different studied cases with the baseline case (case 1) shows that using 100% oxy-gas burners (case 2) or oxygen lances achieving a 50% oxygen volume fraction oxidant (case 10 and 12) allows an important decrease of particle emissions, a reduction of NO_x emissions and an increase of energy efficiency, with an operation cost rise lying between 10% and 30%. However, oxygen injection too close to the glass melt (case 12) may deteriorate its quality. Details about the operation optimization of industrial glass melting furnaces is available from [52].

   Advances in Fusion and Processing of Glass II

Table III. Simulated cases for different combustion conditions

Case	
1	Base case (**air/natural gas**)
2	**100% Oxygen**/natural gas
3	**One oxy-gas** burner (in each side wall) in the burners raw close to the flue gas port
4	**Two oxy-gas** burner (in each side wall) in the burners raw close to the flue gas port
5	**Two oxy-gas** burner (in each side wall) in the burners raw close to the back wall
7	**Oxygen lance** near the burner quarl for a 35% oxygen volume fraction oxidant
8	**Oxygen lance** near the burner quarl for a 50% oxygen volume fraction oxidant
9	Considered burning system architecture
10	Air/natural gas (Baseline case)
11	**Oxygen lance** above the glass surface for a 35% oxygen volume fraction oxidant
12	**Oxygen lance** above the glass surface for a 50% oxygen volume fraction oxidant

Table IV. Prediction of natural gas consumption, operation costs from natural gas and oxygen consumption, NOx emission and particle emission trends for the studied burning system architectures.

Studied Case	Predicted Natural Gas consumption	Predicted NO$_X$ emission	Predicted Particle emission	Predicted Furnace fuel efficiency	Operation cost associated with predicted natural gas and oxygen consumption		
	%	%	%	%	Cost 1	Cost 2	Cost 3
1	100.0	100	very high	36	100.0	100.0	100.0
2	68.0	12	very low	53	132.2	124.2	116.2
3	88.7	90	high	41	102.9	101.2	99.4
4	77.8	82	medium	47	107.9	104.2	100.5
5	88.7	85	high	41	100.7	99.3	97.8
6	90.0	178	medium	40	132.5	127.2	121.9
7	73.7	132	medium	49	121.2	115.3	109.4
8	68.3	106	low	53	120.0	113.6	107.1
9	79.1	79	low	46	129.9	123.6	117.3
10	67.6	26	very low	54	118.8	112.5	106.1
11	80.1	94	low	45	131.8	125.4	119.0
12	63.2	9	very low	57	123.5	116.0	108.5

* The oxygen cost is assumed to be equal to 40%, 35% and 30% of the natural gas cost for "cost 1", "cost 2" and "cost 3" columns respectively.

Advances in Fusion and Processing of Glass II

<u>Model Based Control of a Glass Melting Furnace</u>

R. Bauer et al. [53, 54] showed as the current improvement of the capabilities of mathematical simulation models can be useful for control, specially to improve the product quality. In that work, sophisticated modeling tools have been used to obtain operation rules. That rules were used to form a knowledge-base system part of an innovative control strategy aimed to ensure high levels of glass quality. A fuzzy logic controller was adopted to supervise the furnace operation. A practical example was shown in which the response of a glass-melting furnace to color changes has been optimized through that control strategy.

An innovative procedure for the integration of physically based and model-based knowledge with advanced optimization algorithms has been recently proposed by Carvalho et al. [55]. The proposed concept is based on the use of mathematical modeling capabilities referred in the previous section to predict the combustion equipment behavior and the effect of changes in geometrical dimensions, control parameters and operating set points on the system performance.

The predictive tools to be used are able to deal with a large range of complex phenomena, components and processes. The use of these modeling procedures to test geometrical dimensions, operating set-points and control parameters, requires the presence of optimization algorithms to handle such multi-input/multi-output problem. The complexity of the problem is enhanced by the number of technological constraints which normally have to be considered. The complete exploitation of these capacities requires the development of advanced, and structurally and functionally simple man-machine interface adequate for the use in industrial environment.

A physically-based transient modeling tool coupled with an optimization algorithm was set up to maximize some essential but non-sensored furnace operation parameters such as energy efficiency, molten glass quality, to minimize pollutant emissions and to stabilize the furnace operation.

Computational power requirements limits the application of three-dimensional modeling for dynamic predictions of complex systems as a glass melting furnace. Alternatively, a non-steady zero-one dimensional dedicated modeling procedure is proposed to fill the above referred requirements.

A model-based optimal control strategy was tested for future on-line application in the operation optimization of a regenerative end-port furnace. The furnace is heated by means of the combustion of heavy fuel-oil with a glass melt tank area of 71 m^2 and 200 t/d nominal capacity.

Due to the strong interaction between the several components of the glass furnace system, namely respecting its dynamic behavior, the developed physically-based dynamic model is based on the calculation of the heat and mass transfer between all the furnace considered subsystems. Two regenerative chambers (considered a gas flow and refractories volume division along the regenerator chamber length), combustion chamber, refractories structure, glass-melt volume, batch heating stage, batch melting stage.

The solution procedure is based in the calculation of non-steady heat and mass for each one of the above referred components. Details of the dynamic model may be found in Carvalho et al. [55].

Two merit functions were calculated from the above described dynamic results:

1.overall furnace performance;

2.furnace output stability.

The performance merit function is continuously calculated from the above described time-dependent model results. An index is computed through heuristic relations which make possible an on-line classification of the furnace performance. The following parameters were considered in the development of this performance merit function: pollutant emissions level, glass quality presented by Carvalho and Nogueira[5].; furnace working life time and the furnace energetic efficiency. For each of these parameters, individual indicators were calculated for a synthesized formulation of the performance index. The computation of the glass outlet temperature deviation respecting to a given target is considered to classify the furnace operation stability.

Two dedicated optimization algorithms were developed to find optimal conditions for the furnace operation. These algorithms are based on the so-called "hill climber" approach. They are designed to maximize the stability and performance index. The modeling tool is used to supply data to the optimization algorithm concerning the classification of near future operation scenarios.

The furnace pull is given by Figure 16. The operating conditions with and without the optimal control algorithm are shown in Figure 17 where the glass quality index is shown with and without optimum control actuation. The proposed control strategy is aimed at embodying a supervision system capable to give to the operator the main operation parameters able to ensure glass optimal production.

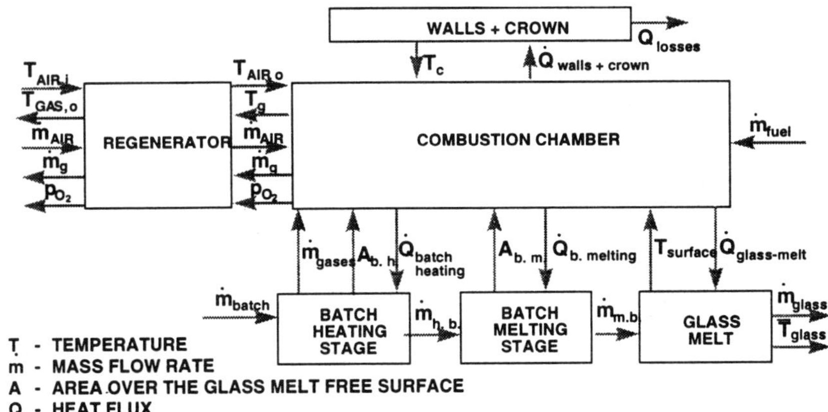

Figure 15. Time-dependent model.

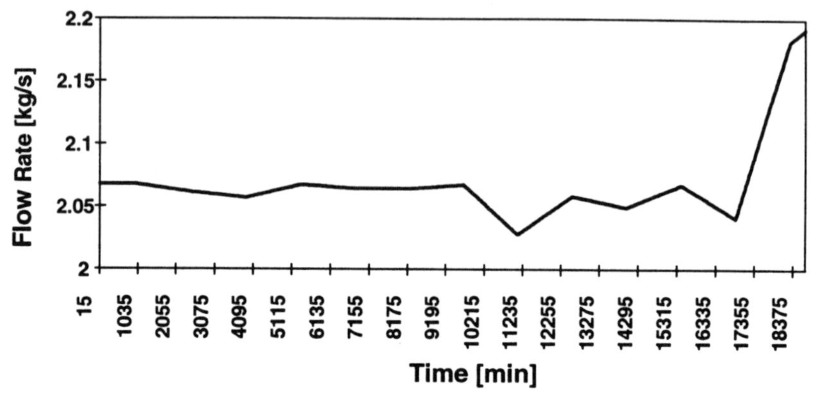

Figure 16. Furnace pull rate (kg/s).

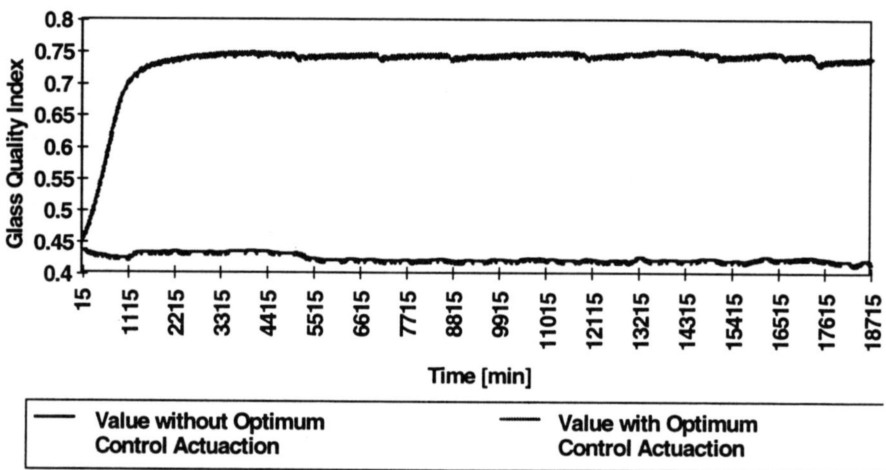

Figure 17. Glass quality index.

An expert systems application to glass melting process was described by Chmelar et al. [56]. Limitations related with computer capacity of calculation of bubbles tracing were possible to solve by implementation of an expert system. This expert system was applied to the blister analysis in glass tank and mathematical model was used for "measurement" of the quality of the process.

Milkulecky et al. [57-58] did an analysis of expert system applications for the glass industry and developed the knowledge-based architecture of the glass quality expert system based on fuzzy control.

Carvalho et al. [59] presented the expert system approach for an industrial glass furnace. The integration of the mathematical models and knowledge-based systems is a key point in the concepts. This system is under development and will include experimental and model-based data using the same knowledge base as it shall be used for integrated design of combustion equipment. This knowledge based operation supervision system will take into account the effect of modifications in the operations set-points on the heat transfer intensity, pollutant emissions, quality and safety. An operation plan will be used instead of the traditional approach of the utilization of set-points.

KBS are primarily useful when the goal is to automate a decision process which is too complex for analytic modeling but which can be, and usually is, addressed by the expertise of domain experts. In such cases, the most common solution is the construction of expert systems: when human knowledge is lacking or unavailable, symbolic learning can be of use. KBS can be used in conjunction with analytic models in two key ways: as a quick means to determine the starting point for model-based optimization and a method of defining merit functions taking into account, heuristically, factors outside the capabilities of current analytic models. The role of the physically-based models is to predict the performance, as measured by an overall merit function, of a given configuration and control strategy. It will be called by the optimization algorithm and must therefore be packaged as a self-contained function or object taking as input parameters configuration settings and returning output variables which can be processed to obtain the merit function. First, the on-line data treatment and reconstruction module prepares data for use in the optimization algorithm. This makes use of simplified models that can be run in "real-time". The off-line detailed module is providing correlations used as input to modules on-line and optimization modules. The optimization algorithm makes use of the hill-climbing method to optimize system as predicted by the models. Finally, the merit function is learned by the Fuzzy expert system based component.

REQUIREMENTS FOR FURTHER RESEARCH AND DEVELOPMENT

The challenges glass industry is facing (reduce environmental impact, offer better quality glass products, introduce innovative glass products, reduce production costs, namely in energy, improve the labor conditions and safety) drove an outstanding Research and Development effort in the last years. The specific aspects of glass-melting process, when compared with process industries in general, required a substantial effort, from glass scientists and technologists, to import to this industry knowledge developed under the scope of outsider R&D working lines. The strong development of specific modeling tools for design and operation optimization, formerly based on more general CFD solving approaches is a clear example. That fact makes the glass industry R&D panorama to be a very specific one and did create a culture of openness and creativity throughout the involved scientific community.

The glass melting industrial process is a large sequence complex physical and chemical phenomena. The expensive cost and the long life cycle of the major involved equipment makes difficult the introduction of experimental technologies. Operator skills, knowledge and experience are still playing a critical role in ensuring production smoothness and quality. These aspects by themselves, together with the large time frame that traditionally involves evolution in the glass fabrication, produced a peculiar current R&D scenario where fundamental investigation aspects are mixed with the integration of novel artificial intelligence technologies in the day-to-day life of glass producing plants. In the authors point of view the current R&D scenarium could be summarized as follows:

The physical understanding of the industrial process is continuously progressing, mainly as far as those aspects that are directly related to quality and refractories degradation.

Development of mathematical models present the following major points.
Advanced models for melting process are established.

- Powerful and integrated numerical solvers (including parallel solvers) are close to be available.
- Integration of more physical phenomena in mathematical/numerical simulator is progressing.
- Levels of complexity are strongly increasing, namely in: solution domain, time-dependency, post-treatment.
- Validation remains scarce

- Process instrumentation is still to be a difficult point mainly due to the harsh glass melting process environment:
    - Very few progresses were noticed in the last years.
    - Glass melting process remains very difficult for direct access and monitoring.
    - Advancements towards the utilization of new sensors remain insufficient.

- New process routes, alternative units, optimization:
    - Very few innovative industrial experiences have been made, with the exception of Oxy-fuel utilization.
    - Non-conventional design local solutions have been tried (partially based in mathematical modeling).
    - Application of advanced control was accelerated due to recent developments in systems engineering.

From the above outlined scenario several aspects may be understood as specially motivating for glass scientists and technologists. Several working lines may be found as interesting topics for next year's R&D work, in order to answer to the following drawbacks

- Full automation of the glass melting process still delayed :
    - Glass melting process is physically complex.
    - Glass melting environment remains inaccessible to standard instrumentation.
    - Standard control and supervision (expert) systems are difficult to be adapted for melting optimization due to the difficulty in accessing to industrial process critical parameters.

- New melting process routes may be required to fit new environmental, quality and productivity requirements:
    - Mathematical modeling is not yet a day-to-day methodology to investigate new concepts.

Advances in Fusion and Processing of Glass II

— Reverse solution methods are not yet available for such a complex problem as designing a glass melter.

Trying to suggested priority topics for next years R&D work, the following topics could be selected:

<u>Making modeling a day-to-day tool</u>. The following aspects may be considered as important steps to achieve this objective:

— Validation of existing modeling structures, sub-models and its coupling loops against relevant industrial data.
— Simplification of modeling procedures, utilization of tools and interfaces, namely trough automatic interpretation of modeling results for wide range of parametric studies.
— Coupling of modeling tools with conventional optimization algorithms for selection of optimal design solution in multi-variable parametric studies (Taguchi method or others).
— Development of stronger numerical solutions for larger modeling flexibility and faster resolution.
— Coupling of modeling tools with day-to-day computational platforms to diminish user
— training effort.
— Enlarging the number of phenomena considered in the modeling tool namely as far as the glass quality is concerned

- Development of <u>new sensors</u> for melting environment.
- Use of <u>on-line physically-based modeling and data reconciliation</u> techniques helping to balance the process inaccessibility (process data reconstruction using physically-based considerations).
- Exploitation of <u>self-learning tools</u> to build, with on-line models, relations between measurable data/process targets.
- Development of a <u>new generation of expert-systems</u> (with built-in physically-based models) for: i) on-line process optimization; ii) reverse optimal design solutions; iii) predictive control and regulation systems. The following aspects may be considered as important steps to achieve this objective:
    — Development of on-line physically-based models for data reconstruction purposes.
    — Development of simplified on-line physically-based models to be coupled with "hill-climber" optimization methods forming an operation optimization cycle.
    — Use of specific physically-based modeling modules for the on-line generation of specific process data usable by self-learning modules
    — Developing quality merit functions to be coupled with built-in models of an optimization expert system
    — The same for pollution and refractory degradation.

CONCLUSIONS

The paper has discussed the basic concepts and research strategies involved in an enhanced utilization of modeling technologies for optimized design, operation and control of glass melting furnaces. Several examples of recent developments and applications were given to illustrate those concepts. Prediction codes capable of computing the main characteristics of the combustion chamber, glass tank and batch region were presented through two aspects: the description of modeling strategy and the reporting of significant industrial applications involved in the optimization of design, operation and control of actual industrial installations.

The state-of-the-art of the comprehensive modeling of the industrial equipment involved in the glass melting process was also reviewed considering the main works recently reported in the literature. Special attention was paid to those cases where an integrated modeling optimization approach was presented.

Based on the current research and development scenario, future priorities in the field of dedicated mathematical modeling, namely as part of a strategy for the development of computer based engineering tools for day-to-day use in industry as part of an effort for control, operation and equipment design optimization, were outlined.

The proposed and discussed concepts and tools may contribute to reduce the environmental impact of the industrial glass production, the competitiveness of glass as a material for new uses and applications, the quality of the current glass products and productivity of the suitably managed glass industries.

ACKNOWLEDGEMENTS

The authors wish to thank the cooperation of the following companies: Praxair, Saint Gobain Recherche, Gaz de France and Santos Barosa. The financial support from JNICT to Mr. Wang through the Programme PRAXIS XXI BD/5765/95, and from the European Commission through the JOULE III programme (project EXLIBRIS, Expert Systems for Energy Saving and Pollution Abatement in Industry – JOE3-ct950040) is acknowledge.

REFERENCES

[1] R. Viskanta, "Review of Three Dimensional Mathematical Modeling of Glass Melting", *Journal of non-Crystalline Solids*, **177** 347-363 (1994).

[2] A. Ungan, "Numerical Simulation of Glass Melting Furnaces, A Review", Proc. of Int. Symp. on Glass Problem, Istambul, Turkey, **1** 351-366 (1996).

[3] A.D. Gosman, F.C. Lockwood, I.E.A. Megahed and N.G. Shah, "The Prediction of the Flow, Reaction and Heat Transfer in the Combustion Chamber of a Glass Furnace", AIAA 18th Aerospace Sciences Meeting, Pasadena, CA, USA (1980).

[4] M.G. Carvalho and F.C. Lockwood, "Mathematical Simulation of an End-Port Regenerative Glass Furnace", *Proc. Inst. Mech. Engrs.*, **199** (C2), 113-120 (1985).

[5] M.G. Carvalho and M. Nogueira, "Glass Quality Evaluation Via 3D Mathematical Modelling of Glass Melting Furnace", Proc. of 1st Conf. of the European Society of Glass Science and Technology, Sheffield, England, 169-171 (1991).

[6] L. Post and C.J. Hoogendoorn, "Modelling of Gas Fired Furnace", 1st European Conf. on Furnaces and Boilers, Lisbon, Portugal (1988).

[7] A.M. Lankhorst, H.P.H. Muysenbeerg and M.P.J. Sanders, "Coupled Combustion Modeling and Glass Tank Modeling in Oxy- and Air-Fired Glass-Melting Furnaces", Proc. of the Int Symposium on Glass Problems, Istanbul, Turkey, **1** 378-379 (1996).

[8]H. Kobayashi, K.T. Wu and W. Richter, "Numerical Modeling of Alkali Volatilization in Glass Furnace and Applications for Oxy-Fuel Fired Furnace Design", Proc. of 4th Int. Conf. on Advances in Fusion and Processing of Glass 119-125, Wurzburg, Germany (1995).

[9]F. Simonis, H. De Waal and R.C.G. Beerkens, "Influence of Furnace Design and Operation Parameters on the Residence Time Distribution of Glass Tanks, Predicted by Computer Simulation", Collected Papers XIV Int. Congr. on Glass, New Delhi, Part III, 118-127 (1986).

[10]A. Ungan and R. Viskanta, "Three-Dimensional Numerical Modeling of Circulation and Heat Transfer in a Glass Melting Tank", *Glastech. Ber.*, **60** 71-78 (1987).

[11]M.G. Carvalho, M. Nogueira and J. Wang, "Mathematical Modelling of the Glass Melting Industrial Process", Proceeding of 17th International Congress on Glass, **6** 69-74, Beijing, China (1995).

[12]P. Schill, J. Chmelar and A. Franek, "Simulation of Large Furnace Using Mathematical Model", Proceedings of the International Symposium on Glass Problems, Istanbul, Turkey, **1** 399-407.

[13]A. Pugh, "A Method of Calculating the Melting Rate of Glass Batch and its Use to Predict Effects of Changes in the Batch", *Glastek. Tidskr.*, **23** 95-104 (1968).

[14] H. Fuhrmann, "Contribution to the Approximate Calculation of the Melting down of Glass Batch Blanket", *Glastech. Ber.*, Vol. 46, pp. 201-208 (1973).

[15] P. Hrma, "Thermodynamics of Batch Melting", *Glastech. Ber.*, **55** 138-150 (1982).

[16]M.G. Carvalho, "Computer Simulation of a Glass Furnace", Ph.D. Thesis, Imperial College of Science, Medicine and Technology, University of London (1983).

[17]R. Viskanta and X. Wu, "Effect of Radiation on the Melting of Glass Batch", *Glastech. Ber.*, **56** 138-147 (1983).

[18]A. Ungan, and R. Viskanta, "Malting Behaviour of Continuously Charged Loose Batch Blankets in Glass Melting Furnaces", *Glastech. Ber.*, **59** 279-291 (1986).

[19]P. Schill, "Batch Melting in Mathematical Simulation of Glass Furnace", Proc. of 3rd Int. Seminar on Mathematical Simulation in the Glass Melting, Horni Becva, Czech Republic, 97-101 (1995).

[20]J. Wang, "The Mathematical Model of Glass Batch Melting in Fired Furnaces", Internal Research Report, Instituto Superior Técnico (1995).

[21]R.R. McConnel and R.E. Goodson, "Modeling of Glass Furnaces Design for Improved Energy Efficiency", *Glass Technology*, **20** 100-106 (1979).

[22]H. Mase and K. Oda, "Mathematical Model of Glass Tank Furnace with Batch Melting Process", *Journal Non-Crystaline Solids*, **38-39** 807-812 (1980).

[23]J. Wang, M.G. Carvalho and M. Nogueira, "Physically-Based Numerical Tool for the Study of Glass Melt Quality", Proceedings of IV International Seminar on Mathematical Simulation in the Glass Melting, 67-76, Vsetin, Czech, June 16-17 (1997).

[24]A.M. Lankhost, G.P. Boerstoel, H.P.H. Muysenberg and K.K. Koram, "Complete Simulation of the Glass Tank and Combustion Chamber of the Former FORD Nashville Float Furnace including Melter, Refiner and Working End", Proc. of 4th Int. Sem. on Mathematical Simulation in Glass Melting, Horni Becva, Czech Republic, 140-153 (1997).

[25]J. Chmelar, P. Schill and A. Franek, "Mathematical Models of Glass Melting Furnaces", Proc. of 4th Int. Conf. on Advances in Fusion and Processing of Glass, Wurzburg, Germany, 63-66 (1995).

[26]B.C. Hoke and R.D. Marchiando, "Using Combustion Modeling to Improve Glass Melt Model Results", Proc. of 4th Int. Sem. on Mathematical Simulation in the Glass Melting, Horni Becva, Czech Republic, 87-95 (1997).

[27]H. Takamuru, Y. Ozeki, E. Kudo and T. Miyamoto, "Evaluation of Combustion Chamber in Furnaces by Numerical Simulation", Proc. of the Int. Symp. on Glass Problems, Istanbul, Turkey, 392-397 (1996).

[28]B.F. Magnussen and B.H. Hjertager, "On Mathematical Modelling of Turbulent Combustion with Special Emphasis on Soot Formation and Combustion". 16th Symp. (Int.) on Combustion, Combustion Institute, 719-727 (1976).

[29]I.M. Khan and G. Greeves, "A Method for Calculating the Formation and Combustion of Soot in Diesel Engines, *Heat Transfer in Flames*", Edited by Afgan and Beer, Scripta Book Company, Washington, 391-402 (1974).

[30]M.G. Carvalho, P.J. Coelho and V. Semião, "Predictions of Nitric Oxide Emissions from an Industrial Glass-Melting Furnace", *Journal of Institute of Energy*, 39-47 (1990).

[31]F.C. Lockwood and N.G. Shah, "A New Radiation Solution Method for Incorporation in General Combustion Prediction Procedures", Proc. 18th symp. (Int.) on combustion, Combustion Institute, 1405-1413 (1980).

[32]M.G. Carvalho, J. Wang and M. Nogueira, "Numerical Simulation of Thermal Phenomena and Particulate Emissions in an Industrial Glass Melting Furnace", Proceedings Fundamentals of Glass Science and Technology, Växjö, Sweden, June 9-12, 416-421 (1997).

[33]M. Cable and D. Martlew, "Effective Binary Diffusivities for the Dissolution of Silica in Melts of the Sodium Carbonate — Silica System", *Glass Technology*, **26** [5] 212-216 (1985).

[34]M. Mühlbauer and L. Nemec, "Dissolution of Glass Sand", *American Ceramic Social Bulletin*, **64** [11] 1471-1475 (1985).

[35]L. Bodalbhai and P. Hrma, "The Dissolution of Silica Grains in Isothermally Heated Batches of Sodium Carbonate and Silica Sand", *Glass Technology*, **27** [2] 72-78 (1986).

[36]P. Hrma, J. Barton and T.L. Tolt, "Interaction Between Solid, Liquid and Gas during Glass Batch Melting", *Journal of Non-Vrystalline Solids*, **84** 370-380 (1986).

[37]M.K. Choudhary, "The Effect of Free Convection on the Dissolution of a Sphericle in a Viscous Melt", *Glass Technology*, **29** [3] 100-102 (1988).

[38]M.K. Choudhary, "Free Convection Effects on the Dissolution of a Sphericle", Advances in the Fusion of Glass, Proceedings of 1st International Conference, Alfred, NY, USA, 11.1-11.23 (1988).

[39]R. Beekens, H. Muijsenberg and T. Heijden, "Modelling of Sand Grain Dissolution in Industrial Glass Melting Tanks", *Glastech. Ber.*, **67** [7] 179-188 (1994).

[40]L. Nemec, "The Behaviour of Bubbles in Glass Melts, Part I: Bubble Size Controlled by Diffusion; Part II: Bubble Size Controlled by Diffusion and Chemical Reaction", *Glass Technol.*, **21** [3] 134-144 (1980).

[41]P. Onorato, M.C. Weinberg and D.R. Uhlmann, "Behaviour of Bubbles in Glass Melts, III Dissolution and Growth of a Rising Bubble Containing a Single Gas", *J. Am. Cer. Soc.*, 64 [11] 667-672 (1981).

[42]J.L. Ramos, "Behaviours of Multicomponet Gas Bubbles in Glass Melts", *J. Am. Cer. Soc.*, **69** [11] 149-154 (1986).

                            Advances in Fusion and Processing of Glass II

[43]R. Beerkens, "Chemical Equilibrium Reactions as Driving Forces for Growth of Gas Bubbles during Refining", *Glastech. Ber.*, **63** 222-242 (1990).

[44]F. Simonis, "Estimation of the Local Redox Distribution in the Melt by Numerical Flow Modeling", *Glastech. Ber.*, **63** 29-38 (1990).

[45]A. Paul, *"Chemistry of Glasses"*, Chapman and Hall, New York (1982).

[46]J. Cassiano, M.V. Heitor and T.F. Silva, "Combustion Tests of an Industrial Glass-Melting Furnace", *Fuel*, **73** 1638-1642 (1994)..

[47]H. Barklage-Hilgefort and W. Sieger, "Primary Measures for the NOx Reduction on Glass Melting Furnaces", *Glastech. Ber.*, **62** 151-157 (1989).

[48]T. Nakamura, J.P. Smart and W.L. Van de Kemp, "The Effects of Fuel Air Mixing on NOx Reduction and Heat Transfer in High Temperature, Gas Fired, Glass Melting Furnaces", The Institute of Energy, First International Conference on Combustion and Emissions Control, Cardiff, U.K. (1993).

[49]M. Costa, M. Mourão, J. Baltasar and M.G. Carvalho, "Combustion Measurements in an Industrial Glass Melting Furnace", Journal of the Institute of Energy", **69** 80-86 (1996).

[50]J. Chmelar, M. Novackova, I. Safarik and P. Budik, "Mathematical Modeling of Furnace Design", Proc. of 4th Int. Sem. on Mathematical Simulation in the Glass Melting, Horni Becva, Czech Republic, 140-153 (1997).

[51]P. Bruchet and E. Lepoutre, "Conversion des Fours de Verrerie à la Fusion Oxygène. Collaborastion Verrier/Fournisseur d'Oxygène pour une Réalisation Industrielle/Convertimg to Oxygen in Glass Tanks: A Necessary Collaboration Between Glass Manufacturer and Oxy-fuel Technology", *Verre*, **1** [3] 20-25 (1995).

[52]M.G. Carvalho, M. Nogueira, J. Wang, T. Ferlin and H. Malvos, "Model-Based Study of a Glass Melting Furnace for Reduced Particle Emission", Proceedings of International Symposium on Glass Problems, Istanbul, Turkey, September 4-6, 398 (1996).

[53]R.A. Bauer, R. Beerkens, H.P.H. Muijsenberg and F. Simonis, "Advanced Control of Glass Tanks by Use of Simulation Models", Proc. of 17th Int. Congr. on Glass, Beijing, **7** 1-8 (1995).

[54]R.A. Bauer and H.P.H. Muijsenberg, "Advanced Control of Glass Tanks using Simulation Models and Fuzzy Control", Proc. of 4th Int. Sem. on Mathematical Simulation in Glass Melting, Horni Becva, Czech Republic, 163-169 (1997).

[55]M.G. Carvalho, A.P. Xeira and M. Nogueira, "Model-Based Operation Optimisation of Glass Melting Furnaces", *Int. Glass Journal*, **89** 31-37 (1996).

[56]J. Chmelar, P. Schill and A. Franek, "Expert Systems and Mathematical Modeling in Glass Melting Technology", Proc. of 4th Int. Conf. on Advances in Fusion and Processing of Glass, Wurzburg, Germany, 63-66 (1995).

[57]P. Milkulecky, R. Bodi and J. Chmelar, "Towards Total Glass Quality Management", Proc. of 4th Int. Sem. on Mathematical Simulation in Glass Melting, Horni Becva, Czech Republic, 172-181 (1997).

[58]P. Milkulecky, D. Ponce and J. Hynek, "Knowledge-Based Projects in Industrial Applications", Proc. of 4th Int. Sem. on Mathematical Simulation in Glass Melting, Horni Becva, Czech Republic, 191-201 (1997).

[59]M.G. Carvalho, M. Nogueira and N. Speranskaia, "Expert System and the Use of Mathematical Models in On-line Glass Melting Optimisation", In Workshop on Advanced Control for industrial glass melters, Maastricht (1997).

USING COMPUTATIONAL FLUID DYNAMICS MODELS TO EXPLAIN OPERATING
CONDITION EFFECTS ON GLASS QUALITY

Bryan C. Hoke, Jr., and Robert D. Marchiando
Air Products and Chemicals, Inc.,
7201 Hamilton Blvd.
Allentown, PA 18195

ABSTRACT

Two operating conditions which produce very different glass quality results were modeled. Results from the models are compared to explain the change in glass quality with regard to operating conditions. The case corresponding to improved glass quality showed longer minimum and mean residence times, a higher temperature on the critical trajectory, and spring zone and turning point locations that are moved back towards the batch when compared to the other case.

The models are validated using furnace data. Model results are compared to measured bottom thermocouple temperatures, crown thermocouple temperatures, and tuckstone optics. In addition, return flow from the refiner, a phenomenon known to occur for this furnace design, was predicted by the models.

INTRODUCTION

Sharp changes in glass quality for different pull rates and operating conditions were observed for a TV glass melter. Early in the campaign, acceptable glass quality was obtained at a lower pull rate. Later, unacceptable glass quality was obtained when the pull rate was increased. Two general types of defects occurred depending on operating conditions. Under high temperature conditions, excessive chord and viscous defects attributed to AZS rundown occurred. When the furnace firing rate profile was changed to relieve AZS rundown, seed and blister defects increased. More recently, excellent glass quality was obtained at the higher pull rate with a change of operating conditions. The main objective of this modeling study is to understand what caused these quality changes.

FURNACE DESCRIPTION AND OPERATING CONDITIONS

The furnace under investigation is a cross fired oxy-fuel TV glass melter with air bubbling. The combustion system uses Cleanfire® HR burners in staggered arrangement. A throat connects the melter and refiner. The refiner has three forehearths feeding the shops

Two sets of operating conditions were simulated; case 1 describes the conditions resulting in poor glass quality and case 2 describes the conditions resulting in good glass quality. The pull rate, total firing rate, and bubbler flow rate was the same for both cases. The firing rate distribution between burners was different for each case.

©Air Products and Chemicals, Inc., 1997

CFD MODEL DESCRIPTION

Two CFD modeling codes were used in this study. For the glass melt, the *Glass Model* by Glass Service, Ltd. was used. This model is described by Schill[1]. For the combustion space, the [2]FLUENT® solver was used. In a earlier paper[3], it was shown that results using the coupled model approach were more consistent with furnace data than using a glass melt model alone.

The formulation of the *Glass Model* is based on the assumptions:

    a)   molten glass is an incompressible Newtonian fluid,

    b)   variations in density due to temperature is important only in buoyancy forces (Boussinesq approximation),

    c)   viscous heat dissipation effects are negligible, and

    d)   molten glass is optically thick (Rosseland approximation).

The following assumptions are made for the combustion space model:

    a)   operation is steady state,

    b)   the combustion model is coupled to a glass melt model by sharing energy balance information at the glass/gas interface,

    c)   gases evolved from the batch do not affect the flow patterns in the combustion space,

    d)   flow in the furnace is turbulent. The standard two-equation $k - \varepsilon$ turbulence model is used[4],

    e)   radiation heat transfer is well described by the discrete transfer radiation model (DTRM) by Shah[5].

BOUNDARY CONDITIONS

Boundary conditions are required everywhere along the edge of the computational domain. They describe the thermal and mass flow conditions at the boundaries.

The total mass flow rate is specified in the batch region. Through the energy balance, the distribution of mass entering the glass melt from the batch region is calculated by the model. The mass flow rate at the forehearths are specified according to shop usage rate.

A zero shear boundary conditions is applied to free surface glass. Since walls are included in the Glass Model, zero velocity is applied implicitly on external boundaries.

Cold face refractory temperatures were measured using a Minolta Cyclops 300AF optical pyrometer. These temperatures were used as temperature boundary conditions on the refractory in the model.

At the glass surface, temperatures and heat fluxes are transferred between the glass melt and combustion models.

In the combustion space model, an overall heat transfer coefficient which includes the heat transfer resistance of the refractory was used for refractory wall boundary conditions. Conservation of mass fixed the inlet velocities for the burners and the temperature of the inlet gas was set to 27°C. A fixed pressure of 101325 Pa (14.7 psia) was used as the boundary condition for the exhausts. Since the flow of glass is very slow relative to the velocity of combustion gases, the combustion model uses a zero velocity boundary condition at the glass surface.

GLASS PROPERTIES

Thermophysical properties of glasses are determined by laboratory measurements and fitting the values to temperature functions suitable for use in the Glass Model. Properties required by the model include density, effective thermal conductivity, specific heat, and viscosity.

RESULTS & DISCUSSION

Five features are chosen to describe the glass melt motion in the tank, validate the model, and compare results between the cases.

1. Bottom temperatures
2. Combustion space temperatures
3. Return flow from the refiner
4. Location of the spring zone and turning point
5. Residence times

To prevent disclosure of confidential information, a reduced temperature, T_r, is defined as

$$T_r = \frac{\left(T - T_{ambient}\right)}{\left(T_{ref} - T_{ambient}\right)}$$

where T is the predicted or measured temperature, $T_{ambient}$ is 25°C, and T_{ref} is some reference temperature.

Bottom Temperatures

In Figure 1, predicted bottom temperatures from the two simulations are compared to bottom thermocouple temperature data measured for case 1 conditions. It is difficult to compare the model results with measured temperatures because of the unknown quality of the thermocouple readings. However, thermocouples 1 and 3 were within 12°C of optical pyrometer measurements in the thermocouple hole. Thermocouple 2 hole was inaccessible to the optical pyrometer.

As seen in this figure, model results match the verified thermocouple data.

The model shows a higher bottom temperature for the case 2 conditions. Furnace supervisors confirmed this trend.

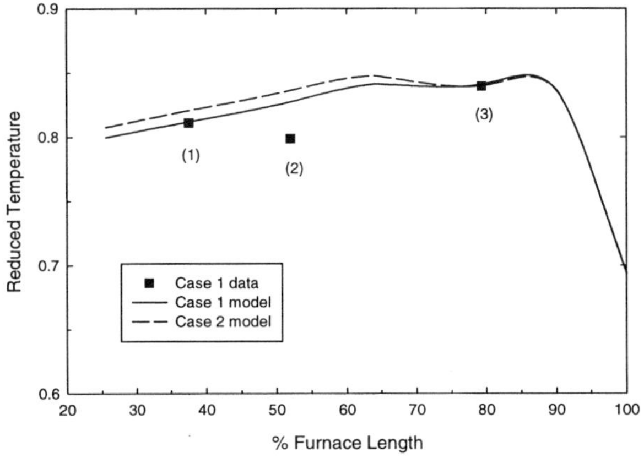

Figure 1. Predicted and measured temperatures in the bottom refractory.

Combustion Space Temperatures

Figure 2 shows plots of hot face superstructure temperature at the crown centerline for both cases. Also plotted are crown thermocouple temperatures. Crown thermocouples are embedded in the refractory resulting in a temperature lower than at the hot face. Agreement is excellent in the melt region. In the batch region, the model predicts temperatures higher than measured. Direct

Advances in Fusion and Processing of Glass II

comparison at the same refractory depth is not possible because the crown refractory is modeled with an effective heat transfer coefficient.

Comparing thermocouple data for cases 1 and 2 in the melt region, the temperature is about the same at the hot spot but lower at the front wall. The model results are in agreement with this trend.

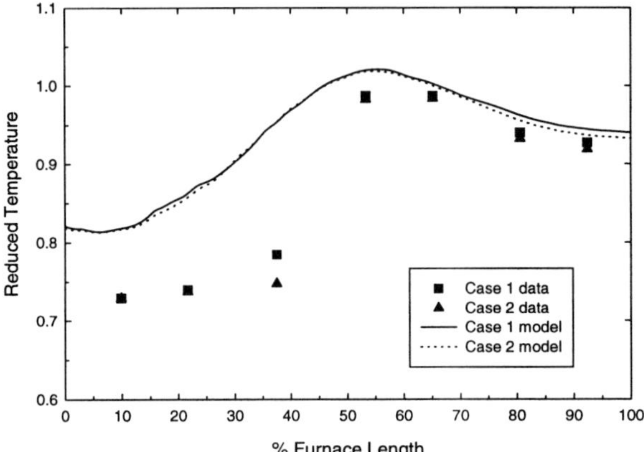

Figure 2. Crown temperature vs. furnace length.

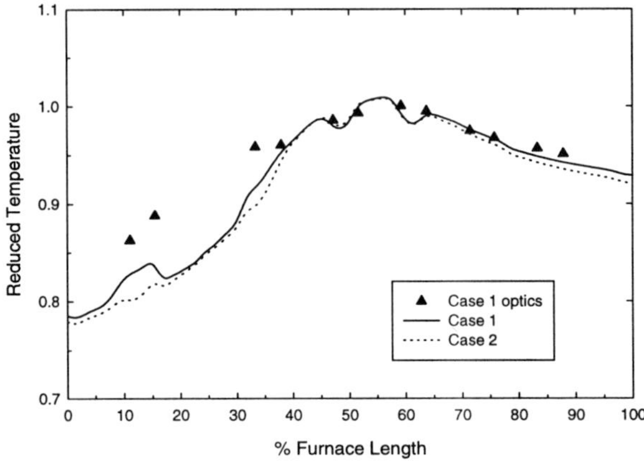

Figure 3. Tuckstone temperature vs. furnace length.

Tuckstone temperatures were measured using a Minolta Cyclops 52 with the emissivity set to 0.9. Figure 3 shows plots of tuckstone optic temperatures measured for case 1 and results of modeling cases 1 and 2. Again, there is excellent agreement in the melt region. In contrast to the comparison at the crown centerline, the model predicts temperatures lower than measured in the batch region.

Advances in Fusion and Processing of Glass L

The measured and predicted tuckstone temperature at the hot spot is about the same for both cases. At the front wall, measured and predicted temperatures are lower for case 2 than case 1.

Return Flow from the Refiner

For this furnace design, return flow is known to occur. Inspection of the throat at the end of a furnace campaign reveals wear patterns consistent with return flow. Return flow makes a very strong and stable barrier and prevents short circuiting of newly formed glass from the batch region to the throat.

Return flow from the refiner was observed in both modeling cases presented here.

Location of Spring Zone and Turning Point

The spring zone is defined as the centerline position on the melt surface where surface glass separates and flows towards the batch and throat. The turning point is the place on the bottom centerline position where forward flow on the bottom of the melter is stopped. From the turning point, the flow is either turned back along the surface or at least lifted upward.

The positions of the spring zone and turning point are given in Table I. The general location of the spring zone is consistent with that observed in the furnace. The spring zone and turning point are moved towards the doghouse for case 2 (the case with better glass quality). Moving the spring zone and turning point back suggests stronger backward currents and circulation which may be better for glass quality because new glass from the batch is lifted upwards to higher surface temperatures sooner. This high temperature region is usually associated with higher oxygen concentrations which are good for refining of bubbles.

Table I. Positions of spring zones and turning points

Case	Spring zone (% of furnace L)	Turning Point (% of furnace L)
1	73	78
2	70	73

Residence times

Residence time information is given in Table II. Massless particle tracing studies were performed using the model where the maximum residence time was two times the volumetric mean residence time. The volumetric mean residence time is defined simply as the volume of the melter divided by the volumetric flow rate. For the mean residence time to have meaning, Levenspiel[6] recommends that the maximum residence time is at least two times the volumetric mean residence time.

Table II. Residence time information

Case	Minimum residence time (% of volumetric mean residence time)	Mean residence time (% of volumetric mean residence time)	Reduced mean temperature on the critical trajectory
1	12.3	71.6	0.866
2	15.3	75.7	0.870

These results suggest why case 2 operating conditions are better for glass quality. The minimum and mean residence times are greater for case 2 and the mean temperature on the critical trajectory is slightly higher for case 2. The critical trajectory is the trace for the minimum residence time particle. The mean temperature on the critical trajectory is expected to be higher for case 2 since the predicted bottom temperature is higher.

Unfortunately, there is no measured residence time data from the furnace to compare with the predictions.

CONCLUSIONS

Bottom glass temperatures predicted by the model compare well to the measured data. The model predicted an increase in the bottom temperature for case 2 operating conditions.

In the region of the melt, predicted superstructure temperatures agree with measurements. In the batch region, the model predicts temperatures higher than measured for the crown and lower than measured for the tuckstone. Consistent with measurements, the predicted hot spot temperature is about the same for both cases and the frontwall temperature is lower for case 2.

Consistent with known wear patterns in the throat at the end of a furnace campaign, the model predicted return flow from the refiner for both cases.

Based on product packing rates, glass quality was observed to be affected by furnace operating conditions. Consistent with the glass quality measurements, modeling results show the case 2 operating conditions to have more favorable features for melting performance. Main features are:

- movement of the spring zone and turning point back towards the doghouse,
- longer minimum and mean residence times, and
- higher mean temperature on the critical trajectory.

This study illustrates how models can be used for furnace understanding and operating improvements.

REFERENCES

[1] P. Schill, "Models of Glass Melting Furnaces," II. International Seminar on Mathematical Simulation in the Glass Melting, Vsetin, Czech Republic, 1993.

[2] *FLUENT User's Guide*, Version 4.3, Fluent Inc., Lebanon, NH, March, 1995.

[3] B.C. Hoke and R.D. Marchiando, "Using Combustion modeling to Improve Glass Melt Model Results," IV. International Seminar on Mathematical Simulation in Glass Melting, Horni Becva, Czech Republic, 1997.

[4] B.E. Launder and D.B. Spalding, "The Numerical Computation of Turbulent Flows," Imperial College of Science and Technology, London, England, NTIS N74-12066, January, 1973.

[5] N.G. Shah, *A New Method of Computation of Radiant Heat Transfer in Combustion Chambers*, Ph.D. Dissertation, Imperial College of Science and Technology, London, England, 1979.

[6] O. Levenspiel, *Chemical Reaction Engineering*, John Wiley and Sons, Inc., New York, 1962.

                                    Advances in Fusion and Processing of Glass II

INVESTIGATION OF GLASS MELTING AND FINING PROCESSES BY MEANS OF COMPREHENSIVE MATHEMATICAL MODEL

M. G. Carvalho and J. Wang
Instituto Superior Técnico/Technical University of Lisbon - Mechanical Engineering Department
Av. Rovisco Pais, 1096 Lisboa Codex, Portugal

M. Nogueira
IrRADIARE, Research & Development in Engineering and Environment, Lda
Tagus Park, Núcleo Central, 2780 Oeiras, Portugal

ABSTRACT

The present paper is aimed at describing a mathematical modelling structure aboe to predict the redox reaction and the fining and refining processes. A predictive evaluation of the molten glass quality is possible using that modelling structure which is part of a comprehensive numerical simulation tool for whole the glass melting furnace. The combustion chamber, the glass melting region and the glass melting tank are considered, in a coupled way, as part of that simulation tool. This procedure was applied to a conventional industrial end-fired glass melter. A significant set of results is presented in the paper as an illustration of that tool's ability to assist the optimization of the produced molten glass quality. These results have shown a good agreement to the qualitative observations done in the industrial environment.

INTRODUCTION

The improvement of furnace thermal efficiency and glass quality and the increase of furnace lifetime are strong requirements of the glass industry. In order to fulfil these requirements the use of mathematical models for better understanding thermal phenomena occurring in glass melting furnaces is necessary.

Recent developments in the comprehensive mathematical model of glass melting furnace offer the possibility of studying glass furnaces and predicting whole furnace performance [1]. In the present paper an industrial glass tank furnace is simulated using the comprehensive mathematical model and a fining model. The furnace performance and refining process are predicted.

MATHEMATICAL MODELLING

Mathematical Modelling

The comprehensive model of glass melting furnace is composed of three main sub-models: the combustion chamber, the batch melting, and the glass tank. The sub-models have been described in detail in previous publications [2,3]. Only the main points of the models are briefly outlined here.

The heat transfer, fluid flow, batch melting and combustion phenomena within the furnace are predicted by solving the governing partial differential equations set in its steady state time-averaged form. Mass continuity, transport of momentum and turbulent quantities, conservation energy, and concentration of combustion species are calculated in the combustion chamber domain and in the glass tank domain by solving the following general conservation equation:

$$div(\overline{\rho u}\,\phi) = div(\Gamma_\phi\, gard\phi) + S_\phi \tag{1}$$

where ϕ represents the transported scalar, Γ_ϕ the diffusion coefficient and S_ϕ denotes the source term. The quantities ϕ, Γ_ϕ and S_ϕ, which are taken to each differential equation, are referred in [2]. ρ denotes the density and $\overline{u}$ velocity.

Combustion Chamber Model

The combustion model incorporates physical modelling for the turbulent diffusion flame, soot formation and oxidation, NO_X formation and dissociation and radiative heat transfer. The flow in the combustion chamber is turbulent. A Reynolds decomposition has been followed to obtain a time averaged solution of the considered variables. An ideal gas assumption was considered to predict the thermodynamic behaviour of the gaseous mixture present in the combustion chamber. The Bousinesq approximation was considered applicable to gases flow which presents a Newtonian behaviour. The combustion chamber model includes several sub-models such as: turbulence sub-model; combustion sub-model; soot sub-model; NO_X sub-model; radiation sub-model [2].

Batch Melting Model

The glass batch, which is fed into tank by batch charger and floats on the surface of the molten glass to form the batch blanket, is heated from top surface by heat radiation from the combustion chamber and heated from bottom surface by heat transfer from hot molten glass underneath. The batch melting model incorporates physical modelling for the heat transfer and melting down process. The batch will become molten glass and will melt down into glass tank when its temperature arrives at the molten temperature. The temperature field, the melting down process, the shape of batch blanket and the coverage area of batch blanket on the surface of molten glass may be predicted by the batch melting model [3].

Glass Tank Model

The glass melt is heated by heat flux from the combustion space to the free surface of the molten glass and cooled by heat transfer to the batch blanket. The glass tank model incorporates physical modelling for the flow and heat transfer of the molten glass. The main assumptions for the glass tank model are: the molten glass is a homogeneous, incompressible Newtonian viscous fluid, neglecting bubbles and chemical reactions, the flow and heat transfer of the molten glass in tank are steady-state, the radiation heat transfer inside the glass melt was handled by using an effective thermal conductivity in the energy equation [2]. Mass continuity, transport of momentum and energy were solved.

Redox and Fining Model

When the glass batch has been melted down and mutual solution of the liquid phases has taken place, there remains a large amount of gases in the glass melt. The gases can present in three forms in the glass melt: in bubbles, physically dissolved and chemically dissolved. Bubbles are undesirable in the glass product and must be removed from the glass melt. Redox reactions take

                                    Advances in Fusion and Processing of Glass II

place in the molten glass when chemical fining agents are used in the glass batch which normally are Na_2SO_4, As_2O_5 and Sb_2O_5, etc. The polyvalent ions convert from higher valence state into lower valence state during the temperature increase in the glass melt. This is accompanied by gases release in the glass melt. These gases will diffuse into bubbles that will grow rapidly, ascend up easily and escape out of the melt from glass surface The fining efficiency is mainly determined by the redox reactions occurring in the glass melt. The equilibrium constant K(T) of redox reaction in the glass melt can be written as follows [4,5]

$$K(T) = \frac{\left[M^{x+}\right]_r Po_2(T)^{n/4}}{\left[M^{(x+n)+}\right]_r} = \exp\left(-\frac{\Delta H^*}{R_g T} + \frac{\Delta S^*}{R_g}\right) \tag{2}$$

where
$[M^{x+}]$:   concentration of polyvalent ion at x+ valence state (mol/m^3)
$[M^{(x+n)+}]$:   concentration of polyvalent ion at (x+n)+ valence state (mol/m^3)
$Po_2(T)$:   partial equilibrium pressure of oxygen in glass melt   (bar)
n :   number of electrons that is transferred when the polyvalent ion M is converted from one valence state to another
ΔH^*:   standard enthalpy of redox reaction   (J/mol)
ΔS^*:   standard entropy of redox reaction   (J/(mol.K))
R_g :   molar gas constant   8.314   (J/(mol.K))

The partial equilibrium pressure of oxygen in the glass melt is expressed [4]:

$$Lo_2(T)Po_2(T) - Lo_2(T1)Po_2(T1)$$
$$= \frac{1}{4}C_M\left(\frac{K(T)}{K(T) + Po_2(T)^{n/4}} - \frac{K(T1)}{K(T1) + Po_2(T1)^{n/4}}\right) \tag{3}$$

where $Lo_2(T)$ and $Lo_2(T1)$ are the physical solubility of oxygen in the glass melt at temperature T and T1 respectively. The concentration of the polyvalent element, C_M, is assumed to remain constant during the melting and fining processes. $Po_2(T1)$ can be measured, K(T) and K(T1) can be calculated using equation 2. $Po_2(T)$ can be calculated using equation 3.

The gas transfer rate into the bubble can be written as [4,6]:

$$\frac{dn_i}{d\theta} = 4\pi R^2 h_i\left(C_{i,gl} - C_{i,bub}\right) \tag{4}$$

where
n_i :   mole of gas i in a bubble   (mol)
θ :   time   (s)
R: bubble radius   (m)
h_i: mass transfer coefficient   (m/s)
$C_{i,gl}$:   concentration of gas i dissolved in the glass melt (mol/m^3)
$C_{i,bub}$:   concentration of gas i in equilibrium with the partial pressure of gas i in the bubble according to Henry's Law (mol/m^3)

Mass transfer coefficient can be calculated as following:

$$h_i = \frac{D_i}{2R} Sh_i \tag{5}$$

where D_i is the diffusivity of species i in (m^2/s), and Sh_i is the Sherwood number of species i.

The bubble velocity in the glass tank $\vec{V}_B$ is calculated as following:

$$V_{BO} = \frac{2\rho_{gl} gR^2}{9\mu} \qquad\qquad T > 1250\,^{\circ}C \tag{6a}$$

$$V_{BO} = \frac{\rho_{gl} gR^2}{3\mu} \qquad\qquad T \leq 1250\,^{\circ}C \tag{6b}$$

$$\vec{V}_B = \vec{V}_{BO} + \vec{V}_{gl} \tag{7}$$

where $\vec{V}_{BO}$ is the velocity of bubble ascension in the glass melt that is calculated using equations 6a and 6b, $\vec{V}_{gl}$ the velocity of glass melt in the tank that is calculated by the glass tank model.

Coupling Algorithm and Numerical Solution

The comprehensive model is composed of three main sub-models: the combustion chamber, the batch melting, and the glass tank. The simulation of the whole glass furnace is performed in an integrated method. The three main sub-models are coupled by a cyclical iterative way, matched by the relation between the heat flux from the flame to the molten glass surface and the top surface of batch blanket, and the heat flux from the molten glass to the bottom surface of batch blanket. Interface temperatures between flame and batch, flame and molten glass, molten glass and batch were calculated by a cyclical iterative way. The whole procedure is calculated until "convergence" of the coupled process is achieved. The finite difference method is used to solve the model equations. As the upstream conditions determine the downstream ones on the batch blanket, a matching downstream technique in its finite difference form is used to the batch model [3]. The redox and fining model can predict the growth and ascension processes of gas bubbles in the glass melt. As a post processing-model, the redox and fining model is able to be coupled with glass tank model and be used to study the effect of fining process of glass melt in the industrial glass tanks.

RESULTS AND DISCUSSION

The above referred comprehensive model is applied to an end-fired container glass melting furnace. The pull rate of furnace was 166 ton glass each day. Air bubbling and a weir were used in the tank. The flow rate of fuel oil was 827 kg per hour. The heat losses from tank bottom and from other tank walls were taken as 1.5 kW/m^2 and 4.0 kW/m^2 respectively. The heat losses from the crown and the combustion chamber walls were taken as 2.5 kW/m^2 and 3.0 kW/m^2 respectively. For glass batch, raw materials constitutes 75 % by weight and cullet 25 % by weight. Fining agent in the batch is 1.0 weight-% Na_2SO_4 and assumes the equilibrium concentration of total sulphur content in the glass melt, Cs, equals to 10 mole in each m^3 glass. Thermal physical properties of the glass, the batch and the fining used in the model are available in [3,4].

Advances in Fusion and Processing of Glass II

Numerical results for the combustion chamber, the batch melting and the glass tank are presented in figure 1 to 8 and table 1. The results obtained from the model are in good agreement with those obtained from the measurement on an operating furnace. Details about the results comparison between model and measurement is available from [7].

Figure 2 shows the predicted gas temperature field. The higher temperature region, which is near the front flame, is well visible. A significant feature of the calculated field is the location and magnitude of temperature gradients, apparent in the air/fuel flow shear layer. Flame was pushed to flow near the top surface of batch blanket and the free surface of glass melt, which is great help to the heat transfer from the combustion space to the batch and the molten glass. The gas flow field, mass fractions of O_2, CO_2, NO, SO_2 and particulate emissions in the hot gases inside the combustion chamber can be predicted also.

The heat flux from the combustion chamber to the free surface of the molten glass and the top surface of the batch blanket is shown in figure 3. For numerical simulation of the comprehensive model, all these interface temperatures and heat fluxes are calculated by a coupling way. The distribution of heat flux shown by figure 3 can only be obtained when three main sub-models are coupling calculated by a cyclical iterative way.

The batch melting rate on the surface of the molten glass is shown in figure 4. The batch is fed into glass tank by batch charger from the doghouses that are located at both side walls. When the temperature is over molten temperature, the batch is converted to molten glass and melting down into glass tank.

The temperature distribution on the surface of glass melt shown by figure 5 could only be obtained when the interface temperatures and heat fluxes between the combustion space, the batch blanket and the molten glass are calculated in a coupling way.

Figure 6 to 8 show the numerical results of the redox model and the fining model. Results include the partial equilibrium pressures of O_2 and SO_2, bubble's flow trajectories. For sulphate fining agent, both SO_2 and O_2 are fining gases. As the partial equilibrium pressure of SO_2 is higher than that of O_2, SO_2 is more powerful than O_2 at fining.

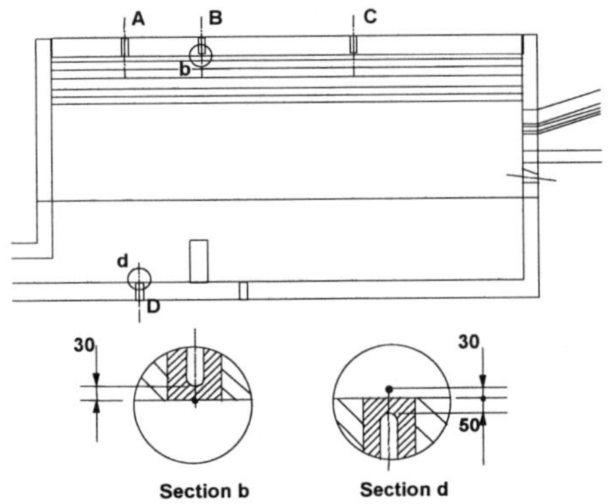

Figure 1. Sketch of end-fired glass furnace

Point	T_m	T_c	(T_c-T_m)
A	1532	1550	18
B	1502	1548	46*
C	1495	1497	2
D	1249	1301	52*

* A displacement between the location of measurementsand calculations exists, see detail in Figure 1.

Table 1 Measured and calculated temperatures (T in °C)

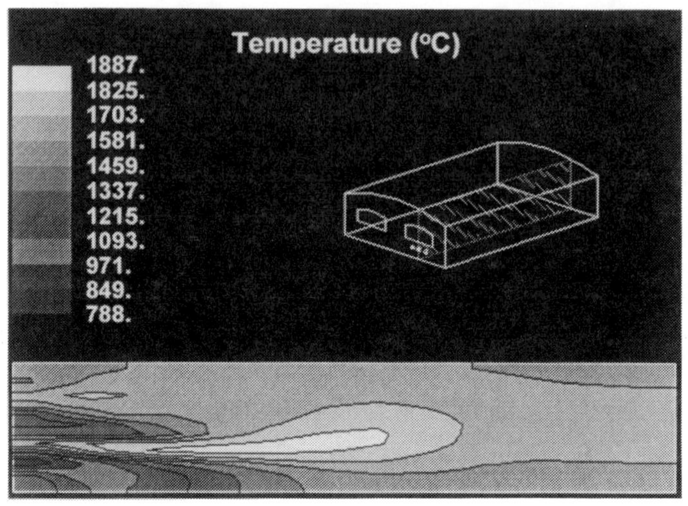

Figure 2. Predicted temperature field in the combustion chamber

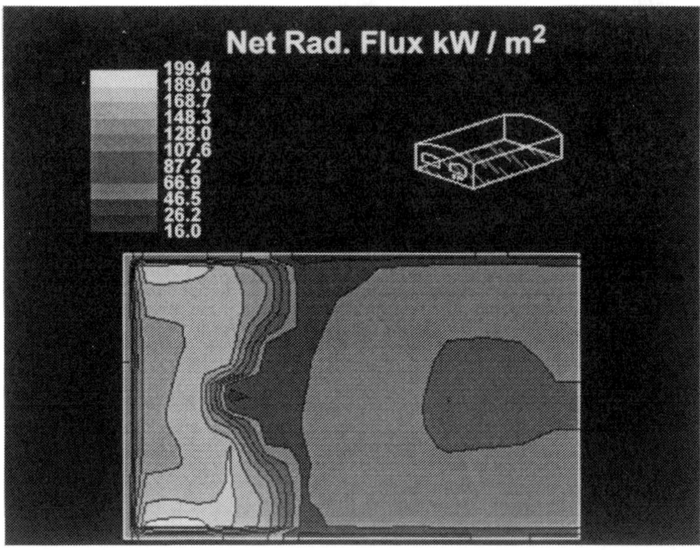

Figure 3. Predicted heat flux from the combustion space to the surfaces of batch and glass melt

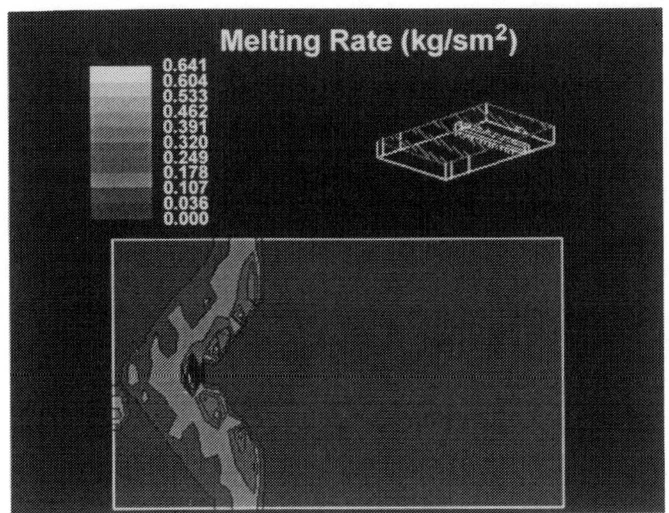

Figure 4. Predicted batch melting rate on the surface of glass melt

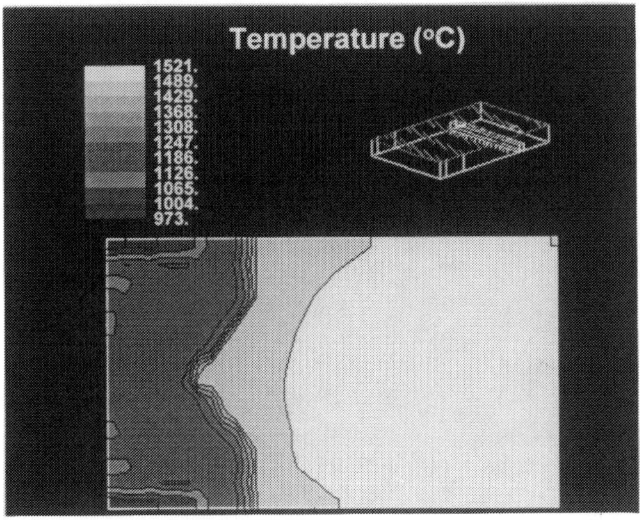

Figure 5. Predicted surface temperature of glass melt

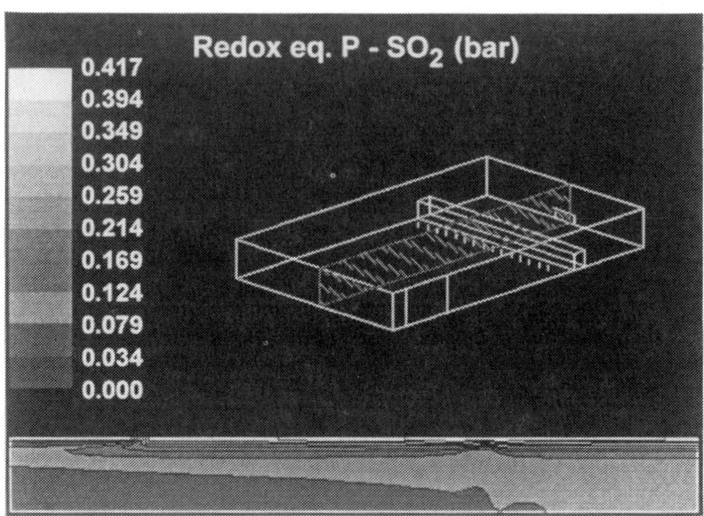

Figure 6. Distribution of partial equilibrium pressure of O_2 in the tank for the glass melt that contains 1.0 weight-% Na_2SO_4

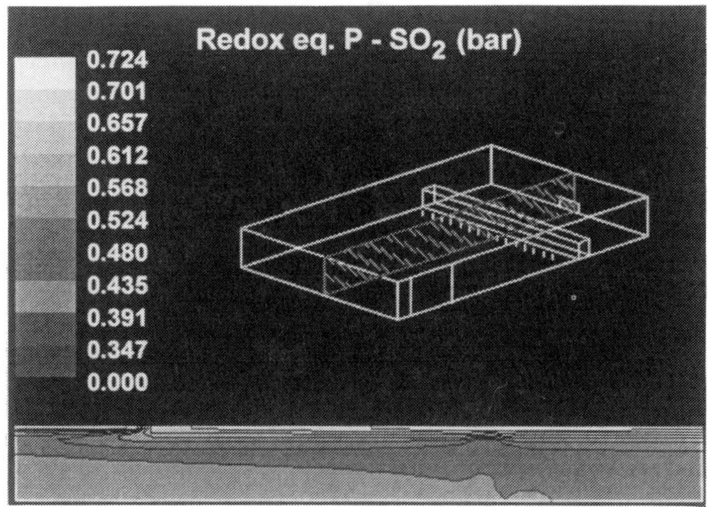

Figure 7. Distribution of equilibrium concentration of Sb^{3+} in the tank for the glass melt that contains 1.0 weight-% Na_2SO_4

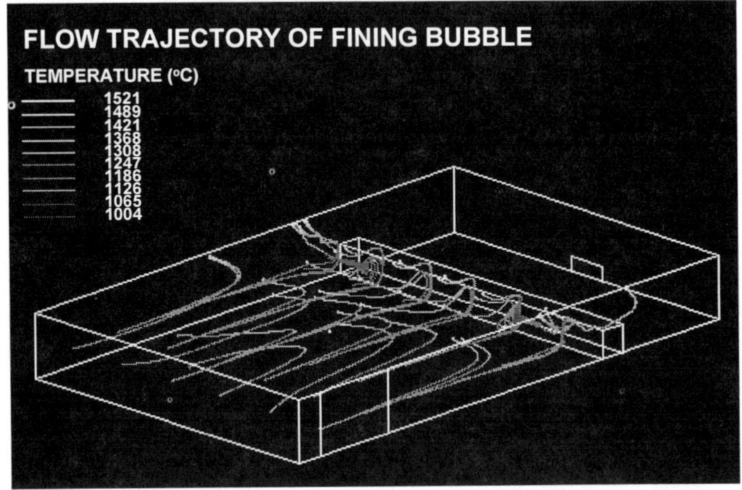

Figure 8. Bubbles flow trajectories in the tank for the glass melt that contains 1.0 weight-% Na_2SO_4

CONCLUSIONS

This paper presents a comprehensive mathematical model of glass melting furnace that can simulate the main thermophysical phenomena occurring in an industrial glass melting furnace and predict the redox and fining process in the glass melt. The interface temperatures and heat fluxes between the combustion space, the batch blanket, and the glass melt were calculated by a coupled cyclical iterative way by the comprehensive model. The accuracy of numerical simulation is improved greatly. Very useful information about whole glass furnace may be supplied by the model as long as the inlet and the boundary conditions are given. Analysis of the furnace performance and prediction of product quality are therefore possible. The comprehensive model constitutes a useful tool for understanding the thermophysical phenomena occurring in a glass furnace, improving furnace design, furnace operation and quality control of production.

ACKNOWLEDGEMENTS

The authors acknowledge JNICT for the financial support of Mr. Wang Jian through the scholarship PRAXIS XXI BD/5765/95.

REFERENCES

[1] M. G. Carvalho, M. Nogueira, Wang Jian, "Mathematical modelling of the glass melting industrial process", *Proceeding of 17th International Congress on Glass*, **6** 69-74, Beijing, China, October 9-14, 1995.

[2] M. G. Carvalho, M. Nogueira, "Glass quality evaluation via 3-D mathematical modelling of glass melting furnaces", *Proceedings of the first conference of the European Society of Glass*

Science and Technology, pp. 169-177, Sheffield, England, Sept. 9-12, 1991.

[3] J. Wang, "The mathematical model of glass batch melting in fired furnaces", *Internal Research Report*, Instituto Superior Técnico. October, 1995.

[4] J.Wang, M. G. Carvalho, M. Nogueira, "Physically-based numerical tool for the study of glass melt quality", *Proceedings of IV International Seminar on Mathematical Simulation in the Glass Melting*, pp.67-76, Vsetin, Czech, June 16-17, 1997.

[5] A. Paul, "Chemistry of glasses", Chapman and Hall, New York, 1982.

[6] J. I. Ramos, "Behaviours of multicomponent gas bubbles in glass melts", *Journal of the American Ceramic Society*, **69**, [11] 149-154 (1986).

[7] M. G. Carvalho, N. Speranskaia, J. Wang, M. Nogueira, "Modelling of glass melting furnaces – Application to control, design and operation optomization", *Proceedings of 5th International Conference on the Advances in Fusion and Processing of Glass*, Toronto, Canada, July 23-27, 1997.

                                   Advances in Fusion and Processing of Glass II

PROCESS PROTOTYPING -
A LOW TEMPERATURE MODEL OF A GLASS PRESSING PROCESS

Prof. Dr. Gunther Reinhart
Hans-Jürgen Trossin

iwb
Institute for Machine Tools and Industrial Management
Munich University of Technology
85747 Garching near Munich

ABSTRACT

The Institute for Machine Tools and Industrial Management (iwb) has a long tradition of investigating Rapid Prototyping technologies. Recently the field of these prototyping technologies was extended with Process Prototyping in order to build not only prototype parts or tools, but also to set up prototypes of complete technical and industrial processes. Process Prototyping involves modeling and investigating into technical processes. In the following an industrially applied glass pressing process is used as an example to create a process model. Dimensional Analysis was chosen out of the variety of methods. In this paper below the fundamental principles and boundary conditions related to the problem of modeling a real glass forming process are discussed.

THE FORMING OF GLASS AT INDUSTRIAL SCALES

Technological and economic circumstances surrounding the industrial production of glassware, particularly tableware, lead to a conflicting situation in most of the companies when it comes to scientific investigation of production processes. Unlike other glass products, tableware shows relatively short life cycles. Therefore frequent designing of new products is vital to a company's economic success. Strongly related to these new products is the designing of their production. The general structure of glass production plants is characterized by a centralized melting process in few but enormous furnaces, which are the source of glass for as many production lines as possible in the available space. There, installations do not allow one to carry out experiments in order to test and pre-optimize new tools or moulds and their corresponding machine parameters before they are used in the actual production. As a consequence, moulds, as soon as their manufacturing is completed, are installed in production lines and adjusted there in an iterative process until the production line shows a satisfactory functionality. This iterative procedure of tuning new moulds with the existing installation usually consumes a considerable amount of time as the installation as a whole has to be brought into thermal equilibrium for each run to make assessments possible.

According to our institute's experiences, gained in cooperation with glass producing enterprises, the average value of production loss due to introducing e. g. one new mould into a typical production line of a pressing process amounts to approximately 50 000 US \$. Depending on the a company's size, at least several dozens of new products have to be successfully brought to market each year which leads to a total annual loss easily exceeding the amount of one million US \$.

In order to find a possibility to reduce the loss caused by introducing new products the iwb concentrates first of all on the process of glass pressing because it is governed by the least amount of parameters among all glass-forming processes.

DIMENSIONAL ANALYSIS

Setting up a digital model (e. g. by FEM) is usually very difficult because of the complex material behaviour of glass combined with the problems of heat transfer and fluidic flow [1]. Therefore we decided to design a physical model, and selected Dimensional Analysis as method of investigation.

Dimensionless Parameters of Interest for a Glass Pressing Process

In dimensional analysis the key base to operate from is a dimensionless description of the object of investigation. All processes sharing the same values for the same dimensionless parameters represent only different aspects of the same dimensionless process. Therefore, at first we deduced a set of dimensionless parameters describing the process of glass pressing by the so called transformation of a relevance list.

The relevance list contains all parameters of significant influence on the examined process. Regarding the pressing of glass, phenomenons as listed in column 1 of Table I occur and can be sufficiently described by the corresponding parameters of column 2.

Table I: Occurring Physical Phenomena and Corresponding Describing Parameters (index G: glass; index M: mat. of the mould)

Phenomena	Parameters
Heat Transfer between Mould and Glass	$\lambda_G,\ \lambda_M,\ \rho_G,\ \rho_M,\ c_{pG},\ c_{pM}$
Viscous Flow of Glass	$\eta_0,\ T_0,\ \gamma$

In Table I λ, ρ and c_p represent the thermal conductivity, density and specific heat respectively. As it will be shown, the three parameters η_0, T_0, γ describe the temperature dependence of the viscosity η of the glass. η_0 and T_0 represent an arbitrarily chosen point of reference of the $\eta(T)$-function of the glass, while γ indicates the 'sensitivity' of the $\eta(T)$-function at the point of reference T_0. γ is defined as

$$\gamma = \frac{1}{\eta_0}\left(\frac{d\eta}{dT}\right)_{T=T_0} \qquad\qquad \text{(Eq. 1)}$$

In addition to defining the physical phenomena, which are related to the materials involved, it is necessary to describe the process-related parameters as well as the process' boundary conditions. Table II contains a list of the parameters found in this group:

Advances in Fusion and Processing of Glass II

Process Entity	Corresponding Parameter
Applied Pressure	p
Characteristic Length	l
Characteristic Period of Time	Δt
Temperature of the Glass at the beginning of a Cycle	T_{Gin}
Temperature of the Outer Surface x of the Mould	T_{OSx}

Corresponding to the industrial reality the investigated process is considered to be cyclic. The characteristic period of time mentioned in Table II can be chosen arbitrarily. Nonetheless it should refer to the observed process, and should be constant and easily reproducible, as e. g. the cycle period or the duration of the pressing of the hot glass.

Simplifications
Considering the subject of investigation the set of parameters can be further reduced. Regarding the thermal properties of glass it is reasonable to assume that during the short period of pressing the glass into shape no convection of heat occurs, and that therefore conduction is the only mechanism of heat transfer. In this case the so called thermal diffusivity a can be taken into account instead of the three quantities Heat Conductivity λ, Density ρ and Specific Heat c_p.
Internal radiation is taken into account by using the so-called effective thermal conductivity instead of the true thermal conductivity of the glass [2].
Furthermore the temperatures of the outer surfaces of the moulds can be considered constant. They are indeed either forced to be constant by cooling systems or are, in case of free surfaces, constant by nature as it was proved by W. Trier's measurements [3].

Based on the consideration above, the set of dimensionless parameters shown in Table III is obtained.

Table III: Dimensional and Dimensionless Parameters Describing the Pressing of Glass

Process Entity	Dimensional Parameters	Dimensionless Parameters	
Characteristic Period of Time	Δt	$\dfrac{\Delta t \cdot a_G}{l^2}$	(Fourier-Number)
Applied Pressure	p		
Characteristic Length	l	$\dfrac{p \cdot l^2}{a_G \cdot \eta_0}$	
Viscosity at T_0	η_0		
Temperature of the Glass at the beginning of a Cycle	T_{Gin}	$T_{Gin} \cdot \gamma$	
Temperature Sensitivity	γ		
Temperature of the Outer Surface x of the Mould	T_{OSx}	$T_{OSx} \cdot \gamma$	
Temperature at Reference Point	T_0	$T_0 \cdot \gamma$	(Arrhenius-No.)
Thermal Diffusivities of the Glass and the Mold	a_G a_M	$\dfrac{a_G}{a_M}$	

Table III further shows that by changing the system of description of the process the number of relevant parameters was reduced by four from ten to six.

Later on, the same set of dimensionless parameters could be obtained directly from the involved differential equations. The set of dimensionless parameters from Table III can thus be considered reliable.

Scale Equations

The equations necessary to design a model process can be deduced directly from the dimensionless parameters in Table III by demanding that those parameters must have the same value in the model as in the original process. The following example shows how a time scale for the model can be deduced from the Fourier-number (s. also Table III):

$$_{Mod}Fo = {}_{Orig}Fo \quad \Leftrightarrow \quad {}_{Mod}\left(\frac{\Delta t \cdot a_G}{l^2}\right) = {}_{Orig}\left(\frac{\Delta t \cdot a_G}{l^2}\right) \qquad \text{(Eq. 2)}$$

$$\frac{_{Mod}\Delta t}{_{Orig}\Delta t} = \frac{_{Orig}a_G}{_{Mod}a_G} \cdot \left(\frac{_{Mod}l}{_{Orig}l}\right)^2 \qquad \text{(Eq. 3)}$$

According to these model laws represented by the scale equations, an experimental model for the production plant can be designed and set up. Theoretically one has an infinite number of possibilities to realize such a model although not all of them might be of practical interest due to scale values exceeding the limits of practicability.

The results gained in the experiments with the model process can be re-calculated, i. e. re-transferred, to the original process.

A MODEL LIQUID

The main objective in creating a scale model of a pressing process was to transfer the necessary experiments from the production line to a laboratory. Therefore the process temperature had to be reduced to values that could easily be handled. Thus, a material had to be found that showed at lower temperatures the same characteristic thermo-viscous behaviour as glass.

The lack of such a substance had so far been the main obstacle in realizing an experimental installation of a glass forming process. Compared to other quantities involved in the process the viscosity of the glass varies most. The region of working of the glass reaches from starting at about 10^6 dPas to approx. 10^9 dPas [4], which indicates a change by the factor of 1000.

In the following a substance is proved to be able to represent the thermo-viscous property of an industrial soda-lime glass. The Vogel-Fulcher-Tamann-Equation of the soda-lime glass is

$$\log(\eta) = -0.2932 + \frac{2898.3}{T - 621.77} \qquad \text{(Eq. 4)}$$

Variable Properties

In order to reproduce the material behaviour of glass the treatment of variable properties in Dimensionless Analysis has to be taken into account. Here a formulation introduced by Pawlowski [5] with regard to chemical apparatus engineering was used to dimensionlessly describe the thermo-viscous behaviour of glass.

The $\eta(T)$-function has to be standardized to a dimensionless function $\Phi(u)$ in a way that the following conditions will be met:

$$\Phi(0) = 1 \qquad \text{(Eq. 5)}$$

$$\left(\frac{d\Phi}{du}\right)_{u=0} = 1 \qquad \text{(Eq. 6)}$$

This is achieved by selecting an arbitrary reference point (η_0 / T_0) and applying the following transformation:

$$u = \frac{(T - T_0)}{\eta_0}\left(\frac{d\eta}{dT}\right)_{T=T_0} \Leftrightarrow (T - T_0) \cdot \gamma \qquad \text{(Eq. 7)}$$

$$\Phi(u) = \frac{\eta(u)}{\eta_0} \qquad \text{(Eq. 8)}$$

According to the so-called "standard transformation" explained above, η_0, T_0 and γ must belong to the relevance list in order to define the material properties of the glass. Two materials can be considered sufficiently similar concerning a certain property, if the material's dimensionless descriptions of the property are identical. In the case of the standard transformation, the shape of the dimensionless function depends to a high degree on the chosen point of reference. However, there is one class of functions that allows a transformation regardless of the point of reference. This practically means that the Arrhenius-number (s. also Table III), which contains exclusively values related to the point of reference, can be neglected and hence the set of dimensionless parameters can be further reduced. In order to achieve this the thermo-viscous properties of glass have to be described by Eq. 9:

$$\log(\eta) = \log(\eta_0) \cdot \left[1 + \mu \cdot \frac{(T - T_0)}{a}\right]^{1/\mu} \qquad \text{(Eq. 9)}$$

Generally Eq. 9 will not provide the same accuracy in describing the thermo-viscous behaviour of glass over a wide temperature range as the VFT-Equation. However, it is sufficient for Eq. 9 to provide correct values in the region of working of glass, i. e. 10^6 to 10^9 dPas. Table IV shows that

Eq. 9 is capable of displaying the thermo-viscous property of the examined glass with satisfactory accuracy. The constants used in Eq. 9 were a=-294.72; log(η_0)=9.32, T_0=923 K and μ=-0.9039.

Table IV: Comparison between the VFT-Eq. and Eq. 9 in the area of $\eta=10^6$ to 10^9 dPas

Temp. [K]	log(η) acc. to VFT	log(η) acc. to Eq. 9	$\Delta\eta/\eta_{Eq.9}$
913	9.6569	9.6592	-2.38 10^{-4}
953	8.4578	8.4580	-2.36 10^{-5}
1003	7.3102	7.3100	2.74 10^{-5}
1043	6.5882	6.5881	1.52 10^{-5}
1083	5.9915	5.9919	-6.68 10^{-5}

As the transformation function described above is single valued and mathematically sufficient, it is clear that only amorphously solidifying materials are able to replace glass in the model process. Any form of (even partial) temperature-dependent crystallization, in contrast to an amorphous solidification, is characterized by a sudden change of viscosity at the melting point. This sharp increase of viscosity, when cooling a substance below its critical temperature of crystallisation, will also be transformed to the dimensionless description of its thermo-viscous property. Thus, the graph of the thermo-viscous material behavior of a (partially) crystallizing substance can never match with that of glass, i. e. a basic requirement of Dimensional Analysis could not be met.
A known amorphously solidifying system with a sufficiently low T_G is a composition of $Ca(No_3)_2$ and KNO_3 [6]. Figure 1 shows a comparison of the temperature-dependent viscosities of a soda-lime glass and the system of nitrates.

Figure 1: Viscosity of the Model Glass and an industrially worked Soda-Lime Glass in dependence on the temperature

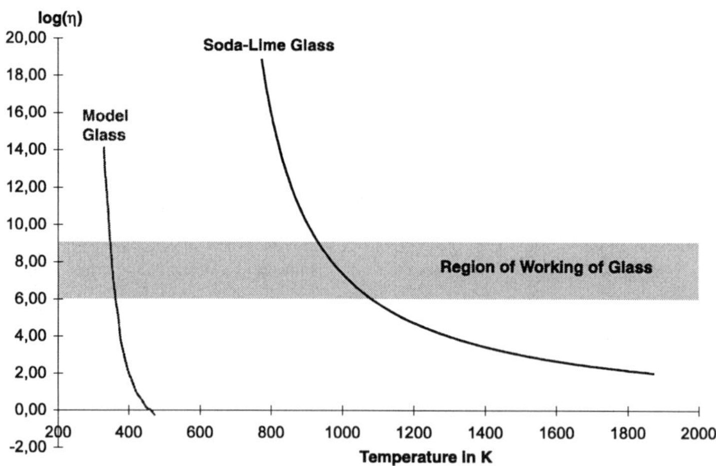

The modeling abilities of the nitrate system regarding the soda-lime glass are obvious, if the working region of both glasses are displayed dimensionlessly as shown in Figure 2.

　Advances in Fusion and Processing of Glass II

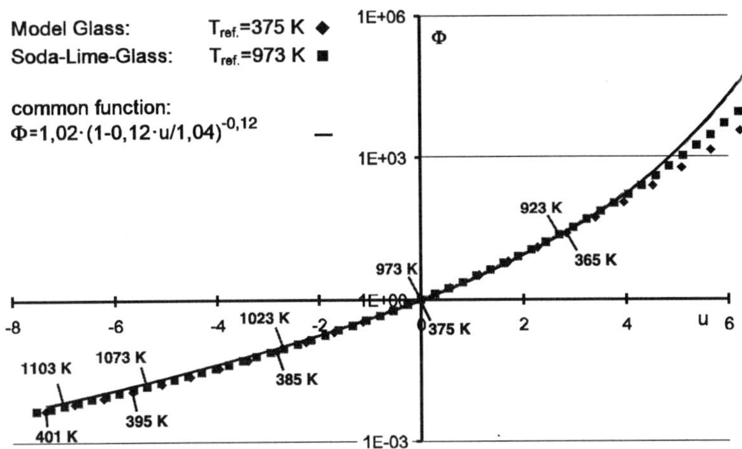

As Figure 2 shows, the parts of the glass pressing process concerning the thermo-viscous properties of glass can be investigated by a model process using the nitrate system mentioned above as a model glass. The temperatures of the model process run through a region from approx. 400 K down to 360 K, which leads to a reduction of the temperature of more than 600 K. By shifting the dimensionless values of viscosity of the model glass along their graph to higher temperatures (a raise of approx. 10 K) the necessary pressures to form the model glass can also be reduced. This can be deduced from the dimensionless parameter containing a characteristic pressure p of the process, in which η_0 can be interpreted as the viscosity of the model glass at the origin (s. Table III). Shifting the dimensionless values of viscosity itself without affecting the shape of their dimensionless function is possible by using Eq. 9 to describe the thermo-viscous property of the glass.
Of course the whole model process has to be set up according to the derived scale equations.

So far many amorphously solidifying substances, e. g. borosilicate glasses, pure SiO_2, GeO_2, BeF_2, and B_2O_3 have been compared with each other in the same theoretical way. They all show sufficient similarity concerning their thermo-viscous properties. That means, that the nitrate system is generally able to display a temperature-dependent change of viscosity of a amorphously solidifying substance.

CONCLUSION
Within the region of 10^6 to 10^9 dPas the thermo-viscous properties of a soda-lime glass used in industrial production of tableware can be represented by a modell glass. A dimensionless description of these properties is necessary to derive the model laws for a model process. On condition that all previously mentioned laws according to the theory of Dimensional Analysis are followed, the temperatures in the model process are roughly 600 to 700 K lower than the one in the original production. Experiments can thus be carried out regardless of production installations, and the results obtained can be transferred to the process of interest by the scale equations. Besides the soda-lime glass used as an example above, all known industrially worked types of glasses could be correlated with the model glass.

REFERENCES
[1] H.-J. Trossin: "So Ähnlich"; *Die Neue Fabrik*, p. 52, verlag moderne industrie, Landsberg, Germany, 1997
[2] H. Scholze: "Glass"; p. 360, Springer, New York, 1991
[3] W. Trier: "Temperaturmessungen in Glasformen", *Glastechnische Berichte* **28** [9] pp. 336-351 (1955)
[4] H. Scholze: "Glass"; p. 160, Springer, New York, 1991
[5] J. Pawlowski: "Veränderliche Stoffgrößen in der Ähnlichkeitstheorie"; Verlag Sauerländer, Aarau, Switzerland, 1991
[6] H. Tweer et al.: "Inadequacies of Viscosity Theories for a Vitreous KNO_3-$Ca(NO_3)_2$ Melt"; *Journal of the American Ceramic Society*, **54** [2] 121-123 (1971)

Advances in Fusion and Processing of Glass II

EXPERIMENTAL INVESTIGATION OF FLOW AND HEAT TRANSFER IN AN ELECTRICALLY HEATED VISCOUS LIQUID

Kari A. Van Dyke
Avery Dennison
250 Chester Street
Painesville, OH 44077

Manoj K. Choudhary
Owens Corning
2790 Columbus Road
Granville, OH 43023-1200

ABSTRACT

The paper describes an experimental set-up for investigation of viscous flow and heat transfer phenomena in melts which are heated electrically by means of electrodes immersed in the melt. The set-up involved measuring velocity and temperature distribution in a glycerin-5% LiCl solution contained in a cylindrical tank and heated by a pair of graphite electrodes. The velocity distribution was measured by tracking the displacement of particles in the fluid using a video camera and a large memory buffer frame grabber board (the Particle Displacement Tracking [PDT] technique). The PDT technique allowed the measurement of velocity distribution in an entire plane of the apparatus illuminated by a He-Ne laser. Results are reported to show the effects of power distribution and the electrode immersion on flow and heat transfer in the system. In addition to providing insights into flow and heat transfer in an electrically heated melt, the results from this study may also be used to validate mathematical models for such a system. Preliminary findings from a mathematical model show good agreement between the predicted and the measured temperature distribution.

I. INTRODUCTION

Electric heating in melting operations is used widely in the materials processing industry, a prominent example being the electric melting of glass. Electric melting as used in the glass industry typically involves the passage of electric current through the glass melt. The passage of current results in a spatially non-uniform release of joule heat in the melt. The resulting non-uniform temperature distribution gives rise to natural convection in the glass melt. The flow and heat transfer phenomena in electric glass furnaces have been studied extensively using mathematical modeling techniques [1-10]. However, experimental data needed for the validation of mathematical models have been lacking. Not only is it difficult to obtain detailed temperature and flow data from operating furnaces, but also uncertainty in specifying some of the key boundary conditions (e.g. at the glass melt/batch interface where the term batch refers to unmelted mixture of raw materials) makes a rigorous validation of the mathematical models difficult.

The principal objective of the present study was to set up an experimental system where flow and heat transfer phenomena in electrically heated fluids may be investigated under well defined boundary and operating conditions. To this end, velocity and temperature were measured in a glycerin-5% LiCl solution heated by a pair of graphite electrodes. The results obtained here are important in their own right as they provide valuable insights and may also be used for the validation of mathematical models for flow and heat transfer in electrically heated viscous fluids. Another important contribution of this work is that it has demonstrated the use of a flow measurement technique (the Particle Displacement Tracking [PDT] technique) which offers considerable advantages over the techniques commonly used (Exposure Photography, Laser Doppler Anemometry) to measure flow in physical models of glass furnaces.

Regarding the organization of this paper, Section II describes the experimental set-up, Section III describes the velocity measurement technique used in this study, and Section IV presents results showing the effects of electrode penetration and electric power on the velocity and temperature distributions in the system investigated here. Finally, some concluding remarks are given in Section V.

II. EXPERIMENTAL SET-UP

The experimental set-up and techniques used in this study are described in detail in Reference 12.

Figure 1 shows the apparatus and the principal dimensions. The apparatus consisted of a 0.15-meter diameter cylinder kept inside a 0.38-meter square thermal bath. The thermal bath and the tank were constructed of 6.35-mm thick acrylic. The cylinder was filled with a glycerin-5% LiCl solution to a height of 0.23 m. The cylindrical shape was chosen so that, with axial symmetry, a two-dimensional mathematical model may be used to compute velocity and temperature distribution in this system. As shown in Figure 1, two electrodes (graphite, 6.35-mm diameter) were used to pass electric current through the fluid. Mineral oil was chosen as the thermal bath fluid.

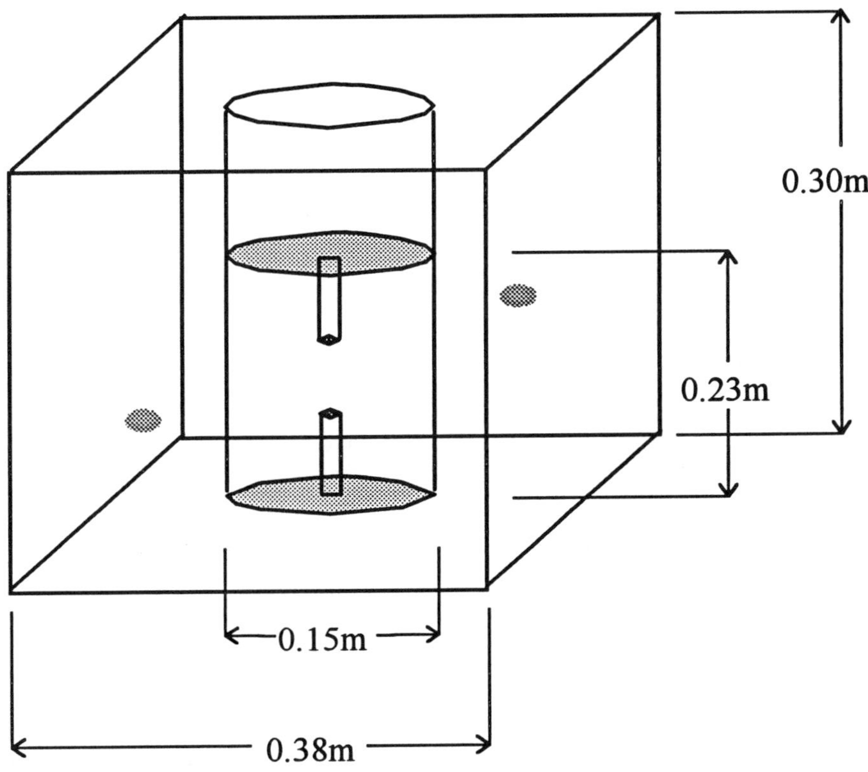

Figure 1: Schematic of the Apparatus

The dimensions and the operating conditions were chosen to give the following approximate values for the Grashof (Gr) and Ralyeigh (Ra) numbers:

$$Gr = \rho^2 g L^3 \beta \Delta T / \mu^2 \cong 2.3 \times 10^4 \quad , \quad Ra = Gr.C_p \mu / k \cong 1.6 \times 10^7$$

where;

C_p	=	specific heat
g	=	acceleration due to gravity
k	=	thermal conductivity
L	=	length scale (radius)
ΔT	=	characteristic temperature differential
β	=	coefficient of volumetric expansion
μ	=	viscosity
ρ	=	density

The values given above for the Grashof and Rayleigh numbers are roughly in the range encountered in electrically boosted furnaces (where a part of the melting energy requirement is met by electricity and the rest by fossil fuel combustion) used by Owens Corning for melting E-glass compositions.

Figure 2 shows a block diagram of the experimental set-up.

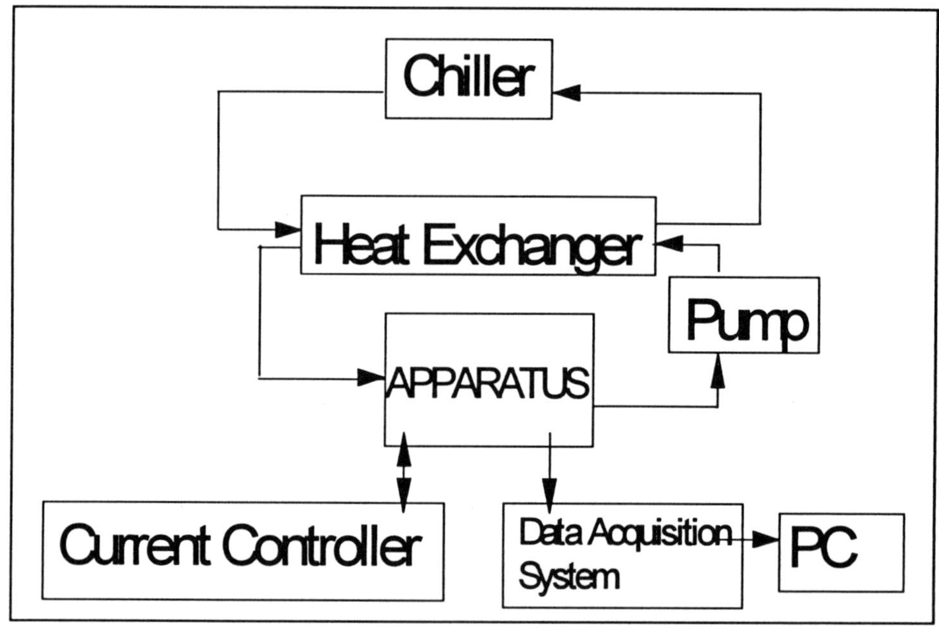

Figure 2. Block Diagram of the Experimental Setup

 Advances in Fusion and Processing of Glass II

A chiller (Neslab Model HX-100) in conjunction with a heat exchanger was used to maintain the temperature of the thermal bath fluid (mineral oil). Mineral oil was circulated, by means of a 11.2 $\times 10^{-6}$ m^3 Zenith positive displacement pump, through the heat exchanger (triple-pass, unbaffled and tubular). The temperature of the cooling fluid (distilled water) was controlled by the chiller. The current controller (Silicon Control Rectifier) regulated the current passed between the electrodes in the apparatus. Fifteen Type J thermocouples, permanently installed in the apparatus, provided information on temperature. The thermocouple readings were collected by a scanning digital voltmeter and recorded on a PC. The details of the thermocouple locations are provided in Figure 3.

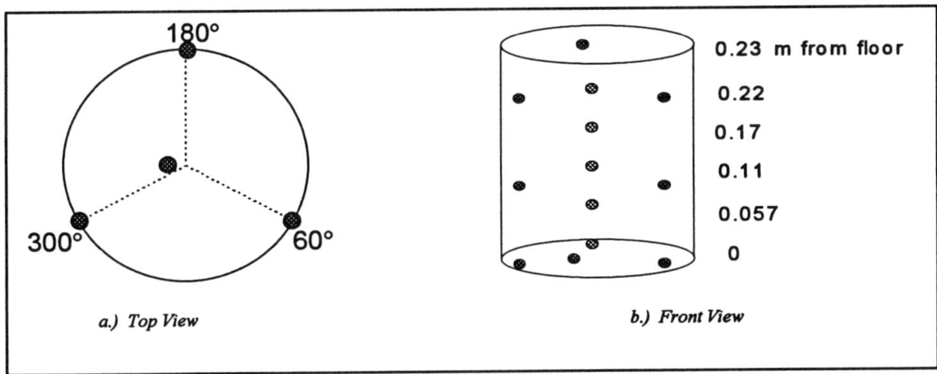

Figure 3: Locations of thermocouples; (a) top view, (b) perspective view. The locations are indicated by the dots. In addition to the 13 thermocouples shown in (b), 2 more were used to measure the inlet and the outlet temperatures of the mineral oil.

Two thermocouples were installed to measure temperature in the vicinity of the electrodes. These were located 6.35 mm from the electrodes on either the floor or the lid of the cylindrical tank. The inlet and the outlet temperatures of the mineral oil (thermal bath fluid) were measured by two additional thermocouples. The remaining 11 thermocouples were used to measure temperature at several locations in the glycerin-5% LiCl solution. As seen in Figure 3, these thermocouples were located at 60°, 180°, and 300° angles. At all three positions, thermocouples were installed at 0, 0.11 m, and 0.22 m from the floor. At the 180° position, thermocouples were also located at 0.057 m and 0.19 m from the floor. These 11 thermocouples measured temperature of the fluid at a distance of 6.35 mm from the inside face of the cylinder wall.

Experimental conditions used in this study are summarized in Table I. As seen here, experiments were conducted to investigate the effect of power input (80, 100, 140 W) and electrode penetration (0.051, 0.076, 0.01 m). The use of three levels for the two variables meant running nine experiments. Two additional experiments were incorporated to determine the reproducibility of the measurements. Thus, a total of eleven experiments were conducted.

Table I: Details of Experimental Conditions

Run Order	Electrode Penetration (m)	Power (W)
1	0.076	100
2	0.076	140
3	0.076	80
4	0.010	80
5	0.010	140
6	0.010	100
7	0.076	100
8	0.051	100
9	0.076	100
10	0.051	80
11	0.051	140

III. VELOCITY MEASUREMENT TECHNIQUE

The velocity distribution in the electrically heated glycerin-5% LiCl was measured using the Particle Displacement Tracking (PDT) technique. Specifically, we used the Particle Image Velocimetry (PIV) data acquisition and analysis system, developed by Dr. Mark Wernet of the NASA Lewis Research Center (11). This system tracks the displacement of particles in the fluid using a video camera and a large memory buffer frame grabber board. This technique allows one to measure the complete velocity distribution in a plane and thus offers considerable advantage over a Laser Doppler Anemometer, which measures velocity at one point at any given time.

The essence of the Particle Displacement Tracking technique is that it tracks particles through a series of single exposure images to determine particle velocities in fluid flow. A simple He-Ne laser source (5 mW) is used to generate a light sheet in the apparatus, and a video camera and frame grabber board are used to record particle image data. The frame grabber board is used to acquire five video fields, equally spaced by a time interval, Δt as set by the operator, through which individual particles are tracked. The time interval, Δt, is limited at a minimum to 1/60 of a second, and virtually unlimited on the maximum. The time interval is set according to the velocities of interest: the slowest velocities require the largest time intervals, while the fastest velocities require the shortest time intervals.

The principal components of the PDT system included

- Metrologic 5 mW He-Ne laser (Model G61,314)
- Edmund Scientific straight line generator lens accessory (Model G43,499)
- COHU RS-170 monochrome CCD camera (Model 4915-2100)
- PC with 4 MB frame grabber board
- Sony Trinitron color video monitor (Model PVM-1270Q)
- Software for data analysis

The fluid contained in the cylinder was seeded with 50-60 micron diameter alumina (Al_2O_3) particles. The two primary considerations in choosing the seed particles are the amount of light the particles scatter and their terminal velocities (i.e. their ability to accurately represent the flow). The scattering of light by the particles is what is actually observed when studying particle movement. The amount of light that a particle scatters is a function of both the particle diameter and the difference in refractive index of the particle and the fluid. Therefore, the refractive index of the particles should be very different from that of the fluid. As for the diameter of the particles, it should be large enough so that scattering is significant, but the particles should still be able to track the flow. For the alumina particles chosen, the index of refraction is different from that of glycerin by 0.21, and the terminal velocity of the particles in the fluid is approximately 4×10^{-5} m/s, which is about 0.4 to 1% of the maximum velocities measured in the present study.

The error in the measurement of the maximum velocity in the system was estimated to be about 10% (or in the range of 1 x 10^{-3} to 2 x 10^{-3} m/s).

IV. EXPERIMENTAL RESULTS

As mentioned earlier, the cylindrical geometry was chosen so that, with axial symmetry, flow and heat transfer in the system could be described by a two-dimensional mathematical model. In order to check axial symmetry, temperature data along the circumference at the base of the cylinder and at the mid-depth (i.e. at 0.11 m) position were examined. The temperature data were generated by fixed thermocouples located at 60°, 180°, and 300° positions (see Figure 3) and 6.35 mm away from the cylinder wall and into the fluid. For the 11 experiments described in Table I, the largest temperature variation in the circumferential direction was found to be about 6 K. This is small compared to the radial temperature differences measured in the system (e.g. for a case with 100 W power and 0.076 m electrode penetration, the difference between the maximum temperature in the center and the average temperature on the inside cylinder wall was about 35 K). When possible, velocity distribution was measured for an entire vertical plane (90°- 270°, see Figure 3) passing through the cylindrical axis. The electrodes, however, blocked the laser light, leaving the area behind them dark. The vector plots showed the flow to be essentially symmetric about the center line.

As mentioned before, in order to evaluate the reproducibility of the experiments, three experiments were conducted at 100 W power and 0.076 m electrode penetration. The maximum velocities reported for the three experiments were within the ± 0.001 m/s velocity measurement error (the maximum velocity was about 0.01 m/s). The temperatures measured at the lower electrode tip were found to be within 4 K. Furthermore, the vector plots for the three cases were indistinguishable from one another.

Figure 4 shows the measured velocity distribution for a case with 100 W power and 0.076 m electrode penetration. The vectors are shown in a portion of the 90°-270° plane mentioned above. The portion of the plane shown here is between the two electrodes. The x- and the y- axis numbers used in this figure refer to the pixel numbers. The upper and the lower regions on the left side of this figure do not show any vectors because laser was blocked by the electrodes, and these regions were therefore not illuminated. As expected, the fluid goes up in the center and falls down near the wall. The velocities are relatively large in the central part of the fluid (i.e. in cylindrical volume between the two electrodes) and near the wall. The rest of the fluid is relatively stagnant.

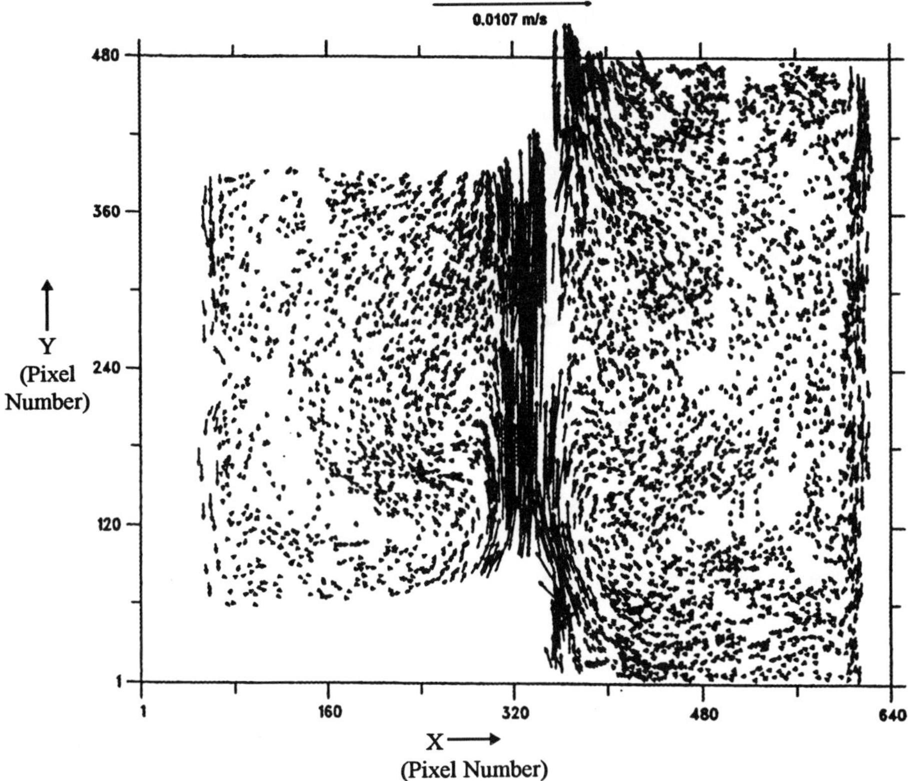

Figure 4: Velocity distribution in a plane containing the two electrodes for the case with 100 W power and 0.076 m electrode penetration.

For all the cases studied, as expected, the maximum velocity was observed in the central "core" (cylindrical volume comprising the space between the two electrodes), which also had the highest temperature in the system. The maximum values of the velocity and temperature increased with the increasing electrode penetration and power. The rise in temperature and velocity with increasing power is rather obvious. As the electrode penetration is increased, the distance between the two electrodes decreases, and hence, for the same power the heat release density (W/m^3) in the space between the two electrodes increases. This results in fluid in the space between the two electrodes getting hotter and more mobile. In the experiments reported here, both the temperature in the central "core" and the radial temperature differences (maximum temperature in the central core - temperature near the wall) were found to increase with the increasing power and increasing electrode penetration. The largest velocity was observed when both the power input and the electrodes penetration were at their maximum values (140 W and 0.10 m, respectively).

Figure 5 shows the variation in the maximum velocity with the power level and electrode penetration. Except for the discrepancy observed at 0.051 m electrode penetration, the maximum velocity magnitude increases with power. This discrepancy is probably an experimental error and needs to be examined critically. The maximum velocity also increases with increasing electrode penetration.

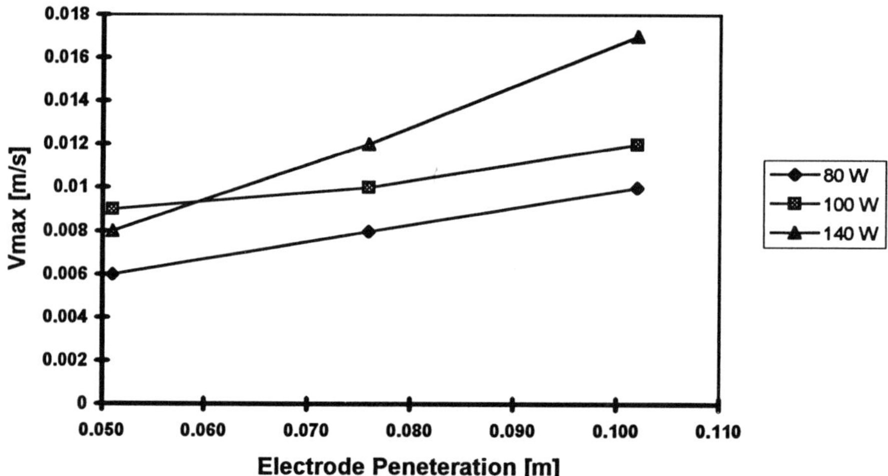

Figure 5: Maximum velocity as a function of power input and electrode penetration

Figure 6 shows the axial distribution of temperature at the mid radius position (i.e. r = 0.0381 m) for 80, 100, and 140 W power levels and for a fixed electrode penetration of 0.076 m.

                    Advances in Fusion and Processing of Glass II

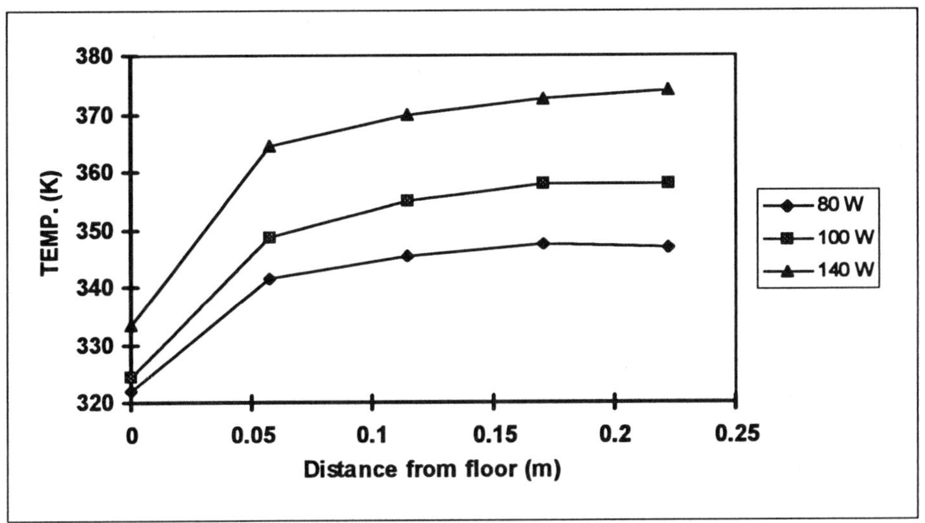

Figure 6: The axial variation of the mid-radius temperature for 80, 100, and 140 W power levels and a fixed electrode penetration of 0.076 m.

At all power levels the temperature of the fluid rises rapidly within a short distance from the floor (0.05 m). The temperature then increases slowly. At the fluid-lid interface itself (0.23 m from the floor), the temperature will decrease because the outer surface of the lid is exposed to air. As mentioned earlier, an objective of this work was to generate data which may be used to validate mathematical models for flow and heat transfer in electrically heated viscous melts. Some limited results of a mathematical model to be described in a subsequent publication are given below in Table II for the case with 100 W power and 0.076 m electrode penetration.

Table II: Predicted and measured values of temperature at the mid radius position for 100 W power and 0.076 m electrode penetration

Axial position (distance from floor, m)	Measured value (K)	Predicted value (K)
0	325	321
0.057	349	349
0.114	355	356
0.171	358	359
0.222	358	360

The agreement between the predicted and the measured values is remarkably close considering that no adjustable parameters were used in the mathematical model.

The measured and the predicted values for the maximum velocity were 0.01 m/s and 0.006 m/s respectively. The agreement was also good between the measured and the predicted voltage drops in the system (70 V and 71 V respectively). The maximum temperature in the fluid was measured to be 377 K. The corresponding predicted value was 364 K. Further details on the mathematical modeling results for this system will be provided in a subsequent publication.

V. CONCLUDING REMARKS

The principal objective of the work reported here was to design an experimental system for investigating flow and heat transfer phenomena in melts which are heated electrically by means of electrodes immersed in the melt. Electric heating and melting are used in many materials processing systems and an understanding of how design and operation variables (e.g. spacing between the electrodes, power input) affect flow and heat transfer is of considerable theoretical and practical interest. The system designed in this study consisted of a glycerin-5% LiCl solution contained in a cylindrical tank and heated by a pair of graphite electrodes. The velocity distribution was measured using a Particle Displacement Tracking technique . This technique allowed the measurement of velocity distribution in an entire plane illuminated by a He-Ne laser. This offers considerable advantage over some other techniques which either only give qualitative information (e.g. flow visualization through dye tracing) or measure velocity at one point at a time (e.g. Laser Doppler Anemometry). Temperature distribution was measured by means of thermocouples.

The velocity and temperature distribution were measured for three levels of power (80, 100, 140 W) and three levels of electrode penetration (0.051, 0.076, and 0.10 m). As is to be expected for thermal convection, the hot fluid in the space between the two electrodes moved upward and the relatively cooler fluid near the wall moved downward. The maximum velocity (0.005-0.02 m/s) was found to increase with both an increase in power and an increase in the electrode penetration (i.e. a decrease in the spacing between the electrodes).

Finally, it is noted that the experimental system reported here is characterized by well defined boundary conditions and cylindrical symmetry. This was done by design so that, in addition to providing insights into flow and heat transfer in an electrically heated melt, the results from this study may also be used to validate mathematical models for such a system. Such an effort is currently underway and

Advances in Fusion and Processing of Glass II

the preliminary findings reported here show a reasonably good agreement between the predicted and the measured temperature distributions.

REFERENCES

(1) R. L. Curran, "Mathematical Model of An Electric Glass Furnace: Effects of Glass Color and Resistivity," *IEEE Trans. Ind. Appl*, vol. 1A-9, pp 348-356, May/June 1973.

(2) T. S. Chen and R. E. Goodson, "Computation of Three-Dimensional Temperature and Convective Flow Profiles for an Electric Glass Furnace." *Glas. Tech.*, vol. 13, pp 161-167, Dec. 1972.M. J. Austin and D. E. Bourne, "A Mathematical Model of an Electric Glass Furnace," *Glas. Tech.*, vol. 14, no. 3, pp 78-84, June 1973.

(3) M. J. Austin and D. E. Bourne, "A Mathematical Model of an Electric Glass Furnace," *Glas. Tech.*, vol. 14, no. 3, pp 78-84, June 1973.

(4) M. K. Choudhary; "Three-Dimensional Mathematical Model for Flow and Heat Transfer in Electrical Glass Furnaces," *Heat Transfer Engineering*, vol. 67, no. 4, pp 55-65 (1985).

(5) A. Ungan and R. Viskanta; "Three-Dimensional Numerical Simulation of Circulation and Heat Transfer in an Electrically Boosted Glass Melting Tank," *IEEE Transaction on Industry Applications*, vol. IA-22, no. 5, pp 922-933 (1986).

(6) M. K. Choudhary; "A Modeling Study of Flow and Heat Transfer in an Electric Melter," *J. of Non-Crystalline Solids*, vol. 101, pp 41-53 (1988).

(7) P. Schill; "Modeling of Electric Boosting," *Proceedings II International Seminar on Mathematical Simulation in Glass Melting," Czech Glass Society, Czech Republic,* pp 140-151 (1993).

(8) I. G. Choi, D. F. Bickford and J. T. Carter, "Thermal Effects of Electrically Conductive Deposits in a Joule-Heated Melter," *Ceramics Transactions* (editor Varshneya et al), vol. 29, pp 645-658, The American Ceramic Society, Westerville, Ohio (1993).

(9) M. K. Choudhary, "A Modeling Study of Flow and Heat Transfer in the Vicinity of an Electrode," *Proceedings of XVII International Congress on Glass, Beijing,* Chinese Ceramic Society, vol., 6, pp 100-107 (1995).

(10) M. K. Choudhary, "Mathematical Modeling of Flow and Heat Transfer Phenomena in Electric Glass Melting Furnaces," to be published in the *Proceedings of the Julian Szekeley Memorial Symposium on Materials Processing,* The Minerals, Metals and Materials Society, Warrendale, Pennsylvania (1997).

(11) M. P. Wernet, "Particle Displacement Tracking for PIV," *NASA Lewis Report,* Cleveland , Ohio (1990).

(12) K. Van Dyke, "Flow and Heat Transfer in an Electrically Heated Fluid," MS Research Report, Case Western Reserve University (1996).

ACKNOWLEDGEMENT

The authors gratefully acknowledge Ms. Cynthia Rarick's help in preparing this manuscript.

                    Advances in Fusion and Processing of Glass II

The Fiber Glass Industry

TIME-TRANSIENT MATHEMATICAL MODELING OF A FIBER GLASS FURNACE

J.P. Fletcher
Fiber Glass Research Center, PPG Industries, Inc.
P.O. Box 2844, Pittsburgh, PA 15230

G. Peters
TNO Glass Technology, P.O. Box 595, 5600 AN
Eindhoven, The Netherlands

B. Shome
Fiber Glass Research Center, PPG Industries, Inc.
P.O. Box 2844, Pittsburgh, PA 15230

INTRODUCTION

Three-dimensional mathematical modeling has been successfully applied towards understanding the effects of glass furnace design and operating changes under steady-state conditions. However, due to the nature of the glass melting process, modeling of time-dependent perturbations represents a further refinement towards improved furnace stability and product consistency. Extension of the time-dependent model can also be applied to the area of sensor development.

This study demonstrates the application of time-transient calculations, using a complete 3D mathematical simulation of an air-fuel fired recuperative fiber glass furnace, subject to a process change involving 3% increase in total fuel. Modeled results are compared with the measured process response using glass temperatures, and are tracked through the transition period.

BACKGROUND

Descriptions of Models

The TNO complete glass furnace model consists of a model for the glass tank and a model for the combustion space. For the glass tank the finite-volume based model TNO-GTM is used which is described by Beerkens et al. [1] In this model the time-dependent mass continuity equation, Navier-Stokes equations and energy equation are solved in three dimensions. A batch blanket model for calculating the raw materials melting process is incorporated in the model. Conduction in the furnace walls is taken into account and highly temperature dependent glass properties are used. Radiation is accounted by using an augmented thermal conductivity. The TNO-GTM model is equipped with additional sub-models for forced bubbling, fining and degassing, melting kinetics, homogenization, tracing of potential glass defects, determination of residence time distribution and the calculation of glass quality indices.

For the combustion space, the finite-volume based model TNO-WISH3D is used. A detailed description of this model can be found in Lankhorst et al. [2] This model describes time-dependent turbulent flow and mixing, flame chemistry as well as radiative and convective heat transfer in the combustion space. Three-dimensional flow is described by the Navier-Stokes equations and the continuity equation. Turbulence is being accounted for by the standard k-epsilon model with wall functions. Radiation is modeled by the discrete transfer method where the gas is assumed to be gray. An appropriate grid structure was used to resolve the large gradients. Combustion is described by the conserved scalar/presumed probability density function (PDF) approach to high temperature non-premixed combustion. The thermochemical variables are functions of a single conserved scalar, the mixture fraction, f. To account for turbulent fluctuations of the thermochemical variables, a double delta PDF for f can be used. Chemistry is described with a one-step global reaction taking into account the effect of dissociation. Models predicting soot have also been incorporated. A postprocessor calculates the thermal NO formation according to the Zel'dovich mechanism [3]. Properties of the gas mixture are composition and temperature dependent and conduction in the refractories are taken into account. The various submodels are described in more detail by Boerstoel [3].

The TNO-GTM and TNO-WISH3D models can be used separately with assumed or measured boundary conditions on the glass surface. The models can also be coupled with a shared interface on the glass surface. In the coupled case, the distribution of the total heat flux on the glass surface is calculated by TNO-WISH3D and is used as a boundary condition for the glass bath. Conversely, the

                    Advances in Fusion and Processing of Glass II

temperature on the glass surface as calculated by TNO-GTM is used as a boundary condition for the combustion space. Details of the iterative coupling procedure can be found in Lankhorst et al. [4].

In the present study, a coupled model was used. The time-transient GTM calculations were performed using the calculated heat flux from the simulated combustion calculations. Since the time scales in the combustion space are much smaller than those in the glass bath, we can assume that when changing gas/air flow rates, the new flow patterns and temperature distributions in the combustion space are instantaneously established and that the temperature distributions in the combustion space follows changes on the glass surface also without delay. Therefore, time-transient calculations for the combustion space can be avoided which results in significant savings in computing time without loss of accuracy.

Description of Glass Furnace

An air-gas, cross fired fiber glass furnace was chosen for this study. This furnace produces continuous strand E-glass fibers at a rate of approximately 30 tons/day. The total energy to the melting system is provided through total gas flow and recuperative pre-heat to the combustion air. The batch is introduced on one side in the rear of the furnace.

A schematic of the furnace is shown in Figure 1. For this study, a series of melter underglass (MUG) bottom thermocouples was monitored. These are located on the melter centerline at different positions along the length.

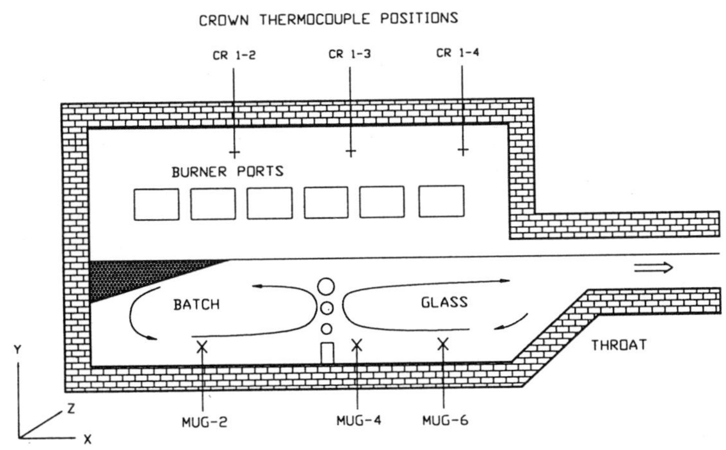

Figure 1. Side View of Fiber Glass Furnace Center Plane

DISCUSSION OF RESULTS

Model Validation

In order to validate the time-transient calculations performed by the TNO-GTM, data were collected every 15 minutes, for a period of 6 hours, from four thermocouples in the glass tank. The location of these four thermocouples is shown in Figure 1. The first thermocouple (MUG2) is located underneath the batch, the second (MUG4) is located near the bubblers, the third (MUG6) is located near the throat of the glass tank, and the last one (PC0) is located in the throat region representing the glass outflow temperature.

The model was validated by comparing the modeled temperature changes relative to a base value with the measured data for a 6 hour period. The base value for the measured data was calculated by averaging the temperatures over a 12 hour period prior to the 3% gas increase, whereas the steady state results were used as the base value for the modeled data. The standard deviation of the measured data during that 12 hour period could then be interpreted as the uncertainty in the measured values and are indicated, as error bars, in Figures 2 through 5.

As seen in Figures 2 through 5, the agreement between the predicted and measured temperature difference for all the monitored thermocouples are mostly within experimental uncertainty. In addition, the predicted rate of increase in the temperature shows excellent agreement with the measured values. The scatter in the measured data could be attributed to small fluctuations in the operating conditions (such as perturbations in throughput, ambient, etc.) and are not seen in the model as they are not taken into account.

Using the time required for a typical temperature change of 3 degrees C, the melter modeled temperature response rate to the increased gas flow can be examined in more detail. The exit glass reacts first, due to the direct forward flow in the top glass layer of the tank. The short return flow under the batch starts to heat up MUG2, thereafter. The reason for MUG6 to react earlier than MUG4 is because MUG4 sees the return flow over the bottom from the front portion of the tank, rather than the forced convective flow from the bubblers.

Application to Sensor Development

The time-dependent furnace responses to the energy increase, as a function of both glass depth and melter length, can be studied in more detail through a series of side view melter centerline planes as shown in Figure 6. These represent the

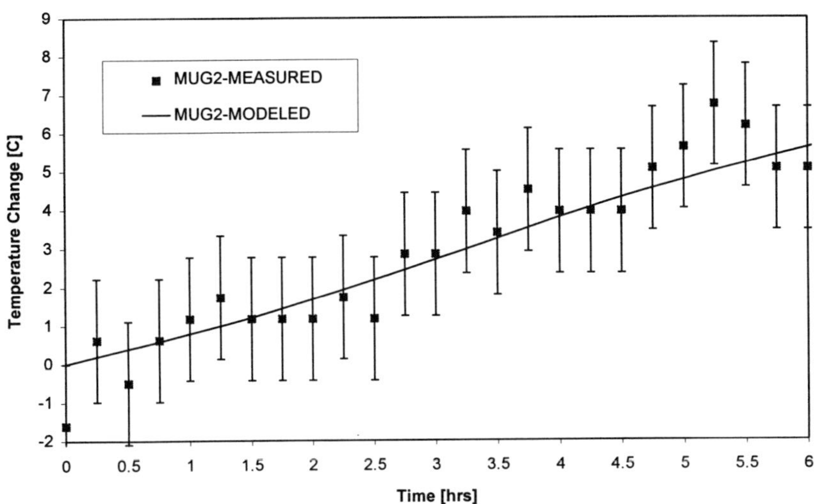

Figure 2. Transient Temperature Change of MUG2

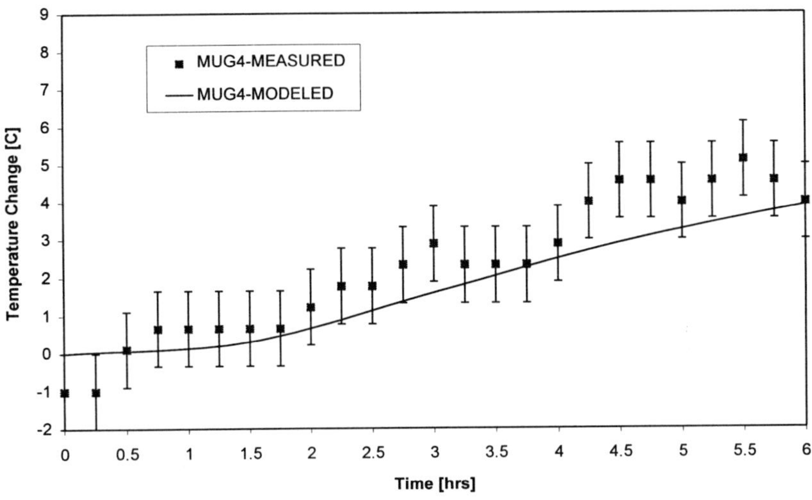

Figure 3. Transient Temperature Change of MUG4

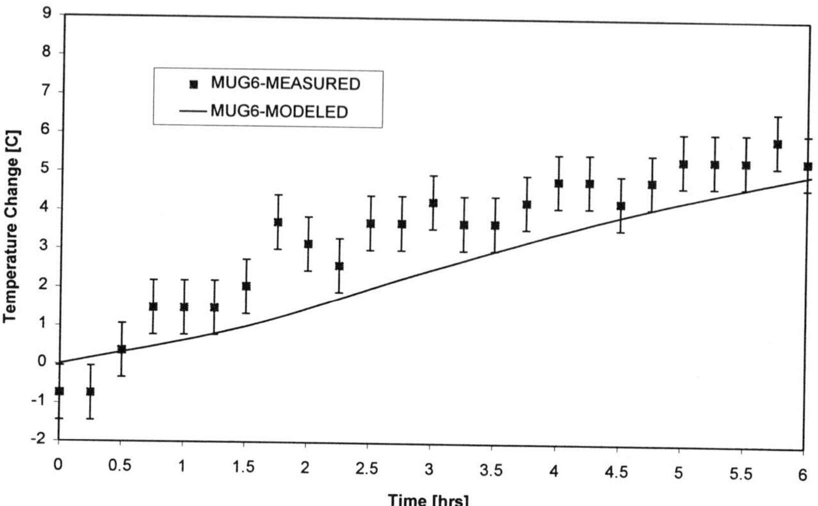

Figure 4. Transient Temperature Change of MUG6

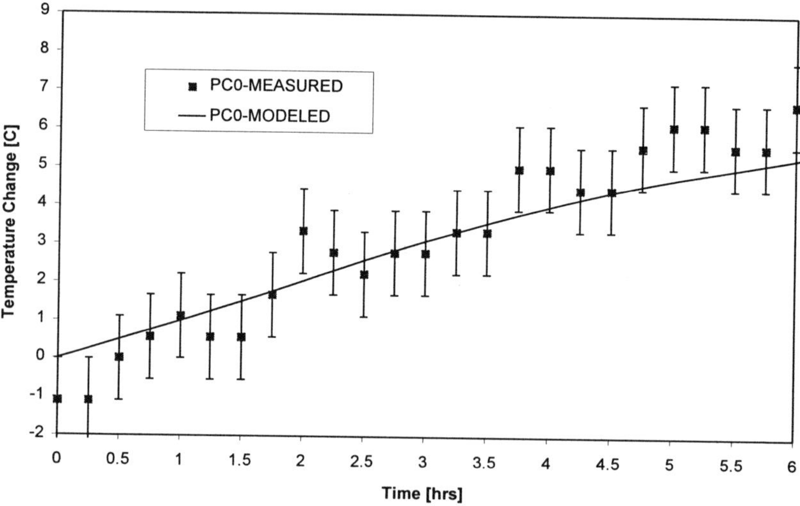

Figure 5. Transient Temperature Change of PC0

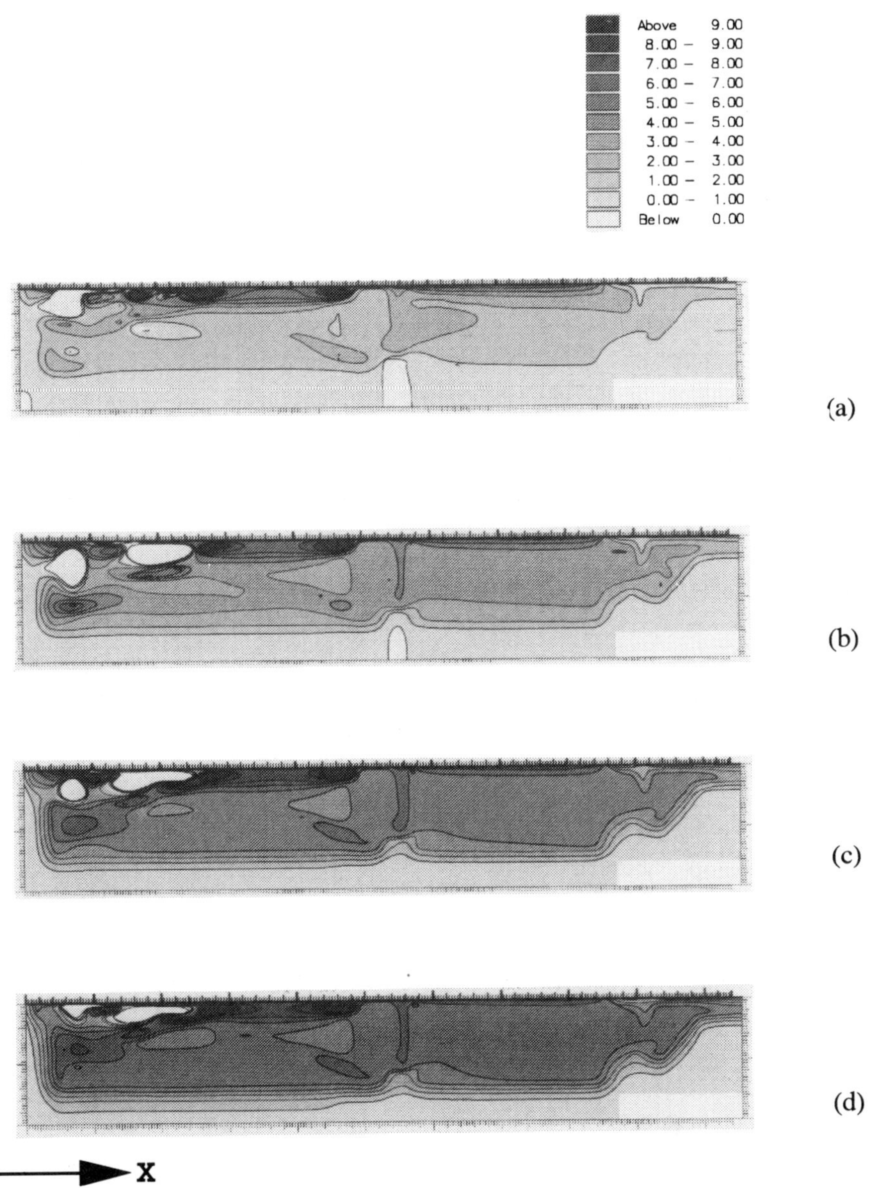

Figure 6. Side View of Melter Center Plane – Temperature Distribution
Change as Function of Time at (a) 1.5 hrs., (b) 3.0 hrs., (c) 4.5 hrs., and
(d) 6.0 hrs.

temperature distribution difference (in degrees C), relative to that of the starting time (base value), at 1.5 hour time intervals for a 6 hour time period.

These plots graphically illustrate which parts of the furnace respond first to the perturbation. Initial melter response is most strongly seen in the burner region between the batch meltout and row of bubblers. Rearward convective flows transfer heat towards the bottom thermocouples in the backend of the tank. The bubbling region also sees a quick response to the higher glass surface temperatures and is transported away by the local convective flows. Within the first 3 hours, the entire furnace experiences the higher temperatures.

The value of time-transient calculations as applied to sensor placement, in this case thermocouples, can readily be seen throughout the glass melt. This can provide a beneficial tool in determining sensor location for process control and in new furnace designs. Since understanding the time-dependent temperature changes allows us a better understanding of the temperature dependent glass properties such as redox, this concept can also be generalized for application in new sensor technologies (e.g. redox probe).

CONCLUSIONS

It is concluded that the coupled TNO GTM and WISH3D models are capable of simulating a time-transient process change for a complete glass melter. Differences in the rate of change and magnitude of the melter temperature response to the increased melter gas flow can be quantified as a function of position. From these results, the coupled models can be used to develop optimum temperature sensor locations in both existing tanks and in the future for new tank designs.

REFERENCES

1. R.G.C. Beerkens, T. van der Heijden, and H.P.H. Muysenberg, "Possibilities of Glass Tank Modelling for the Prediction of the Quality of Melting Processes," *Ceram. Eng. Sci. Proc.*, **14** 139-160 (1993).

2. A.M. Lankhorst, G.P. Boerstoel, H.P.H. Muysenberg, and K.K. Koram, "Complete Simulation of the Glass Tank and Combustion Chamber of the Former Ford Nashville Float Furnace Including Melter, Refiner and Working End," *Fourth Int. Seminar on Mathematical Simulation in the Glass Melting*, (1997).

3. G.P. Boerstoel, "Modeling of gas-fired furnaces," Ph.D. thesis, Delft University of Technology (1997).

                                    Advances in Fusion and Processing of Glass II

4. A.M. Lankhorst, H.P.H. Muysenberg, and M.P.J. Sanders, "Coupled Combustion Modeling and Glass Tank Modeling in Oxy- and Air-Fired Glass Melting Furnaces," *International Symposium on Glass Problems*, Istanbul, Turkey (1996).

ACKNOWLEDGMENTS

The authors are grateful to PPG Industries, Inc. and TNO Glass Technology management for allowing the publication of this work.

EFFECT OF FIBERIZATION PROCESS ON GLASS FIBER SURFACES

Jon F. Bauer
Johns Manville Technical Center
Littleton, Colorado   USA

ABSTRACT
Surface properties such as chemical durability are of critical importance to the behavior and performance of fine diameter, discontinuous glass fiber. In addition to bulk composition, these properties are influenced by the method of fiberization and the specific process variables employed in production. Controlled dissolution studies combined with SEM and surface chemical analysis were used to evaluate process effects, particularly as they relate to durability and persistence of fibers in the lung. Results indicated that fibers produced by pure flame attenuation exhibited a unique concentric structure (core / shell) with an outer surface that can be more durable than the bulk composition would indicate. Fibers produced by air attenuation in either continuous draw or rotary processes are more uniform with surfaces that are more open and reactive to moisture or simulated body fluids. A variety of fiber types are formed when air and flame are mixed in a rotary process, ranging from more durable "biphasic" fibers as produced by flame attenuation to highly soluble, presumably low density fibers that are even more reactive than those formed by air attenuation alone. Results allow a better projection of likely behavior of fine glass fibers in aggressive environments such as the lung.

INTRODUCTION
Commercial glass fiber is produced by a number of processes that differ principally in the methods used for final attenuation of the melt stream. Key process variables that affect the quality and end properties of the fiber produced include the rate of attenuation, initial melt temperature, temperature distribution during forming, and the ambient atmosphere or environment in which the fiber is drawn. Within a given process, control of these variables is used principally to change the diameter and length distributions of the fibers produced. However, these variables also have an important impact on the nature of the fiber surface, imparting properties that can impact forming as well as the fiber's performance in the final product.

Fine diameter, discontinuous fibers used in insulation and filtration products, in particular, exhibit unique properties attributable to surface effects induced by process. These fibers can have quite high specific surface areas for non-porous solids (up to 7 m^2 / g), and may be rapidly quenched upon formation creating surfaces that are quite different from those of bulk glass or even fibers of much larger diameter such as those used in reinforcement applications. As such they might be expected to possess tensile strengths, chemical durabilities, and adhesive and dispersive properties that are somewhat unique.

One of the most important properties of discontinuous glass fibers yet to be fully understood is their behavior in the lung upon inhalation, specifically with respect to chemical durability which determines biopersistence. Biopersistence in turn is directly correlated with incidence of respiratory disease (*1*) – the longer fibers endure within the deep lung, the greater the chance for induction of both malignant and non-malignant reaction. There have been a number of studies relating glass fiber durability to bulk composition (*2, 3, 4*). No information, however, has been available on the impact or "overprint" created by process that might be expected based on the effects on fiber characteristics noted above.

This study attempts to look at fiberization process effects on the nature of glass fibers and their surfaces, particularly as they relate to the chemical durability of the fiber in the physiological environment of the lung. To accomplish this, several commercial glass fiber compositions were produced by three separate processes, using conditions and operating variables similar to those used in actual manufacture. The fibers were then exposed to simulated conditions of the deep lung for periods up to four months, during which their dissolution progress was monitored using leachate analysis. Following exposure, samples were recovered and characterized to determine the morphological, chemical, and physical changes that occurred. Results were used to project the likely biopersistence of each of the fibers, and also to infer some of their properties relative to utility and performance in end use.

MATERIALS AND METHODS
Glass compositions evaluated are provided in Table I. All three have been used in the commercial production of glass fiber for at least 15 years. They have also been employed in numerous toxicological studies, so that their response in laboratory animals is reasonably well known (if not completely understood). Results of such tests have shown these fibers to be non-toxic to laboratory animals by inhalation, with the exception of recent results giving a positive response in hamsters for JM475 (*5, 6*).

Table I. Glass fiber compositions (mole %)

	JM753	JM475	JM901
SiO_2	65	64	58
Al_2O_3	3	4	3
B_2O_3	5	11	8
$CaO + MgO$	10	2	14
$Na_2O + K_2O$	17	13	17
BaO	-	3	-
ZnO	-	3	-

Fiberization was accomplished using methods representative of those used in actual production. Processes used were (1) continuous drawing (CD), (2) flame attenuation (FA), and (3) rotary spinning. Continuous drawing involves mechanically pulling and attenuating a melt stream, in air, through single or multiple orifices in a crucible or bushing. The flame attenuation process incorporates a secondary attenuation step in which fibers or melts produced by primary attenuation are further fragmented and drawn out by high velocity gas jets or flames to produce finer diameter, discontinuous fiber. The rotary process also produces discontinuous fiber, but does so by spinning of primary melt streams through small holes in the walls of rapidly spinning "baskets." More details on each of these processes is provided elsewhere (*7, 8*).

Durability studies were performed by exposing representative aliquants of the fibers to simulated extracellular fluids (SEF) held at pH 7.4 and 37° C. For these studies, fiber samples were initially elutriated in water to remove any residual non-fibrous particulate present, and, for some

                                        Advances in Fusion and Processing of Glass II

experiments, may have been further processed to select only certain diameters, e.g., those in the respirable range. The preparative procedures have been discussed previously (*9*).

Experiments were performed using two system designs. The first, and most frequently used was a continuous flow system in which SEF was percolated through a thin web of fibers at constant ratio of fluid flow rate to surface area (F / S) - in these studies, $0.04 \pm .003$ cm / hr. Aliquots of fluid were collected at fixed time points and analyzed by ICP to accurately determine the concentrations of dissolved glass components which were then used to construct mass loss curves. After conclusion of each run, residual fiber samples were removed, rinsed in de-ionized water for up to one hour to remove any remaining fluid, and then dried to constant weight. These latter values were used as a check on mass loss derived from leachate analyses. Details of this procedure have also been described previously (*10, 11*). In the second system, dissolution was allowed to proceed under conditions of near infinite dilution, thought to more closely approximate those of the deep lung (*12*). In this system (termed NIDS), fiber dissolution occurred in a 100 litre bioreactor containing SEF maintained at 37° C and pH of 7.4. Fluid was circulated internally at rates just below the maximum permeability of the filters holding the samples, creating not only high V_{fluid} / S, but also F / S conditions of approximately 3000 cm / hr. Here gravimetric data, as opposed to leachate data was used to construct mass loss curves so that the NIDS system also served as a check on results from the continuous flow system. Details have also been presented earlier (*13*).

Rates for total mass loss (dissolution rates) for each fiber sample were derived by fitting the experimental data from the continuous flow system to results predicted by known models for fiber dissolution. Differences between the experimental and predicted masses were then minimized over the entire range of data by adjusting the respective rate constants until a "best fit" value was obtained.

The method employed here was a "constant velocity" (CV) model in which fiber dissolution proceeds by constant rate of reduction in thickness, or in this case, diameter. This is the most commonly used model for characterizing fiber dissolution both *in vitro* and *in vivo*. It begins with a fundamental relationship - i. e., that the rate at which mass is dissolved from a surface is directly proportional to the total area of that surface presented to the attacking fluid. Mathematically, this is expressed as

$$\frac{-dm}{dt} = kS \tag{1}$$

where m represents the mass of the sample at time t, S represents the total surface area of the sample, and k is the dissolution rate constant, often referred to as k_{cv} or k_{dis}. If the variation of S with t is known or can be approximated, then the expression may be integrated and k calculated from the experimentally-derived mass loss.

The constant velocity model assumes that S(t) can be described by a simple geometric progression based on continuous diameter reduction. For cylindrical fibers with initial diameter d_o, and density ρ, this leads to the following relation between k_{cv} and the mass fraction <u>remaining</u>:

$$k_{cv} = d_o \rho [(1 - (m_t/m_o)^{1/2}) / 2t] \tag{2}$$

where m_o and m_t are respectively the initial mass and the mass left at time t. Details of the derivation are given elsewhere [3,*11*].

An appealing feature of this model, based on the initial assumptions, is that fiber diameter (d_t) may be predicted for any time t by the relation

$$d_t = d_o - (2k_{cv}t / \rho) \tag{3}$$

By setting d_t to 0 one can then predict the time to disappearance or total fiber "lifetime". This means that this model may be readily employed in the interpretation of diameter distributions and clearance by dissolution of fibers recovered either from *in vitro* dissolution studies or from the lung itself, if time of initial exposure is known.

Given the above models and assumptions, rate constants may be calculated for each fiber type once a compilation of individual fiber diameters comprising each sample is known. The calculations are best handled by treating the dissolution process separately for each fiber in the distribution, allowing the fiber to dissolve independently according to the chosen rate law as described above, and then summing results for all of the fibers at each experimental time point. In this study, fiber mass loss and dissolution rate calculations were performed in this manner. For a chosen k_{cv}, a mass loss curve (dissolution profile) was computed for the particular diameter distribution characterizing each sample. This simulated data was then fit to the experimental mass loss curve with the optimum k_{cv} found iteratively by varying k_{cv} until χ^2 was minimized over the total data set. Somewhat similar methods have been described by others (*14*).

Samples recovered from the *in vitro* test cells were then characterized by SEM / EDS to determine diameter distributions at a given time of exposure (also a good check on the CV model), and physical, chemical, and morphological changes to the fibers and their surfaces. Pristine and exposed fiber samples were also evaluated by bulk chemical analysis, and XPS to determine surface chemistry.

RESULTS

Comparisons of mass loss profiles and dissolution rates for the three glass types showed significant differences depending upon type of process used for fiberization. The differences were most noticeable between the continuously drawn fiber which underwent only a single attenuation in air and the flame attenuated fiber formed by secondary attenuation in a hot gas jet. The variation is illustrated in Figure 1 for JM753 glass. As seen in Figure 1a, the experimental mass loss data for the continuously drawn fiber can be fit quite well with the CV model giving an overall rate constant of just over 100 ng / cm^2 hr. In such cases, dissolution has been shown to be largely congruent, and likely controlled principally by the rate of breakdown of the aluminosilicate network of the glass comprising the fiber (*10*). Little difference exists between surface and bulk properties of the fiber, at least with respect to durability. In contrast, Figure 1b reflects a different behavior for the FA fiber, which is similar in mean diameter to the CD fiber sample but contains a significantly broader distribution of diameters. Here the dissolution profile is quite non-linear, showing pronounced positive curvature, and the experimental mass loss data cannot be easily fit to

 Advances in Fusion and Processing of Glass II

a simple CV model. In addition, the best fit to the data gives a rate constant of only 28 ng / cm^2 hr, approximately ¼ that for the CD fiber.

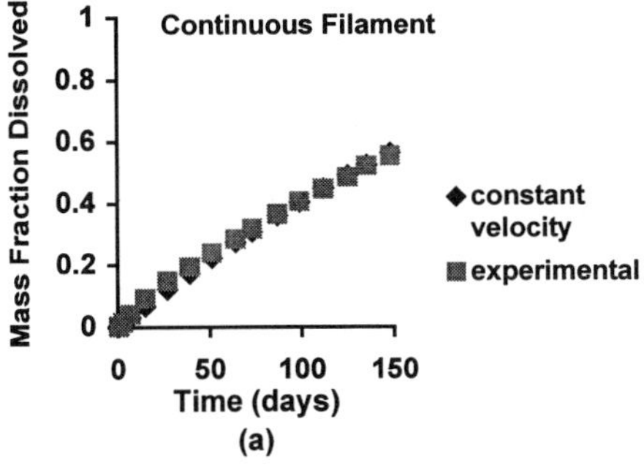

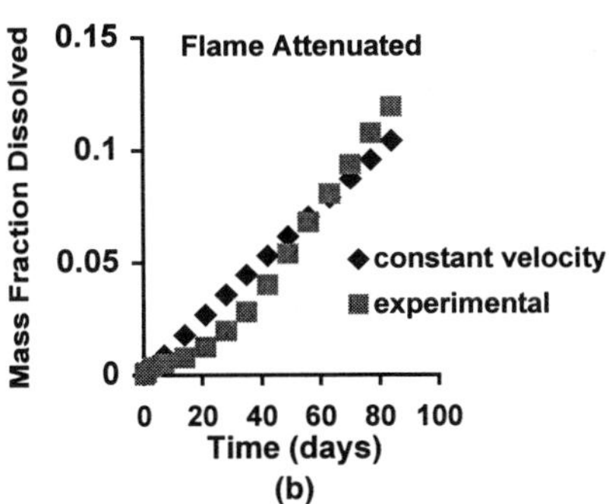

Figure 1. Dissolution profiles for JM753 glass fiber in SEF, 37º: (a) continuously drawn (CD) filament, mean fiber diameter = 7 µm, k_{cv} = 108 ng / cm^2 hr, (b) flame attenuated (FA), mean fiber diameter = 6.5 µm, k_{cv} = 28 ng /cm^2 hr.

Attempts were made to better reproduce the experimental mass loss for the flame attenuated JM753 using several different models. The data was initially modeled assuming that the sample was actually comprised of two or more different types of fiber, each dissolving with a different rate constant. Individual attempts were made where k_{cv} values were assigned randomly across the initial diameter distribution and also selectively, e.g., assuming that only certain diameter fractions dissolved at a given k_{cv}. The dissolution profiles were also modeled assuming that there was mass present in the initial sample that was unaccounted in the original fiber diameter distribution; such would be the case if the sample contained sufficient residual non-fibrous particulate or "shot" to influence the mass loss. None of these attempts proved successful.

Similarities were, however, observed between the dissolution profiles for flame attenuated JM753 and fibers which were coated with hydrophobic or otherwise durable coatings. Using this concept, experimental mass loss curves were reproduced reasonably well using a simple model which assumes that each fiber in the sample is essentially "concentrically biphasic", comprised of a moderately soluble inner core and a more durable outer shell. A better fit was obtained when core and shell were discrete, each with its own k_{cv} value. Initially, a constant shell thickness was assumed for all fibers. The best results, however were obtained when the shell thickness was allowed to vary according to fiber diameter with a 1/d dependence, assuming that, similar to the CV model for dissolution, surface reactions or interactions producing the shell would vary with specific surface area of each fiber. Results are shown in Figure 2a. For the flame attenuated JM753, the "best fit" predicts a k_{cv}(shell) of 12 ng / cm^2 hr, a k_{cv}(core) of 33 ng /cm^2 hr, and a shell thickness that averages 82 nm, but varies considerably from about 10 nm for the thickest fibers in the distribution to over 200 nm for the thinnest fibers. This has some important implications for the issues discussed above. First of all it suggests that not just the surface, but the also "bulk" of the FA fiber is distinctly different from that of the continuously drawn JM753. Long term persistence of this fiber would not be well predicted using the CD fiber as a surrogate with k_{cv} of 108 ng / cm^2 hr, as had been thought previously. Secondly, it implies that thinner fibers should be intrinsically more durable than thick ones, and have dissolution profiles that are more linear and closer in accord with the CV model. This hypothesis can be easily tested by determining mass loss for either a fiber sample which has a much smaller mean diameter or for a fine fiber fraction separated from the original sample. Results of this exercise are illustrated in Figure 2b for a fine fiber sample exposed to SEF for almost four months. Here a considerable portion of the mass loss profile has been captured, so that dissolution should have proceeded well into the interiors of most fibers. Notice that the profile is much more linear than that of Figure 1b, with a calculated uniform fiber k_{cv} very close to that predicted by the core / shell concept. For these finer diameter fiber samples, biopersistence may then be reasonably well approximated by a single value for k_{cv}.

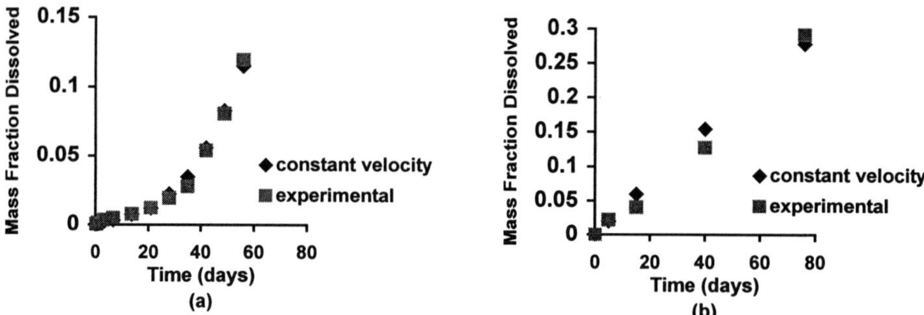

Figure 2. Dissolution of flame attenuated JM753 fiber samples in SEF, 37°: (a) experimental from Figure 1b, fit to a core / shell model with shell thickness varying as 1 / d, k_{cv} (core) = 33 ng / cm^2 hr, k_{cv} (shell) = 12 ng / cm^2 hr, avg. shell thickness = 82 nm, (b) experimental data from a fine fiber sample (0.8 μm mean dia.) from a NIDS run, fit to a single k, CV model: k_{cv} = 11 ng /cm^2 hr.

JM475 fiber was found to show similar behavior with regard to dissolution of the CD and FA fiber. Dissolution profiles constructed for the CD version were highly linear, conforming well to the CV model with a rate constant of 70 ng / cm^2 hr. Results for the FA fiber, however were very close to those seen for JM753 glass. Figure 3 shows the mass loss profiles for fiber samples with mean diameters of 1.4 um, evaluated using both the continuous flow and the NIDS system. Both gave similar dissolution profiles (only slight positive curvature – as predicted) with rate constants of 12 and 11 ng / cm^2 hr, respectively. The reproducibility further supports that flow independent conditions have been achieved in the continuous flow system and that the effects described above are not artifacts of the test system.

 Support for the core / shell concept for FA fibers was also derived from SEM images of fibers recovered from the test cells after prolonged exposure to SEF.

Advances in Fusion and Processing of Glass II

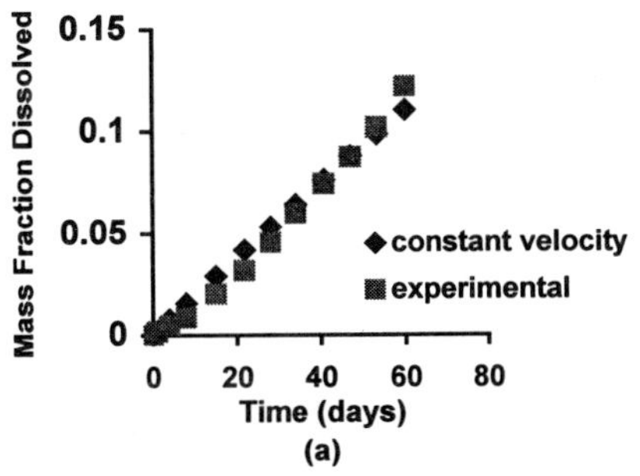
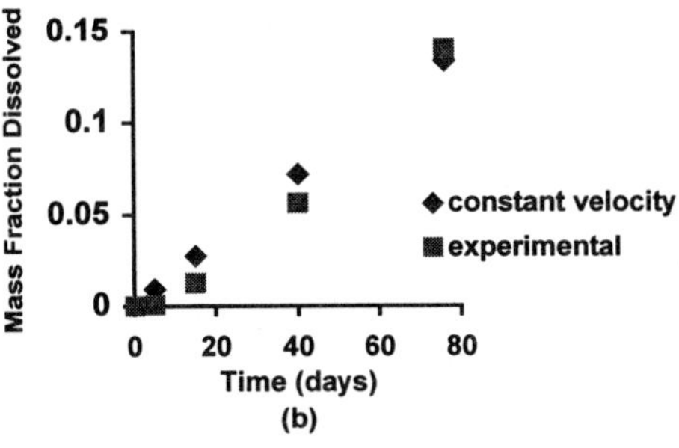

Figure 3. Dissolution of flame attenuated JM475 in SEF (1.4 μm mean diameter): (a) continuous flow system, k_{cv} = 12 ng / cm^2 hr, F / S = .04 cm / hr., (b) NIDS system, k_{cv} = 11 ng / cm^2 hr

While morphological distinctions between fiber surface and interior are not visible in the pristine fibers, they are revealed upon reaction with the fluids. Figure 4 shows representative end views of continuously drawn JM475, and flame attenuated JM475 and JM753. The image of CD JM475 was taken after over 40% mass loss and reveals a distinct core of residual glass surrounded by a highly porous leached layer that, based on the mass loss profiles, has not yet become diffusion-limiting. In contrast FA JM475 and JM753 appear to have dissolved quite differently, primarily by local penetration of an otherwise durable outer layer. While such "pit corrosion" was not specifically accounted for in the core / shell approximation, the images support the idea that dissolution over time progresses by increased contribution of glass dissolved from the fiber interiors – hence, the positive curvature to the mass loss curves.

The similarity in surface features, dissolution profiles, and rate constants between flame attenuated JM475 and JM753 raised the issue of whether the flame attenuation process produced the same type of surface regardless of the initial bulk composition. Glass compositions that were initially less durable were of particular interest, since these would give a good indication as to whether bulk composition or process was dominant in determining the fiber's ultimate surface properties. JM901was chosen as a good candidate for evaluation primarily because, as a CD fiber, its k_{cv} is at least four times greater than that of either JM753 or JM475. Unlike these two glass types, JM901 fibers also dissolve incongruently resulting in leaching and subsequent breakage *in vitro* (*13*).

A dissolution profile for flame attenuated JM901 glass is shown in Figure 5, compared with fiber of identical mean diameter produced by air attenuation in a rotary process. As discussed later, this process produces fiber which is quite similar to that obtained by continuous drawing; it was used here to achieve a 1 μm mean diameter fiber for comparison, which is not attainable by the CD process. Mass loss is quite different in both cases, indicating that flame attenuation had a major effect upon the glass, producing a fiber with a k_{cv} value only about ¼ that of the CD fiber. The profiles also show slight positive curvature, similar to those of FA JM753 and JM475 indicating the generation of a core / shell structure. This is in direct contrast to the negative curvature shown by CD JM901 (a consequence of incongruent dissolution noted above). The measured dissolution rates, however, for FA JM901 are still much greater than those for the previous two fibers in the same diameter range, suggesting that, while flame attenuation does produce a more durable fiber, it is sensitive to initial glass composition and does not modify the glass in such a way that the same type of surface or shell is produced every time. This has been confirmed for other compositions of fiberizable glasses as well.

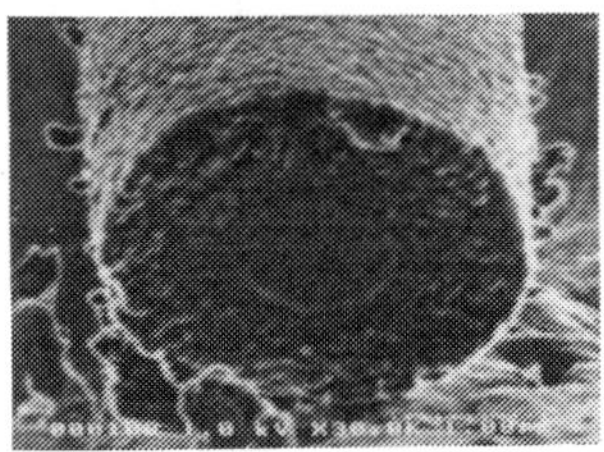

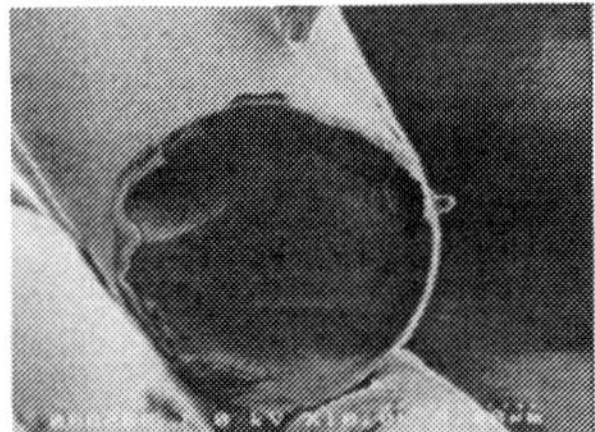

Figure 4. Fibers exposed to SEF, continuous flow system: (a) CD JM475, 72 days, (b) FA JM475, 72 days, (c) FA JM753, 30 days

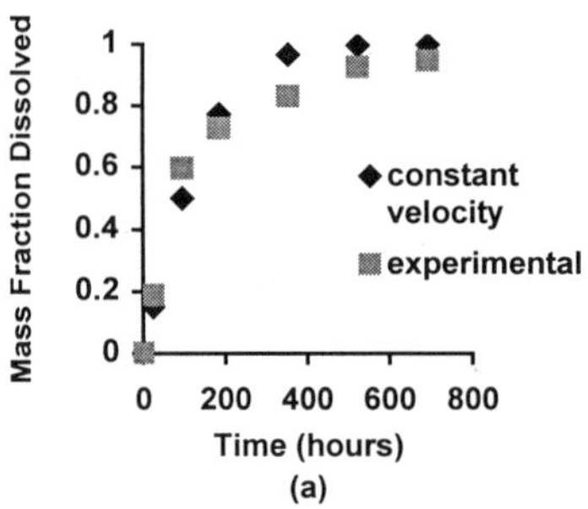

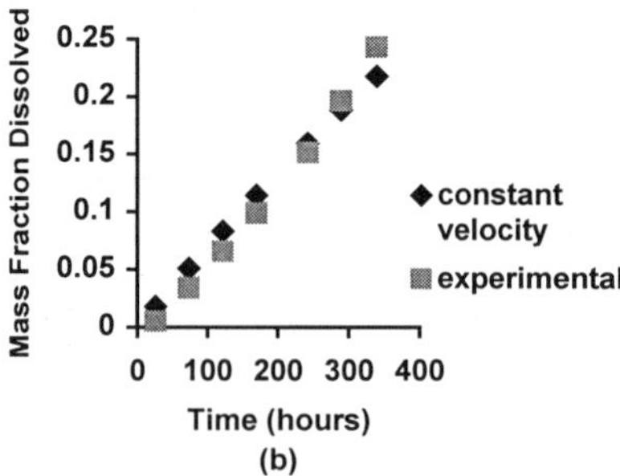

Figure 5. Dissolution of JM901 fiber in SEF, mean diameter = 1 μm: (a) rotary, air attenuated, k_{cv} = 560 ng / cm^2 hr, (b) flame attenuated, k_{cv} = 101 ng / cm^2 hr.

Advances in Fusion and Processing of Glass II

A key question remaining is whether the flame-attenuated surface represents primarily a structural difference in the glass relative to the bulk glass or to the CD fiber, or if it is essentially a chemical difference. Structural changes resulting in increased durability often accompany density increases such as those produced upon annealing of the glass, where relaxation occurs resulting in the elimination of high energy, reactive bonds. To test this concept, samples of CD JM475 were subjected to different time-temperature heating and cooling conditions, all at temperatures at or below the strain point of the glass. Durability in SEF was then evaluated as above. Results showed that k_{cv} for CD JM475 could be lowered from an average value of 70 ng / cm^2 hr (as produced) to 26 ng / cm^2 hr, suggesting that structural changes alone could account for a significant fraction of the difference between CD and FA fibers.

It was believed, however, that formation of a glass surface with a k_{cv} as low as 11 or 12 ng / cm^2 hr would require an additional change that was primarily chemical. The choices here included both loss of volatile glass components from the surface as well as possible incorporation of "atmospheric" components into the glass itself. To determine if such reactions had occurred during the manufacture of the flame attenuated fiber, bulk samples of pristine FA JM475 and FA JM753 were analyzed by XPS. Results normalized to atomic concentration of Si are given in Table II.

As seen here, data for JM475 shows no good evidence for a significant depletion of volatile glass components and may even suggest a relative enrichment, especially in F. In contrast, JM753 appears to be completely depleted in B albeit potentially enriched in Na (data for Na is somewhat suspect based on inconsistent results obtained from analysis of bulk glass fresh fracture surfaces). Similar results have been found in other fiber samples of different initial diameter. Any depletion of volatiles or rearrangement of glass components at the fiber surface as a result of flame attenuation may well be specific to the original glass composition.

Table II.   XPS analysis of pristine fiber surfaces

Atomic Ratio	JM475 nominal	JM475 Flame attenuated	JM753 nominal	JM753 Flame attenuated
Na / Si	.36	.42	.47	.87
B / Si	.34	.33	.14	0
F / Si	.03	.09	.05	.03
Ca / Si	.04	.04	.08	.07
Al / Si	.12	.09	.09	.08

XPS data was also useful in evaluating the potential for reaction of the newly-forming fiber surface with the ambient atmosphere. In Table III, a comparison of data from CD and FA JM475 shows apparent differences in composition, largely attributable to a major difference in the amount of carbon adsorbed onto (or reacted with) the surface. At the least, this indicates that the surface chemistry of the CD and FA fibers cannot be directly compared quantitatively due to the difference in adsorption by the carbon layer(s) (*15*). When considering the distinct difference in specific surface area between the two samples, however, there is a strong indication that the CD and FA fibers differ appreciably in their affinity for carbon – the highest carbon values being found on the lower surface area CD fibers. In both cases fiber samples were stored open to ambient atmosphere and room conditions (23° C, 25% relative humidity) prior to analysis – no special precautions were taken to avoid hydrocarbon or other surface "contamination" which typically forms adventitious carbon layers on surfaces.

To determine if the carbon contents were influenced by process, FA JM475 and JM753 fibers were produced using both oxidizing and reducing conditions in the final attenuation burner. Results indicated that carbon values averaged nearly three times greater in fibers formed in the reducing flame. This indicates that much of the surface carbon on FA fibers is likely formed during fiberization and is not the result of subsequent atmospheric contamination.

Table III. XPS surface chemistry – JM475 (atom %)

	Continuously Drawn (CD)	Flame Attenuated (FA)	Nominal
Si 2p	12.2	15.6	20.0
Al 2p	1.0	1.3	2.3
B 1s	3.7	7.1	6.7
Ca 1s	0.3	0.3	0.7
Ba 3d5	0.2	0.3	0.7
Zn 2p3	0.3	0.8	1.0
Na 1s	3.4	13.6	7.1
K 2p	0.7	1.0	1.0
F 1s	0.7	0.9	0.7
O 1s	37.8	57.7	59.7
C 1s	39.0	1.3	-
Specific surface area (m2 / g)	0.6	2.5	

With a better understanding of the effects of flame attenuation in hand, efforts were focussed upon fibers and fiber surfaces produced by the rotary spinning process which is perhaps the most complicated of all commercial processes employed today. There are many variations of rotary processes that differ principally in design and in the methods used for secondary attenuation of the primary melt stream spun from the rotating "basket" or spinner. Typically these employ high velocity jets of either air or a combination of air and flame. In this regard, it was anticipated that fiber surface properties could resemble those seen in CD or FA fibers, depending upon the specific operating conditions used.

Typical dissolution profiles for JM901 produced by rotary spinning with air attenuation (RA) have been shown in Figure 5a. They appear nearly identical to those obtained from CD fiber. Both the shape of the profile and the derived rate constants at 50% mass loss are similar, indicating that even though the time-temperature cooling regimes may be different in both cases, the conditions responsible for formation of the glass surface appear to be similar. As such CD fibers may serve as useful surrogates for unbonded discontinuous fiber or wool in evaluations of their potential biopersistence via *in vitro* dissolution in SEF. This observation was further supported by studies on several experimental glasses produced by both CD and RA methods.

The environment in which the fiber is formed also appears to be rather consistent from one location to another in the veil of discontinuous fiber formed by rotary spinning in air. This is suggested in Figure 6, showing very uniform corrosion surfaces developed on fiber exposed to

Advances in Fusion and Processing of Glass II

SEF, especially fiber to fiber. This is quite similar to the type of corrosion surface developed on CD JM901and is perhaps somewhat surprising considering the rapid fiber formation and high production volumes and in most commercial rotary fiberization operations.

The introduction of flame as a component of the attenuation process in rotary spinning, however, can introduce a good deal of complexity both in the forming environment and in the nature of the fiber produced. At first approximation it might be expected that the bulk and surface properties of the fiber formed would be influenced by exposure to both air and flame during formation and, hence, values for k_{cv}, for example, might lie somewhere between those obtained for CD or RA fibers and those obtained from FA fibers. For JM475, this appears to be valid. Typical product produced by rotary spinning using both air and flame in the secondary attenuation (RAF) show k_{cv} values around 40 ng / cm^2 hr with dissolution profiles showing slight positive curvature for the coarser fibers (> 2 μm mean diameter), similar to those seen for FA fibers.

Figure 6. SEM image of fibers of JM901 glass – rotary spun with air attenuation, recovered from *in vitro* test cells – 21 day exposure to SEF @ 37°.

However, rotary spinning with contributions from both air and flame during attenuation can also create many local microenvironments which can affect the newly forming fibers and their surfaces, particularly upon cooling from peak temperatures during initial (secondary) attenuation down through T_g. In addition, there are many variations upon design, configuration and operating conditions (mostly proprietary) that can further increase variability, so that it may be difficult to compare surface properties such as durability on fiber produced form one production unit to another. This can create a significant "overprint" on initial glass composition in determining values such as k_{cv}.

An example of this is shown in Figure 7 for JM901 fiber produced by rotary spinning with both air and flame components in the final attenuation. Here differential "etching" resulting from exposure to SEF has revealed distinct types of fiber formed in the same product – morphological distinctions being independent of fiber diameter and, hence, not a function of extent of mass loss. In this sample, four different types of fiber could be recognized. Type 1 fibers appeared quite similar to RA or CD JM901 fiber with rather uniform corrosion surfaces. Type 2 fibers clearly resembled flame attenuated fiber with well-formed core / shell structure. Types 3 and 4 however, were unique and found only in RAF fiber product. They are both characterized by their extensively leached structures, being comprised entirely of residual alumina and silica – the original network formers in the glass. The distinction between type 3 and 4 is purely morphological: type 3 showing a rather uniform, highly porous, skeletal structure, and type 4 characterized by the development of concentric layers in the now residual gel. Although difficult to confirm, it is believed that these morphological distinctions probably reflect real initial differences in the structure of the glass comprising the fibers and are not simply artifacts formed in the leached hydrogel or the subsequently-desiccated fiber.

SEM / EDS analysis was also employed to determine if the fiber to fiber variability seen in RAF samples was due to an initial difference in fiber chemistry, produced by local variations in melt homogeneity. Results indicated, at least to the sensitivity of the technique, that this was not the case and that the differences were more likely the result of structural differences in the glass itself – very likely a reflection of the microenvironment of formation. Types 3 and 4, for example, might represent fibers comprised of glasses of relatively low density and open structure formed by very rapid quenching in one of these microenvironments. Not all samples produced by rotary spinning with air and flame attenuation show such gross distinctions among fibers – the types of fibers produced depend strongly on design and operating parameters. RAF fibers do, however, in general show a greater degree of fiber to fiber variability than CD, FA, or RA fibers of the same bulk composition.

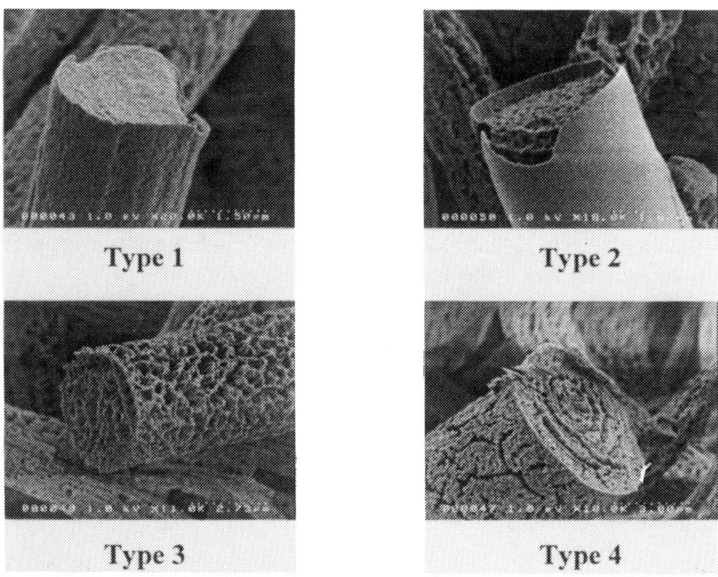

Figure 7. Morphologies of fibers developed by exposure of RAF JM901 fibers to SEF, 21 days

The types of fiber formed by rotary spinning with air and flame attenuation, not surprisingly influence measured physical properties obtained from bulk samples. k_{cv} values at 50 % mass loss for JM901, for example, were found to range from 530 to 1110 ng / cm^2 hr, depending on the particular operating conditions used in their formation. Increase in measured k_{cv} was also correlated with an increase in type 3 and 4 fibers in the sample; type 2 fibers, however, were

                    Advances in Fusion and Processing of Glass II

present throughout, albeit in very low percentages and did not significantly influence the k_{cv} values obtained for the collection of fibers comprising each sample.

Fiber to fiber variability in RAF samples was also reproduced using acid leaching (8 % H_2SO_4) at various fluid volume to sample surface area ratios. Although morphological distinctions were not as well defined as those seen from SEF exposure, results indicated that the variability was in fact real and not simply an artifact of the *in vitro* test system.

Another measure of sample variability can be obtained from BET surface area measurements on fiber samples of similar mean diameter exposed to elevated temperature and high humidity for periods of several days. Table IV shows results obtained for JM901 RA and RAF fiber after 72 hours exposure at 50° and 90% relative humidity. These correlate well with the measured k_{cv} values – fibers that are less durable in SEF also show the greatest development of surface porosity upon exposure to water vapor, again suggesting a more open, lower density structure to the glass comprising the individual fibers.

Table IV. Increase in specific surface area (BET- m^2/g)
of rotary-spun JM901 fiber following exposure at
50°, 90% relative humidity, 72 hours.

	% Increase	kcv in SEF
Air attenuated	21	526
Air + flame attenuated #1*	91	928
Air + flame attenuated #2*	159	1110

* different operating conditions used in production

DISCUSSION

Dissolution studies on fibers formed by different processes have provided insights into many of the properties that are important to their usefulness and performance in product applications. The importance of process on fiber durability relevant to persistence in the lung lies primarily in the effects upon k_{cv}, which is believed to be a reasonable predictor of the longevity of the more toxic long (> 15 to 20 μm) fibers (*16*). This is best represented graphically as shown in Figure 8.

Following equation 3, time to complete dissolution (persistence) can be seen to vary non-linearly with k_{cv}. The most critical dependence occurs when k_{cv} is less than about 50 ng / cm^2 hr where persistence increases sharply with decreasing k_{cv}. It is in this region that accuracy of measurement becomes most critical. The results of this study clearly show that process effects, particularly those resulting from flame attenuation are capable of creating fine fibers and fiber surfaces with measured k_{cv}'s in this critical region – even if results obtained from the bulk glass composition or from air attenuated fiber surrogates would predict otherwise.

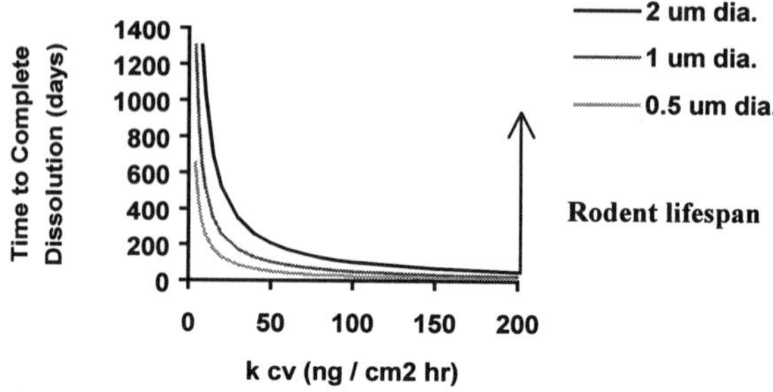

Figure 8. In vitro persistence of vitreous fibers of various diameters as a function of measured k_{cv}.

Persistence, however, becomes particularly important when considering the lifespan of the animal in question (*16*). Most experimental inhalation studies have employed either rats or hamsters with lifespans of only 2 to 2 ½ years. As seen in figure 8, the most durable flame attenuated fibers of 1 μm diameter (k_{cv} as low as 11ng / cm^2 hr have been measured), if sufficiently long, would be predicted to persist in the lung for nearly half of the animal's life. Relative to the lifetime of a human, however, this would be comparatively short. This may help to explain why such fibers have shown no evidence of inducing respiratory diseases in humans (*17*), but have produced positive responses in inhalation experiments with some laboratory animals as noted above.

CONCLUSIONS

Commercial fiberization processes used in the production of discontinuous glass fiber create fiber forming conditions that can be quite distinct from those present in the drawing of continuous filament. Operating conditions (especially those affecting time - temperature) and ambient environment during secondary attenuation are responsible for controlling not only final fiber dimension, but also the properties of the fiber itself. Both bulk and surface properties of fibers are affected; changes may be chemical, physical or both. Commercial secondary attenuation processes can produce significant fiber to fiber variability; the greatest variability occurs when both air and flame are part of the attenuation environment (creation of microenvironments). Flame attenuated fibers are frequently formed with a unique structure ("concentrically biphasic") which is responsible for enhanced chemical durability and likely strength. Such process effects can significantly impact persistence of fine respirable glass fiber in the lung - particularly important for interpretation of biological responses in laboratory animals.

Acknowledgements

The author would like to give special thanks to B. D. Law for work on the durability studies and F. D'Ovidio for scanning electron micrography.

     Advances in Fusion and Processing of Glass II

REFERENCES

(1) Davis, J.M.G., " The Role of Clearance and Dissolution in Determining the Durability or Biopersistence of Mineral Fibers", *Environmental Health Perspectives*, **102** (Suppl. 5) 113-117 (1994).

(2) Bauer, J. F., Law, B. D., and Roberts, K. A.,. "Solubility and Durability of Man-made Mineral Fibers in Physiological Solutions", Presented at TAPPI Nonwovens Conference, Nashville, (1988). Available from the author.

(3) Potter, R. M. and Mattson, S. M., "Glass Fiber Dissolution in a Physiological Saline Solution", *Glasstechnische Berichte*, **64** 16-28 (1991).

(4) Mattson, S. M., "Glass Fibers in Simulated Lung Fluid: Dissolution Behavior and Analytical Requirements", *Annals of Occupational Hygeine*, **38** (6) 857-877 (1994).

(5) Hesterberg, T.W., Miiller, W.C., Musselman, R.P., Kamstrup, O., Hamilton, R.D., and Thevenaz, P., "Biopersistence of Man-Made Vitreous Fibers and Crocidolite in Rat Lung Following Inhalation", *Fundamental and Applied Toxicology*, **29** 267-279 (1996).

(6) Hesterberg, T. W,. Chase, G., Axten, C., Miiller, W.C., Musselman, R. P., Kamstrup, O., Hadley, J., Morscheidt, C., Bernstein, D., and Thevenaz, P., "Biopersistence of Synthetic Vitreous Fibers and Amosite Asbestos In the Rat Lung Following Inhalation, A Joint Study of NAIMA and EURIMA", In preparation. (1997).

(7) Mohr, J. G. and Rowe, W. P., *Fiber Glass*, Van Norstrand Reinhold, New York, 1978.

(8) TIMA, *Man-made Vitreous Fibers: Nomenclature, Chemistry, and Physical Properties*, North American Insulation Manufacturers Association, Alexandria, Va. 1991.

(9) Hesterberg, T.W., Miiller, W.C., Thevenaz, P., and Anderson, R., "Chronic Inhalation Studies of Man-made Vitreous Fibers: Characterization of Fibers in the Exposure Aerosol and Lungs", *Annals of Occupational Hygiene*, **9** 637-653 (1994).

(10) Bauer, J. F., Law, B. D., and Hesterberg, T. W., "Dual pH Durability Studies of Man-made Vitreous Fiber (MMVF)", *Environmental Health Persp*ectives, **102** (Suppl. 5) 61 - 67 (1994).

(11) Zoitos, B., DeMeringo, A., Royer, E., Thelohan, S., Bauer, J., Law, B., Boymel, P., Olson, J., Christensen, V., Guldburg, M., and Koenig, A., "*In Vitro* Measurement of Fiber Dissolution Rate Relevant to Biopersistence at Neutral pH: An Interlaboratory Round Robin", *Inhalation Toxicology*, **9** 525 - 540 (1997).

(12) Kanapilly, G. M., Raabe, O. G., Goh, C. H. T., and Chimenti, R. A., "Measurements of *In Vitro* Dissolution of Aerosol Particles for Comparison to *In Vivo* Dissolution in the Lower Respiratory Tract After Inhalation", *Health Physics* **24** 497-507 (1973).

(13) Hesterberg, T.W., Miiller, W. C., Hart, G.A. , Bauer, J. F., and Hamilton, R.D., "Physical and Chemical Transformation of Synthetic Vitreous Fibers in the Lung and *In Vitro*", *Journal of Occupational Health and Safety – Australia, New Zealand*, **12**(3) 345-355 (1996).

(14) Förster, H. and Tiesler, H., "Contribution to Comparability of *In Vitro* and *In Vivo* Man-made Mineral Fibre (MMMF) Durability Experiments", *Glasstechnische Berichte*, **66** (10) 255 - 266 (1993).

(15) Pantano, C. G., "X-ray Photoelectron Spectroscopy of Glass"; pp.129-160 in *Experimental Techniques of Glass Science*, Edited by C. J. Simmons and O. H. El-Bayoumi , The American Ceramic Society, Westerville, Oh., 1993.

(16) Eastes, W. and Hadley, J. G. "A Mathematical Model of Fiber Carcinogenicity and Fibrosis in Inhalation and Intraperitoneal Experiments in Rats", *Inhalation Toxicolog,y* , **8** 323-343 (1996).

(17) Marsh, G. M., Enterline, P. E., Stone, R. A., and Henderson, V. L., "Mortality Among a Cohort of US Man-made Mineral Fiber Workers: 1985 Follow-up", *Journal of Occupational Medicine*, **32** (7) 594-604 (1990).

Float Glass

GLASS-TIN INTERACTIONS DURING THE FLOAT GLASS FORMING PROCESS

Amarendra Mishra and George A. Pecoraro
PPG Industries, Inc.
P.O. Box 11472
Pittsburgh, PA 15238-0472

Thomas E. Paulson and Carlo G. Pantano
Department of Materials Science and Engineering
The Penn State University
123 Steidle Building
University Park, PA 16802-5005

ABSTRACT

The thermochemical interactions that occur between tin and molten glass during the float forming process were investigated using experimental techniques and theoretical calculations. The major glass composition variables were iron, sulfur and the atomic Fe^{+2}/Fe^{+3} redox ratio. The iron, sulfur and tin concentration profiles were determined in commercially formed surfaces using XRF, EPMA and SIMS. These were compared with the profiles generated by floating the identical glasses under near isothermal conditions in the laboratory. The characteristic bump in the tin diffusion profile was reproduced in the laboratory and was related to redox interactions with Fe and S in the glass. The calculations suggest that a steep oxygen activity gradient at the glass/tin interface is fundamental to the mass transport and depth-distribution of the variable-valent Fe, S and Sn ions.

INTRODUCTION

Float glass is formed by pouring molten glass onto the flat surface of a liquid tin bath. The glass is cooled as it progresses along the tin bath, and after solidification, is lifted off the tin and finally annealed in the lehr. The float glass process produces high quality, optically flat, distortion-free glass. During this process, however, tin is incorporated into the glass surface at the glass/tin interface to a depth of approximately 40-50µm. The SnO_2 concentrations in the outermost 20 µm are of the order of 4-10%. This phenomenon has been studied by many authors. [1-11] A desirable effect of the tin-enriched bottom surface is that this glass surface has increased chemical durability. However, there are several undesirable effects associated with the tin/glass interaction. Some deteriorate the glass quality or process yields, while others deteriorate the bath refractories, generate defects in the glass, or influence secondary processing.

A well-known consequence of tin diffusion into the bottom surface of float glass is a condition referred to as "tin bloom." [7] During the reheating of glass in a bending process, for example, the tin surface develops an iridescence. Examination of the surface at high magnification reveals a wrinkled appearance.

The formation of a calcium-rich scum layer on the tin-side of float glass having high tin concentrations is another manifestation of tin diffusion into glass. The scum is believed to be the result of a calcium ion exchange with tin and iron in the tin-side surface. [8,9]

The diffusion of oxygen and sulfur into the tin accompanies the diffusion of tin into the glass to help satisfy electron transfer balance. The resulting reaction produces SnO and SnS which can vaporize from the tin and condense onto the colder regions of the bath superstructure. Reaction of these species with hydrogen can then produce elemental tin that drips onto the top of the ribbon causing defects and downgrading of glass quality. There is also an out-diffusion of Fe from the glass to the tin.[10]

Subsequent coatings applied to the glass can also be effected by the increase in tin and iron at the bottom surface. Coatings are applied to the glass substrate to produce products having improved spectral properties, i.e., low IR and UV transmission for improved creature comfort, energy savings , decreased fabric color fading resistance and personal safety considerations. Excessive levels of tin and iron at the surface can have a detrimental effect on the coatings applied to the glass surface.

Deterioration of the tin bath's basin refractories has been caused by the reaction with sodium originating from the glass.[12,13] The sodium that is freed from the glass because of tin migration into the glass, and subsequent ion exchange, diffuses through the tin to the alumino-silicate refractory liner. If the alumina content of the refractory is less than 40%, a molten product can result from the reaction, at the hot end of the bath, between sodium and the refractory. This low viscosity glass forms into viscous nodules. These float up through the tin and lodge themselves against the bottom surface. A "bottom dropper" defect can result from this situation. For blocks that have an alumina concentration exceeding 40% alumina, the product of reaction between sodium and the refractory is nephelite, which is a solid at the bath temperatures. There is a large volume expansion and stress build-up associated with the formation of Nephelite, which can result in delamination of the refractory liner. Solid particles that spall off can stick to the bottom surface of the ribbon and can reduce production yields.

The reported investigations of tin/glass melt interactions and mutual diffusion have been limited to the more conventional glass compositions having total iron concentrations ranging from 0.1 to 0.5% and about 0.20% SO_3. Recently, new glass formulations have been developed that broaden the range of total iron concentration from less than 0.01% up to 2% for the heat-absorbing glasses. Some of these glasses also have less than 0.005% SO_3 concentration. Our level of knowledge and understanding of the interactions of tin and molten glass can be increased by studying this wider range of commercial float glass compositions.

Although a direct measurement of the composition profiles in the bottom surface of float glass can be used to evaluate the mechanisms and kinetics of the tin-glass reaction, theoretical models are required to interpret these observations. One approach to modeling the reactions of complex high-temperature liquids, like molten glass, is that developed by Hastie, et al.[14] The model is referred to as Ideal Mixing of Complex Phases (IMCP). It accounts for negative non-ideal mixing energies by the inclusion of complex phases in the liquid solution. Thus the mole fraction of a particular liquid system component may decrease after equilibrium is reached due to the formation of complex phases which include the liquid components. At equilibrium, the thermodynamic activities of the system component species are equivalent to their calculated mole fractions. Hastie and Bonnell[15] demonstrated the ability of the IMCP model to predict the phase equilibria of high-temperature multi-component, liquid-phase systems. Specifically, the authors investigated glasses composed of various quantities of Na_2O, K_2O, CaO, MgO, Al_2O_3 and/or SiO_2, and compiled a list of the "significant" and "insignificant" complex phases of the glass, based on the equilibrium mole fractions of each complex phase. In that study, the model calculations were compared with experimentally measured activity data for various species of the glass. The results showed excellent agreement, validating the model predictions. Spear[16], et al., also used the IMCP approach to model interface reactions in carbon fiber-glass matrix composites. Eriksson[17] performed calculations were using the computer program SOLGASMIX™, the predecessor to CHEMSAGE™, which was used in the present study.

Advances in Fusion and Processing of Glass II

In this study, the effect of sulfur concentration was evaluated using glass compositions where nearly all the other constituents in the glass were held constant. The glasses were floated in the laboratory to separate glass composition effects from the specifics of the commercial float glass process. The Sn, Fe, and S profiles were experimentally measured in both the commercial and laboratory float surfaces, and were compared with the calculated thermodynamic driving forces.

EXPERIMENTAL PROCEDURE

Two samples of commercial float glass having approximately the same levels of iron, but two different levels of sulfur, were used. Table I summarizes their bulk chemical composition. Note that there is a difference of a factor of two in the tin count between these two compositions.

Table I. Chemical Analyses of the Float Glasses

in weight %	Float Glass 1	Float Glass 2
Al_2O_3	0.14	0.19
SO_3	0.22	0.007
Fe_2O_3 (total)	0.53	0.49
FeO	0.14	0.16
Fe^{+2}/Fe^{+3} (atomic)	0.41	0.53
Tin Count (XRF)	55.1	27.5

Figure 1 shows a schematic of the lab-scale float bath module used to equilibrate the top-side of the float glasses with molten tin. The graphite boat containing the glass and tin was placed in a fused quartz glass tube having the highest quality fittings and connections to insure a leak-free atmosphere inside. As the tin melts, it flows into the space under the plate of glass through an opening in the vertical partition between the chambers. The dross and impurities remain on the tin side. The tube and its contents were placed into a furnace that heated the glass, tin and graphite module.

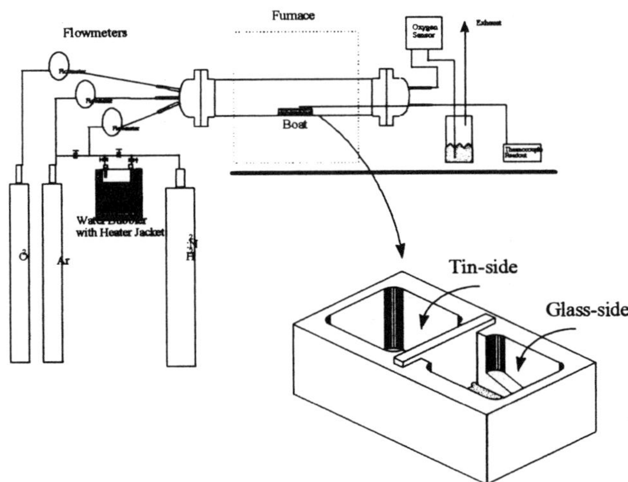

Figure 1. Schematic of experimental apparatus and tin/glass experimental boat.

The as-received tin (20 mesh A.C.S. reagent grade Aldrich Chemical Co.) was refined by premelting the tin in a graphite boat under reducing conditions. Since the solubility of oxygen in tin decreases with decreasing temperature, the tin was refined at 600°C for 72 hours while subjected to a 4%H_2 / 96%N_2 gas flowing at a rate of 150 sccm. These conditions were sufficient to remove the visible oxide layer on the tin. All subsequent glass-tin equilibration experiments were carried out at temperatures above 600°C. This insured that the oxygen concentration of the tin is below the solubility limit at the test temperature.

The commercial glasses were floated on their top-surfaces in the laboratory apparatus. (The top surface of float glass is very low in tin concentration and near-bulk in composition.) It needs to be emphasized that there was no expectation to reproduce the float process in these experiments. It was more desirable to extend the times of tin-glass interaction so that transport phenomena could be clearly observed. In some cases "near"-isothermal reactions were performed. ("near" in the sense that it is necessary to cool the sample to room temperature from the isothermal hold temperature.) Therefore, in some cases, the isothermal hold time was designed to significantly exceed the cool-down time. In other cases, two step experiments were conducted wherein the glass and tin were equilibrated at a higher temperature (1050°C), and then cooled and held for an extended period at a lower temperature. This is the manner in which the temperature dependency of the glass-tin interaction was explored.

Both the tin and air surfaces of each commercial float sample were subjected to quantitative depth profiling for Fe, Sn and S using the SIMS (0.2-1.5µm depth range) and EPMA (1-125µm depth range). The SIMS (CAMECA IMS/3f) depth profiles were measured by rastering a 250nA, 14.5keV $^{18}O^-$ primary beam over a 250x250µm area of the sample and collecting the positive secondary ions from a 10µm diameter area from the center of the sputtered crater. The relative sensitivity factors used to quantify the SIMS depth profiles[18] were calculated from the depth profile of a fracture surface in a glass with a known bulk composition. Concentration profiles extending more than 1 micron deep into the glasses were measured using EPMA (CAMECA SX-50, 100nA, 10keV) cross-sectional profiling. The sensitivity factors for quantifying the EPMA profiles were calculated from Sn metal, ZnS, and Fayalite standards.

Thermodynamic modeling of the tin bath and float glasses listed in Table 1 was performed using CHEMSAGE,™ and included calculating the equilibrium activity of oxygen in the glass and in the tin. The activities of oxygen in the glasses were compared with the activities of oxygen in the tin bath to establish gradients in chemical potential, i.e. the thermodynamic driving force for chemical interactions, across the tin/glass reaction interface. The two glasses chosen for this study represent the range of sulfur concentration present in today's commercial float glasses.

RESULTS AND DISCUSSION
Concentration Profiles

The Fe and Sn SIMS profiles (which extend from the surface to a depth of one micron) of the as-received tin-sides of Float Glasses 1 and 2 are presented in Figures 2 and 3. The tin-side surfaces of Float Glasses 1 and 2 are enriched in Fe relative to the bulk concentrations by factor of 2.5 to 4.5, respectively. The peak concentration of iron occurs at a depth of approximately 0.075µm. The tin-side surfaces of both glasses are also enriched in Sn. The peak concentration of tin occurs at the surface. The depth of iron and tin enrichment are approximately the same (about 0.3µm). The ratio of tin concentration between the two glasses is comparable to the ratio of tin counts derived from XRF as shown in Table I.

The Fe and Sn EPMA profiles from 1 µm below the surface to a depth of 25 µm of the as-received tin-sides of Float Glasses are presented in Figures 4 and 5, respectively. It is important to recognize that the first data point in the EPMA profiles represents the average surface composition

Advances in Fusion and Processing of Glass II

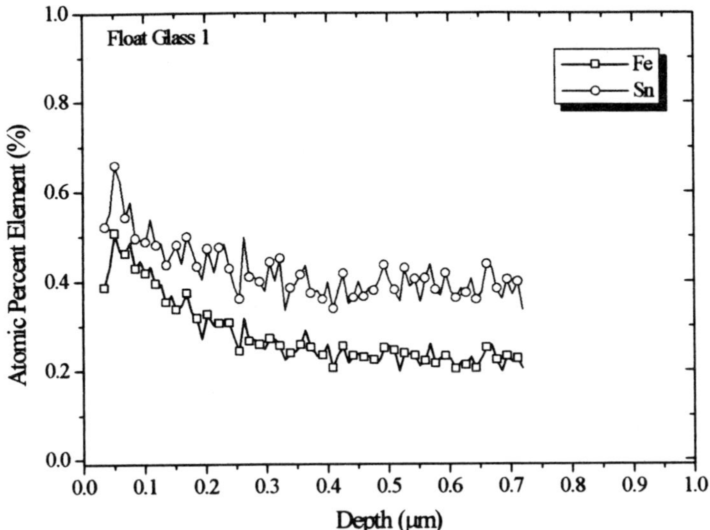

Figure 2. SIMS depth profiles in the bottom (as-received) surface of Float Glass 1.

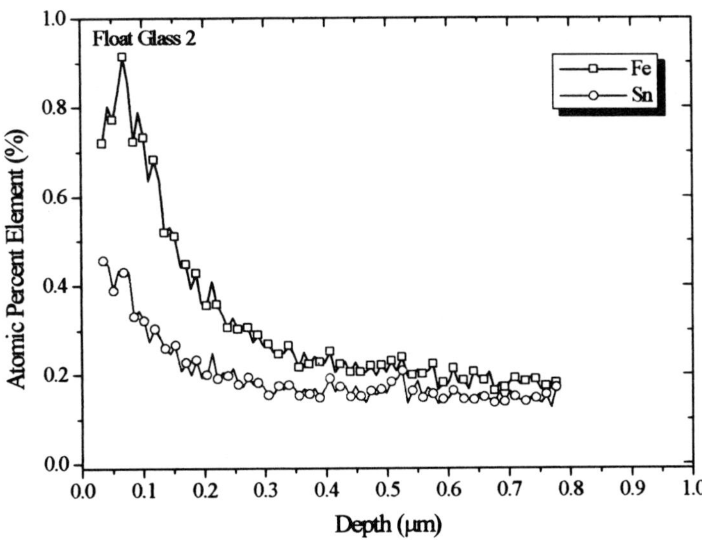

Figure 3. SIMS depth profiles in the bottom (as-received) surface of Float Glass 2.

over ~ 1 μm; i.e., the SIMS profiles presented in Figures 2 and 3 characterize the first micron where steep gradients in the concentrations of Fe and Sn exist. The gradients are especially steep in the case of Fe.

As shown in Figure 4 for Float Glass 1, the Fe concentration decreases from 0.15% to 0.11% over a depth of 10 μm where it begins to gradually increase to a level of 0.18%. The Sn concentration profile exhibits a very different pattern. The level decreases sharply from 0.36% to 0.20% from 1 to 7 μm from the surface. The characteristic bump in the tin profile is seen from 7 to 10 μm from the surface.

The Fe profiles of Float Glass 2 are similar to those of Float Glass 1. However, the Sn profiles are different. As can be seen in Figure 5, in the former, the Sn concentration "bump" is absent. The differences in these tin profiles suggest that the presence of the bump relates strongly to the bulk sulfur concentration of the glass. Altogether, these observations confirm a dependence of the bump upon the in-depth redox ratio since Sn, Fe and S are the only variable valent species in the glass.

The Fe and Sn concentration profiles of the air-side of Float Glass 1 floated on tin in the laboratory, for ten minutes at 1050°C (initially) and then for 1 hour at 950°C, are shown in Figure 6. The Fe concentration is depleted to a depth of 75 μm. The Sn profile shows a sudden bump at 25 μm. In contrast, the Sn bump occurred at 10 μm in the commercial glass.

The Fe and Sn profiles of the air-side of Float Glass 2 floated on tin in the laboratory, for ten minutes at 1050°C (initially) and for 1 hour at 950°C, are shown in Figure 7. The concentration profiles of Fe and Sn are significantly different for Float Glass 1 compared to Float Glass 2. The latter contains low sulfur; the former contains high sulfur. The sudden increase in Sn concentration seen for Float Glass 1 lab melt is not observed with Float Glass 2, presumably because of the low sulfur concentration of the latter as discussed above. In both glasses, Fe concentrations decrease as the profiles approach the surface.

The commercial glasses show much steeper tin and iron concentration profiles from the surface into the glass, whereas the lab floated glasses of the same compositions do not. There are at least four possible reasons for these differences:

1. The lab melts are equilibrated with pure tin. The commercial glasses are formed on tin that contains iron and other impurities that can limit the diffusion of tin into the bottom surface of the glass ribbon.

2. The lab melts are formed in a system that eliminates relative motion between the glass and the tin. The commercial glasses are formed in a continuous process whereby there is relative motion between the glass and the tin, as well as attenuation of the ribbon thickness and extension of its surface.

3. The lab melts are annealed in contact with the tin, under reducing conditions. The commercial glasses are annealed in air after they are removed from the tin.

Advances in Fusion and Processing of Glass II

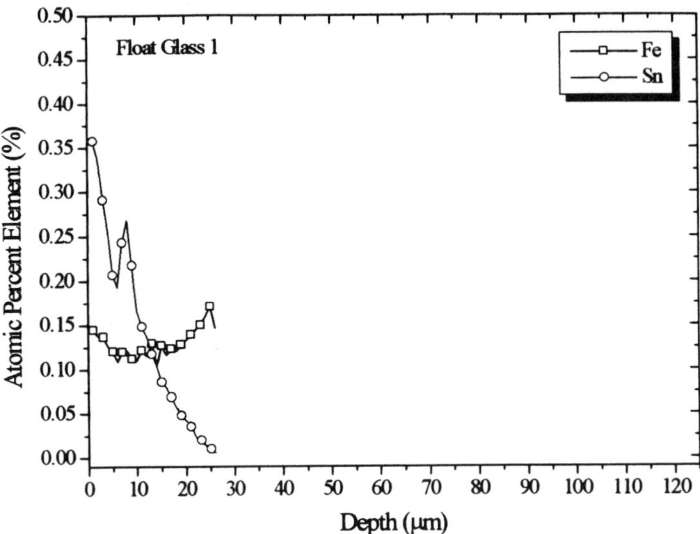

Figure 4. EPMA depth profiles in the bottom (as-received) surface of Float Glass 1.

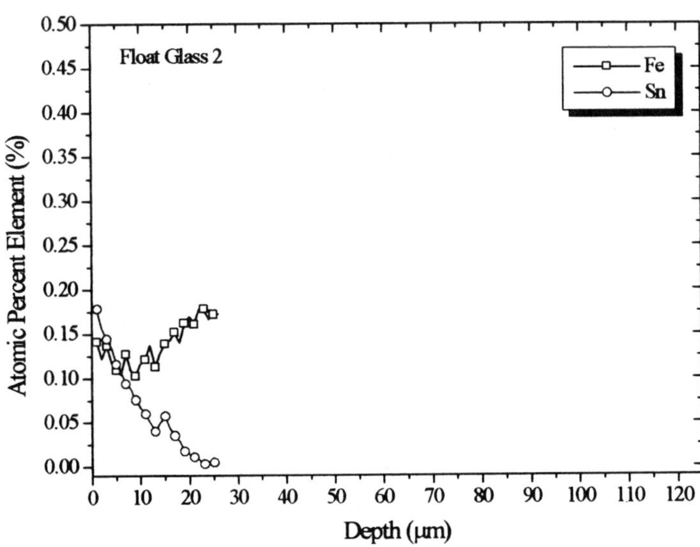

Figure 5. EPMA depth profiles in the bottom (as-received) surface of Float Glass 2.

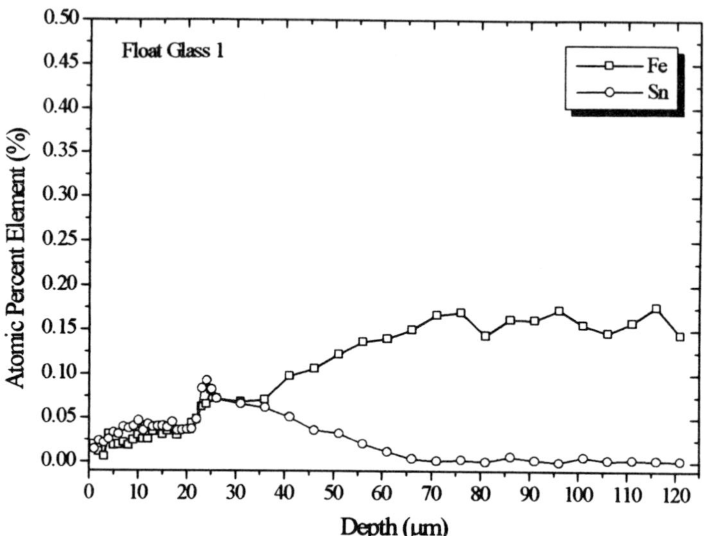

Figure 6. EPMA depth profiles in the bottom surface of Float Glass 1 floated in the lab at 950°C-1h. Glass-tin contact time is 10 minutes at 1050°C.

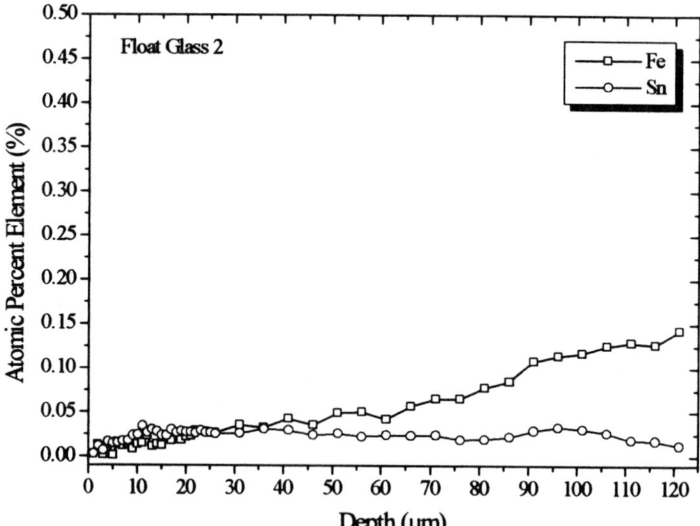

Figure 7. EPMA depth profiles in the bottom surface of Float Glass 2 floated in the lab at 950°C-1h. Glass-tin contact time is 10 minutes at 1050°C

Advances in Fusion and Processing of Glass II

4. The lab melts are cooled to near room temperature before they are removed from the tin and are subjected to oxygen. The commercial float glasses are exposed to oxygen shortly after they are cooled into the transformation range.

The tin bump, on the other hand, is a fundamental characteristic of the high sulfur glass whether it is made on a float bath in a continuous process or in the lab in a small boat of tin. It appears to be an intrinsic property of the glass that is related to the glass composition, and not only to external factors associated with the commercial float bath and ribbon forming process.

Thermodynamic Modeling

<u>Tin Bath.</u> Assuming that the tin bath achieves and maintains equilibrium with the gas above it, the oxygen activity in the bath should be equal to the p_{O2} in the gas. The p_{O2} in this gas atmosphere is governed by the H_2:H_2O equilibrium in the gas phase. The atmosphere above the lab-scale tin bath was 5%H_2:95%N_2 gas that contained 1ppm each of O_2 and H_2O. Argon was substituted for nitrogen in the calculations to simplify the gas composition with respect to the kinetically-limited formation of NH_x and NO_x gases. Therefore, the input gas composition for the calculations was 0.95 moles Ar, 0.05 moles H_2, 10^{-6} moles O_2 and 10^{-6} moles H_2O, and the total system pressure was kept fixed at one atmosphere. The gas was equilibrated over a solution of tin which contained iron, and then the gas and solution phase equilibria were calculated. Carbon was included in a second set of calculations because a graphite boat was used for the lab float experiments. Comparing the results with and without carbon in the calculations, an upper and lower limit on the calculated oxygen activity in the tin bath for the lab float experiments was determined.

The gas phase equilibria for the calculations relating to the tin bath without and with carbon are presented in Figures 8 and 9, respectively. In the case where carbon was not included in the calculation (Figure 8), the incoming 1 ppm O_2 reacted with hydrogen to form water. Thus, the equilibrium composition of the gas phase was 5% H_2, 3 ppm H_2O and 95% Ar. This composition remained essentially constant for all temperatures and indicates that the amount of oxygen in the gas was not effected by incorporation of small amounts of oxygen into the tin bath solution. The oxygen activity, therefore, was defined by the H_2:H_2O ratio in the gas at all temperatures.

When carbon was included in the calculation (Figure 9), however, the oxygen activity was no longer a simple function of the H_2:H_2O ratio. At the lowest temperature, the incoming gas equilibrated with the excess carbon forming primarily CO, as seen in Figure 9. Again, the sum of the CO and H_2O partial pressures remained essentially constant, indicating that the all the oxygen introduced into the system was remaining in the gas phase. As temperature increased, the H_2O partial pressure continually decreased while the CO partial pressure increased and leveled off at about 3 ppm, indicating that the $C(s)+\frac{1}{2}O_2 \blacklozenge CO$ equilibria was ultimately fixing the p_{O2} at the higher temperatures. The effect of including carbon in the calculations was a reduction in oxygen partial pressure by up to 8 orders of magnitude at 1100°C, relative to the oxygen activity defined by the H_2:H_2O ratio in the carbon-free system at that temperature.

<u>Glass Compositions.</u> The two float glass compositions listed in Table 1 (including the other major elements) were allowed to equilibrate in an atmosphere of 10^{-3} moles of Ar. This was done to keep the total volume of the gas phase low and reduce changes in glass composition due to volatilization. No H_2, H_2O or O_2 was included in the input gas phase because it was desired to calculate the intrinsic activities of species in the glass; i.e. those fixed by the glass composition and

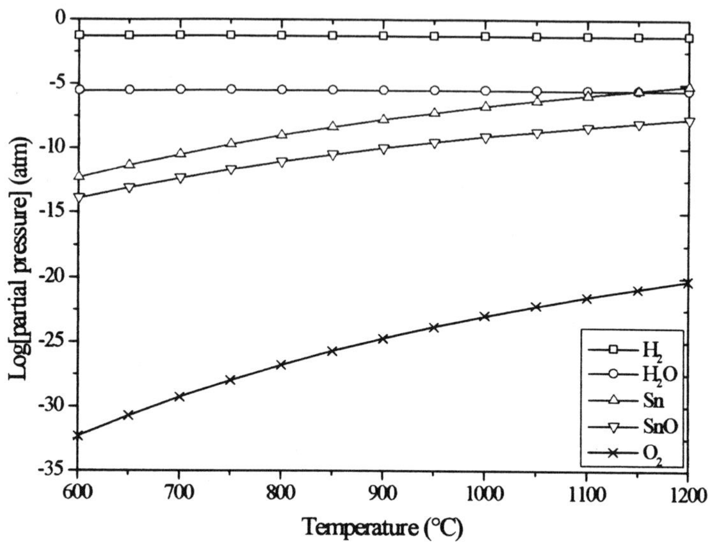

Figure 8. Calculated gas phase equilibria above the pure lab-scale tin bath.

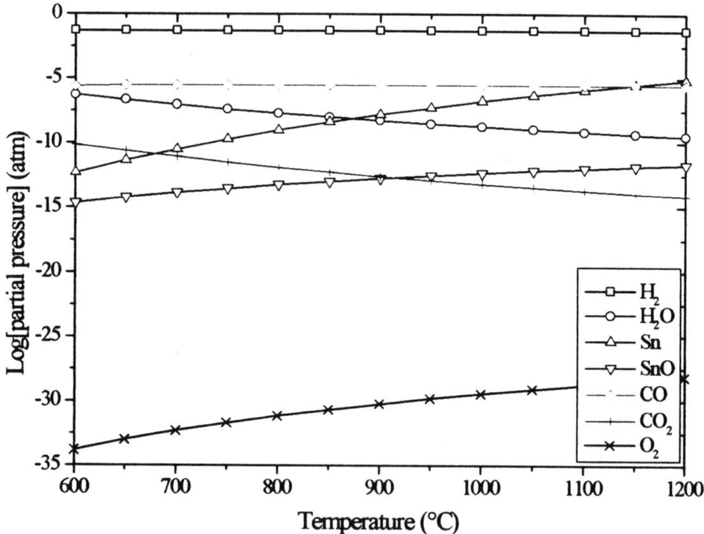

Figure 9. Calculated gas phase equilibria above the lab-scale tin bath which includes carbon.

     Advances in Fusion and Processing of Glass II

not by the external gas phase. The SO_3 content of the glass composition was included as the equivvalent molar amount of Na_2SO_4, but keeping the total Na_2O content constant according to the bulk composition. The corresponding oxygen activities calculated for Float Glasses 1 and 2 are shown in Figure 10. The calculated oxygen activities in two hypothetical glasses with the same bulk composition as Float Glass 1, except forcing the redox to extreme values of 0.1 and 2.0, are also shown in Figure 10, to determine the effect of Fe redox on the oxygen activity.

It is not surprising that the low-sulfur glass resulted in a glass with a lower oxygen activity than the high-sulfur glass; even though the oxygen activity in a glass is a function of both the batch composition and melting conditions, these calculations were performed to isolate only the composition effect of SO_3. The effect of Fe redox on the oxygen activity in Float Glass 2 can be seen in Figure 10 where the redox values were set to 0.1 and 2. As expected, higher redox values resulted in lower calculated oxygen activities.

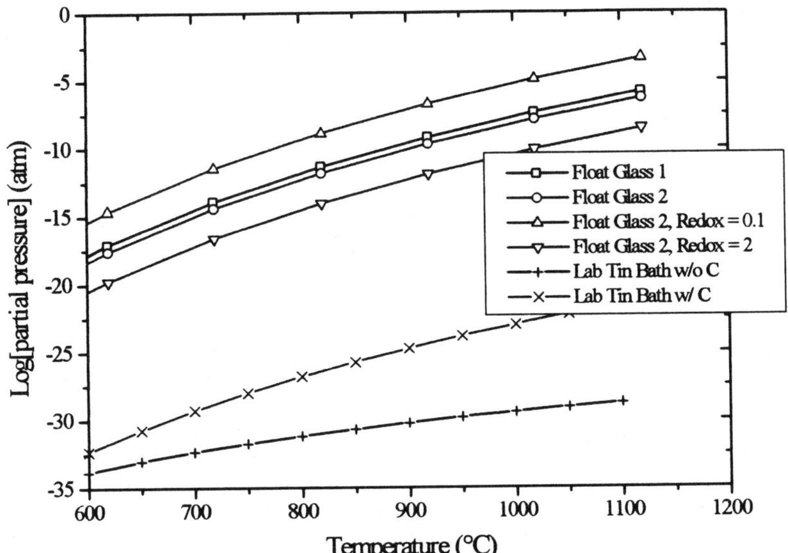

Figure 10. Calculated oxygen activities in the lab-scale tin bath and in Float Glasses 1 and 2, and in Float Glass 2 fixing the redox at 0.1 and 2.

The most important point to note in Figure 10 is that regardless of the sulfur content or redox, the calculated oxygen activity in float glass is at least 10 orders of magnitude higher than that calculated for the tin bath. This oxygen potential difference is believed to play an important role in the transport of metal ions across the tin-glass interface, and consequently in the final tin distribution profile. The significance of the oxygen activity gradient in the glass becomes even more apparent when considering the distribution of Fe^{2+}, Fe^{3+}, Sn^{2+} and Sn^{4+} in the near-surface region of the float glass because the relative proportions of these species is a direct function of the local p_{O2}.[10] Assuming the glass and tin reach a local equilibrium state in the lab-float experiments, the oxygen activity in the glass will define the relative proportions of polyvalent species present in the glass; of particular interest in this study is the Sn^{2+}:Sn^{4+} concentration ratio. The tin that enters the glass from the bath is initially in the Sn^{2+} state. Once in the glass, it experiences a higher oxygen potential, relative to the tin bath, and if that oxygen potential is sufficient, the tin will oxidize to the Sn^{4+} state. The diffusion coefficient for Sn^{4+} in glass is very low,[19] and so it becomes essentially immobilized at the point of oxidation. Since additional Sn^{2+} can continue to enter into the glass from the tin bath, an accumulation of Sn^{4+} can occur. This is consistent with the presence of a bump in the total Sn profile and is precisely what is observed in the two glasses examined in this study, as well as a recent study based on Mössbauer spectroscopy.[11] Float Glass 1 has a higher intrinsic oxygen activity than Float Glass 2, and presumably it is high enough to oxidize the incoming tin to create the observed bump in the tin depth profile. Float Glass 2, with the lower oxygen activity, does not exhibit the distinct bump. The fact that these general factors were observed in <u>both</u> the commercial and lab-scale floats shows the specific effect of glass composition (versus the float bath operation). The next phase of this work is to continue lab-scale glass floating experiments to provide experimental concentration profiles for Fe and Sn after reactions under fixed conditions; this data can be used to test the thermochemical modeling predictions. This work is currently underway and will be published in the future.

CONCLUSIONS

It has been shown that the concentration of sulfur in float glass influences the final concentration profile for tin after float processing. This was verified using a lab-scale float furnace where the external effects of the commercial process could be eliminated. The thermochemical modeling showed that the presence of sulfur (as well as the Fe redox) influences the driving force for the reaction, but only to a limited extent relative to the much larger difference in oxygen potential between the glass and tin. This suggests that the role of S (and Fe) in the tin-glass reaction relates more to the mechanism of equilibration (rather than to establishment of the driving force). In the presence of oxidized S and Fe species in the glass, the steep oxygen activity gradient can be reduced by transport of S and Fe to the tin bath (as well as by transport of Sn into the glass).

REFERENCES

1. Brückner,R. and Navarro,J.F., "Spannungen und Ionentransport im Floatglas, *Glastech. Ber.* 44 361368 (1971).
2. Sieger,J.S., "Chemical Characteristics of Float Glass Surfaces," *J. Non-Crys. Sol.*, **19**, 213-220 (1975).
3. Colombin, L., Jelli,A., Riga,J., Pireaux,J.J., and Verbist,J., "Penetration Depth of Tin in Float Glass," *J. Non-Crys. Sol.*, **24**, 253-258 (1977).
4. Baitinger,W.E., French,P.W., and Swarts,E.L., "Characterization of Tin in the Bottom Surface of Float Glass by Ellipsometry and XPS," *J. Non-Crys. Sol.*, **38&39**, 749-754 (1980).
5. Suscavage,M.J. and Pantano,C.G., "Tin Penetration into a Soda-Lime-Silica Glass," *Glastech. Ber.*, **56 K** [1] 498-503 (1983).

     Advances in Fusion and Processing of Glass II

6. Verità,M., Geotti-Bianchini,F., Hreglich,S., Pantano,C.G., and Bojan,V.J., "EMPA, RBS and SIMS Analyses of Tin Profiles in Commercial Float Glasses," *Bol. Soc. Esp. Ceram. Vidrio*, **31c** [6] 415-420 (1992).

7. Pantano,C.G., and Bojan,V.J., "Tin Profiles in the Bottom Surface of Float Glass: Manufacturing and Heat Treatment Effects," *Fundamentals of Glass Science and Technology*, ESG 1993, Venice.

8. Verità,M., Geotti-Bianchini,F., Guadagino,A., Stella,A., Pantano,C.G., and Paulson,T.E., "Chemical Characterization of the Bottom Side of Green Float Glass," *Glastech. Ber.*, **68** [C1] 251-258 (1995).

9. Franz,H., "Surface Chemistry of Commercial Glasses," *Ceram. Eng. Sci. Proc.*, **16** [2] 221-227 (1995).

10. Paulson,T.E., Spear,K.E., and Pantano,C.G., "Thermochemical Modeling of High-Temperature Interactions Between Glass and Tin," Proceedings of the Ninth International Conference on High Temperature Materials Chemistry, May 19-23, 1997, in press.

11. Williams,K.F.E., Thomas,M.F., Johnson,C.E., Tilley,B.P., Greengrass,J, and Johnson,J.A., "Oxidation States of Tin and Iron in Clear and Tinted Float Glass by Mössbauer Spectroscopy," *Fundamentals of Glass Science and Technology*, 1997, Sweden.

12. Müller,F., Lim,S.K., Gebhardt,F., and Küstner,D., "Physico-chemical Investigation of the Behavior of Sodium in the Float Glass Process, Part 1. Distribution of Sodium and Tin Between the Glass Melt and the Bath of Molten Tin," *Glastech. Ber.*, **62** [11] 369-376 (1989).

13. Müller,F., Lim,S.K., Gebhardt,F., and Küstner,D., "Physico-chemical Investigation of the Behavior of Sodium in the Float Glass Process, Part 2. Reactions of Sodium and Oxygen Dissolved in the Bath of Molten Tin," *Glastech. Ber.*, **62** [12] 417-421 (1989).

14. Hastie,J.W., Horton,W.S., Plante,E.R., and Bonnell,D.W., "Thermodynamic Models of Alkali Vapor Transport in Silicate Systems," *High Temp.-High Press.*, **14**, 669-680 (1982).

15. Hastie, J.W., and Bonnell,D.W., "A Predictive Phase Equilibrium Model for Multicomponent Oxide Mixtures, Part II. Oxides of Na-K-Ca-Mg-Al-Si," *High Temp. Science*, **19**, 275-306 (1985).

16. Spear,K.E., Benson,P.M., and Pantano,C.G., "Thermochemical Modeling for Interface Reactions in Carbon-fiber Reinforced Glass Matrix Composites," <u>High Temperature Chemistry IV</u>, Z.A. Munir, D. Cubicciotti, and H. Tagawa, eds., The Electrochemical Society, Pennington, NJ, 345-354 (1988).

17. Eriksson,G., *Chem. Scripta*, **8**, 100 (1975).

18. Bojan,V.J., Buyuklimanli,T.H., and Pantano,C.G., "Quantification of SIMS Data for Multicomponent Glasses," *Surface and Interface Analysis*, **21**, 87-94 (1994).

19. Russel,C., "Self Diffusion of Polyvalent Ions in a Soda-Lime-Silica Glass Melt," *J. Non. Crys. Sol.*, **134**, 169-175 (1991).

COUPLED COMBUSTION SPACE-GLASS BATH MODELING OF A FLOAT GLASS
MELTING TANK USING FULL OXY-COMBUSTION

Carol Schnepper
Air Liquide
Chicago Research Center
5230 S. East Avenue
Countryside, IL 60525

Benjamin Jurcik and Christel Champinot
Air Liquide
Claude-Delorme Research Center
Les Loges-en-Josas, B.P. 126
78350 Jouy-en-Josas
France

Jean-François Simon
Air Liquide
75 quai d'Orsay
75321 Paris
France

INTRODUCTION
Motivation

Full oxy-firing has been accepted as a cost-effective process for melting many types of glass with the benefit of reducing emissions such as NO_x and particulates, while improving the operational flexibility. For a variety of reasons, there has been only limited use of oxy-firing in float glass melting tanks. One of the reasons for the limited use of oxygen is the question of how to control the oxy-combustion system, including firing rate profile, burner location, and burner characteristics, so that the glass quality is not adversely affected [1]. As float glass tanks are typically quite large, and the glass product has extremely stringent quality demands, the potential economic consequences of a glass tank producing low quality product has prevented the use of full oxy-combustion technology. This occurs despite the fact that many glass makers observe an increase in the quality of glass product from full oxy-fired tanks. Possible explanations for the observed quality improvement include increased water content in the glass due to the higher water concentration in the atmosphere and hence decreased viscosity, better thermal control of the furnace, and better control of the furnace atmosphere and the fining process.

Challenges of Modeling Float Glass Melting Tanks

While Air Liquide has extensive capabilities in coupling internal ATHENA™ combustion space simulation code with the glass batch/melt simulations of the GTM code from TNO, the simulation of float glass tanks with air combustion represents a technical challenge. A brief description of our coupled calculation capabilities was presented in [2]. The first limitation is simply the grid size that must be used, given the size of the float glass tank. As the power and capacity of computer resources increase and become increasingly affordable, this limitation can be overcome.

The second limitation is more fundamental in nature and is related to regenerator firing-side reversal. For the simulation of only the glass batch/melt, the reversal process is not taken into account since the time scale of the reversal process is much faster than the pertinent time scales for the glass melt. As a result, a line of symmetry down the centerline of the glass tank is frequently assumed. There is no natural extension of this approach to coupled calculations of oxy-combustion with ATHENA™ because the combustion process by its very nature is not symmetric down the centerline of the glass tank. We have made proprietary calculations of side-fired regenerative furnaces in the past which show that the profile of energy flux into the glass/batch is not symmetric about the centerline.

In the coupled-calculation approach, iterations are made in both the combustion space and the glass melt until convergence is reached in both domains and there is continuity of energy flux and temperature at the interface between the two domains. A time-averaged energy flux could be estimated from a series of asymmetric steady-state ATHENA™ calculations, however, simply averaging the temperatures to obtain the average glass temperature profile is not physically correct since the predominant mode of heat transfer to the glass surface is radiation, which has a nonlinear dependence on temperature. The correct method for averaging the glass surface temperature is unknown. As a result, fully coupled ATHENA™-GTM calculations for regenerative glass tanks remain an area of active research at Air Liquide. Preliminary results related to the coupled combustion space/glass melt simulation of the TC21 tank were recently presented [3].

Present Scope

Modeling, both coupled and batch/glass only, is used to study the effects of conversion to full oxy-fuel firing on the behavior of a float glass melting tank. The float glass melting tank under consideration is patterned after the TC21 tank, not because the TC21 tank necessarily represents optimal float tank operation, but because the data required for this modeling study are easily and publicly available. Because of the difficulties associated with the coupled modeling of regenerative float glass tanks, a model of only the glass batch and melt was performed to provide references for the glass temperature profile and the circulation pattern.

Coupled calculations were used to model the conversion of this glass tank to full oxy-fuel combustion. The basic geometry was maintained in terms of the dimensions of the tank, but the regenerators and regenerator ports were removed and the exhaust ports were relocated. The variable parameters were exhaust location, burner location, firing rate profile, and burner design. Two different burner designs were considered: 1) the rapid mixing between oxidant and fuel achieved in the conventional pipe-in-pipe ALGLASS™ burner and 2) the second generation, luminous, low NO_x, high efficiency ALGLASS FC™ burner. The goal was to adjust these parameters until the batch profile, glass melt temperature profiles, and glass melt flow patterns closely approximated those calculated for the reference case. Careful attention was also paid to the temperature profiles on the crown.

FURNACE CHARACTERISTICS AND REFERENCE GLASS TEMPERATURE AND FLOW PATTERN: AIR-FUEL OPERATION

While detailed information about the TC21 test case is available, a brief description is given here so that the reader may understand the case without consulting additional references. The float tank is a side-fired regenerative furnace operating at a pull-rate of 453 metric tons per day. It consists of a melting and refining end measuring 34.69 m long by 10.05 m wide and a working end measuring 14.93 m long by 9.15 m wide. The melter/refiner is separated from the working end by a narrow waist measuring 3.56 m long by 4.27 m wide. The waist is equipped with a 4-1-4 cooler which circulates water at 3.5×10^{-3} m^3/s and presents a 0.25 m blockage to the glass flow. The glass melt is 1.17 m deep throughout, except where it exits the tank. The thermal conductivities of the refractory in different regions of the tank were set to values consistent with the available heat fluxes and the utilized refractory thicknesses.

The inner dimensions of the combustion space length and width match those of the melting tank. The maximum crown height is 2.25 m above the glass surface in the melter/refiner, 2.40 m in the working end, and 1.10 m in the waist. Five ports operate on each side, the last port on each side being located 15 m from the back wall.

The values for the physical properties of the glass melt are consistent with values reported in both public and confidential literature for a colored float glass [4, 5]. The batch contains 50% cullet.

The computational domain of the glass melting tank was discretized into $171 \times 50 \times 26$ (length × width × height) elements for a total of 222,300 cells. The heat-flux profile to the batch/glass surface was calculated from the measured crown temperature profile, assuming a constant emissivity for the gas atmosphere. In other words, a rather simple and steady-state model of the heat transfer from the combustion space to the glass is implemented (similar in nature to Hottel's method).

Results from the batch/glass melt simulation are shown in Figures 1, 2, and 3. The temperature profile on a vertical line centered side-by-side and 16.74 m from the back wall is shown in Figure 1, and Figure 2 shows the plan view of the batch and glass melt. Velocity vectors on a vertical plane centered side-by-side in the melter/refiner are shown in Figure 3. The spring zone is evident where the two major circulation zones separate at approximately 16.5-20.0 m from the back wall. In Figure 3, the vertical axis is scaled by a factor of 5 to improve legibility.

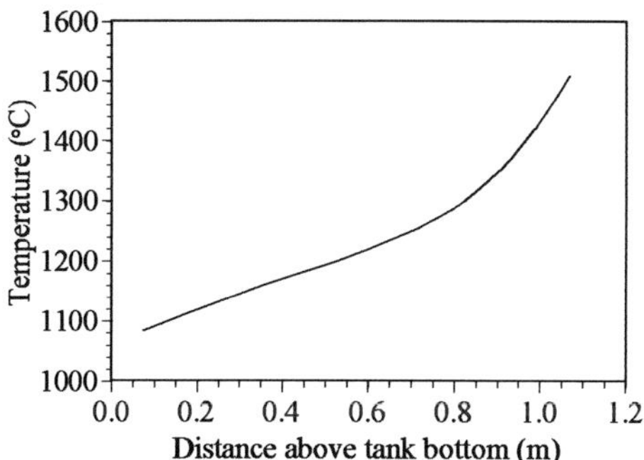

Fig 1. Temperatures on a vertical line centered side-to-side and 16.74 m from the back wall for the reference case.

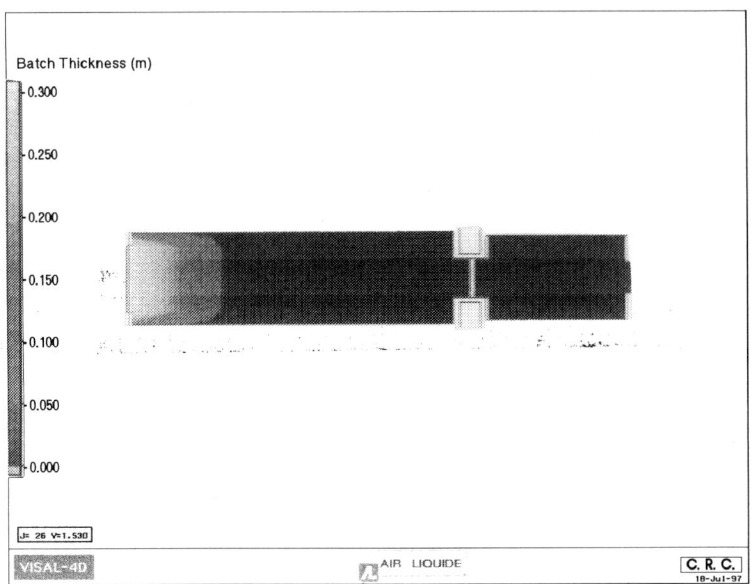

Fig. 2. Plan view of the batch and glass for the reference case.

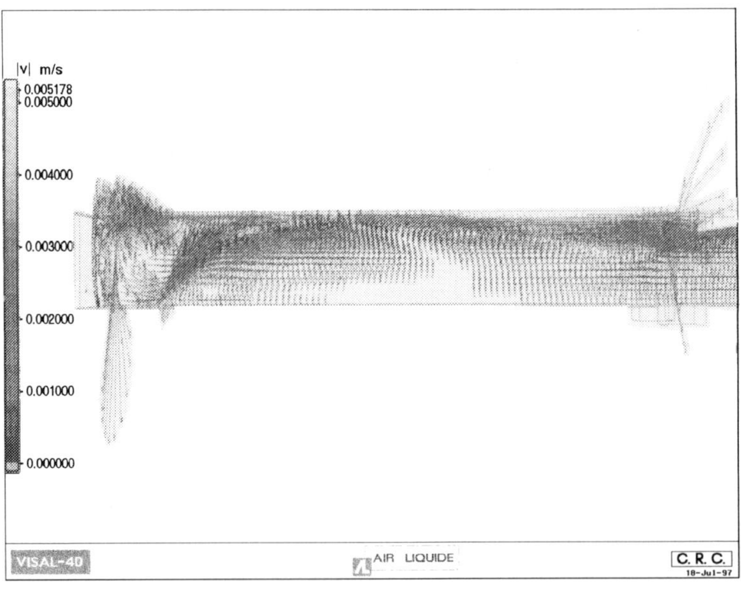

Fig. 3. Velocity vectors on a vertical plane centered side-by-side in the melter/refiner for the reference case. The vertical axis is scaled by a factor of 5.

CONVERSION TO FULL OXY-FUEL FIRING
Case 1

For the first conversion to oxy-fuel firing, ten conventional ALGLASS™ burners were placed along each side with each burner directly opposed to the corresponding burner on the opposite side of the tank. The first three burners on each side were placed 3.0, 4.5, and 6.0 m from the back wall, and the remaining burners were placed at 1.5 m intervals starting at 10.5 m from the back wall. To estimate the total firing rate of the burners, an energy balance was performed over the combustion space. The amount of energy transfered to the batch/glass surface was taken from the reference case simulation, an average wall loss from the combustion-space refractory of 5 kW/m^2 was assumed, and an exhaust temperature of 1800K was assumed based on 2% oxygen in dry flue gases. The firing rate distribution among the first four burners on each side was 11.8%-13.6%-8.0%-9.0% with the remaining energy divided evenly among the other burners. An exhaust port 0.70 m wide by 0.79 m high was placed 8 m from the back wall on each side of the tank.

The design of the combustion space for the conversion to oxy-fuel was based upon the results from the reference case. The exhaust ports were placed immediately downstream from the leading edge of the melting batch in an effort to minimize the mixing of suspended particulates and gaseous reaction products from the batch with the atmosphere over the refining zone. The burner placement was conceived to create high heat flux to the batch surface while adding energy downstream to drive circulation in the refining zone.

For this case, the combustion space was discretized into 186 × 54 × 28 (length × width × height) elements for a total of 281,232 cells. Note that the spatial discretization of the combustion space does not have to match the discretization of the batch and molten glass.

The resulting profile of the batch and molten glass is sketched in Figure 4, along with the profile from the reference case. This case was not considered adequate because the batch blanket had migrated quite far down the length of the tank in comparison with the reference case. Two problems were identified: 1) the energy from the burners did not penetrate into the tank far enough to melt the batch blanket, and 2) the exhaust location caused a long section where the heat transfer to the batch surface was decreased.

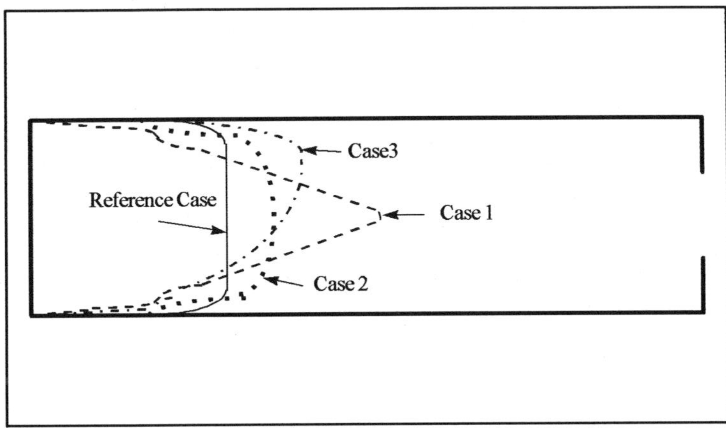

Fig. 4. Plan view sketch of the melter/refiner showing batch profiles for the refernce case and Cases 1-3.

Cases 2 and 3

To correct the problems identified from Case 1, the exhausts were located farther downstream (in the refining zone), and higher power burners were utilized to give longer flame lengths. Space limitation prevents a full description of the various cases, which are described in detail in the presentation. Five burners were placed on each side at 3.44 m intervals starting at 5.6 m from the back wall on the left side and at 3.9 m on the right. The exhaust ports were relocated to 29.4 m from the back wall in order to accommodate the new burner positions. The power distribution between the burners was 31.2%-33.4%-18.4%-10.2%-6.8%.

For the third conversion, second generation ALGLASS FC™ burners were utilized. The ALGLASS FC™ burner produces a long, wide, luminous flame for maximum coverage of the load. The flame sweeps close to the glass surface with low flame temperature for high efficiency heat transfer. The same firing rate distribution was used for the ALGLASS FC™ burners as for Case 2.

The calculated batch profiles for Cases 2 and 3 are sketched in Figure 4. The batch profile for Case 2 most closely resembles that of the reference case whereas the profile is asymmetric in Case 3. The asymmetry is most likely due to the high power of the ALGLASS FC™ burners in the batch region. As a result of the heat transfer efficiency of the ALGLASS FC™ burner, the batch is melted faster than with ALGLASS™ burners, and the staggered arrangement of the burners gives rise to the asymmetric batch profile. A reduction in the power of the burners near the batch is necessary to eliminate this effect. For the moment, therefore, Case 2 can be considered the most representative example of a full oxy-combustion float glass tank which produces glass temperatures and flow patterns similar to those of the regenerative glass tank.

CONCLUSIONS

A study has been performed using coupled combustion space-glass bath modeling to evaluate the behavior of a float glass tank after conversion to full oxy-fuel firing. For the furnace studied here, a conversion using five ALGLASS™ burners on each side of the tank results in glass temperature and circulation profiles very similar to those resulting from the regenerative reference case. Further calculations are underway to determine whether different exhaust locations and power distributions provide better results.

REFERENCES

[1] R.G.C. Beerkens, T. van der Heijden, and E. Muijsenberg, "Possibilities of Glass Tank Modeling for the Prediction of the Quality of Melting Processes"; pp. 139-160 in *A Collection of Papers Presented at the 53rd Conference on Glass Problems*. American Ceramic Society, Westerville, OH, 1992.

[2] B. Jurcik and C. Schnepper, "The Use of Coupled Combustion Space and Glass Melt Models for Oxy-Fuel Fired Glass Tanks," presented at the American Ceramic Society's Glass & Optical Materials Division Fall Meeting, San Antonio, TX, 1996.

[3] C. Champinot, TC21 Round Robin Discussion, IV. International Seminar on Mathematical Simulation in Glass Melting, Vsetin, Czech Republic, 1997.

[4] R. Beerkens, E. Muijsenberg, G. Peters, and O. Verheyen, "Modeling Explores Bubble Behavior in Molten Glasses," *The Glass Researcher*, 6 [2] 19-23, 25 (1997).

[5] O.V. Mazurin and O.A. Prokhorenko, "High-Temperature Studies of IR-Absorption Spectra of Silicate Glasses (The Second Part)"; pp. 175-301 in the Semiannual Research Report of the NSF Industry-University Center for Glass Research, Alfred Univ., Alfred, NY, June 1996.

Advances in Fusion and Processing of Glass II

Special Topics in Glass

Dynamic Expert Systems in Glass

Atique Malik (APCI),P.B. Eleazer (APCI),   J. Chmelař (Glass Service)

ABSTRACT

A closed-loop expert system has been commissioned and has proven to be be robust and effective in dealing with job changes. The principles behind the system are outlined The system has been implemented at a container glass plant, and has consistently maintained a high efficiency , and reduced transition time, during job changes. The system is based on a novel hybrid technique of conventional multivariable control, neural networks and fuzzy logic control.

{Corresponding author maliksa@ttown.apci.com}

1   Introduction

It is a competititive advantage in container glass manufacturing to make gobs of uniform temperature, under variable operating conditions. This goal is a desirable one, but often difficult to achieve in practice. The difficulty lies in the fact that process equipment is often designed for steady state operation , and has to operate in a dynamic environment. This simply means that the furnace and forehearth designer cannot solve process operating problems on the drawing board through furnace design, as most disturbances occur in unplanned ways [10]. These departures from ideal operating conditions are due to :

1. Process Characteristics

    (a) Nonlinearity. This means that the behaviour of the forehearth at a gob temperature of 1150°C is appreciably different to that behaviour observed at around 1180°C. Similiarly, for the melter. A symptom of this problem is the constant retuning that seems to go on at container glass sites, or having reached exhaustion point, the operators leave the control loops on manual mode.

    (b) Constraints. Sooner, or later, every process reaches a limit, in terms of heat addition or cooling. The method of operation has to be changed , for example, if heating is called for, and the conventional means of heat addition have been exhausted. Some other point in the process now has to be used to provide "double-duty".

2. Instrumentation.

    (a) Drift .There are a large number of thermocouples in a glass melter and forehearth. These are expected to fail, over the life of the furnace, as they are subjected to mechanical erosion and high temperatures. The good thermocouple reading has to be consistently differentiated

Advances in Fusion and Processing of Glass II   227

from the bad. Bad thermocouples have to be tagged, and brought to the attention of the maintenance people. Algorithms have been developed to detect, and substitute alternate measurements for the deleted thermocouples [4].

(b) Failure .While thermocouples generally deteriorate slowly, emitting a particular statistical signature, an often noted characteristic of some thermocouples is to go "bad", and then return to normal operation. Quite often conventional control systems will not catch this, and by taking the wrong action, they can magnify the problem.

Therefore, a system is needed to optimise, and regulate, glass furnaces for day to day operation. This system has to provide more than that achievable by standard first-level process control in a demanding environment [6]. The next section briefly considers the systems on the market, and the unique aspects of the system that we have implemented,

2 Literature Survey

A large number of process control systems exist for the regulation of glass furnaces. They rely largely on technology that has matured in the chemical process industry over the last two decades [5], [9]. Some of the larger proprietary systems in place in the industry are essentially DDC systems, of an older generation. Recently, work has started on using fuzzy logic for glass furnaces also, but only simulation work has been done, and no applications have been reported for a real furnace, [2], [1]. A study underway will provide a detailed comparison of the various approaches at both the instrumentation and the mathematical level of the various systems available,[8], [7].

3 Operational Requirements for Glass

A distinguishing characteristic of all glass plants, relative to the chemical industry, is the low level of technical staffing. Whereas a typical oil refinery or petrochemical plant will have control engineers and process engineers to maintain and design new ways to optimise the manufacturing process on-site, the glass plant will generally leave these functions unattended. Conventional advanced control and optimisation methods are very maintenance intensive , and it is therefore not very surprising that these approaches have withered on the vine at glass plants. An approach we took to tackle the maintenance problem, was to design robust intelligent control algorithms with the following functionality [3] :

1. Adaptation. The approach used is adaptive in the modelling process , and self-tuning via fuzzy logic, in the controller. There is sufficient data available on present computer data bases to develop approximate nonlinear statistical dynamic models, as a starting point.

2. Robustness. By incorporating fault detection, we increase the on-line time of the controller and minimise the upsets that a controller can inadvertently cause to the process by taking the wrong action.

The individual components of the dynamic expert system are therefore :

1. A fault detection and reconfiguration module

                                   Advances in Fusion and Processing of Glass II

2. A closed-loop model updating module, based on neuro-fuzzy principles.

3. A multivariable fuzzy controller, augmented with rules

4   Results

This section deals with the performance of the expert system. The system has been operational at a container facility , since October 1996. It has proven to be efficient in the handling of job (or pull) changes. Average efficiency over 99.5% has been achieved. Previous efficiencies were at or below the 94% level. A weakness of the present system is that some plant testing is required. The next version under development will overcome this requirement through the use of a true adaptive fuzzy system. Initial tests have been promising. The dynamic expert system is also being applied to a TV glass plant, but the system has not been commissioned.

5   Conclusion

An expert system has ben developed and demonstrated for controlling glass quality in the closed-loop. It has been shown to extend the performance of an old furnace, by giving higher efficiencies than those achievable with newer designs. This is a direct economic benefit, and in some circumstances, can delay furnace rebuild if deterioration in glass quality is a problem. The algorithm uses intelligent control methods of fuzzy logic and neural networks to make the traditional controls more robust, and does not require extra instrumentation over that available at the typical furnace and forehearth.

6   REFERENCES

[1] S. Aoki and S. Kawachi. Application of fuzzy control for dead time processes in a glass melting furnace. *Fuzzy Sets and Systems*, 38:251–265, 1990.

[2] R. Bauer and H. Muijsenberg. Advanced control of glass tanks using simulatiom models and fuzzy control. In *Fourth Int. Seminar on Mathematical Simulation in Glass Melting*, pages 163–171, June 1997.

[3] J. Chmelar and S. A. Malik. Expert systems in intelligent control of continuous glass melting processes. In *Workshop on Advanced Control for Industrial Glass Melters*, MECC Exhibition and Congress Centre, Maastricht, Netherlands, May 1997.

[4] D. M. Himmelblau. *Fault Detection and Diagnosis in Chemical and Petrochemical Processes*. Elsevier, 1978.

[5] W. Lee and V. Weekman. Advanced control practice in the chemical process industry: A view from industry. *AIChE J.*, (22):27–38, 1976.

[6] S. A. Malik. Applications of neural networks to the control of reactors and oil fractionators in a refinery. In *IEEE Conference on Intelligent Control*, Chicago, U.S.A, Aug. 1993. IEEE.

[7] S. A. Malik. A review of nonlinear process control by henson and seborg. Technical report, Int. J. Of Robust and Nonlinear Control, 1997.

[8] S. A. Malik. Theoretical and practical comparison of methods of controlling glass quality. Technical report, Air Products and Chemicals, 1997.

[9] M. Morari. Three critiques of process control revisited a decade later. *CPC IV Conference, San Padre Island, 1991*, pages 306–321, 1987.

[10] W. Trier. *Glass Furnaces. Design, Construction and Operation.* Springer Verlag, 1984.

Advances in Fusion and Processing of Glass II

TV Glass

THE EFFECTS OF ATMOSPHERES ON VOLATILIZATION AND SURFACE TENSION

Haochuan Jiang, William C. LaCourse
NYS College of Ceramics at Alfred University
2 Pine Street
Alfred, NY 14802, USA

ABSTRACT

The effects of atmospheres on the volatilization and surface tension of TV panel and other silicate glasses were studied by measuring the weight loss and wetting angle of small glass samples heat treated in an atmosphere-controlled furnace. Water vapor in the furnace highly increased the weight loss rate. The wetting angle of TV glass on Pt increased with time in O_2, N_2, and air, due to changes in the glass surface composition, but exhibited a maximum with time in dry and wet CO_2. The volatilization data was also used to estimate the compositional and viscosity changes which might occur in thin walled foams over a glass melt.

INTRODUCTION

Reactive vapors such as H_2O, SO_2, and SO_3 in the furnace play an important role during the volatilization by altering the reaction mechanisms at the interface between the glass melt and atmosphere [1,2]. Among the reactive atmospheres, H_2O vapor is particular interest due to its increased presence in oxy-fuel fired furnace. The objective of the present work was to investigate potential effects of atmosphere on the viscosity and surface tension of the glass comprising the foam which often appears during glass melting reactions. This foam, which reduces heat transfer from the combustion space into the melt, can be a particular problem in some oxy-fuel fired furnaces, and methods to prevent or eliminate foam are needed. Since the cell walls comprising the uppermost layer of foam is thin (on the order of 10's of microns) its composition and properties (viscosity and surface tension) can change with time, due to volatilization of alkali, and other species. At the same time, gasses from the environment may be dissolved. These changes can have important influences on the "foam stability", and information regarding such changes is needed. For this reason, very small samples were employed so that volatilization would cause changes in the glass composition, and therefore the viscosity. No attempt was made to alter flow conditions of the atmosphere, or to create an "equilibrium" environment. The results are therefore of limited use for studies of volatilization from melts. On the other hand, the results should show how the properties of the foam may be altered by environment.

EXPERIMENTAL

Glasses were pre-melted in a platinum crucible under air at 1450°C for two hours. A small piece of glass (about 10 mg) was placed on a platinum substrate which was pre-heat-treated at 1000°C for 15 minutes and weighed. Then both the glass and platinum substrate were heated at 1000°C for 10 minutes to remove the adsorbed moisture and weighed to measure the original weight of the glass sample. The sample was then inserted into an atmosphere-controlled furnace for high temperature heat treatment. After the treatment, the glass was quenched in air and the weight loss and wetting angle were measured. Water vapor was generated by bubbling a water bath with the required atmosphere. Precautions were taken to prevent recondensation of water vapor. Water content in the furnace atmosphere was controlled by controlling the temperature of the water bath. All data were corrected to account for weight loss of the Pt substrate in different environments.

RESULTS AND DISCUSSION

TV Glass: The major volatile constituents in this glass are K_2O and Na_2O. Their original contents are 6 mol% and 9 mol% respectively. Fig.1 shows the normalized weight loss of this glass in different atmospheres. It is evident that water vapor accelerates volatilization, in agreement with previous investigations[1,2].

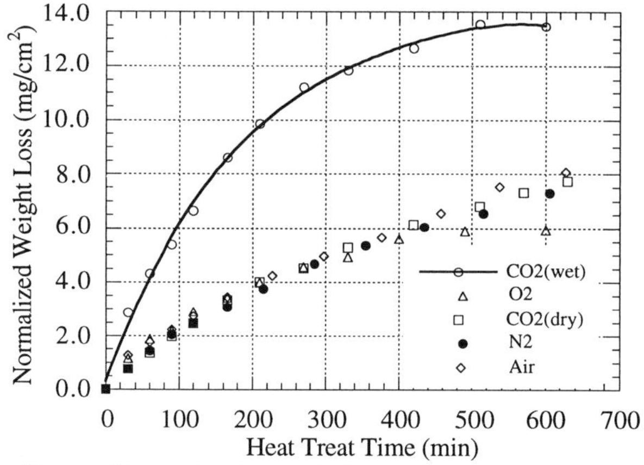

Fig. 1. Normalized weight loss of TV glass at 1450°

H_2O vapor in the furnace reacts with alkali oxides to form alkali hydroxides. Because of the high vapor pressure of alkali hydroxides (e.g. 760 mm Hg for NaOH)[4], the reaction rate is high. If volatilization, rather than diffusion in the glass is rate controlling, the reaction mentioned above will increase the weight loss rate. Another effect of H_2O is to reduce the viscosity of the melt. This factor could increase the volatilization rate if diffusion-controlled.

Fig. 2 shows the effect of temperature on the volatilization from TV glass in dry CO_2. From 1400 to 1450°C, the higher the temperature, the faster the weight loss rate. But the weight loss at 1475°C is much slower than that at lower temperatures.

We assume when the glass was treated at 1475°C, a SiO_2 rich layer (verified by Secondary Ion Mass Spectroscopy of the sample) evidently formed quickly because of the fast surface depletion of alkali, creating a barrier to diffusion of alkali ions and slowing volatilization. When H_2O vapor was incorporated with CO_2, the volatilization rate was increased by almost 3 times during the initial stage, and the weight loss rate at 1475°C was faster than that at lower temperatures. This can be attributed to the effect of H_2O in reducing the viscosity of the SiO_2 rich layer as discussed by LaCourse[5].

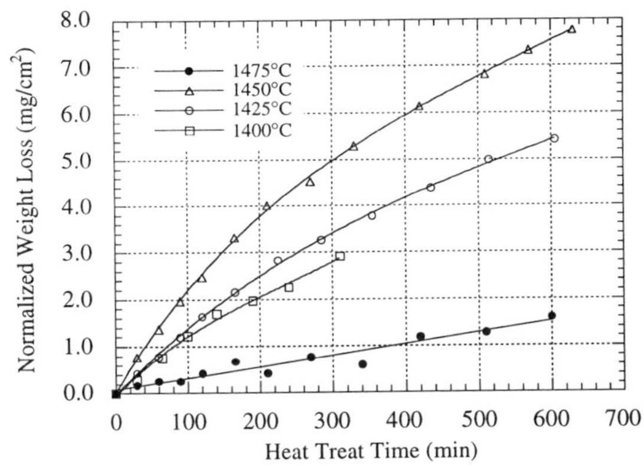

Fig. 2. The normalized weight loss of TV glass in CO_2.

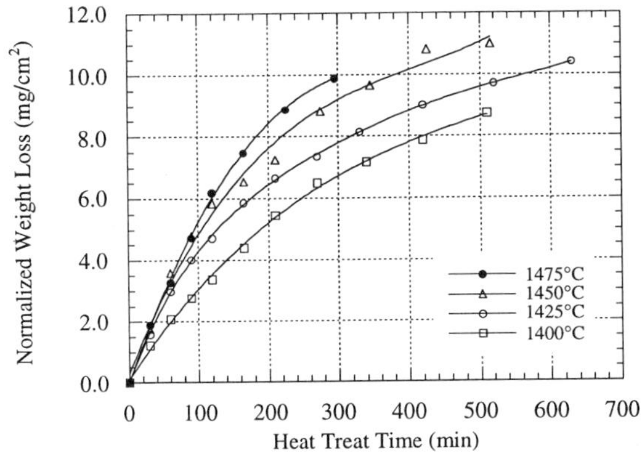

Fig. 3. Normalized weight loss of TV glass in CO_2 + 30% H_2O.

<u>$25Na_2O$-$75SiO_2$ Glass:</u> As shown in Fig. 4, the weight loss in CO_2 increases with increasing temperature. At 1450°C or lower temperatures, the weight loss was nearly linear, implying a volatilization controlled process. At 1475°C, because

surface reaction rate is faster, diffusion becomes more important. It should be noted, however, that the total concentration of Na_2O in the glass is also decreasing with time, and separation of these effects has not yet been attempted. Fig. 5 shows a dramatically increased weight loss in CO_2 with 30% H_2O vapor, especially at lower temperatures. This is also an evidence of the surface controlling mechanism for the volatilization process. The linear time dependence disappeared at all temperatures when water vapor was present.

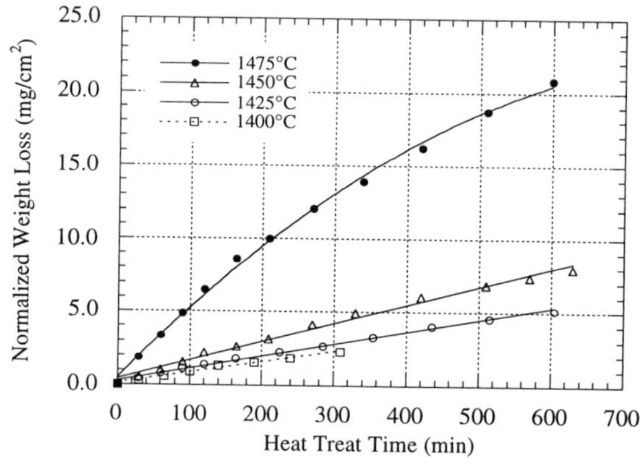

Fig. 4. Normalized weight loss of $25Na_2O-75SiO_2$ glass in CO_2.

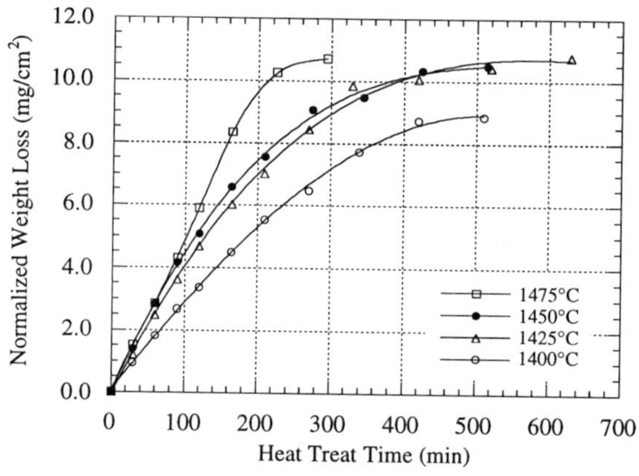

Fig. 5. Normalized weight loss of $25Na_2O-75SiO_2$ glass in CO_2 with 30% H_2O.

<u>$25K_2O-75SiO_2$ Glass</u> : Fig. 6 and 7 show the weight loss of the glass in CO_2 and CO_2 with 30% H_2O respectively. This glass also showed lower weight loss rate at 1475°C than at 1450 and 1425°C due to the SiO_2 rich layer formation on the surface. The weight loss rate was considerably increased by H_2O. In Fig. 7 the weight of the glass became nearly constant after 200 minute treatment. According to

Advances in Fusion and Processing of Glass II

the relative weight loss calculation in which alkali oxides were considered to be the only constituents lost, almost all the K_2O was lost after 200 minutes.

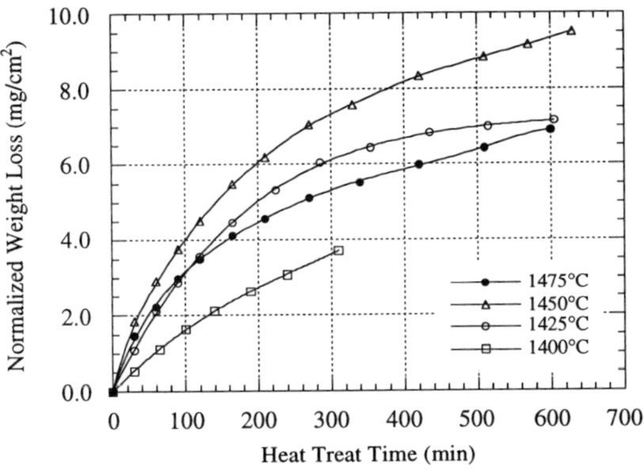

Fig. 6. The normalized weight loss of $25K_2O$-$75SiO_2$ glass in CO_2.

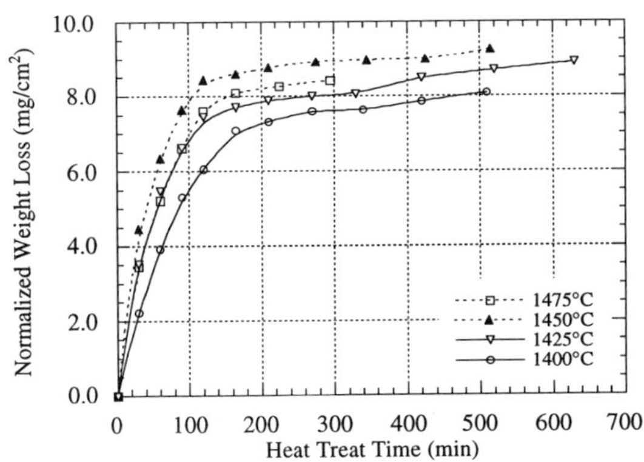

Fig. 7. Normalized weight loss of $25K_2O$-$75SiO_2$ glass in CO_2+30% H_2O.

<u>Wetting Angle and Surface Tension</u>: As shown in Figure 8, the wetting angle of TV glass in O_2, N_2 and air increased linearly with time, but showed a maximum when treated in CO_2 or CO_2 with 30 % H_2O. The high value of wetting angle in CO_2 30% H_2O in the initial stage was due to the extremely high rate of volitilization, but the specific reason for the maximum is unknown. The wetting angle is higher in CO_2, N_2 thatn that in O_2 and air, in agreement with surface tension measurements dtermined using the maximum bubble pressure technique[6.] Generally, H_2O vapor decreases the surface tension and wetting angle, but here we find higher values, probably due to the rapid compositional changes in wet environmnets.

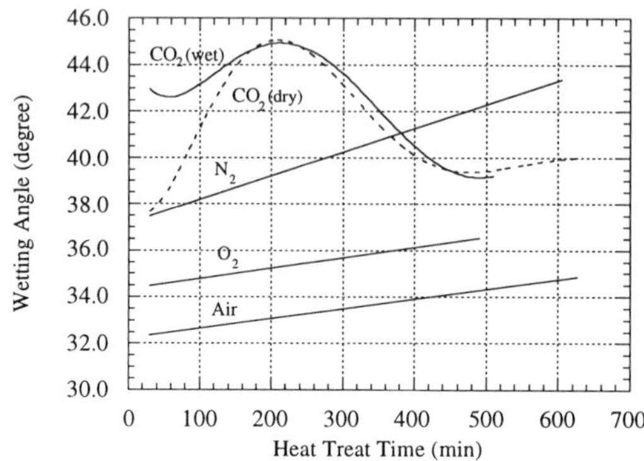

Fig. 8. Wetting angle of TV glass in various atomospheres.

CONCLUSIONS:

Furnace atmosphere has very important effects on the volatilization and surface tension or wetting angle of glasses. Water vapor enhances the volatilization by reacting with alkali in the glass melt and reducing the viscosity. Surface SiO_2 rich layer formation can be a barrier to ion diffusion, thus reducing the volatilization rate. The wetting angle of TV glass increases linearly with heat treat time in N_2, O_2, and air, but shows a maximum when treated in CO_2 or CO_2 with water vapor. The weight loss rates can be used to estimate the viscosity change in the bubble walls of high temperature glass foams.

ACKNOWLEDGMENT

This work was supported by NSF Industry-University Center for Glass Research at Alfred University and US Department of Energy.

REFERENCES

[1]D.M. Sanders and H.A. Schaefer, "Reactive Vaporization of Soda-Lime-Silica Glass Melts," J. Am. Ceram. Soc., **59**[3-4]96-101(1976).

[2]M.H.V. Fernandes and M. Cable, "Reactive Vaporization of Sodium Tetraborate With water Vapor," Glass Technology, **34**[1]26-32(1993).

[3]H. Jiang and W.C. LaCourse, "Foaming Behavior of Glass Melts - Volatilization," Presented at the 99th Annual Meeting of American Ceramic Society, Cincinnati, Ohio, May 4-7, 1997.

[4]A.M. Kruithof, C.M. La Grouw, and J. de Groot, "Volatilization of Glass," pp. 515-527, in the Proceedings of the Symposium Sur La Fusion Du Verre, Bruxelles, October, 1958.

[5] W. C. LaCourse, Surface Viscosity Measurements via Fiber Elongation, this volume.

[6] H. Jiang and W.C. LaCourse, unpublished work.

SIMULATION OF A TV GLASS REFINER USING COUPLED COMBUSTION AND GLASS FLOW MODELS

Richard Bergman, Corning Incorporated, Corning, NY, USA
Dimitri P. Tselepidakis, Ph.D., Fluent Incorporated, Lebanon, NH, USA

ABSTRACT

Fluid flow and heat transfer in both the combustion space and glass bath of a circular TV panel refiner was simulated to evaluate the effect of combustion space heating and cooling on glass flow and temperatures. FLUENT software was used to simulate both the glass and the combustion space.

Two separate models, one for the combustion space and one for the glass, were coupled by passing boundary conditions between the models. The results of the modeling are: 1) temperature correspondence with glass surface opticals is improved over simplified assumptions, and 2) surface glass flow is influenced by the placement of burners and cooling air.

Computationally, it is necessary to relax the passing of boundary conditions between the two models to achieve convergence.

INTRODUCTION

With the demand for ever higher quality glass for color TV panels has come the necessity to more precisely understand the process of melting glass and the sourcing and processing of defects. The refiner has increasingly become an important area to understand as a significant source of blisters, knots, and cord.

Simple modeling of the refiner has provided a general understanding of the flow and has aided in the understanding of melter operations. However, this kind of modeling has not done justice to the sensitivity of refiner operations with respect to defects.

Operators are well aware of the effect burners have on moving surface glass in a melter. The study of these effects on glass processing is not possible without modeling the details of burner flames and cooling air jets.

PROBLEM DESCRIPTION

The process for melting and delivering color TV panel glass to a press is composed of a melter, a refiner, and a forehearth. Generally speaking, the melter is where the glass is melted, homogenized, and fined. The highest temperatures in the process are found in this section. Glass flows through a throat near the bottom of the front wall of the melter and into a refiner. This refiner is circular and has an open space above the glass for conditioning. The refiner is used as a thermal conditioner to cool the glass down to a temperature suitable for the forehearth. Glass flows out of the refiner through forehearth connections located near but a little below the glass line of the refiner. The forehearth is used to cool the glass still further to the right forming temperature and to restore thermal homogeneity to the process glass.

This paper focuses on simulations of the refiner. The firing conditions above the glass are particularly complex and to this point have not been studied in detail. The refiner is a critical area for glass quality, since defects in the forward part of the process have little opportunity, both because of time and temperature, to be eliminated by the process.

Figure 1 shows the grid of the combustion and glass flow model. Glass enters from the left through the throat and exits at the right through the 2 forehearths and the overflow. In the combustion space there are 3 burner pairs. The rear burners (1, 2, 5, and 6) are cooling while the front burners, 3 & 4, are heating slightly.

MODEL RESULTS

Figure 2 shows the flow of gases in the plane of the burners. The rear -most burners drive the convection in this set-up. Gas flow is forward along the centerline. At the nose of the refiner the gases turn back along the side to the vents.

Advances in Fusion and Processing of Glass II

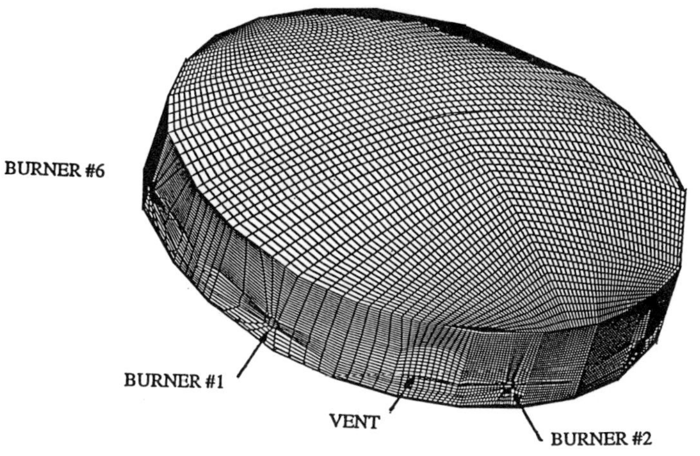

COMBUSTION MODEL

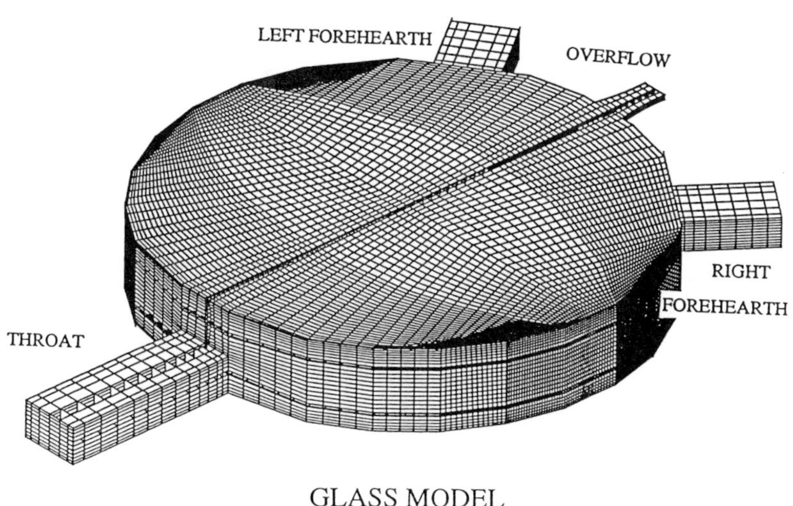

GLASS MODEL

Figure 1: Grid of 2 Models

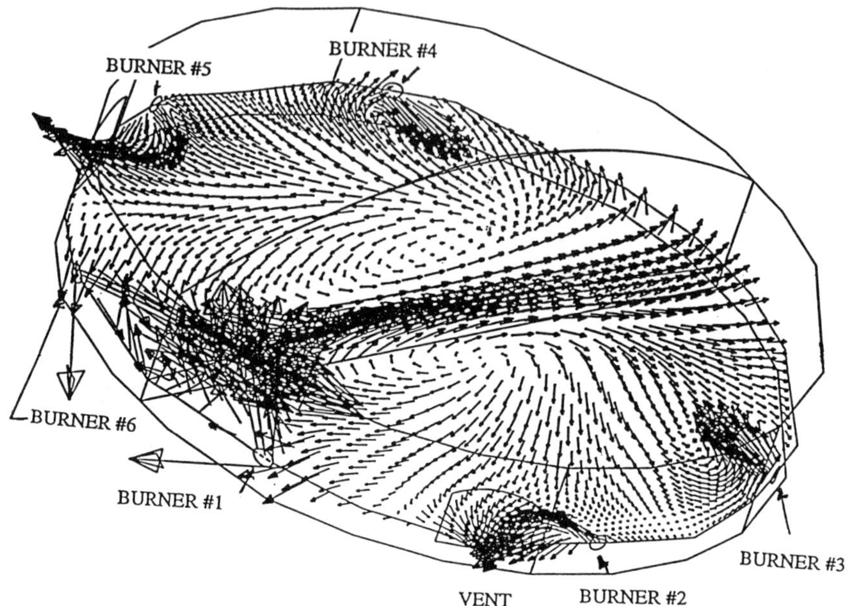

Figure 2: Combustion Gas Flow in Burner Plane

COMPARISON OF SURFACE FLOW vs. MODEL COUPLING

Figure 3a shows contours of temperature on the glass surface with a simple crown assumption - radiation to a fixed, constant temperature.   Figure 3b shows the surface temperatures with a model coupled to the combustion space.

The glass surface temperature gradients are more realistic in the coupled model case. Burners #1, #2, #5, and #6 supply cooling air. Burners #3 and #4 in the front supply fuel and air.   Strong cooling by the rear burners causes the glass to lose temperature rapidly in the rear of the refiner. Heating in the front causes 2 hot spots to form near the forehearth connections. Temperature correspondence is improved as well, as seen in Table I.

                                Advances in Fusion and Processing of Glass II

Location	Measured	Simplified Model	Coupled Model
Rear Optical	1311	1340	1300
Center Optical	1222	1260	1210
Front Optical	1211	1215	1190
Side Temperature	1190	1223	1195

The coupled model gives better temperature agreement than the simplified model. Just as importantly, the coupled model calculates surface temperature gradients, which influence surface flow.

Figures 4a and b show the surface pathlines. The simplified model shows very regular motion out to the sidewalls and to the front of the refiner. The coupled model shows that surface flows are directed away from the surface hot spots. This phenomena could be very important to understanding how to control surface glass defects.

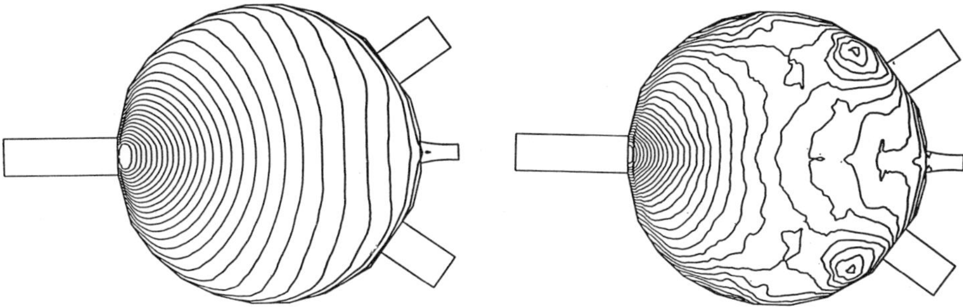

Figure 3a:Simplified Model-Temperatures   b) Coupled Model-Temperatures

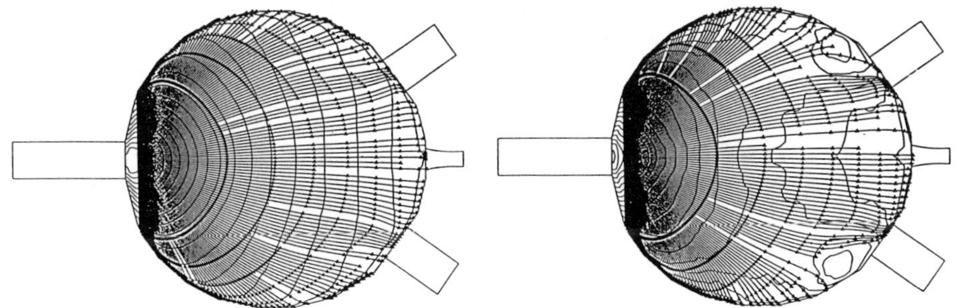

Figure 4a: Simplified Model – Pathlines   b) Coupled Model – Pathlines

CONVERGENCE ISSUES

It was learned very early on that simply passing boundary conditions from one model to another would not yield a converged solution. Figure 5 is a comparison of surface temperatures at the front of the refiner when numerical relaxation is applied.

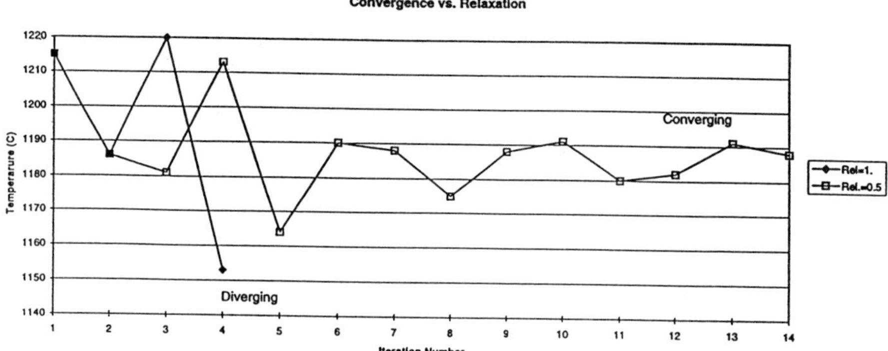

Figure 5: Refiner Front Temperature vs. Iteration

Various methods can be used with success. In this case, temperatures are passed to the combustion model. The resulting heat fluxes are averaged with the heat fluxes of the previous iteration and applied to the glass surface. In this way the heat fluxes are relaxed. Other researchers avoid this particular relaxation issue by alternating iterations of the combustion model and the glass model within one master program.

FUTURE WORK

The coupled model has been successfully applied to provide a more realistic simulation of the refiner of a glass melting tank. The challenge is to use this model to explain in a direct way furnace observations and quality data.

This kind of modeling is very resource intensive. As with any modeling the right balance must be struck between simplicity and speed versus complexity and accuracy.

Using coupled models provides new insights into how furnace operation affects glass convection. These models have the potential to show how to correct set-up problems and more optimally design the glass melting units.

SIMULATION OF BUBBLE REMOVING PROCESS IN A TV FURNACE

Shinji Kawachi
Nippon Electric Glass Co.,Ltd.
7-1, Seiran 2 Chome, Otsu, Shiga   520
Japan

1. Purpose of research

Growth and dissolution of bubbles generated during the batch reaction into glassy state is a very complicated physico-chemical phenomenon. Assuming that bubbles are removed by two mechanisms, floatation or absorption, we developed a simulator to evaluate the overall influence of (1) glass and batch composition, including refining agents, (2) geometrical tank design and (3) furnace operating conditions. For verification, the simulated results were compared with actual data from our TV panel furnaces.

2. Procedure of simulation

The simulation is conducted according to the procedure shown in Fig. 1. The first step "Thermal fluid calculation of molten glass in tank furnace" computes convection and temperature distribution in a furnace. The second step "Computation of gas concentra-

tion in glass melts" reveals gas distribution in the melts. The third step "Computation of (re)fining process of bubbles" gives the number of (re)fined bubbles that they are generated beneath the batch blanket, travel in the molten glass and disappear because of floatation or absorption. The fourth step "Overall evaluation of bubble removing performance of furnace" yields clues to pursue better bubble quality.

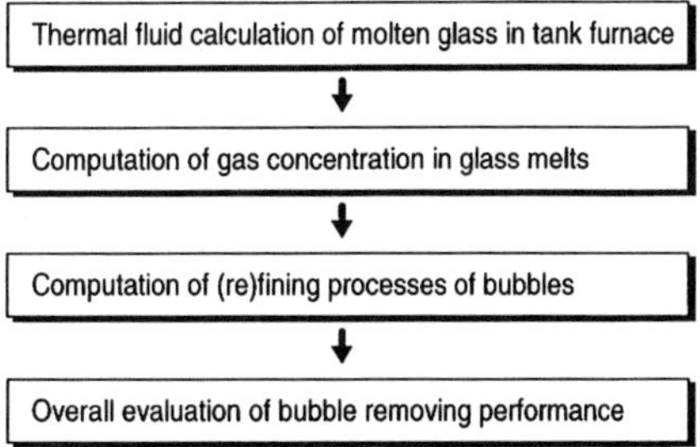

Fig. 1   Sub-models and computation procedure

Advances in Fusion and Processing of Glass II   245

3. Basic equations of floatation and absorption

(1) Floatation:

Ascending velocity of a bubble Vs is assumed to follow Stokes' law given in Equation (1):

$$Vs = 2\, g\, r^2\, \rho/(9\, \eta) \qquad (1)$$

where, g: gravitational acceleration, r: bubble radius, ρ: glass density, η: glass viscosity.

(2) Absorption:

Mass of gas that moves in or out of a bubble per unit time I_j may be described by Equation (2).

$$I_j = 4\, \pi\, r^2\, k_j\, (Cb_j - Ca_j) \qquad (2)$$

where, j: subscript of gas, k: mass transfer coefficient, Cb: Gas concentration in glass melts, Ca: Gas concentration of bubble/glass boundary. Equation (2) denotes that the difference between gas concentration in a bubble and that in glass melts will become the driving force for gas migration.

4. Example of simulation

4.1. Thermal fluid calculation of molten glass in tank furnace:

The convection and temperature in the furnace are obtained by solving the equation of continuity, Navier-Stokes equation and the energy equation si-

multaneously. Fig. 2 is the calculated results shown at the furnace center section.

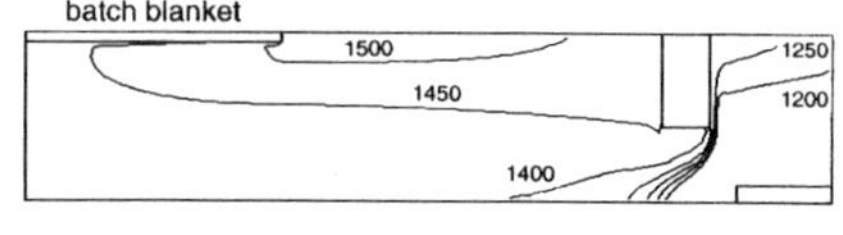
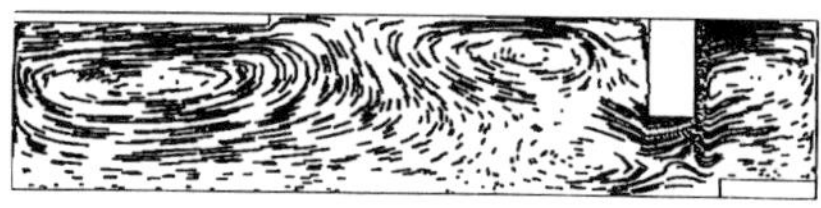

Fig. 2 Temperature(upper) and flow pattern (lower) at furnace center section

4.2. Computation of gas concentration in glass melts

The factors which influence the gas concentration in glass melts are as follows;

(a) amount of oxygen evolved by the refining agent, (b) gas concentration at the batch/glass boundary, and (c) gas concentration at the glass-melt/combustion-space boundary. Fig. 3 schematically shows the relationship of these three factors. Substituting the condition (a) to the source term of the diffusion equation, and conditions(b) and (c) to the boundary condition of the same equation, we can obtain the gas distribution in glass melts.

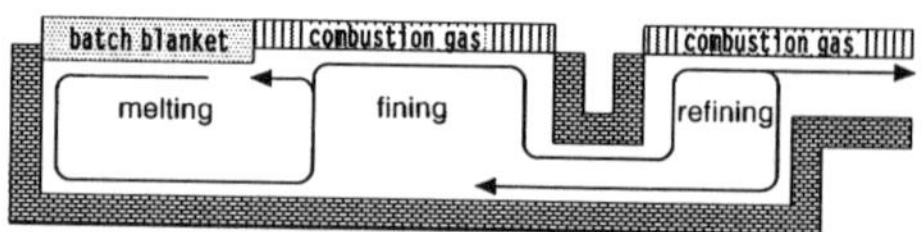

Fig. 3 Schematic diagram of refining gas evolution and boundary conditions

   Advances in Fusion and Processing of Glass II

Fig. 4 illustrates the situation of O_2 generated by the decomposition of refining agents traveling with the convection current in the furnace. Oxygen seems to evolve violently beneath the batch blanket but its decomposing rate becomes gradually stable as it descends toward the furnace bottom where the temperature is lower. However, once the glass is raised to the surface in the vicinity of furnace center, the gas is generated again actively.

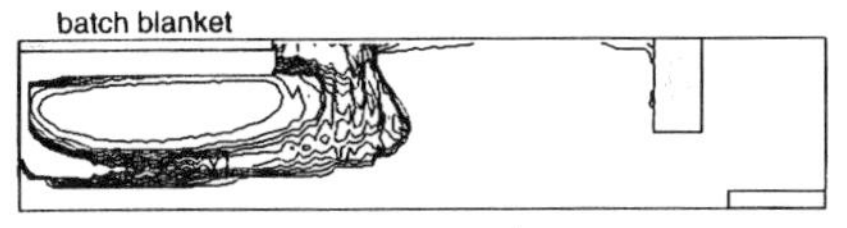

Fig. 4 O_2 gas evolution by refining agent

Fig. 5 shows the distribution of gas concentration in the furnace. The value is converted from a concentration to partial pressure.

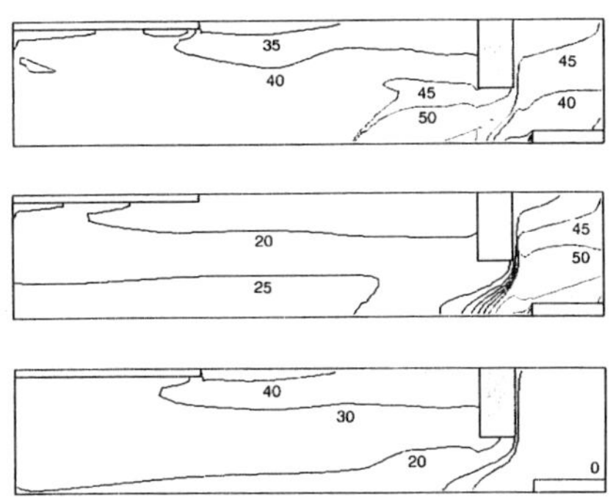

Fig. 5   Partial gas pressure [%]
(upper): CO_2, (middle): N_2, (lower): O_2

4.3. Computation of (re)fining process of bubbles:

Within 30 minutes after the batch charged into our TV furnaces, it is observed to become initial glass containing a huge number of bubbles.

In order to know the properties of the initial glass, we conducted a test. We melted the batch in a crucible at 1400℃ which is supposedly equivalent to the temperature of batch charging area. Measurement of bubble diameter with an image processor showed the distribution in Fig. 6. The maximum size was 0.4 mm and its frequency of appearance became higher as the size got smaller. The average gas composition of the bubbles were as follows: CO_2: 15.7 vol%, N_2: 0.3vol%, O_2: 84.0vol%.

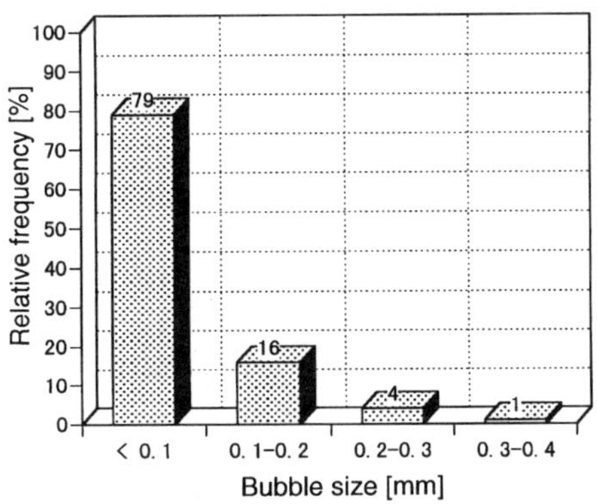

Fig. 6   Size distribution of initial bubbles

Fig. 7 illustrates three common trajectories of bubbles which are generated beneath the batch blanket. The bubbles are 0.4 mm in diameter and have the same gas composition as above. Fig.7 represents three typical cases: (1) the case of absorption, (2) the case of floatation and (3) the case of flowing-out.

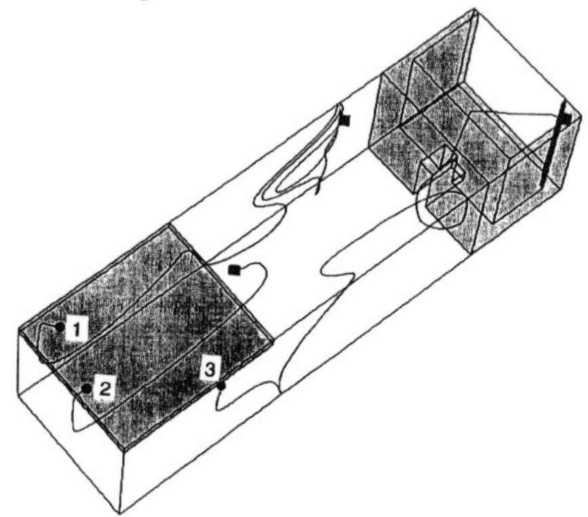

Fig. 7   Trajectories of three kinds of bubbles
1: absorbed, 2: floated, 3: flowed out
● : starting point,   ■ : terminal point

In Fig.8, the change of bubble size and composition with time is illustrated for the case that the bubble is removed due to floatation in Fig.7. As the bubble moved toward the furnace bottom, the bubble diameter tended to shrink, but the bubble diameter increased again when it approached the glass surface. In accordance with the movement, O_2 content in the bubble almost halved and CO_2 content roughly tripled once, but the bubble which reached the surface became rich in O_2 , and poor in CO_2 and N_2. Our experience that the glass scooped from the melter surface contains O_2 rich bubbles may well be explained by the simulation result.

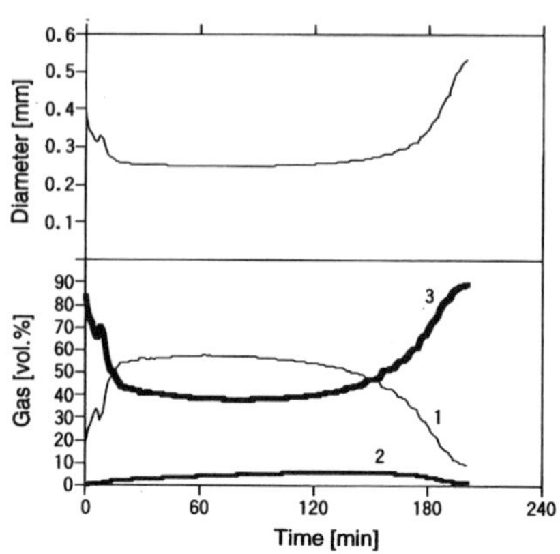

Fig.8   Change of diameter and gas composition of floated bubble

1 : CO_2,   2 : N_2,   3 : O_2

4.4. Overall evaluation of bubble removing performance of furnace:

Fig. 9 is the fate of 2000 initial bubbles with four sizes, 0.1, 0.2, 0.3, 0.4 mm in diameter. They were placed beneath the batch blanket and their trajectory was pursued for 24 hours. The graph shows that the bubble disappears by absorption, as the dimension gets smaller. Bubbles also disappear by floatation, as the size becomes larger.

As a criterion to evaluate bubble removing performance of furnace, we hypothesize that the more the number of bubbles is decreased because of absorption or floatation, the better the furnace performance is. The product of initial bubble number shown in Fig.6, and bubble removal ratio due to floatation or absorption shown in Fig.9 is defined as "weighted disappearance

                          Advances in Fusion and Processing of Glass II

ratio". In our case it becomes 86.5. It means that 86.5% of the initial bubbles are (re)fined within 24 hours.

This simulation is judged to sufficiently reflect the conditions of real furnaces.

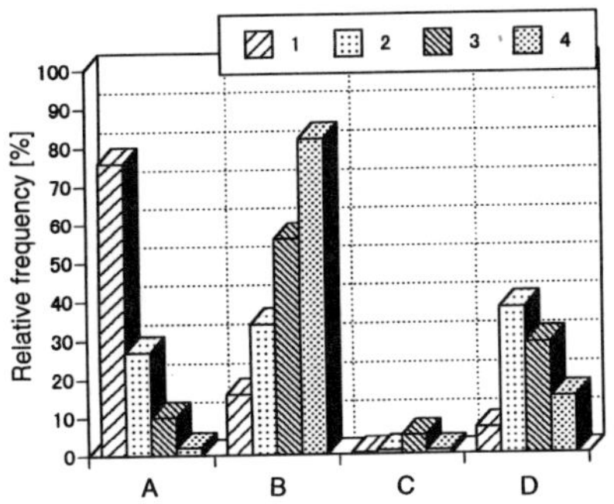

Fig. 9   Fate of bubbles after 24 hours
Initial size 1: 0.1, 2: 0.2, 3: 0.3, 4: 0.4 [mm]
Fate A: absorbed, B: floated, C: flowed out, D: remaining

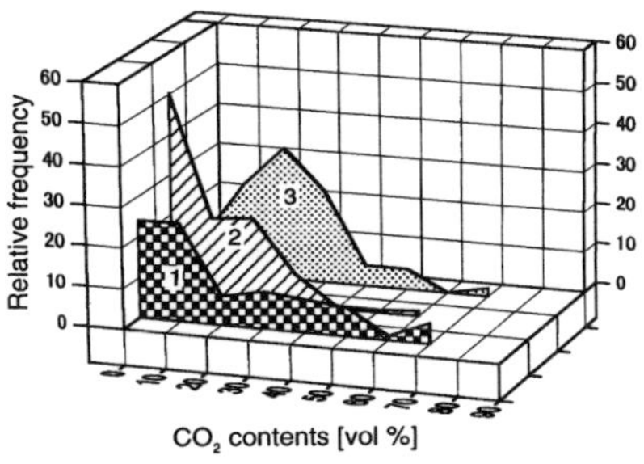

Fig. 10 CO_2 gas distribution in bubbles
1: predicted,   2: measured-1,   3: measured-2

5. Conclusion

By paying attention to CO_2 content in flow out bubbles, we can draw a histogram, resulting in Fig. 10. O_2 in the bubbles which reach the outlet of refiner is almost completely absorbed and the bubble gases consist of CO_2 and N_2. Therefore, the characteristics of a bubble can be explained by CO_2. Curve No.1 indicates the predicted results by the simulation, and Curves No.2 and 3 are the analytical results obtained from our real furnaces. These data were gathered from different furnaces and on different dates. However, it was found that the predicted CO_2 were reasonably within the range of fluctuation of measured TV glass data.

APPLICATION OF BUBBLING IN A CRT PANEL FURNACE

Victor O. Aume and Bin Hong (Thomson Consumer Electronics , 24200 U.S. Route 23 South, Circleville, Ohio, USA 43113)

ABSTRACT

Thomson has successfully applied forced bubbling as an operating technique to optimize the quality produced from furnaces melting glass for the production of CRT faceplates. Through a combination of mathematical and physical modeling of the bubbling and melting processes, the production results were predicted and then verified in operation. In addition to the expected improvement in defect levels, benefits also include improved process robustness.

PROJECT GOAL

This project was a small portion of a larger effort to understand the driving forces for glass quality in the melting of panel glass, and to test the various approaches to impact it ahead of four major furnace rebuilds and expansions that Thomson was scheduling for 1997 and 1998. Mathematical modeling, physical modeling, and high-temperature laboratory testing were all included as part of this effort.

Out of these efforts, came the conclusion that forced bubbling should have a significant impact on the glass convection patterns, and so result in a significant improvement in glass quality. It was decided to install bubblers on-the-fly in Circleville's C Tank to demonstrate the technology, and also to improve a nagging problem with viscous knots that had not been significantly impacted by other operational approaches.

BACKGROUND

Most glass manufacturers operate with an intuitive "model" of how their melting furnaces operate, the driving forces involved, the resulting glass convective flow patterns, and Thomson is no different in that account. What was different in this case was the fact that the actual operation did not typically respond to process changes consistent with expectations, leading to the conclusion that the furnace operation deviated significantly from expectations.

Thomson's math modeling efforts began over five years ago to address this problem specifically, but also to look for future approaches necessary to support business expansions and the desire to improve manufacturing efficiencies.

The model results for C Tank were quite unexpected. The traditional expectation that the glass witnessed a hotspot upwelling and a resulting pair of convective cells was not verified, nor could it be produced with any chosen operating temperature profiles. In fact, the model predicted that a single large convective cell existed, and that newly melted glass flowed slowly along the bottom of the furnace, picking up refractory corrosion products, and then flowed directly through the throat, into the refiner and forehearth (See Figure 1). The result was that the normal refractory corrosion products that would be easily assimilated into the melt with a "normal" flow pattern and glass time/temperature history, were in our case prevented from reaching assimilation temperatures by the poor flow pattern, and were instead observed as defects in our finished product.

Figure 1 -- Vector Plot Of C Tank Convective Flow Pattern -- Base Case

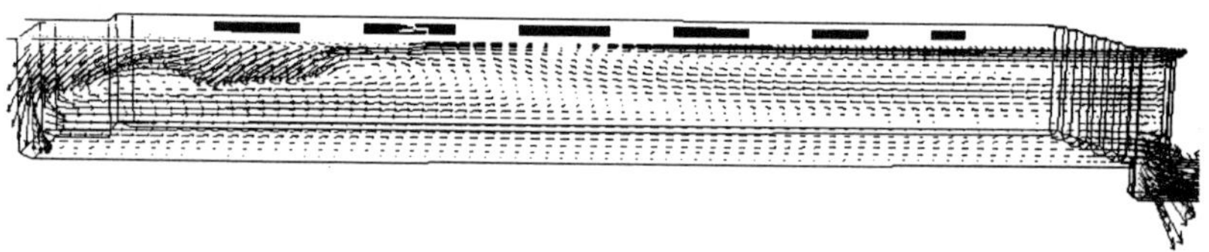

APPLICATION OF BUBBLING

Through continued math modeling efforts, forced bubbling was identified as a technique that could be applied to alter the flow pattern, as well as the only significant process change that could be applied to the existing furnace. Figure 2 shows the dramatic change to the convective flow patterns as a result of bubbling, which creates more of the expected upwelling in the hotspot region.

Unfortunately, bubbling did not have all of the impact we would have desired, as it did not change the bottom flow near the forward portion of the melter near the throat. From the model predictions, the best we could hope for is a 50-60% reduction in defects, assuming of course that we were correct in our assumption that the source is corrosion products from the melter bottom refractory. However, this improvement was enough to warrant further investigation of bubbling.

                                    Advances in Fusion and Processing of Glass II

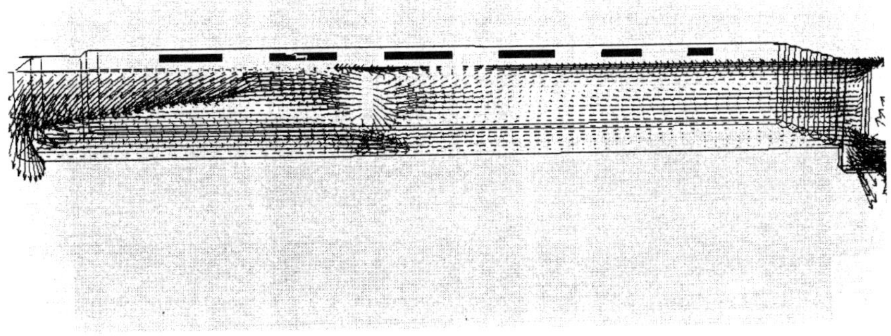

ANALYSIS OF BUBBLING

The task of gaining a more detailed understanding and confidence in bubbling, lead us to build a full scale physical model. It was not intended to achieve rigorous similarity criteria, but rather give us some operational experience with the parameters of bubbling with the goal of directional, qualitative guidance once bubblers were installed in the operating furnace. These included : bubbler tip design, continuous gas flow, pulsed gas flow, pulsing parameters, and their influence on fluid flows within our model.

In one series of experiments varying the gas flow rate, our laboratory clearly demonstrated that the continuous-flow approach would most likely be limited by the increase in entrapped seeds caused by bubble collisions. There was some concern that the seed level would be unacceptably high for most flow rates, but fortunately this was not the case in practice.

It was also desired to increase our confidence in the furnace math model predictions for bubbling, and so a math model was constructed of our bubbling physical model to compare bubbler inputs to predicted and measured fluid velocities (See Figure 3). Since the math model input for bubbling is simply a gas flow rate (from which a resulting buoyancy force is calculated and applied to the grid cells above the bubbler location), the model doesn't know if the gas is introduced as a pulsed bubble or as continuous flow. Obviously, the resulting bubble size and frequency depend on the specifics of the application setup, but it was unknown how these parameters would influence the flows in our physical model, or the math model's (volume flowrate) ability to predict their influence on the melter flows, which is the ultimate goal.

Figure 3a -- Vector Plot Of
Bubbler Math Model

Figure 3b -- Particle Trace Plot Of
Bubbler Math Model

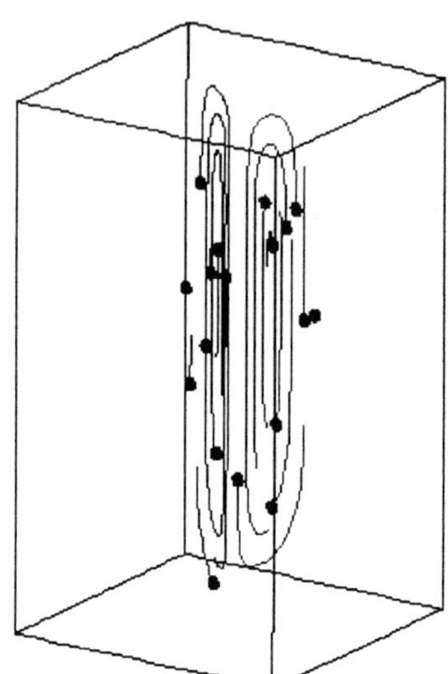

Our melting process laboratory undertook an experiment to measure the fluid velocities
resulting from various bubbler inputs. From the pulsed bubbling experiments, we
calculated the volume of gas in the bubbles, and then used this volume flow as an input
to the math model. Resulting velocities were found to be within approximately 10-15%
of the predicted values, which is thought to be quite satisfactory. (See Figure 4)

Figure 4 -- Comparison Of Bubbler Physical Model Velocities To Bubbler Math
Model Predicted Velocities

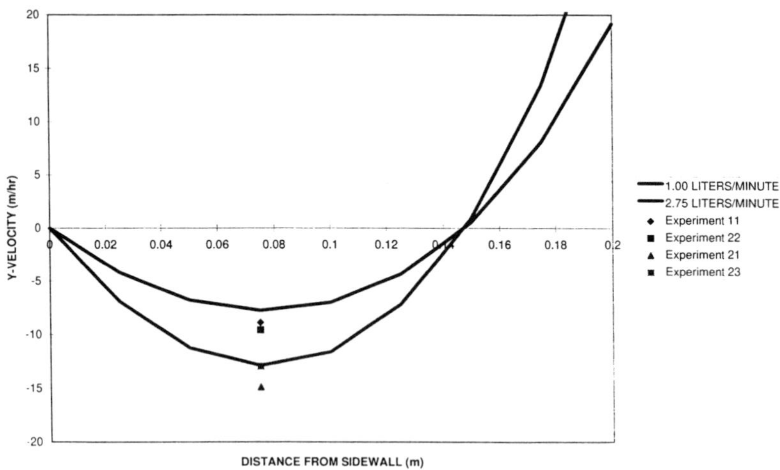

Advances in Fusion and Processing of Glass II

Another element of the experiment was to verify the hypothesis that introducing many smaller bubbles resulted in the same overall fluid flow as fewer large bubbles, assuming that the volume flow rate was the same. The experiment was successful in verifying this, which gave us more latitude in choosing our bubbling operating setpoints.

FURNACE APPLICATION

Thomson's first application of forced bubbling was in C Tank in February 1996. The installation consisted of 15 non-cooled, continuous-flow bubbler pipes, installed to reinforce the location of the furnace hotspot, as was show in the earlier modeling plot.

The initial operation was at a nominal gas flowrate of 1.5 SCFH, and matched the transient model expectations extremely well in terms of the reaction time of the process and the predicted change in temperatures to within 5 DegC in most instances. A chemical tracer study was also performed and compared with a similar study prior to the bubbler installation. The change to the minimum residence time and the overall mixing characteristics matched the model's predictions extremely well.

Soon after the installation, higher flowrates were attempted to achieve further temperature increases and a stronger influence on the convective flow pattern. However, the increase in seed count soon exceeded the quality requirements and demonstrated the limitations of a continuous-flow bubbler. All subsequent applications within Thomson have been pulsed bubblers, which are now operating in three different panel furnaces.

Figure 5 -- Furnace Temperature Response To Bubbler Installation And Adjustments

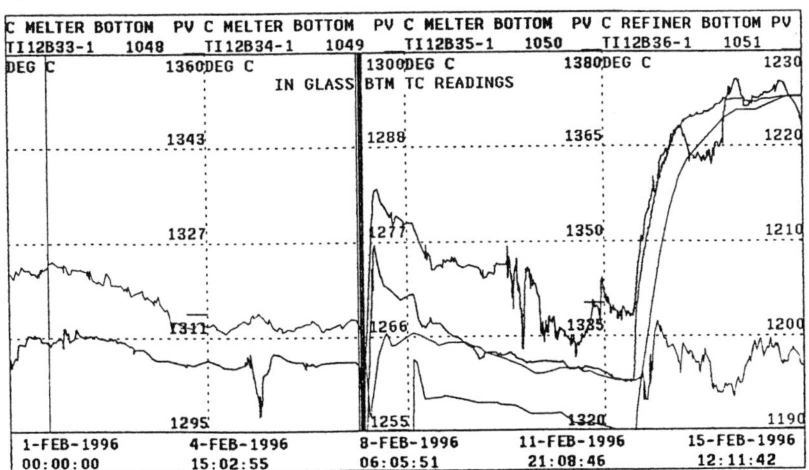

GLASS QUALITY

The ultimate judge of process performance is not the ability to predict response times, nor temperature changes, but simply glass quality. In the case of C Tank, a 50-60% reduction in viscous knot inclusions was predicted, and in fact observed!

While the results were not instantaneous, and an upset in glass quality with a duration of approximately 7-10 days occurred while the process re-established itself while stagnant bottom and surface glass was forced to mix into the process, the overall result was very positive.

Glass losses for viscous knots have decreased by almost 50%, and perhaps more interestingly, have stabilized as well. In previous operation of C Tank, minor variations or upsets in batch feeding, pull changes, batch wetting, etc., would have been enough to de-stabilize the furnace for periods of several days, resulting in increased losses due to knots. The convective flow driving force in our furnace configuration was apparently so weak, that any disturbance in the furnace operation also resulted in disturbing the refractory corrosion boundary layer, which was then swept directly into the product stream, resulting in elevated knot and cord losses, until a new steady state condition was established. Now with operating bubblers, these minor disturbances have virtually no impact on glass quality, and if there is a large enough disturbance, the previous quality level is re-established within a day.

This new process robustness offsets the initial losses caused by the bubbler installation by a factor of at least two in the first 18 months of operation, and the process improvement from the bubblers easily justifies the expense of the engineering, analysis, and installation of a bubbling system.

CONCLUSION

The use of bubbling in Thomson's CRT panel furnaces has resulted in a process improvement due to an improved convective flow pattern, and is easily justified economically by an improvement in manufacturing efficiencies. In several different applications, an improvement in viscous knots has been demonstrated when the limiting factor was the ability to control and manipulate the convective flow pattern. It has also been found that bubbling has a major impact in stabilizing the melting process, which results in additional improvements to yields.

ACKNOWLEDGEMENTS

The authors would like to also acknowledge the work of Ken Robb and Dennis VerDow of Thomson's Melting Technology Laboratory and Ruud Beerkens, Erik Muysenberg, and Olaf Op Den Camp of TNO for their contributions to the success of this project.

FRICTION BEHAVIOUR OF HOT GLASS ON METAL SURFACES

N.M. Keulen and M.G.F.C. Sanders
Philips Components B.V.
P.O. Box 218
5600 MD Eindhoven
The Netherlands

ABSTRACT

Friction experiments between hot television glass and a commercial 13% Cr-steel, uncoated and coated with Cr and NiW have been performed with a pressing-friction apparatus on a laboratory scale. Also the influence and friction reducing effects of several commercial lubricants have been studied. The results showed that friction coefficients measured at contact temperatures above 540°C, which is about 80 °C below sticking temperature, already start increasing sharply. For NiW the lowest friction coefficients were obtained. The effective life-time of the lubricants has been tested as a function of the number of pressing-friction cycles.

INTRODUCTION

The contact behaviour of hot viscous glass towards tool materials is a critical factor in the production of television glass. During the processing of television glass, the employed press tool and glass surfaces are subjected to extremely high thermal, corrosive and abrasive loads. To ensure the quality of the product surfaces and to prevent fracture during television glass and subsequent tube processing, microdefects and flaws have to be avoided in the glass surfaces. Also worn tool surfaces leave marks on the formed glass surfaces, the tools have to be reworked (milled, repolished, shotblasted, replated) and lubricated several times during their lifetime and finally have to be replaced.

The available and necessary information and basic knowledge of the tribological system formed by the hot viscous glass, metal and/or lubricant are far from sufficient at present. Therefore the tribological properties of this system have to be understood better in order to improve and/or control the response of tool and glass surfaces under different tribological loads. Friction and wear tests are often applied to assess these tribological properties.

The authors have carried out friction experiments between hot viscous glass and metallic substrates, with and without coatings, and with and without lubricants. Experiments have been performed using laboratory equipment reflecting the production tribosystem. Both static and dynamic friction coefficients were determined. Such a laboratory tribosystem is not a substitute for industrial tests, but serves as a guideline for industrial developments.

EXPERIMENTAL

Friction apparatus

Figure 1 presents a schematic view of the friction apparatus constituted by two main parts. The left part is formed by two steel rollers, heated and rotating, and two gas burners, to produce small glass gobs of about 1030 °C, with a typical mass of about 8 gr. In order to get spherical shaped glass gobs from cubic glass blocks one of the rollers also is moved in a direction perpendicular to the plane of the drawing. After a pre-determined period of time the hot gob is dropped into a mould by moving one of the rollers, see figure 1. After dropping the gob into the mould, the mould is moved to the press-friction position, right part of figure 1.

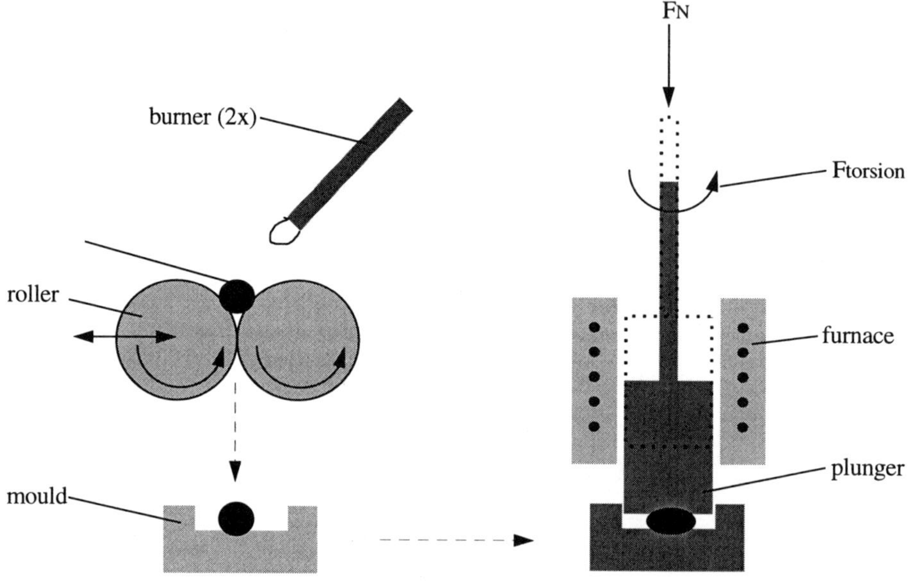

Figure 1. Schematic view of the press-friction apparatus.

The mould temperature can be monitored between room temperature and 650 °C. The right part of figure 1 consists of an interchangeable plunger which can be heated in a furnace. Furnace temperatures can be monitored between room temperature and 650 °C. Press or normal forces in the glass can be adjusted in the range 0.1 - 1 MPa. Glass surface temperature is measured with a pyrometer and is used as a control signal for starting the press-friction measurement cycle. With the plunger glass disks (r = 1 cm) can be pressed and after pressing the plunger can be rotated. The applied torsion force and normal (press) force are measured with standard Kistler® cells. The total apparatus is controlled by a Philips P8 PLC and data are collected by a personal computer. Press and friction measurement cycles can be programmed. The apparatus can be operated by hand or automatically. The total size of the apparatus is about 50x30x40 cm^3.

Experiments

The materials tested were a commercial 13% Cr-steel, and two galvanic coatings (Cr and NiW) on 13% Cr-steel. The friction coefficients were measured without lubricants and with some commercial lubricants: Kleenmold 99, Kleenmold 30, Renite CNM, Renite CH, Renite F9, Renite Checkstop 10, Bel-Ray 65417, No Swab and BCS 362. The glass used in the tests was the standard Philips screen glass.

After 15 seconds the glass bulk temperature is low enough and accompanying glass viscosity high enough to give a reliable measurement. The contact temperature during pressing and

 Advances in Fusion and Processing of Glass II

friction measurements is measured by a thermocouple mounted inside the plunger at 1 mm distance from the plunger surface. So the contact (interface) temperature is exactly known at the moment of the friction measurement. About 15 pressing-friction measurements are done to determine a friction coefficient of a pair of materials. The friction coefficient is calculated by the formula $\mu = F_{torsion}$ / F_N. The standard error in this kind of measurements is typically in the order of 20%.

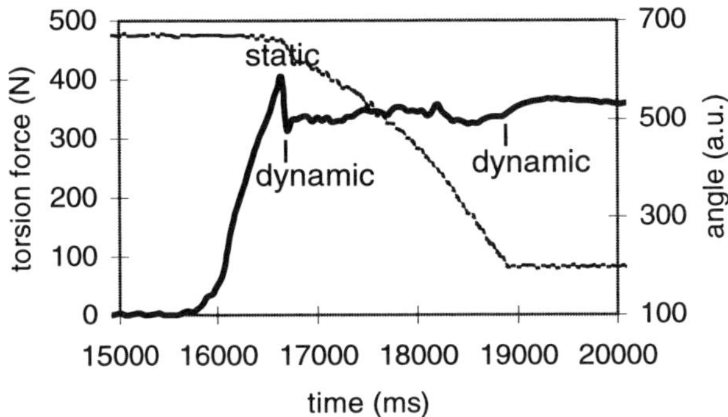

Figure 2. Typical friction measurement output curve, indicated are the points characteristic for static and dynamic friction behaviour. Solid line: torsion (friction) force. dashed line: angle displacement.

Figure 2 shows an example of a typical friction measurement output curve. Indicated are the points where torsion forces, characteristic for static and dynamic friction, can be deduced from these curves. The dashed line shows the angle displacement used for determination of the end of the torsion measurement curve. The equipment is constructed in such a way that glass and metal surfaces slide over each other only once, so no repeated contacts between adjacent glass and metal surfaces. The total angle of rotation is about 15°. Almost all systems reported in literature[1,2,3,4] are designed for measuring dynamic friction, in which case glass and metal surfaces slide over each other numerous times. So the friction measurement system described in this paper reflects the practical television press situation in a more realistic way, friction during filling and at plunger retraction.

RESULTS AND DISCUSSION

In figure 3 the friction behaviour as a function of plunger temperature is plotted. The plotted plunger temperature is the measured contact temperature during the torsion measurement, not the programmed plunger temperature. The plunger temperature (contact temperature) during the torsion force measurement is higher than the programmed temperature, because the metal has been heated up by the hot glass contact for 15 seconds. Shown in figure 3 are the static and dynamic friction. Two values for the dynamic friction are given, see also figure 2. As the friction experiments are done by rotating only over a small angle of about 15 °, this is probably not enough to reach a stationary value for the dynamic friction. Another explanation may be that at the moment of start of the movement the metal slips for a very short moment and then again makes contact and continues with a slip/stick character. Depending on the metal or coating surface and/or lubricant used the two dynamic friction values may differ. In literature only dynamic friction coefficient values are reported for hot soda-lime glass on metal surfaces[1,2] or on ceramic surfaces[3]. Metal temperatures up to 440 °C were measured by von Trier[2]. The dynamic friction values found at 440 °C ranged from 0.75 to 1.55 depending on contact time and angular velocity. Cox and Gee[3] measured in the range 20 - 500 °C and obtained dynamic friction values ranging from 0.35 to 1.0. The authors have measured dynamic friction values at 440 °C ranging from 1.07 to 1.60, which is very similar to the values reported by von Trier[2] , and Cox and Gee[3].

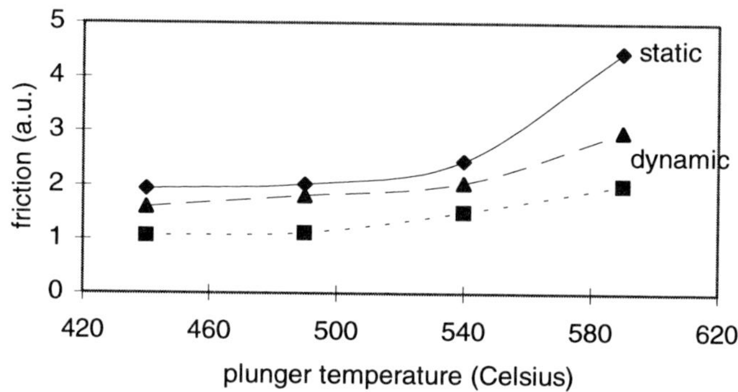

Figure 3. Friction as a function of temperature; both static and dynamic friction are shown. Plunger roughness R_a~1 µm and press force about 0.5 MPa.

In figure 3 one clearly observes that dynamic friction increases with plunger temperature in the range 440 - 590 °C. The same has been observed by von Trier[2] in the metal temperature range 25 - 440 °C. Especially for plunger temperatures greater than 540 °C a large increase in both static and dynamic friction is observed. Going from 440 °C to 590 °C static friction is doubled. Especially from 540 °C to 590 °C an increase of 70% is observed. Thus small temperature changes or differences in temperature over press tool surfaces in the range 540 - 590 °C have large influence on (local) static friction behaviour.

The large increase of static and dynamic friction above contact temperatures of 540 °C may be linked with the sticking temperature. According to Falipou et al.[5] sticking occurs when the temperature at the interface remains sufficiently high, so that the glass viscosity remains low enough to enhance the real contact area and thus, to induce physicochemical interactions between substrate and glass. The sticking temperature of the television glass in contact with 13% Cr-steel is about 620 °C. So the large increase in friction above 540 °C may be an indication of an enhancement of physicochemical interactions near the sticking temperature.

The friction coefficients of 13% Cr-steel uncoated and coated with Cr and NiW were measured. Further some commercial lubricants were measured. Swabbing with the lubricants was done after five pressing-friction cycles without lubricant in order to better see the effect of lubrication. Figure 4 gives an overview of the static friction results obtained for the three substrate materials and the lubricants tested. One clearly observes the low friction value obtained for NiW and the high values for 13% Cr-steel and Cr-coating, without lubrication. Similar results were obtained by Fairbanks[6] which showed high adhesion for Cr-steels and Cr, and low adhesion for Ni and W. The friction reducing effects of most of the tested lubricants are also clearly demonstrated

　　　　Advances in Fusion and Processing of Glass II

in figure 4 for all three substrates. The lubricants Renite CH and Bel-Ray 65417 hardly showed any friction reducing effect. With Kleenmold 99 the largest reduction in friction was obtained.

The plungers were only lubricated once for these tests, so static friction as a function of the number of pressing-friction cycles could be observed. In figure 5 the result for 13% Cr-steel lubricated with Renite CNM is shown. One clearly observes the drop in friction after lubrication; 6th pressing. Friction increases with each pressing-friction cycle returning to the unlubricated situation. So the effective life-time of lubricants can also be tested with this equipment. The friction values presented in figure 4 are the average values of these increasing friction values. However, as this reflects the practical production situation, in which case at regular intervals one has to swab, the authors took this procedure for determining friction coefficients of lubricated plungers.

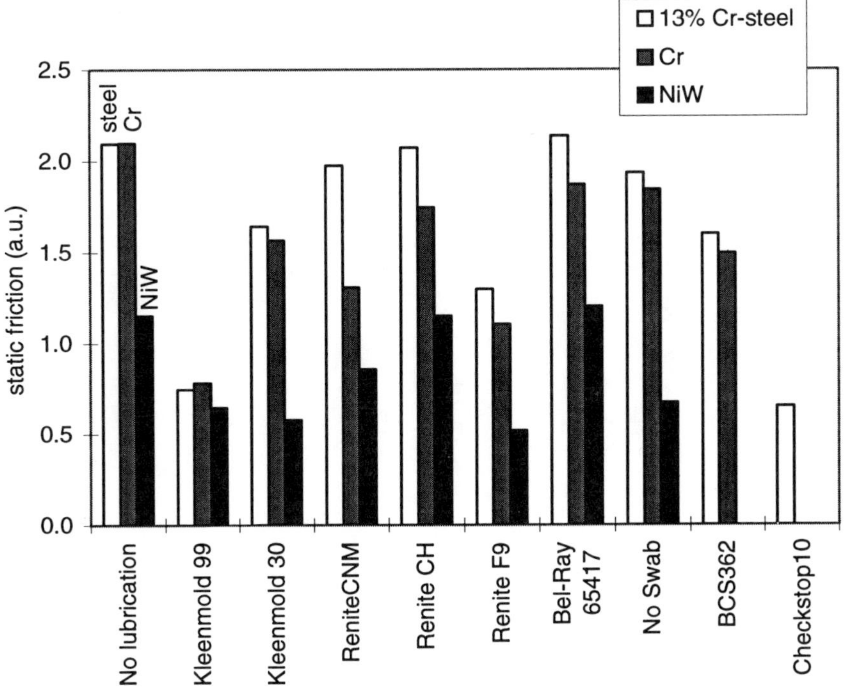

Figure 4. Overview of static friction results, without (no lubrication) and with different lubricants.

CONCLUSIONS
Equipment:
- The experimental set-up for measuring the friction behaviour between hot television-glass and metal or coating surfaces operates very satisfactorily and reliably.

- Both static and dynamic friction can be measured under process conditions (temperature, press force) using fresh glass gobs for each measurement.
- Various process phenomena related with friction can be studied: plunger materials and coatings, lubricants, effective life-time of lubrication, method of lubrication, erosion/wear and corrosion, glass surface damage (flaws, draglines).
Measurements:
- Static and dynamic friction are temperature dependent, especially a large increase appears above a contact temperature of 540 °C.
- NiW showed the lowest static friction coefficient.
- Influence and effects of several lubricants were measured, the effective life-time of the lubricants as a function of pressing-friction cycles was obtained.
- Results of dynamic friction are in agreement with literature data.

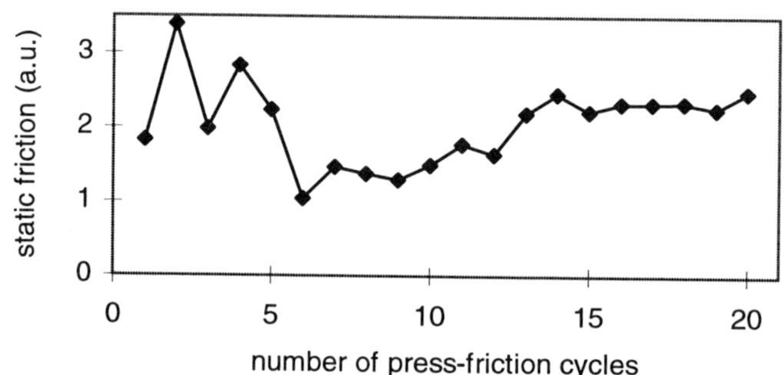

Figure 5. Static friction as a function of the number of press-friction cycles.

REFERENCES

[1] W. von Trier and F. Hassoun, "Mechanik des Gleitens heissen, zähflüssigen Glases auf Metalloberflächen," *Glastech. Ber.* **6** 271-276 (1972).

[2] W. von Trier, "Gleitverhalten von heissen zähflüssigen Glases auf Metalloberflächen," *Glastech. Ber.* **9** 240-243 (1978).

[3] J.M. Cox and M.G. Gee, "Hot friction testing of ceramics," *Wear* **203-204** 404-417 (1997).

[4] D.H. Buckley, "Friction and wear behavior of glasses and ceramics," pp. 101-126 in *Surfaces and Interfaces of Glass and Ceramics*, 1973.

[5] M. Falipou, C. Donnet, F. Maréchal and J.-C. Charenton, "Sticking temperature investigations of glass/metal contacts - Determination of influencing parameters," *Glastech. Ber. Glass Sci. Technol.* 70 [5] 137-145 (1997).

[6] H.V. Fairbanks, "Effect of surface conditions and chemical composition of metal and alloys on the adherence of glass to metal," pp.573-595 in "Symposium sur le contact du verre chaud avec le metal," Scheveningen, The Netherlands, 26-29 May 1964.

          Advances in Fusion and Processing of Glass II

Fatigue and Aging Behavior of Active Matrix Liquid Crystal Display Glasses

John D. Helfinstine and Suresh T. Gulati
Corning Incorporated
SP-FR-4-1, Corning, NY 14831

ABSTRACT

Flat panel displays for both automotive and nonautomotive applications are exposed to mechanical, thermal and vibrational loads during transient and steady-state operating conditions. Their long-term durability is dictated by the stress/time history experienced by Active Matrix Liquid Crystal Display (AMLCD) glass substrate which, in turn, depends on its fatigue, aging and fracture properties. This paper will present aging and fatigue data for five different AMLCD glasses which are necessary for evaluating their relative performance under prescribed loads, boundary conditions and operating environment. These data also help estimate the safe allowable stress/time history, using Power law fatigue model, to ensure their long-term durability. Both static and dynamic tests show that Corning's Code 1737 glass substrates have the highest fatigue resistance and can be made lightweight without compromising their mechanical integrity.

INTRODUCTION

Flat panel displays for both automotive and nonautomotive applications are exposed to mechanical, thermal and vibrational loads during transient and steady-state operating conditions. These loads generate stresses which should be controlled to ensure the long-term reliability of display panels.

While the standard physical properties of active matrix liquid crystal display (AMLCD) glasses are important (see Table I for a summary[1,2]), the long term durability of AMLCD glasses is primarily governed by their fatigue, aging, and fracture characteristics.

Table I. Physical Properties of AMLCD Glasses

Glass Code	Composition	Density (g/cm^3)	Coef. of Thermal Expansion $(10^{-7}/^{o}C)$	Strain Point (^{o}C)
1737	alkaline-earth aluminoborosilicate	2.54	37.8	666
AN635	alkaline-earth aluminoborosilicate	2.76	48.3	643
NA35	alkaline-earth aluminosilicate	2.48	36.5	650
NA45	alkaline-earth aluminoborosilicate	2.76	47.0	610
OA2	alkaline-earth-zinc aluminoborosilicate	2.72	46.7	650

AMLCD glasses, like most silicate glasses, exhibit stress-enhanced corrosion, or, in glass parlance, fatigue and stress-aging[3]. This paper presents aging and fatigue data for five different AMLCD glasses which are necessary for evaluating their relative performance under prescribed loads, boundary conditions and operating environment.

Both static and dynamic fatigue tests show that Code 1737 glass substrate has the highest fatigue resistance and can be made lightweight without compromising its mechanical integrity. These data also help estimate the safe allowable stress/time history, using Power law fatigue model, to ensure its long-term durability.

FATIGUE BEHAVIOR

Fatigue or stress corrosion in silicate glasses results from subcritical growth of surface flaws. It is modeled by the power law fatigue model whereby the crack-tip velocity, V, is related to stress intensity, K_I, i.e.

$$V = A K_I^n \tag{1}$$

where

$$K_I = YSc^{1/2} \tag{2}$$

In Eqn. (1), n is the stress corrosion coefficient, and A is a material/environment constant. In Eqn. (2), S is the applied stress, c is the flaw depth and Y is the flaw shape factor. If the stress, S is constant (static fatigue), or continuously increasing at a stress rate R (dynamic fatigue), then Eqns. (3) and (4) follow from Eqn. (1)[4-6]:

$$S^n t = \text{constant} \qquad \text{(static fatigue)} \tag{3}$$

$$S^{(n+1)}/R = \text{constant} \qquad \text{(dynamic fatigue)} \tag{4}$$

Experiments must be carried out to determine the stress corrosion coefficient for estimating the lifetime, t, corresponding to stress S. When K_I reaches the critical value, K_{Ic}, fast fracture occurs. The critical stress intensity K_{Ic} is related to flaw depth c_f and failure stress, S_f:

$$K_{IC} = YS_f c_f^{1/2} \tag{5}$$

Only the mode I stress intensity factor is needed for describing flaw growth under tensile stress. Table II lists the critical stress intensity or fracture toughness for the glasses being studied[2,7]. The above equations can be combined to calculate the safe allowable stress S_s that the glass can sustain for a long period of time[8]

$$S_s = S_f\{q[(n-2)/(n+1)](S_f/R)^{1/n}\} \tag{6}$$

where q is an inverse time term that is a function of the acceptable safe limit for crack growth. As in Gulati[9,10], we choose the conservative condition

Table II. Mechanical Properties of AMLCD Glasses

Glass Code	Fracture Toughness (MPa m$^{1/2}$)	Young's Modulus (GPa)	Poisson's Ratio
1737	0.83	72.1	0.22
AN 635	0.82	75.4	0.20
NA 35	0.80	68.8	0.22
NA 45	0.83	74.0	0.23
OA2	0.82	70.3	0.17

Advances in Fusion and Processing of Glass II

of 1% crack growth in 40 years. The value of q corresponding to this criterion is found to be 3.7 x 10^{-12} sec^{-1}. We can see that by combining the results of strength test for glass, with typical flaws, with those of fatigue test, one can determine the safe allowable stress the glass can endure for a specified lifetime.

However, it should be noted that the flaws under discussion are neither simple nor single-valued. A statistical distribution is needed to properly describe the flaws being stressed. The distribution that best describes typical surface flaws is the Weibull distribution given by

$$P = 1 - \exp[-(S/S_0)^m] \qquad (7)$$

where P is the cumulative failure probability, S is the applied stress, S_0 is the location parameter (analogous to the mean but at the 63.2 percentile), and m is the shape parameter or "Weibull modulus." This distribution is discussed in more detail in references[11,12]. Thus the determination of a safe allowable stress must be carried out at an acceptable level of failure probability. Similarly the fatigue coefficient n has an associated confidence band. However, since we are interested in fatigue and aging here, we will use mostly the mean value or the Weibullian location parameter.

The dynamic fatigue test consists of breaking a series of monotonically loaded specimens at several rates. The resultant strength/rate data are fitted to the fatigue model, Eqn. (4), and then extrapolated to the desired stress rates. The critical coefficient for this model is the stress corrosion constant n. The data used in this study[9] were measured in the ring-on-ring flexure test[13] (ROR) using 50 mm x 50 mm square specimens. The uniformly stressed region, inside the inner ring, was sandblast abraded[14] to simulate real-life flaws and normalize strength distribution. The stressed region was exposed to 100% R.H. at room temperature. Table III summarizes the results of dynamic fatigue tests.

By combining these stress corrosion coefficients with the corresponding K_{IC} values (Table II) and mean abraded ROR strengths (Table IV), the safe allowable stress can be calculated using Eqn. (6):

STRESS AGING BEHAVIOR

To verify the comparative fatigue resistance of these AMLCD glasses, a stress aging test was performed. In brief, stress-aging data are generated by first statically loading the specimens for a

Table III. Stress Corrosion Constant for AMLCD Glass Substrates with Sandblast Abrasion Tested at Room Temperature in 100% R.H.

Glass Code	Stress Corrosion Coefficient, n (Mean Value)	(95% Conf. Band)
1737	28.8	22.7 - 38.9
AN 635	19.3	16.6 - 23.0
NA 35	28.0	22.3 - 37.3
NA 45	18.1	14.1 - 25.1
OA2	20.0	16.6 - 25.1

Table IV. Estimate of Safe Allowable Stress from ROR Strength Data

Glass Code	ROR Strength (Mean Value, MPa)	Safe Allowable Stress (MPa)
1737	39.5	18.7
AN 635	39.3	12.8
NA 35	36.1	16.6
NA 45	31.0	9.6
OA2	35.3	12.0

fixed period, say 30 days, in 100% RH. Then, any surviving specimens are loaded to failure at a typical strength test rate. Specifically, these tests used the ROR fixture in a water-vapor saturated air. The AMLCD specimens were sandblast abraded, as discussed above, and then zero-stress aged in water for one hour to normalize timing effects between abrading and the static load initiation. The specimens were statically loaded to their safe allowable stress (Table IV) until failure or for 30 days. Unfortunately, through a miscalculation, the static stress applied on the 1737 glass was 20.3 MPa, almost 9% higher than the calculated value of 18.7 MPa. Specimens that survived the aging period were loaded to failure in about 30 seconds at a standard crosshead rate similar to the fastest dynamic fatigue rate.

The test results were analyzed using the Weibull distribution. Figure 1 shows the strength distribution of pre-aged glasses. It is clear that each of the pre-aged glasses exhibits a different response to the standard abrasion process. Some of the differences are only in the Weibull locator, some only in the Weibull modulus, and some are different in both statistics. High strength values do not necessarily imply that the strength at lower failure probability is also comparatively higher. A summary of these statistics is given in Tables 5A, 5B and 5C.

It should be noted that 1737 glass demonstrates the highest combination of S_0, m and n values, both under pre-aged and post-aged conditions, compared with other glasses. These characteristics, which represent excellent resistance to sandblast abrasion, are critical for ensuring long-term reliability of AMLCD substrates, notably at the ultralow failure probabilities, as discussed in the next section.

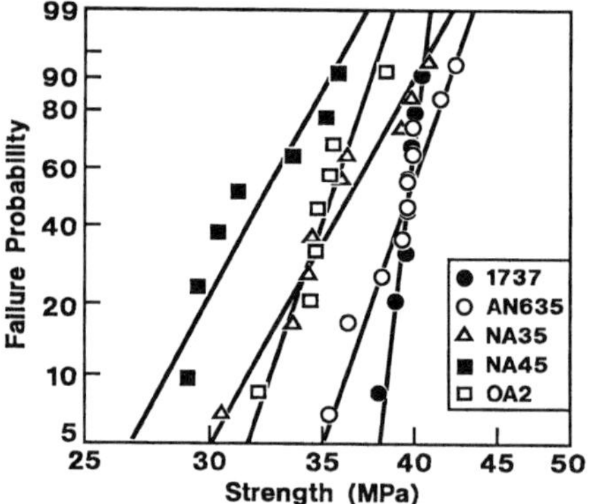

Figure 1. Weibull distribution of biaxial strength of AMLCD glasses prior to stress-aging (courtesy of Corning Incorporated)

Table V. Summary of Stress-Aging Data for Abraded AMLCD Glasses using Weibull Statistics

Table 5A. Strength Data Prior to Aging

Glass Code	S_0 (MPa)	m	No. of Specimens
1737	39.8	68	8
AN 635	40.2	21	10
NA 35	37.4	13	10
NA 45	32.0	14	9
OA 2	36.1	22	8

Table 5B. Stress-Aging Conditions

Glass Code	Static Stress (MPa)	No. of Specimens Survived / Total
1737	20.3	22 / 32
AN 635	12.8	24 / 24
NA 35	16.6	7 / 7
NA 45	9.6	23 / 23
OA 2	12.0	22 / 22

Table 5C. Strength Data After Aging

Glass Code	S_0 (MPa)	m	% Change in S_0
1737	40.2	15	1
AN 635	42.0	15	4
NA 35	41.5	10	11
NA 45	38.9	20	22
OA 2	40.6	16	12

Advances in Fusion and Processing of Glass II

Figure 2 compares the Weibull distributions of biaxial strength of NA35 glass before and after stress-aging at 16.6 MPa for 30 days. Let us note in Table 5C the beneficial aging effect of approximately 11% which shows up in the higher S_0 value albeit at the expense of lower m value. Indeed, the slope of stress-aged distribution in Figure 2 is lower than that of unaged specimens which is undesirable from long-term durability point of view.

Finally, Figure 3 compares the Weibull distributions of biaxial strength of surviving 1737 and NA35 glasses* after stress-aging at 20.3 and 16.6 MPa respectively. The former glass did not enjoy significant aging benefit (see Table V) due, primarily, to 9% higher aging stress than the safe allowable value of 18.7 MPa. And yet, its Weibull slope is significantly higher than that of NA35 glass which will impact its long-term durability positively. We can assess the effect of higher aging stress on expected lifetime of 1737 glass by resorting to Eqn. (3). Substituting $S_1 = 20.3$ MPa, $S_2 = 18.7$ MPa and n = 28.8 we obtain

$$\frac{t_2}{t_1} = \left(\frac{20.3}{18.7}\right)^{28.8} = 10.6 \tag{8}$$

Thus, 1737 glass specimens would have enjoyed an order of magnitude longer lifetime during the stress-aging test if they had been stress-aged at 18.7 MPa instead of 20.3 MPa. This could explain the premature failure of 10 of the 32 specimens during the 30-day stress aging test.

ANALYSIS OF SAFE DESIGN STRESS AND ITS IMPACT ON SUBSTRATE THICKNESS

We will use the fatigue and aging data for estimating the safe design stress corresponding to an acceptable failure probability. To this end, we recall the high reliability requirement for AMLCD substrates over their functional life of, say, 40 years. Thus, we must allow for long life and ultralow failure probability in arriving at the safe design stress. The life factor is readily

The remainder of this paper focuses on Code 1737 and NA35 glasses due to their superior stress corrosion coefficient.

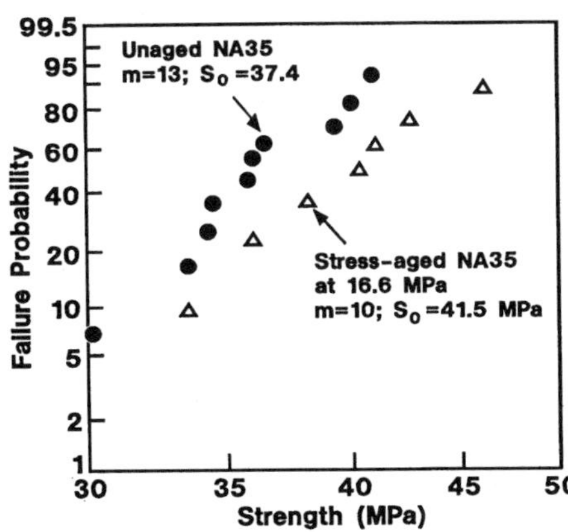

Figure 2. Weibull distribution of biaxial strength of NA35 glass before vs. after aging at 16.6 MPa (courtesy of Corning Incorporated)

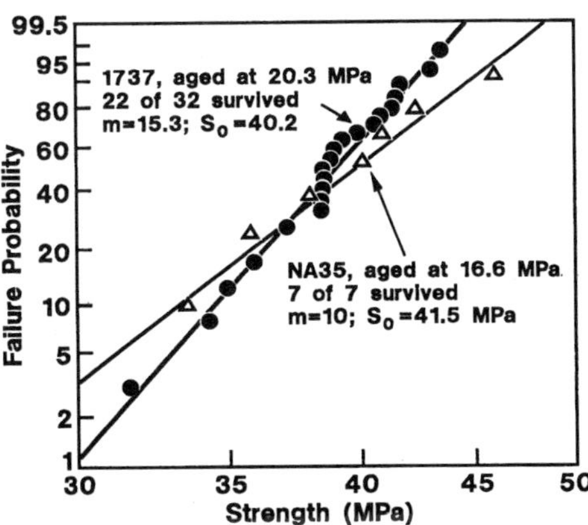

Figure 3. Weibull distribution of biaxial strength of 1737 vs. NA35 glasses after aging (courtesy of Corning Incorporated)

derived from Eqn. (3), namely

$$f_\ell = S_\ell/S_o = (t_o/t_\ell)^{1/n} \tag{9}$$

where

t_o = equivalent static time during measurement of strength $S_o = \dfrac{S_o}{R(n+1)}$,

t_ℓ = 40 yrs. = 1.26 x 10^9 sec. , and

S_ℓ = static stress to be applied for time t_ℓ

With R = 0.3 MPa/sec. and S_o values shown in Table V, we obtain

f_ℓ = 0.51 for 1737 glass (n = 28.8)
    = 0.50 for NA35 glass (n = 28.0)

Next, we use Eqn. (7) for estimating the probability factor, namely

$$f_p = S_p/S_\ell = \left[\ln\left(\frac{1}{1-P}\right) \right]^{1/m} \tag{10}$$

where P may be taken as low as 1 x 10^{-6} or 1 x 10^{-9} for ultralow failure probability. The final expression for safe design stress for AMLCD substrates is simply

$$S_d = f_\ell f_p S_o = f_p S_\ell \tag{11}$$

Table VI summarizes S_d values for 1737 vs. NA35 glasses in unaged vs. aged condition. It is clear from this table that 1737 glass can be stressed to considerably higher values than NA35 glass, without compromising long-term reliability, due to the former's superior fatigue and aging properties. Since the glass substrates are subjected to bending stress, their thickness is inversely proportional to the square root of safe design stress. Denoting the substrate thickness by d_{1737} for 1737 glass and d_{35} for NA35 glass we obtain

$$\frac{d_{1737}}{d_{35}} = \left(\frac{S_{d35}}{S_{d1737}} \right)^{1/2} \tag{12}$$

Substituting S_d values from Table VI, we conclude that 1737 glass substrates can be 20 to 50% thinner than NA35 substrates which for certain applications like laptop computers is highly desirable.

Table VI. Safe Design Stress for 1737 vs. NA35 Glasses for Two Different Failure Probability Levels (MPa)

AMLCD Glass (sandblast abraded surface)	Safe Design Stress S_d	
	$P = 1 \times 10^{-6}$	$P = 1 \times 10^{-9}$
1737 glass (unaged) S_o = 39.8, m = 68	16.6	15.0
1737 glass (stress-aged at 20.3 MPa) S_o = 40.2, m = 15	8.2	5.0
NA35 glass (unaged) S_o = 37.4, m = 13	6.5	3.8
NA35 glass (stress-aged at 16.6 MPa) S_o = 41.5, m = 10	5.2	2.6

SUMMARY

The fatigue and stress-aging behavior of five different AMLCD glasses, with sandblast surface abrasion, has been studied. These data help understand how each glass responds to identical abrasion treatment. Code 1737 glass, for example, with similar fatigue coefficient as NA35 glass deflects the abrasion flaws unlike NA35 glass which permits deeper penetration of the same flaws. Other AMLCD glasses have lower fatigue coefficient similar to that of soda-lime glass. Stress aging tests at the respective safe allowable stress level show that glasses with low fatigue coefficient exhibit the highest strength increase due to crack tip blunting which is consistent with their poor chemical durability. Stress aging also introduces high variability in strength data compared with unaged specimens. With low Weibull modulus the long-term design strength, notably at ultralow failure probability level, is affected adversely. This, in turn, reduces the safe design stress value for a specified lifetime. Lower design stress calls for thicker, hence heavier, glass substrate to ensure long-term reliability which from product's weight point of view can be penalizing.

Of the five AMLCD glasses examined, Code 1737 glass offers the best combination of good abrasion resistance, high stress corrosion coefficient, and well-controlled strength distribution following stress-aging tests. Such a combination leads to high design strength at ultralow failure probabilities and lightweight substrate design with no compromise in long-term mechanical reliability. The fatigue and aging data show that 1737 glass substrates can be made 20 to 50% thinner than the next best candidate, namely NA35. Indeed, 1737 glass appears to be the ideal candidate for AMLCD substrate in view of its optimum mechanical, thermal and optical properties.

ACKNOWLEDGEMENT

The authors express their gratitude to Virginia Doud, Nancy Foster, Tim Roe, and Mike Hobczuk of the Science and Technology Division for their valuable assistance in the completion of this paper.

REFERENCES

[1]Lapp, J.C.; Bocko, P.L. and Nelson, J.W.; "Advanced Glass Substrates for Flat Panel Displays;" Proc. SPIE; Vol. 2174 (1994).

[2]Gulati, S.T. and Helfinstine, J.D.; "Mechanical Reliability of AMLCD Glass Substrates;" SID Digest; Vol. 27 (1996).

[3]Gulati, S.T. and Helfinstine, J.D.; "Long-Term Durability of Flat Panel Displays for Automotive Applications;" Vehicle Displays '96, SID Detroit Chapter (1996).

[4]Weiderhorn, S.M.; "Subcritical Crack Growth in Ceramics;" Fracture Mechanics. of Ceramics, edited by R.C. Bradt, D. P. H. Hasselman, and F.F. Lange, Plenum Press, New York; Vol. 2, pp 6134-646 (1974).

[5]Ritter, J.E. and Sherburne, C.L.; "Dynamic and Static Fatigue of Silicate Glasses;" J. Am. Ceram. Soc.; Vol. 54 (1971).

[6]Evans, A.G.; "Slow Crack Growth in Brittle Materials under Dynamic Loading Conditions;" Int'l J. Frac.; Vol. 10 (1974).

[7]Gulati, S.T. and Helfinstine, J.D.; "Fatigue Resistance and Design Strength of Advanced AMLCD Glass Substrates;" Display Manufacturing Technology Conference Digest of Technical Papers, pp. 29-30, (1997).

[8]Glaescman, G.S. & Gulati, S.T.; "Design Methodology for Mechanical Reliability of Optical Fiber;" Opt. Eng. Journal (1991).

[9]Gulati, S.T.; "Crack kinetics during static and dynamic loading," J. Non-Cryst. Solids; Vol. 38-39, pp. 475-480 (1980).

[10]Gulati, S.T.; "Slow Crack Growth in Silicate Glasses;" 10th U.S. National Cong. App. Mech.; Univ. Texas, Austin (1986).

[11]Weibull, W.; "A Statistical Distribution Function of Wide Applicability;" J. App. Mech.; Vol. 18 (1951).

[12]Gulati, S.T.; "Dynamic and Static Fatigue of Silicate Glasses under Biaxial Loading: Application to Space Windows, CRTs and Telescope Mirrors;" 5th Int'l Otto Schott; Jena, Germany (1994).

[13]Ritter, J.E.; Jakus, K.; Batakis, A. and Bandyopadhyay, N.; "Appraisal of Biaxial Strength Testing;" J. Noncryst. Solids; Vol. 38 & 39 (1980).

[14]ASTM Standard C-158; Am. Soc. Testing Materials; Philadelphia (1995).

Structure/Property Relations

GLASSES WITH TRANSITIONAL STRUCTURES

William C. LaCourse and Alastair N. Cormack
NYS College of Ceramics
Alfred University
Alfred, NY 14802

ABSTRACT

In many glass systems a range of compositions exists over which the structure transforms from one in which properties are controlled by continuous alkali-rich channels to one in which the alkali-rich regions become discontinuous, (clusters) or randomly disperse (random network). In the latter case the properties are controlled by the strongly bonded silicate network. Structures between the two limiting structures are termed "transitional". The present paper describes alkali alumino- and borosilicate glasses with transitional structures. In these systems the properties vary rapidly with composition as the structure converts, but properties for glasses with either of the two limiting structures are not as sensitive to composition. Commercial glasses with transitional structures include soda-lime silicates (float and container).

INTRODUCTION

Experimental studies and molecular dynamic simulations of binary alkali silicate glasses tend to agree that the alkali ions do not enter the SiO_2 randomly.[1] At low concentrations isolated, alkali-rich clusters form, and are surrounded by continuous, nearly pure SiO_2 regions. As the alkali content increases the clusters begin to connect, eventually forming thin, continuous alkali-rich channels. In the high alkali limit the SiO_2 portion of the structure can become discontinuous. The dimensions (width) of the alkali-rich regions in both the low- and high-alkali compositional ranges are on the order of 1-2 nm and do not satisfy the definition of phase separation. On the other hand, the compositional range over which the "percolation" of the continuous alkali-rich region occurs is within the immiscibility region and some glasses do exhibit large scale phase separation.

High-alkali structures are similar to those proposed by Bockris and co-workers[2,3] for alkali silicate melts. Independent anionic silicate cages, (silica "islets") ranging in size from $Si_6O_{15}^{6-}$ at 33 % alkali, to $Si_{32}O_{68}^{8-}$ near 11-12 % alkali are assumed to exist. Below about 10 % the melt structure is assumed to convert to "random network".

The properties of alkali silicate glasses are generally interpreted in terms of the of the non-bridging oxygen (NBO) concentration. However, just as important is the distribution of NBO which determines the long range network connectivity. Glasses with similar alkali contents, having different "network" structures, can have significantly different properties. Phase separation provides the classic example. The structures defined above, while not phase separated on normal scales, will strongly influence properties. At low alkali contents, with a continuous silica rich network, and isolated alkali-rich regions, many "dynamic" properties will be controlled by the silica network. On the other hand, if continuous, alkali-rich regions surround small, anionic silica islets, properties such as diffusion, viscosity, modulus etc., will be controlled by the alkali-rich regions.

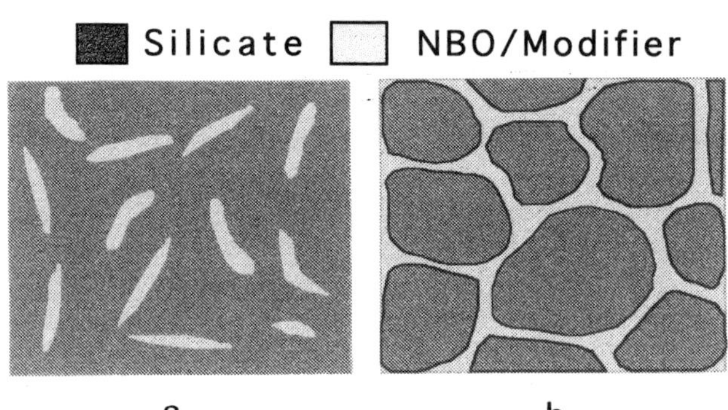

Figure 1. Alkali silicate structures. a. Cluster model for alkali contents below about 10 % alkali. b. Continuous alkali-rich regions.

Examples of this were reviewed by Mackenzie. The thermal expansion coefficient of molten binary alkali silicates is shown in Figure 2. In the compositional range where alkali clusters are isolated (below about 12 % alkali) expansion is controlled by the silica. It remains low and is independent of the type of alkali. At about 10 -12 %, the structure was postulated to undergo a sudden change to the anionic cage structure in which the anionic silicate cages are connected through the $Si-O^-$- R^+- ^-O-Si bonds. It is these bonds which control many of the properties in the high alkali regions. High temperature viscous flow behavior also changes abruptly in the same compositional range, with the pre-exponential constant becoming nearly constant.

While a "sudden" change in structure is postulated in the liquid state, it is not clear that the structure of the solid glass changes so abruptly, and much of the behavior in Figure 2 could be accounted for by a more continuous "transition" from one limiting structure to the other. Differences in expansion amongst the 3 alkali types might be due, in part, to a different rate of structural change. The immiscibiliy boundaries are different for the alkali so the tendency to form differing structures is evident. Effects of phase separation make it difficult to study binary alkali silicates in the 10 - 25 % region, but fortunately phase separation is absent in other compositional systems, where similar changes in network structure occur. In at least some systems the conversion of the network structure apparently occurs over a range of compositions. Many commercial glasses are likely in this structural region since glasses with a fully developed high silica network are highly viscous, while those with continuous alkali or modifier rich regions are not sufficiently

274   Advances in Fusion and Processing of Glass II

durable. The result is a "compromise" between properties and melting ease, leading to glasses with transitional structures. The present paper is concerned with the structures and property variations of glasses with these "transitional" structures.

TRANSITIONAL STRUCTURES

We define "transitional structures" as those in which the overall network is in transition between the two limiting structures. The binary alkali silicate glasses form one such system, but others exist. For present purposes it is convenient to use the 25 Na_2O- 75 SiO_2 composition as the "parent" glass. The structure and properties of this glass has been thoroughly investigated using various analytical techniques as well as molecular dynamic simulations. As noted, it possess a structure in which the Na^+ - NBO rich regions are continuous and provide channels for alkali migration making the glass non-durable, and highly conductive.

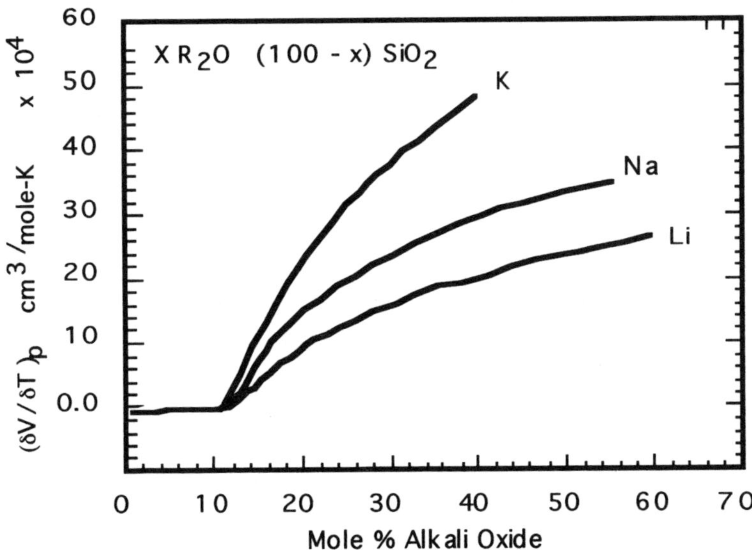

Figure 2. Volume expansion coefficients for molten binary alkali silicate glasses.

Several compositional modifications of this glass could cause a transitional structure.

a. A decrease in the Na_2O (NBO) content.

b. A decrease in the NBO content while holding the 25 % level of Na_2O by addition of B_2O_3, Al_2O_3 or Ga_2O_3.

c. Substitution of alkaline earth compounds for alkali oxides. (Soda-lime silicates.)

d. Combinations of these substitutions; e.g. addition of Al_2O_3 to a glass containing both CaO and Na_2O. Similarly, mixed alkali glasses may respond differently to additions of alkaline earth or Al_2O_3.

The structure resulting from the transition need not be the same as in the case of the alkali silicates. While the "cluster" model is fairly well documented for simple alkali silicates, the alkali (NBO) distribution in soda-lime silicates is not well known. The same is true for alkali aluminosilicates which contain both Si-O$^-$ and (AlO$_4$)$^-$ anionic sites for alkali bonding.

<u>Aluminosilicates:</u> In alumino- and borosilicates with a constant alkali content the concentration of NBO decreases with increasing Al or B. Since the NBO content is critical, it is of interest to know over what compositional range (Al and NBO contents) transitional structures exist. Evidence for transitional structures in 25 Na$_2$O x Al$_2$O$_3$ (75-x) SiO$_2$ glasses is found in Figure 3. As Al$_2$O$_3$ is substituted for SiO$_2$ the modulus first increases slowly between 0 and 5 %, then increases rapidly between 5 and 10 % and again slows between 10 and 25 %. The fracture toughness (K$_{IC}$) increases slightly at first then decreases sharply between 5 and 15 %, then remains low. The fracture surface energy (γ), which is related to E and K$_{IC}$ through

$$K_{IC} \sqrt{2E\gamma}$$

shows a behavior identical behavior to that of K$_{IC}$. Vickers's hardness varies in a manner similar to that for E. It is of interest that, the activation energy for diffusion of Na$^+$ also exhibits a maximum at about 5 % Al$_2$O$_3$.[4]

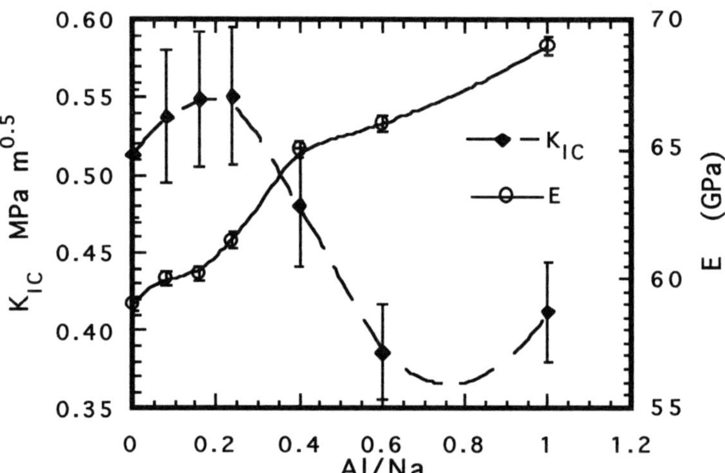

Figure 3. Fracture toughness and Young's modulus of alkali aluminosilicate glasses.

Recent molecular dynamics computer simulations of these glasses[5], as well as the compositional dependence of the properties suggest that the initial Al^{+3} enter the silica rich regions in 4-fold coordination with oxygen, creating AlO$_4$ tetrahedra with 4 bridging oxygen. (Figure 4) The single negative charge on the tetrahedra, are

   Advances in Fusion and Processing of Glass II

balanced by moving Na^+ from the continuous alkali-rich network to the interior of the Al-silicate cages.   In the range from 0 to 5 % Al_2O_3 the alkali-rich region remains continuous, but the concentration of alkali in these regions decreases, and the size of the Si-O-Al islets increases. Isolated Na^+ bonded to AlO_4 tetrahedra do not conduct since the distance between equivalent sites is too large.   The Na^+ remaining in the shrinking alkali-rich channels are less mobile due to their lower concentration, so diffusion decreases and the activation energy increases to that representative of a glass with lower alkali content. Properties which depend on the overall connectivity of the structure (E, γ, K_{IC},H) are less strongly influenced by the initial Al since, high-silica cages remain separated by the alkali-rich regions.

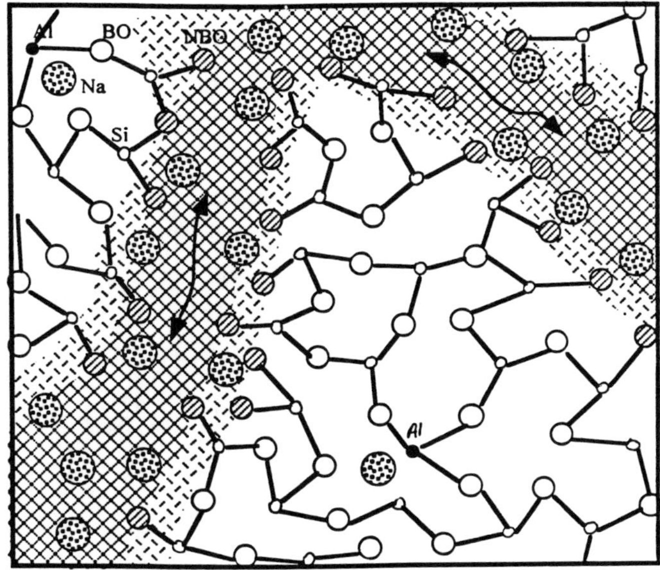

Figure 4. MD simulation of sodium silicate glass structure with small additions of Al_2O_3. Al preferentially enters the silica rich regions.

The more rapid change in properties between 5 and 10% Al_2O_3 (Al/Na = 0.2 - 0.4) can be accounted for by assuming that the continuous alkali-rich regions become increasingly disrupted, with formation of a continuous aluminosilicate network; i.e. glasses in this compositional range possess transitional structures. Molecular dynamic simulations by Cao and Cormack[5] suggest that the structure converts progressively from the "modified random network" to a random network structure. Thus, while the concentration of non-bridging oxygen continues to decrease, a concurrent <u>total rearrangement </u>of the structure is occurring.

As the connectivity of the structure increases H_v and E increase more rapidly but K_{IC} and γ decrease. Evidently the structure is becoming more "brittle". Irreversible local movement (rotation and perhaps limited translation) of the silica cages is possible at low Al contents since the alkali-rich regions permit continuous bonding during deformation. Not unexpectedly, once the conversion to a three dimensional random network is complete, apparently around Al/Na = 0.6 for the 25 % Na_2O glasses, the property changes again slow.

Results on Na_2O-CaO-Al_2O_3-SiO_2 and mixed alkali (Na and Li) alumino-silicates suggest that the transitional structures occur at lower Al_2O_3 contents in both cases.[6] B_2O_3 additions to the parent glass result in a somewhat different effect and illustrate the wide range of possible transitional structures. As shown in Figure 5, the initial addition of B_2O_3 causes a decrease in K_{IC} (and γ) but the modulus increases more rapidly than in aluminosilicates. Molar volume calculations suggest that the initial B enters the structure <u>within</u> the alkali-rich regions, increasing the connectivity of the structure, decreasing the mobility of these regions relative to one another and thus increasing the brittleness. Additionally, at high B_2O_3 contents, the modulus increases above that of pure SiO_2, indicating a more compact network structure with a higher density of strong bonds. A strongly decreasing molar volume with increasing boron supports this, and suggests that the ring sizes in the high boron region are reduced relative to aluminosilicates, and pure SiO_2.

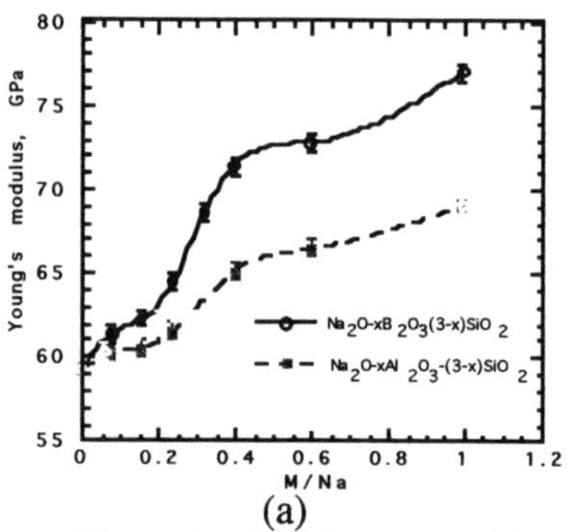
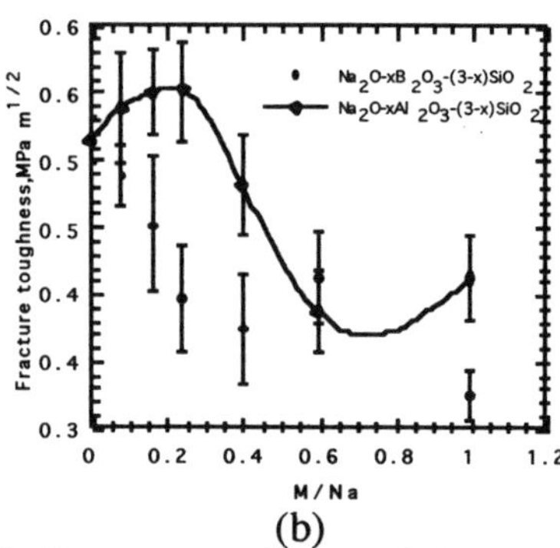

Figure 5. (a) Elastic modulus and (b) Fracture toughness of alumino- and borosilicates glasses.

<u>Soda-lime silicates:</u> Chemical durability is improved while maintaining low melting temperatures when CaO and other alkaline earth cations are substituted for Na_2O. In typical commercial float and container glasses approximately 10 mole % of the Na_2O is replaced by MgO and/or MgO. Property variations are not as dramatic as the substitution progresses but electrical resistivity data suggests possible changes in the structure near 8 % CaO - 18 % Na_2O, and near the same range for a 30 % total modifier glass. (Figure 6).

SUMMARY

Identifying the compositional range over which transitional structures exist can be important since the corresponding glasses can be quite sensitive to compositional variations.   An area of immediate interest is in the ability of this approach to account for effects of Al_2O_3 on commercial soda-lime silicate glasses in which it is well known that small variations in Al_2O_3 (1-3 %) can have significant effects on both durability and surface mechanical properties. Additional interest lies in the possible differences in forming behavior of glasses with transitional structures. Such structures are inherently "unstable" they may be more sensitive to forming rates, redox changes etc..

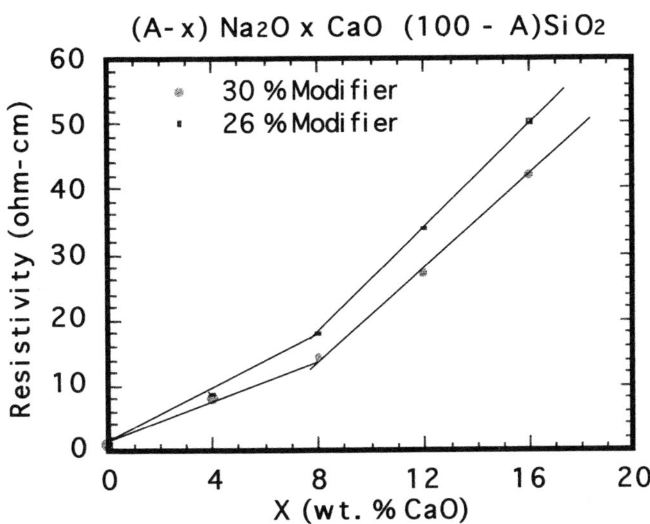

Figure 6. Electrical resistivity of soda-lime silicates at 400 C.

REFERENCES

1.   G. N. Greaves, "EXAFS and the Structure of Glass" J. Non-Cryst.   Solids, **71** 203-217 (1985).
2. J. O'M Bockris, J.D. Mackenzie and J. A. Kirchner, "Viscous flow in silica and binary liquid silicates" Trans. Faraday Soc.**51** , 1734 (1955).
3   J.D. Mackenzie, "Structure of some inorganic glasses from high temperature studies", in *Modern Aspects of the Vitreous State,* ed. J. D. Mackenzie, Butterworths, London (1960) pp 188-218.
4.   J. O. Isard, "Electrical Conduction in the Aluminosilicate Glasses," J. Soc.Glass Tech., **43,** 113T-123T (1959).
5.   Y. Cao and A.N. Cormack,   A Structural Model for Interpretation of an Anomaly in Alkali Aluminosilicate Glasses at Al/Alkali = 0.2 - 0.4 , pp 137-152 in: Diffusion in Amorphous Materials, eds. H. Jain and D. Gupta, TMS, Warrendale, PA (1994)
6. W. C. LaCourse, V. Malot and W. Mason,   "Structural Implications of Elastic Properties, Fracture Toughness and Fracture Surface Energy of R_2O-Al_2O_3-SiO_2 Glasses", Ceramic Trans. **64,** 504-515 (1996)

MD SIMULATED STRUCTURES OF SODA-LIME-SILICA GLASS AND ITS SURFACE

Xianglong Yuan and Alastair N. Cormack
New York State College of Ceramics at Alfred University, Alfred, NY 14802

ABSTRACT

The structures of a series of xNa$_2$O $(1-x)$CaO 3SiO$_2$ glasses and their corresponding glass surfaces have been simulated using molecular dynamics. Mid-range structures were analyzed in terms of Q$_n$ distribution, sizes of largest silicon or modifier rich regions and unique ring size distribution functions. Larger size rings are observed to be associated with modifier rich regions. The surface structures were also investigated and compared to their bulk structures. High content of 3-coordinated silicon (Si$^{[3]}$) was found in the $x = 0.8$ surface and its effect on the distribution of modifiers and non-bridging oxygen in the surface region is discussed.

INTRODUCTION

Silicate glass surfaces, because of their insulative and non-crystalline properties, have not been nearly as well studied as metal and semiconductor surfaces. The experimental difficulties associated with these glassy materials make the analysis of glass surfaces less than ideal and data interpretation much more complicated. Although techniques such as XPS, AES, ESR, ISS, and SIMS have been used to study glass surfaces with varying degrees of success, they do not allow for direct measurement or observation of a particular event or structure. These must be inferred from data. Molecular dynamics (MD) simulation, on the other hand, provides an atomic level view. It allows time and particle averages of specific properties at different temperatures and pressures. These averages can be compared to similar experimental data.

Computer simulation of silica and silicate glasses have been performed using predominantly pair potentials[1] over the past twenty years. More recently, multibody potentials have been developed to more accurately simulate the partial covalency of the silica system,[2-4] and their application to oxide glass surfaces has occurred only within the past several years.[5-8]. In this paper, MD simulations of soda-lime-silica (SLS) glass and its surface are reported and the results are analyzed in terms of ring size distribution (RSD), Q$_n$ species, pair distribution function (PDF), bond angle distribution (BAD), coordination number (CN) and surface depth profile.

SIMULATION PROCEDURE

Simulated silica glass surfaces were created in vacuum from previous studied bulk glasses[9] with 1539-1620 atoms contained in boxes which are around 30 Å at each side. Please refer to Ref. 9 for the potential models and the simulation procedures used to form the SLS bulk glasses. The bulk glasses were relaxed under constant pressure at 879 K and 300 K before they were used to create surfaces. The simulation boxes were then elongated in z direction at 300 K in order to disconnect the periodic boundary condition in that direction and to create free surfaces. One third of atoms in the bottom were frozen as the glass substrate. Using constant volume simulation (NVE ensemble), the system was first heated to 3686 K, then annealed to 1800 K, 879 K and finally cooled back to 300 K. The total timesteps for each temperature were 10,000 with Δt of 1.0 fs for

Table I. Size of largest silicon and modifier rich region of
xNa$_2$O (1-x)CaO 3SiO$_2$ glasses

x	0.4	0.6	0.8	Na$_2$O 3SiO$_2$	Na$_2$O 2SiO$_2$
R_{Si}	5.8	5.1	5.3	5.2	4.3
R_M	7.6	6.3	8.0	5.9	6.7

3686, 1800 K and Δt of 2.0 fs for 879, 300 K. 101 configurations from the last 4000 timesteps were sampled at an interval of 40 to calculate the structural properties.

MID-RANGE STRUCTURE OF SODA-LIME-SILICA BULK GLASS
Ring Size Distribution

There is no significant change in unique RSD for different SLS glasses with the same bridging oxygen to non-bridging oxygen (BO/NBO) ratio, as shown in Figure 1. The criterion for Si-O bond is a cutoff of 2.5 Å (detailed in Ref. 9). The addition of modifiers decreases the number of mid-size rings (5-10) but increases the number of large-size rings (>12). Compared to silica, SLS glasses have several times the number of the large-size rings. By analysis of ring structures, these large size rings were found to be associated with modifier rich regions. Although our algorithm can theoretically search all the rings, at present we are limited to searching for rings up to 14-member ring size because of the cpu time involved. Obtaining the distribution of much larger rings (>30, say) should help to describe more completely modifier rich regions, given adequate cpu time.

Size of Silicon or Modifier Rich Regions

We can describe the size of the largest silicon rich region using the largest distance between each silicon and its nearest modifier, R_{Si}. Similarly, using the largest distance between each modifier and its nearest Q$_4$, R_M, we can describe the size of largest modifier rich region. The results are shown in Table I. R_{Si} suggests that silicon rich regions are relatively small, the largest one being about 11 Å in diameter with the largest modifier rich region being about 12-16 Å in diameter. The addition of calcium increases the size of modifier rich regions, but has less effect on silicon rich regions. The PDFs for Si-Si and Q$_4$-Q$_4$ pairs also suggest that silicon rich regions are around 5.5 Å in radius (Figure 2).

SURFACE STRUCTURE OF SODA-LIME-SILICA GLASS
Coordination of Surface Silicon and Oxygen

Silicon and oxygen ions in the surface regions have different coordination states than those in the bulk glass. As shown in Table II and Figure 3, the BO/NBO ratios in the surface regions are usually lower than those in the bulk, which means there are usually more NBO ions in the surface

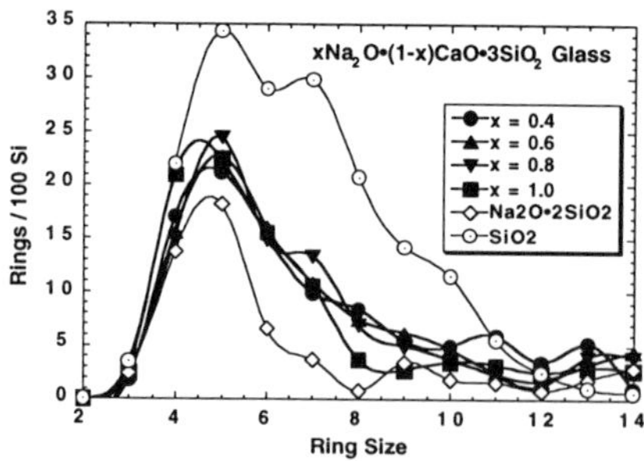

Figure 1. Ring size distribution

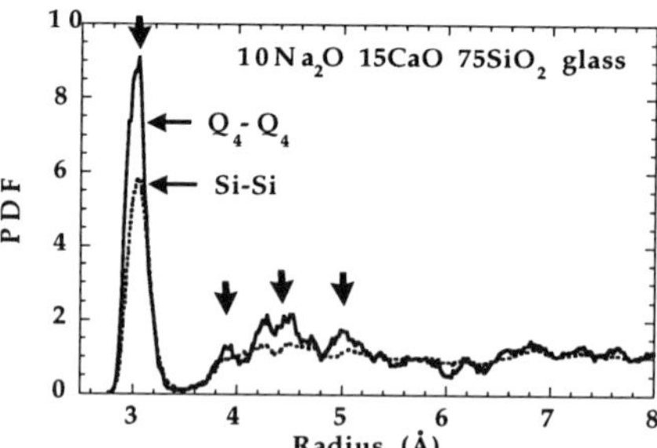

Figure 2. PDF for Si-Si and Q$_4$-Q$_4$ pairs

Advances in Fusion and Processing of Glass II

Table II Comparison between bulk and surface of xNa$_2$O $(1-x)$CaO 3SiO$_2$ glass system

	BO/NBO	Si$^{[3]}$ (%)	Q$_1$	Q$_2$	Q$_3$	Q$_4$
x=0.4						
bulk	2.45	1.0	0.5	10.4	46.4	42.7
surface	1.59	3.4	1.7	24.1	46.4	27.8
x=0.6						
bulk	2.45	1.0	0.2	10.9	46.2	42.7
surface	1.34	2.5	0.0	29.8	41.5	28.7
x = 0.8						
bulk	2.49	0.2	0.5	10.4	44.9	44.2
surface	**2.66**	**15.4**	3.4	12.5	39.2	**44.9**
x=1.0						
bulk	2.50	0.0	0.2	10.9	44.2	44.7
surface	1.58	3.8	1.9	22.6	52.8	22.7
x=1.0 (bulk, exptl.)						
Buckermann, 1992 (NMR)[10]			0	7	45	48
Maekawa, 1991 (NMR)[11]			0	1	61	38

regions. (The surface region here is defined as the region which is above 26 Å along the z-axis, see Figure 3). However, for the x = 0.8 case, the ratio is inverted. The BO/NBO ratio increases from 2.49 to 2.66, as shown in Table II. Correspondingly, a smaller amount of modifier cations in this surface region are observed (Figure 3.c), indicating the association of NBO and modifiers obtains on the surface as well as in the bulk.

Usually, sodium is likely to stay on top of the surface, creating an NBO ion and lowering the surface energy. Our simulation results suggests that the modifier and NBO surface rich regions are

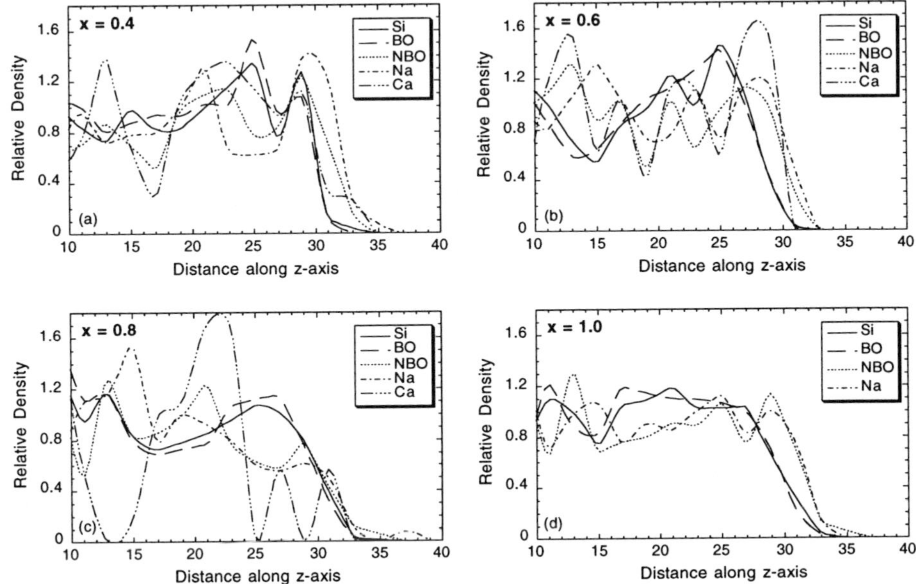

Figure 3. Depth profiles of xNa$_2$O $(1-x)$CaO 3SiO$_2$ glass surfaces

only about several angstroms as shown in Figure 3. However, on the x = 0.8 surface, sodium seems to be stable within the surface and need not stay on top of the surface to lower the surface energy. This observation might be explained by the noticeably smaller amount of oxygen (BO+NBO) ions in this surface region (Figure 3.c) and the abnormal high Si[3] content of 15.4% (Table II). Because the geometry of a [SiO$_3$] is totally different to a tetrahedral [SiO$_4$], a planar [SiO$_3$] can have NBO + modifier ions existing at the same surface level, while a tetrahedral [SiO$_4$] always tends to turn the NBO ion with its modifier cation to the outside surface. Furthermore, the extra charge on Si[3] will also repel the modifiers. Thus, we may say that the suppression of sodium 'leakage' is due to the high Si[3] content, though it is still not clear what causes the abnormal high Si[3] content. Further work is needed to clarify this situation.

Q_n Species Distribution

The Q_n species for both bulk and surface glasses have been calculated and the results are shown in Table II. Experimental NMR results are also listed for comparison. Considering that the BO/NBO ratios remain the same for our SLS system, it is not surprising to find that, for the bulk glasses, there is no significant change in the Q_n distribution as the Na$_2$O/CaO ratio changes. The Q_n species distribution in the surface region, however, is different from that within the bulk. As the BO/NBO ratio decreases from bulk to surface, the number of Q_4 decreases and other Q_n concentrations increase correspondingly, while the trend turns around in the x = 0.8 case. This is expected since the BO/NBO ratio is higher in this surface region.

Bond Angle Distribution

The geometry of silicon tetrahedra in the surface region remains the same as in the bulk since Figure 4 shows that the O-Si-O BADs in the surface regions are identical to that in the bulk except for the x = 0.8 case. Shoulders around 120° result from the content of planar Si[3] in the

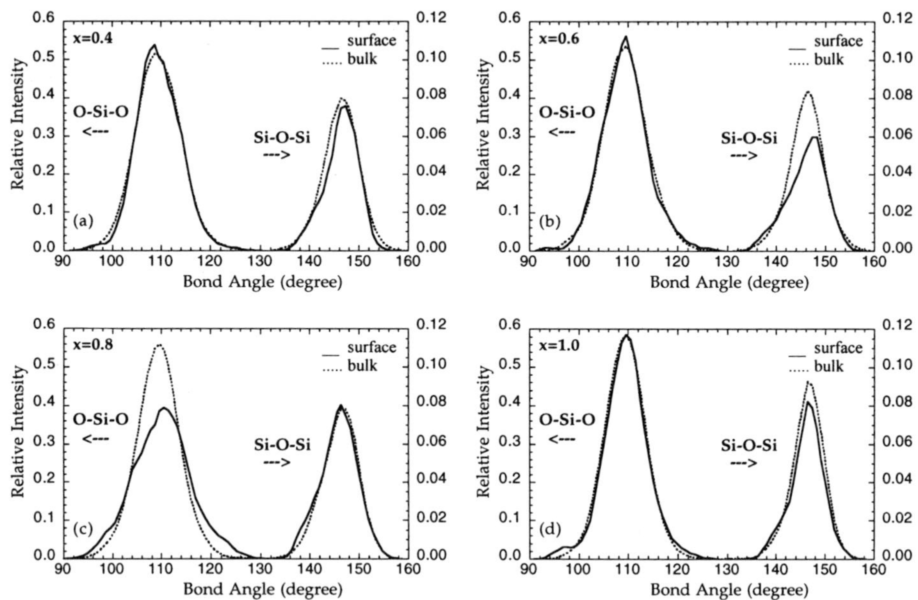

Figure 4. Bond angle distribution of xNa$_2$O (1-x)CaO 3SiO$_2$ glass surfaces

Advances in Fusion and Processing of Glass II

surface regions, whilst the small shoulders around 95° are associated with the increase of 3-member rings (Figure 5). The shoulder of O-Si-O BAD at 120° for the x = 0.8 surface is thus associated with its large portion of $Si^{[3]}$ content.

It should be mentioned that the intensity of BAD is calculated by dividing the total number of bond angles of a certain type with the number of center ions. Thus the area under a peak is the average number of bond angles a center ion has and is proportional to the average CN of a center ion. Thus, for the x = 0.6 surface, the intensity decrease of Si-O-Si BAD is due to its low CN of oxygen, its BO/NBO ratio is only 1.34 (Table II), and for the x = 0.8 surface, the lower O-Si-O peak is attributed to the large portion of $Si^{[3]}$ which lowers the CN state of silicon.

Ring Size Distribution

Figure 5 shows that usually there are fewer large member rings in the surface regions than in the bulk, except for the x = 0.8 surface, which has more 8-member rings. This may be related to the large amount of $Si^{[3]}$ in its surface region. While the increase of 3-member rings is associated with O-Si-O BAD shoulders around 95°, the large amount of 4-member rings in the x = 0.8 surface is echoed by the O-Si-O BAD shoulder around 105°.

SUMMARY

Mid-range structure: No significant change in the unique RSD and Q_n distribution was observed for glasses with the same BO/NBO ratio. The addition of modifiers increases the number of larger size rings which appear to be associated with modifier rich regions. By using R_{Si} and R_M, the largest silicon rich region is about 11 Å in diameter, whilst the largest modifier rich region is about 12-16 Å in diameter.

Surface structure: Increase of the number of NBO ions, modifier cations and 3-member rings was observed in the surface region, whilst the geometry of silicon tetrahedra remains the same as in the bulk structure. The association of NBO and modifiers obtains on the surface as well as in

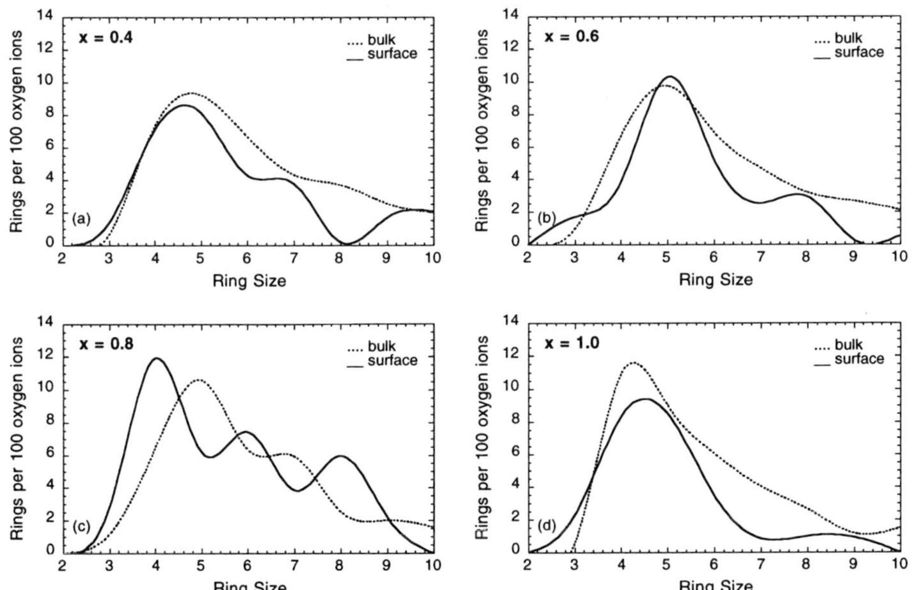

Figure 5. Ring size distribution of xNa$_2$O (1-x)CaO 3SiO$_2$ glass surfaces

the bulk. High Si$^{[3]}$ content found in the x = 0.8 surface region decreases the number of NBO and modifier ions and thus suppresses the sodium 'leakage'.

ACKNOWLEDGMENTS
We gratefully acknowledge the Center for Glass Research, New York State College of Ceramics at Alfred University, for financial support. Some of the calculation was performed at Cornell Theory Center.

REFERENCES
[1]T. F. Soules, "A molecular dynamic calculation of the structure of sodium silicate glasses", *J. Chem. Phys.*, **71**[11] 4570 (1979).

[2]R. G. Newell, B. P. Feuston and S. H. Garofalini, "The structure of sodium trisilicate glass via molecular dynamics employing three-body potentials", *J. Mater. Res.*, **4**[2] 434-9 (1989).

[3]C. Huang and A. N. Cormack, "The structure of sodium silicate glass", *J. Chem. Phys.*, **93**[11] 8180 (1990).

[4]A. N. Cormack and Y. Cao, "Molecular dynamics simulation of silicate glasses", *Molecular Engineering*, **6** 183-227 (1996).

[5]B. P. Feuston and S. H. Garofalini, "Topological and bonding defects in vitreous silica surfaces", *J. Chem. Phys.*, **91**[1] 564-70 (1989).

[6]S. H. Garofalini, "Molecular dynamics computer simulations of silica surfaces structure and adsorption of water molecules", *J. Non-Cryst. Sol.*, **120** 1-12 (1990).

[7]D. M. Zirl and S. H. Garofalini, "Structure of sodium aluminosilicate glass surfaces", *J. Am. Ceram. Soc.*, **75**[9] 2353-62 (1992).

[8]D. C. Athanasopoulos and S. H. Garofalini, "Effect of adsorption on the surface structure of sodium alumino-silicate glasses: a molecular dynamics simulation", *Surf. Sci.*, **273** 129-38 (1992).

[9]X. Yuan and A. N. Cormack, "Local structure of MD simulated soda-lime-silica glass", MRS Symposium Proceedings, Ed by C. A. Angell, 1996, Boston, Materials Research Society.

[10]W.-A. Buckermann, W. Muller-Warmuth and G. H. Frischat, "A further ^{29}Si MAR study on binary alkali silicate glasses", *Glastech. Ber.*, **65** 18 (1992).

[11]H. Maekawa, T. Maekawa, K. Kawamura and T. Yokokawa, "The structural groups of alkali silicate glasses determined from ^{29}Si MAS-NMR", *J. Non-Cryst. Sol.*, **127** 53-64 (1991).

                                        Advances in Fusion and Processing of Glass II

DENSITY AND SURFACE TENSION OF GLASS MELTS AS A FUNCTION OF COMPOSITION AT 1400 °C

Ahmet Kucuk, Alexis G. Clare, Linda E. Jones
NYSCC at Alfred University
2 Pine St, Alfred, NY 14802

ABSTRACT
The density and surface tension of sodium, potassium and soda lime glass melts were measured at 1400 °C using the sessile drop method. The effects of volatilization on these properties were studied since volatilization changes the composition of the melts.

INTRODUCTION
The knowledge of the high temperature physical properties such as density, viscosity, surface tension is important for processing control in glass industry. Although the viscosity data for glass melts are well documented, density and, especially, surface tension data for the glass melts, are sporadically reported.

Density is a necessary input data for computer modeling of the flow in glass tanks. Surface tension is an important parameter in the control of fining, and foaming as well as refractory corrosion in glass making. Surface tension is also effective in the homogenization [1] of glass melts and interaction of glass melts with machinery in the glass making and forming.

Both density and surface tension are composition dependent. One of the problems in glass making is the volatilization of the high vapor pressure components. PbO and CdO, for example, show appreciable volatilization at temperatures as low as 900°C. In addition, alkali oxides such as Na_2O, K_2O start to volatilize at a low rate at 1050 °C and at a high rate at high temperatures(>1300°C). To have a glass product with the desired composition which meets service specifications, the glass industry often adjusts the glass batches to compensate for the volatilization losses. Although, this approach gives the desired composition overall, local heterogeneity due to volatilization may exist. These local alkali-deficient regions with high viscosity can cause problems in glass making and forming.

In this paper, the composition dependence of the density and surface tension of glass melts will be examined, and compositional changes resulting from volatilization will be correlated with the physical properties.

The density of melts can be measured using various methods; Archimedes, maximum bubble pressure, pycnometric, and sessile drop. All have advantages and disadvantages over each other as given in a review by Crawley[2]. The amount of glass used (0.1 to 1.0 gram) makes the sessile drop technique very practical while other techniques need 10 to 50 grams of glass. In addition, measurements with the sessile drop technique can be carried out with a minimum of corrections and assumptions while others need many assumptions to affect analysis. However, there are disadvantages to the sessile drop method such as volatilization from the melt, but this can be alleviated using different experimental procedures as will be discussed in this paper. In addition, the advantage of the small sample size may also be a disadvantage in terms of representing bulk properties.

The surface tension of melts can be measured using various techniques such as maximum bubble pressure, drop weight, dipping plate (or cylinder), fiber elongation, and sessile or pendant drop. Landcastle[3] reviewed all the available techniques with the disadvantages and advantages. The maximum bubble pressure method originally gained attention for metallurgical melts, however, it is limited by viscosity; the technique can give large errors when the viscosity is high such as encountered in glass melts[4]. With the developments in imaging and image processing techniques, the sessile drop method became a very reliable and practical measurement technique. When the wetting angle of melt on the substrate is low (less than 40°), the measurements using the sessile drop can give large errors, however, this limitation can be eliminated by either using an alternative non-wetting substrate such as graphite, BN, Rh-Pt alloys, Au-Pt alloys, or using the pendant drop technique which has the same calculation procedure, but a small change in the experimental procedure.

The sessile drop technique to measure density and surface tension of liquids was developed by Bashforth and Adams[5]. They prepared tables which allow the calculation of volume and surface tension by just measuring the dimension of the stationary liquid drop. The details of this calculation are given in the paper by Ellefson et al.[6]. However, this approach suffers from errors resulting from the dimension measurements, and errors which came from the mathematical iteration used by Bashforth and Adams[5] to solve the differential equation defining the force equilibrium. Also, the calculations are time-consuming.

The developments in computer and imaging technology provide a new approach to the problem. The new techniques include computerized curve fitting and image analysis rather than dimension measurements on the drop shape. The details of this new approach are given by Rotenberg et al.[7]. They successfully applied this approach to room temperature liquids with both low and high wetting angles. Jennings et al.[8] improved the computer algorithm to make it time-efficient. Askari et al.[9] and Weirauch et al.[4] successfully used this approach to determine the surface tension of alkali borosilicate and calcium aluminosilicate glass melts, respectively. In addition to these efforts in the surface tension calculation using image analysis, Clare et al.[10] recently applied image analysis to density calculations by the sessile drop method. In this paper, the approach of Rotenberg et al. [7] for surface tension calculations, and that of Clare et al. [10] for density calculations were used.

EXPERIMENTAL PROCEDURE

The measurements were carried out in the sessile drop experimental arrangement given in Figure 1. The sessile drop was formed at the measurement temperature using a Pt ring attached to a mullite rod. A piece of glass weighing 0.1 to 1 gram was squeezed to the Pt ring. The sample was kept in the cold zone during heating, and pushed into the hot zone at the measurement temperature. The glass flows through the ring onto the substrate at the measurement temperature because of its weight and fluidity. As soon as the drop was obtained, image recording started. For the first couple minutes, depending on the glass composition, bubble formation from substrate-drop interface was often observed. Since this initial period is necessary for drop stabilization, images recorded during this time period was not taken into consideration in the calculations.

The temperature was measured with a 6 Rh/Pt- 13 Rh/Pt thermocouple located in the vicinity of the drop. The final polishing of the graphite substrate was carried out with 1 micron diamond paste and it was cleaned with acetone in a ultrasonic cleaner before being used. During the meaurements, dry argon gas flowed through the tube furnace at a rate of 0.6 l/min. Just before the mullite rod was pushed in, the flow rate was increased to 2.0 l/min and then decreased back to 0.6 l/min to dilute adsorbed oxygen on the surface of the mullite rod.

After the measurement, the furnace was cooled to 1100 °C at a rate of 12 degree/min, 6 degree/min to 900 °C, 3 degree/min to 700 °C and 6 degree/min to 350 °C. The furnace was always kept at a minimum of 350 °C.

The movie recorded on S-VHS video tape was digitized using a Snappy® digitizer[*] and sorfware and converted into series of still pictures with 640X480 pixel resolution in a PC. All dimensions in the pictures were calibrated using grid graphic paper. These images then were processed to determine volume, surface area and the surface coordinates of the drop by using the macros written for Scion Image PC[**] software.

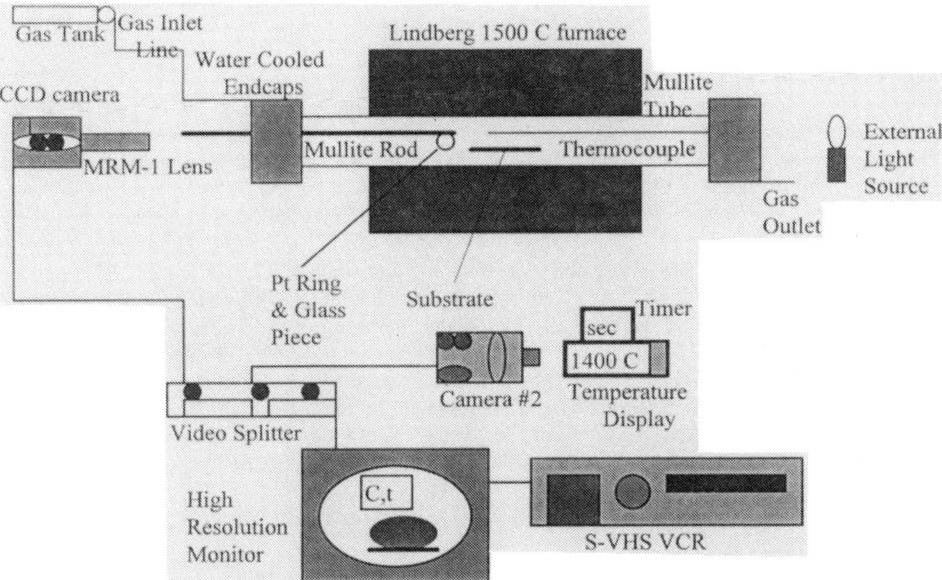

Figure 1. Experimental Setup.

For the volume calculation; the drop was divided into horizontal cylinders of one pixel height, and then the volume of the each cylinder was calculated using $\pi r^2.h$ where r radius and h height of the cylinder. The total volume was found by summing the volume of the individual cylinders. To determine the surface area of the drop, the surface area of the outside of the individual cylinders is summed.

For surface tension calculations; the surface coordinate data were input into a FORTRAN program written with a modified Rotenberg et al.[7] algorithm and surface tension values were extracted.

Glasses were made using reagent grade oxides and carbonates by holding at 1450-1500°C for 2 hours. Batches of 50 to 60 grams were melted in a Pt crucible. The nominal and analyzed composition of glasses are given in Table I.

---

[*] Play Incorporated, Rancho Cordova, CA.

[**] Scion Co., http://www.scioncop.com

Table I. Batch and analyzed compositions of the glasses used The analysis results from ICP emission spectroscopy are given in the parenthesis.

GLASS	mol% Na_2O	mol% K_2O	mol% CaO	mol% MgO	mol% SiO_2
sodium trisilicate (N25S)	25(26.1)				75
potassium trisilicate (K25S)		25(24.0)			75
K30S		30(29.6)			70
N20C10S	20(20.1)		10(11.3)		70
float	12		9	8	70

50 grams of commercial float glass (unknown source) were crushed and re-melted at 1450°C for four hours. The approximate composition of the float glass was $12Na_2O$. $8MgO$. $9CaO$. $70SiO_2$ (mol %) and minority constituents; K_2O, Al_2O_3, Fe_2O_3.

RESULTS AND DISCUSSION

To ensure the reliability of the technique, the sodium trisilicate and potassium trisilicate glasses were taken as the reference materials since many data are available for these systems in the literature. Some similar studies have taken metals or molten salts as their 'standard', however because of the oxidation problem, no metal was found that could be taken as a reference here. Even very inert metals such as gold and silver have an oxygen partial pressure dependent surface tension[18].

The density and surface tension of sodium and potassium trisilicates at 1400 °C are given in Table II and III, respectively. As given, the density of sodium and potassium trisilicates found in this study are in good agreement with other studies within the limit of the experimental error. Although other studies used ostensibly identical techniques to measure the density of the melt, there are discrepancies between them which may be due to using different density, expansion coefficient or surface tension corrections for platinum or platinum alloys. In the current study, the chief error sources are the magnification factor for the camera-lens system and any deviation from symmetry in the drop shape. When a graphite substrate is used, the density values obtained are less scattered.

The surface tension of sodium trisilicate is in good agreement with data given by Shartsis

Table II. Density and surface tension of the sodium trisilicate melts at 1400°C.

Reference	Density(g/cm^3) ± St.Dev	Surf.Tension (mN/m)±S.D.	Method and Conditions
current	2.209±0.009	273±5	Ar atmosp., graphite substrate, 4 runs, 2 operators
current	2.216±0.008	273±4	N_2 atmosp., graphite substrate, 2 runs, 1 operator
current	2.21±0.01		Ar atmosp., 5%Au-Pt alloy subst., 2 runs, 2 opers.
Bockris[11]	2.218±0.004		24.6mol% Na_2O, Archimedean, Pt bob and crucible
Shartsis[12,13]	2.180	273	Interpolated from 19.5 and 30.1 mol% Na_2O data, counterbalance method, Pt sphere
Sasek[14]	2.210		Archimedean method, Pt-Rh bob and crucible
Appen[15]		281	25.4mol% Na_2O, extrapolated from low temp. data, drop weight method
Akhtar et al.[16]		280-320	20mol% Na_2O, max. bubble pressure at 1350 °C, various gas atmospheres
Askari[9]		272	Extrapolated from low temperature data, sessile drop, Ar or vacuum
Lyon[17]		271	calculated from given coefficients

Advances in Fusion and Processing of Glass II

Table III. Density and surface tension of potassium trisilicate melts at 1400°C.

Reference	Density(g/cm^3) ± St.Dev	Surf. Tension (mN/m)±S.D.	Method and Conditions
Current	2.156±0.007	210±2	Ar atmosphere, graphite substrate, 3 runs, 1 operator
Bockris[11]	2.167±0.004		Archimedean method, Pt bob and crucible
Sasek [14]	2.166		Archimedean method, Pt-Rh bob and crucible
Shartsis et al. [12,13]	2.154	211	Interpolated from 23.8 and 26.9 mol% K_2O data, counterbalance method, Pt sphere
Appen et al.[15]		225± 1%	Interpolated from 16 and 36 mol% K_2O, drop weight

et al.[13] and Askari et al.[9], however smaller than the data given by Appen et al.[15] and Akhtar and Cable[16]. Although details of the study of Appen et al.[15] were not published, the maximum bubble pressure method used by Akhtar et al.[16] generally gives higher value than other methods, concluded from the literature data. The surface tension value for potassium trisilicate agrees with the value reported by Shartsis et al.[13], but is lower than Appen's value[15].

Effects of Volatilization

Potassium silicates

The composition dependence of the density and the surface tension of the potassium silicate melts at 1400°C is given in Figure 2. Both density and surface tension decrease with increasing K_2O. Shartsis et al.[12] and Bockris et al.[11] also reported that K_2O addition decreases the density of potassium silicate melts. Bockris et al.[11] explained this decrease with the formation of metasilicate ion rings with various number of silicon depending on the alkali oxide to silica ratio. Each alkali addition creates one non-bridging oxygen to compensate the extra charge from alkali ion. This creates $(Si_9O_{21})^{6-}$ discrete ions when alkali oxide to silica ratio is 0.25, and $(Si_6O_{15})^{6-}$ discrete ions when the ratio is 0.33. The bond strength decreases with decreasing number of silicon in each discrete ion, and therefore melts with higher alkali oxide will have more open structure. i.e. lower density. From Figure 2, it can be seen that the surface tension of

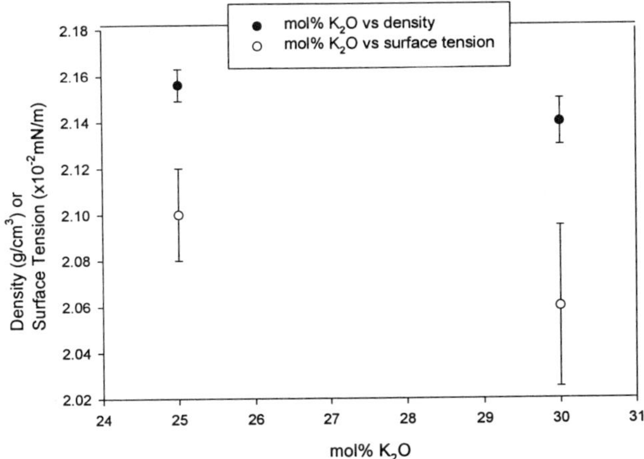

Figure 2. Change of density and surface tension of potassium silicate melts with K_2O concentration at 1400 °C.

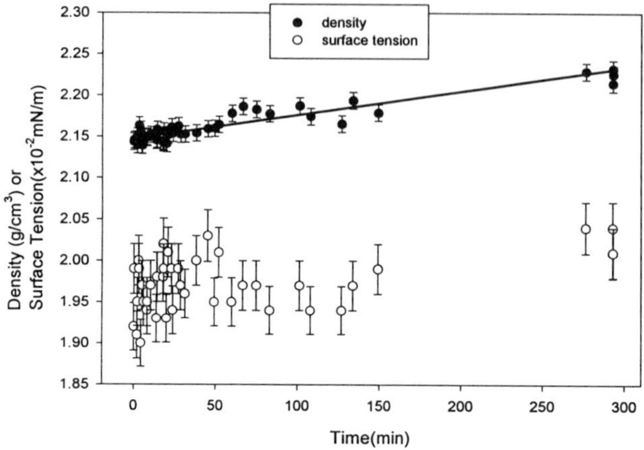

Figure 3. Change of density and surface tension of potassium trisilicate melts with time at 1425 °C.

potassium silicate melts decreases with K_2O addition. Both Shartsis et al.[13] and Appen et al.[15] reported the similar trends. Since surface tension decreases with decreasing average bond strength on the surface, addition of alkali oxide which decreases the bond strength, should decrease surface tension. A decrease in surface tension with increasing K_2O concentration will give a positive excess K_2O concentration according to Gibbs adsorption isotherm [19]. i.e. the presence of K^+ on the surface decreases the Gibbs free energy of the surface.

Figure 3 illustrates the change of density and surface tension of potassium trisilicate melts at 1425 °C with time. As shown, the density increases with the time. This can be interpreted as a decrease in K_2O concentration with time. i.e. volatilization of K_2O. The surface tension of potassium trisilicate, on the other hand, does not change in the limits of error with the time. This is because the K_2O concentration is constant on the surface even though K_2O volatilizes, and the presence of K^+ on the surface decreases the Gibbs Free energy of the system. The K^+ deficiency on the surface of the drop due to volatilization is compensated by the diffusion of K^+ from the bulk, since the temperature is high enough and there is no obstacle in the structure which prevents diffusion of K^+.

Soda Lime Silicates

The change of density and the surface tension of soda lime silicate melts with time at 1400°C is given in Figure 4 and 5, respectively. The density of a glass melt with the batch composition of 20.03 Na_2O-10.13 CaO-69.84 SiO_2 is reported as 2.22 g/cm^3 at 1400 °C [20]. Washburn et al.[21] examined a similar glass and reported the density to be 2.20 g/cm^3, and the surface tension to be 159.4 mN/m at 1454 °C. Although the last two reports [20,21] agree with each other, Washburn's [21] surface tension data jeopardize his density data. The densities calculated in this study; 2.272±0.005 g/cm^3 for run 1, and 2.279±0.009 g/cm^3 for run 2, are higher than the values reported by Cable et al.[20] and Washburn et al[21]. Bottinga et al.[22] statistically examined all available data for the melts, and proposed a model to calculate the density of the melts. According to this model, density of N20C10S glass melt at 1400 °C is 2.29 g/cm^3 which is even higher than our data suggest.

Badger et al.[22] found the surface tension of the glass melt with a composition of

                              Advances in Fusion and Processing of Glass II

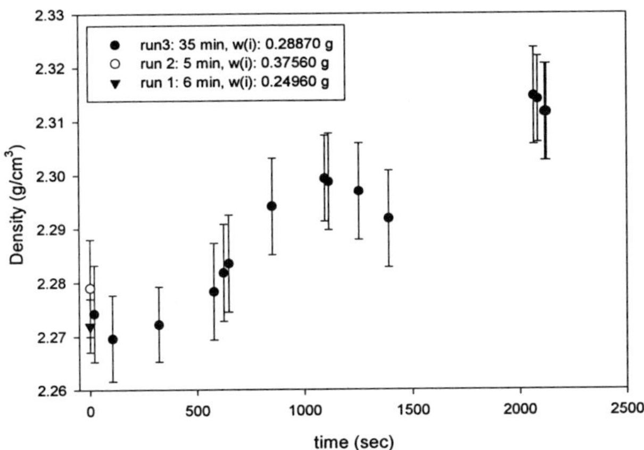

Figure 4. Change of density of 20Na$_2$O-10CaO-70SiO$_2$ with time at 1400 °C.

17.4Na$_2$O-10.1CaO-72.5 SiO$_2$ (mol%) to be 302 mN/m at 1350 °C. Bradley[23] then corroborated this value to be 303 mN/m at 1396 °C. In the current study, surface tension for a N20C10S glass melt was calculated as 299± 4 mN/m from run 1, and 297±3 mN/m from run 2.

The density of the N20C10S melt at 1400 °C increases with time as given in Figure 4. This increase can be explained by volatilization of Na$_2$O; the lower concentration of Na$_2$O creates less non-bridging oxygen and results in a relatively close structure.

The surface tension of Na20C10S melt also increases with time unlike the surface tension of the potassium trisilicate melts which remained constant with time as given above. Higher surface tension is usually sign of stronger bonding. When sodium volatilization from the surface is taken into consideration, the result is reasonable. However, there is contradiction in the literature on the effect of Na$_2$O on the surface tension of silicate melts. Appen et al.[15], and King[25] reported an increase in surface tension of sodium silicate melts with increasing Na$_2$O, while Shartsis et al.[13] reported that Na$_2$O has no significant effect on surface tension of sodium

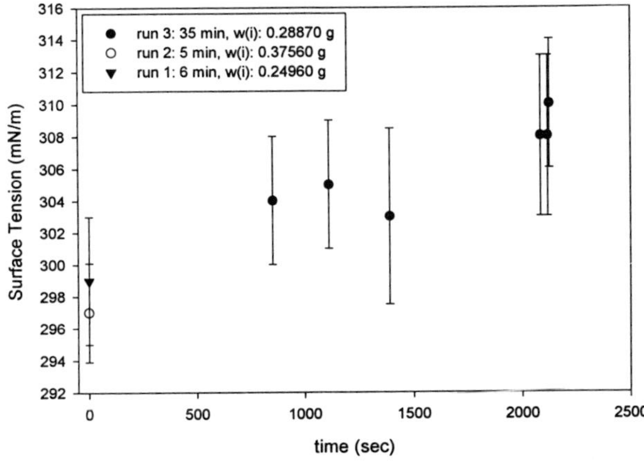

Figure 5. Change of surface tension of 20Na$_2$O-10CaO-70SiO$_2$ with time at 1400 °C.

silicate melts. On the other hand, Weyl et al.[26] explained dealkalization of glassware by a decrease in surface tension with increasing Na^+ concentration on the surface. Our unpublished work showed that Na_2O addition decreases the surface tension of sodium silicate melts. If it is assumed that Na_2O decreases the surface tension of the melt, then a similar trend in surface tension to that observed in potassium trisilicate melt as a function of time (i.e. unchanged) would be expected in soda lime silicate case. However, the increase in the surface tension of N20C10S melts can be explained by an obstacle to the diffusion of Na^+ from bulk to the sodium deficient surface. Divalent calcium ions prevents diffusion of monovalent sodium ion in the melt [20]. Even though diffusion of Na^+ to the surface to decrease surface energy is thermodynamically favorable, kinetics of the process are extremely slow.

Float Glass

The density and surface tension of re-melted commercial float glass melt were 2.37 ±0.01 g/cm^3 and 294±12 mN/m, respectively. Parmelee et al.[27] measured the surface tension of 16.02 Na_2O-7.77CaO-3.02MgO-73.18SiO_2 to be 304 mN/m at 1408 °C. They took an approximate density of 2.4 g/cm^3 in their surface tension calculations.

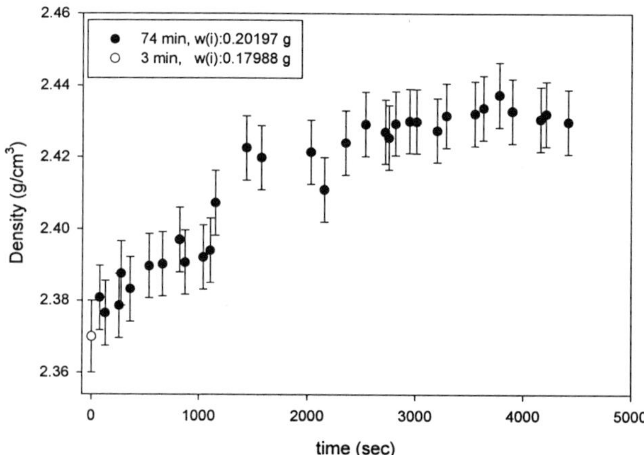

Figure 6. Change of density of float glass melt with time at 1400 °C.

The change of density of float glass melts with time is given in Figure 6. Since the mass of the melt drop is needed to calculate the density from the measured volume, the dynamic mass of the drop at any instant is found using the approach below: The experiments were carried out for three or more different time periods with the various initial glass weights. To normalize the weight loss with respect to various initial weights and instant surface areas, a parameter called normalized percent weight loss (NPWL) was defined:

$$NPWL = \frac{\left(\dfrac{initial\ weight}{initial\ surface\ area} - \dfrac{final\ weight}{final\ surface\ area}\right)}{\left(\dfrac{initial\ weight}{initial\ surface\ area}\right)} x100 \qquad (1)$$

                    Advances in Fusion and Processing of Glass II

Figure 7 gives the normalized percent weight loss (NPWL) for the float glass melt at 1400°C. Although Preston et al.[28] found a logarithmic time dependence for weight loss for a time period of 200 hours, it is possible to fit a linear time dependence at shorter times (less than 20 hours) using their plots. In the current study, a linear relation was found. The negative slope in NWLP vs. time is because the rate of decrease in weight is more than the rate of decrease in surface area. The normalized percent weight loss for the N20C10S glass melt is given in Figure 8. The absolute value of slope of NPWL for N20C10S is greater than that of NPWL for the float

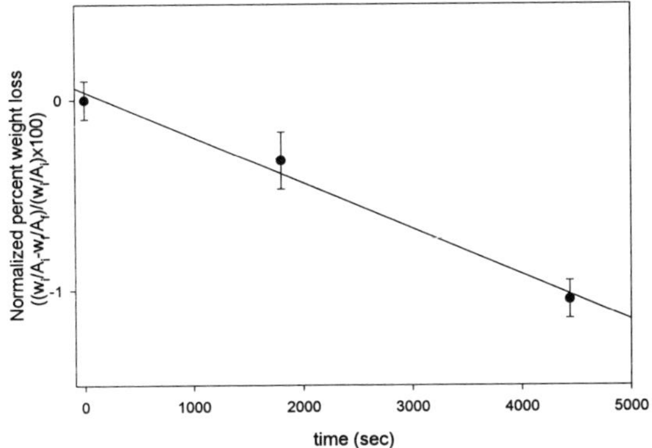

Figure 7. Change of normalized percent weight loss of the float glass melt with time at 1400 °C.

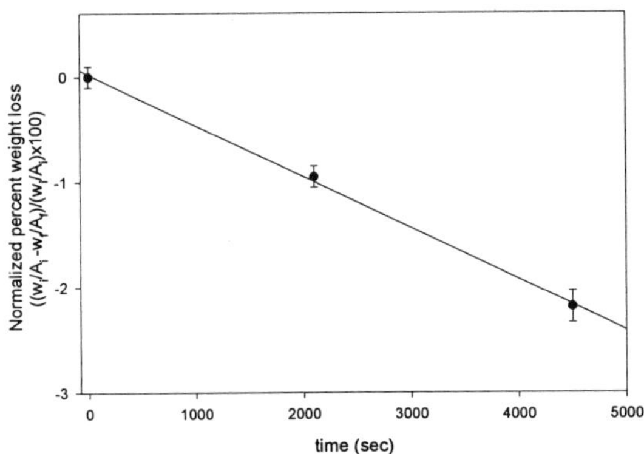

Figure 8. Change of normalized percent weight loss of N20C10S glass melt with time at 1400 °C.

glass melt. This is reasonable since float glass contains other ions such as Mg^{2+}, Fe^{2+}, Fe^{3+}, and so on. These large ions will decrease the rate of volatilization by preventing the diffusion of volatile components. As given in Figure 7, the density of the float glass melt increases and reaches a steady state after certain time. That means a volatile component (alkali) deficient surface is created and no more significant volatilization is observed. The surface tension also shows a similar trend given in Figure 9; at first it increases and then stays steady. Although both observations can be explained by an alkali deficient surface layer and the obstruction of diffusion by large ions, it should be taken into consideration that since this is a multicomponent commercial glass there are many minor components. Phenomena such as reduction or oxidation of multivalent

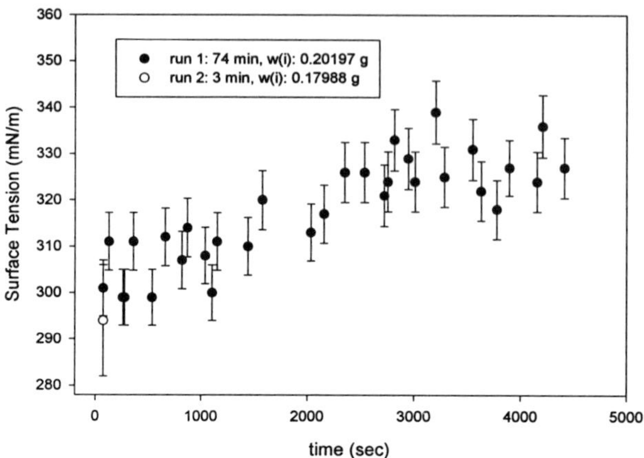

Figure 9. Change of surface tension of float glass melt with time at 1400 °C.

ions such as iron and tin can have unexpected influence on these physical properties.

CONCLUSIONS

The improved sessile drop technique was successfully employed to measure surface tension and density for both simple binary, ternary silicate melts and commercial glass melts. The reference measurements performed on sodium and potassium trisilicate glass melts ensured the reliability of the experimental technique and the calculation methods.

It was shown that volatile components in the glass melts play a critical role on the physical properties. Although an alkali deficient surface layer forms, this layer can be re-enriched with alkali diffusion from bulk depending on the diffusivity of the ion. Since diffusivity strongly depends on temperature and openness of the structure, the diffusion of alkali may not be possible in some cases.

Acknowledgements: The authors gratefully acknowledge the financial support from the NSF University-Industry Center for Glass Research at Alfred University. The suggestions from D.A. Weirauch, Jr. of Alcoa Technical Center, and help from H. Jiang of Alfred University by providing the glass samples are appreciated. Many thanks to chemist C. Edney for the chemical analysis.

REFERENCES
[1]   D. E. Sharp, and A. E. Badger,"Physical Properties of Glasses"; pp. 35-39 in Handbook of Glass Manufacture. Edited by F. V. Tooley. Ogden Pub., New York, 1953.
[2]   A. F. Crawley,"Density of Liquid Metals and Alloys," *Int. Metal. Rev.*, 19 32-48 (1974).
[3]   C. A. Landcastle,"High Temperature Properties of Glass Determined by The Sessile Drop Technique"; M.S. Thesis, Alfred University, Alfred, New York, 1994.
[4]   D. A. Weirauch and D. P. Ziegler," Surface Tension of Calcium Aluminosilicate Glass Using Computerized Drop Shape Analysis," *J. Am. Ceram. Soc.*, 74 [4] 920-26 (1996).
[5]   F. Bashforth and J. C. Adams, An Attempt to Test the Theories of Capillary Action. Cambridge University Pres, Cambridge, U.K., 1883.

[6]   B. S. Ellefson and N. W. Taylor," Surface Properties of Fused Salts and Glasses I", *J. Am. Ceram. Soc*, 21 193-213 (1938).

[7]   Y. Rotenberg, L. Boruvka, and A.W. Neumann,"Determination of Surface Tension and Contact Angle from the Shapes of Axisymmetric Fluid Interfaces," *J. Coll. Inter. Sci.*, 93[1] 169-83 (1983).

[8]   J. W. Jennings Jr, and N. R. Pallas,"An Efficient Method for the Determination of Interfacial Tension from Drop Profile," *Langmuir*, 4[4] 959-67 (1988).

[9]   M. Askari, A. M. Cameron, and J. Oakley," The Determination of Surface Tension at Elevated Temperatures by Drop Analysis," *High Temp. Technol.*, 8 [3] 201-7 (1990).

[10] A. G. Clare, C. A. Landcastle, and L. E. Jones,"The Determination of Density and Surface Tension of Molten Glasses by the Sessile Drop Technique"; pp. 8.1-8.5 in Proceeding Philips-TNO Seminar, April 1997, Netherlands.

[11] J.O'M. Bockris, J. W. Tomlinson, and J. L. White" The Structure of Liquid Silicates: Partial Molar Volumes and Expansivities,"*Faraday Soc. Trans.*, 52 299-310 (1956).

[12] L. Shartsis, S. Spinners, and W. Capps,"Density, Expansivity, and Viscosity of Molten Alkali Silicates," *J. Am. Ceram. Soc.*, 35[6] 155-60 (1952).

[13] L. Shartsis and S. Spinner,"Surface Tension of Molten Alkali Silicates," *J. Res. Nat. Bureau Std.*, 46[5] 385-90.

[14] L. Sasek and A. Lisy(1972); from *"SciGlass" Database* by O. V. Mazurin, M. V. Streltsina, and T.P. Shavaiko-Shvaikovskaya. Scivision, Lexington, Massachusetts 1996.

[15] A. A. Appen, K. A. Shishov, and S. S. Kayalova (1952); from *"SciGlass" Database* by O. V. Mazurin, M. V. Streltsina, and T.P. Shavaiko-Shvaikovskaya. Scivision, Lexington, Massachusetts 1996.

[16] O. V. Mazurin, M. V. Streltsina, and T.P. Shavaiko-Shvaikovskaya;   pp. 343-44 in *Handbook of Glass Data, Part A. Silica Glass and Binary Silicate Glasses*. Elsevier, Amsterdam, Netherland, 1983.

[17] K. C. Lyon,"Calculation of Surface Tension of Glasses," *J. Am. Ceram. Soc.*,27[6] 186-9 (1944).

[18] D. Chatain, F. Chabert, V. Ghetta, and J. Fouletier,"New Experimental Setup for Wettability Characterization under Monitored Oxygen Activity: I Role of Oxidation State and Defect Concentration on Oxide Wettability by Gold,"*J. Am. Ceram. Soc.*, 76[6] 1568-76 (1993).

[19] W. D. Kingery, H. K. Bowen, and D. R. Uhlmann; pp. 177-181 in *Introduction to Ceramics*. John Willey and Sons, New York, 1976.

[20] M. Cable and M. A. Chaudhry, "Volatilisation from Soda Lime Silicate Melts at One Atmosphere and Reduced Pressures," *Glass Tech.*, 16[6] 125-34 (1975).

[21] E. W. Washburn, G. R. Shelton, and E. E. Libman,"The Viscosity and Surface Tension of the Soda Lime Silica Glasses at High Temperatures," *University of Illinois Experimental Station Bull.*, 140 51-71 (1924).

[22] Y. Bottinga, and D. F. Weill," Density of Liquid Silicate Systems Calculated from Partial Molar Volumes of Oxide Components,"*Am. J. Sci.*, 269 169-82 (1970).

[23] A. E. Badger, C. W. Parmelee, and A. E. Williams,"Surface Tension of Various Molten Glasses," *J. Am. Ceram. Soc.*, 20[10] 325-29 (1937).

[24] C. A. Bradley," Measurements of Surface Tension and Viscous Liquids,"*J. Am. Ceram. Soc.*, 21 339-44 (1938).

[25] T. B. King,"The Surface Tension and Structure of Silicate Slags," *J. Soc. Glass Technol.*, 35 241, 241-59(1951).

[26] H. S. Williams and W. A. Weyl,"Surface Dealkalization of Finished Glassware," *Glass Ind.*, 26[6] 275-301 (1945).

[27] C. W. Parmelee and C. G. Harman,"The Effect of Alumina on the Surface Tension of Molten Glass," *J. Am. Ceram. Soc.*, 20[7] 224-30 (1937).

[28] E. Preston and W. E. S. Turner,"The Volatilisation and Vapor Tension at High Temperatures of the Sodium Silicate-Silica Glasses," *J. Soc. Glass Tech.*, 16 331-49 (1932).

Advances in Fusion and Processing of Glass II

REDOX MEASUREMENTS IN CONTAINER GLASS

Thuan TONTHAT and Christophe OLLIER

Research and Development, BSN Emballage (France)

1.Introduction

The amount of recycled cullet used in container glass furnaces has been increased drastically in the last decade. To meet the customer's demands for higher quality, and maintaining stable glass color, BSN EMBALLAGE in cooperation with Heraeus Electro-Nite and TNO-TPD(1), has developed a fast and reproducible method to characterize and quantify the REDOX state of recycled cullet and also the redox state of the glass products. The method is now applied on a routine basis in many of BSN factories. The application and reproducibility are demonstrated in this paper.

2. Suitable sensor

Target of the research program has been the development of a laboratory but also for industrial use sensor which measures the oxidation state directly in the cullet melt. The sensor should be applicable up to 1400°C glass melt temperatures and should give reproducible oxygen activity values.

New equipment for redox sensing, called Rapidox(2), has been designed and developed for the market together with Heraeus Electro-Nite based on an Iridium electrode, and a reference electrode in Ni / NiO, operating with a rapid electrical melting furnace, available for glass samples of about 600 gr. This set up, fully automatized, has been tested and approved by our factories in control of redox state of all types of glass, and mostly used now for the quality control of recycled cullet.

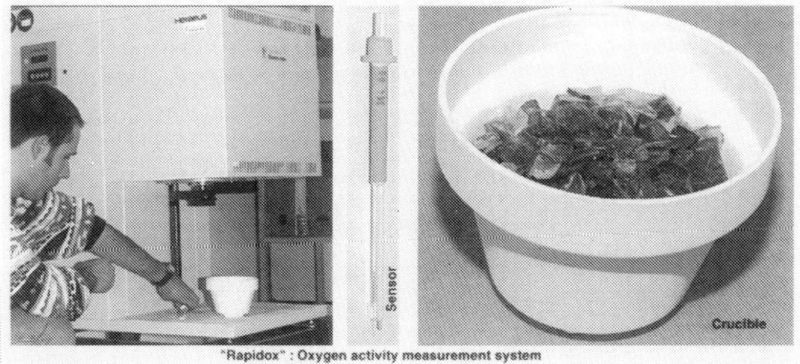

"Rapidox" : Oxygen activity measurement system

The range of measured EMF values at 1300°C are between -600 mV to +400 mV for pO2 ranges of 4,5*E-15 to 3* E-2 bar, following the Nernst law :

$$\text{EMF (volts)} = \text{RT/nF*Lg } pO2(\text{melt})/pO2(\text{reference})$$

The inaccuracy in this range is only 6 mV.The analysis of the redox state of clean glass coming directly from the melter, forehearth, gob or production lines show very precise, stable and reproducible values. On a control on 6 measurements of the same sampling, the results showed :

Average of the sample : 192,3 mV
Standard deviation : 3,44 mV
6 S.D. : 21 mV*

3. Examples of redox analysis by the sensor system

3.1 Measurements of different cullet types :

The cullet usually delivered to the glass plant contains more or less organic contaminants varying from day to day. The redox measurements directly quantify the reducing effect of these organic contaminants on the glass melt. Fig. 1 shows measured EMF values (a high value is very oxidized glass, a low value is reduced glass) for a melt directly prepared from fresh polluted cullet and for the same cullet after preheating at 350° or 500°C. Due to the burning of the organic contaminants, the reducing power of the preheated cullet is much lower than in the case of untreated fresh cullet. The EMF value changes from -490 mV up to +15 mV after preheating up to 350°C and even to +190 mV after preheating the cullet at 500°C

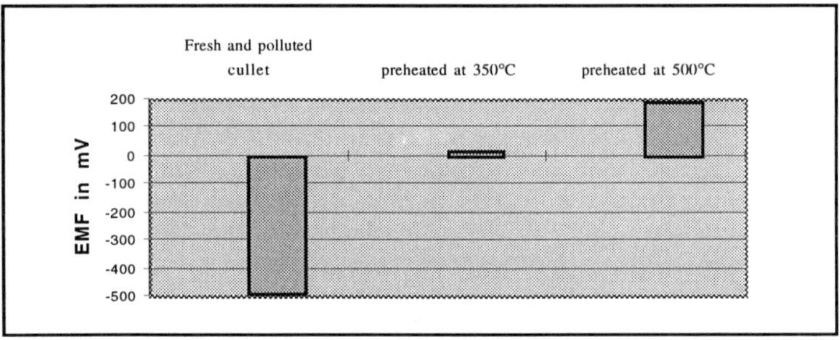

Fig.1 Rapidox values with fresh and preheated recycled cullet

Results for different types of cullet (mixed colors, fine cullet, coarse cullet, flat glass cullet etcetera) in use at BSN are presented hereafter. The results show that pulverized (powder) cullet behaves much more reducing than coarse cullet from the same source. Washing the cullet lowers the reducing power.

Advances in Fusion and Processing of Glass II

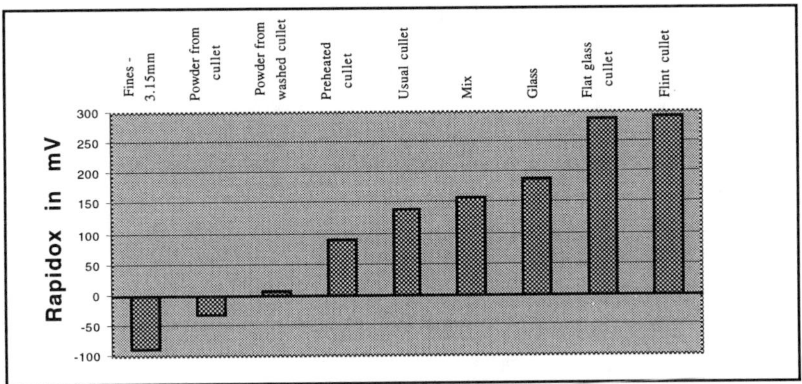

Fig.2- Different cullet types in use in container glass

The granulometry of the cullet has an effect on the redox state of the cullet melt, finer cullet always shows lower EMF-values thus lower oxygen activities.

BSN has found that most of the organic contaminants are often concentrated in the fine grain sizes of the collected cullet from the collection banks. The measurements performed for the cullet sizes below 3.15 mm show a very reducing state of the obtained glass or glass melt, compared to cullet larger than 3.15 mm.

3.2 The powdering of cullet

In some factories, the cullet is partly ground into powder in order to minimise the occurence of glass defects coming from contaminants such as stones, ceramics or china ware. However, this pulverisation leads to a very large change in redox state while melting the powder. In our example the sensor in a coarse cullet melt showed a EMF value of 50 mV, after grinding this coarse cullet, the EMF value of the molten powder became -100 mV at a 1300°C melt temperature. This problem is currently identified and controlled in the BSN factories by the daily use of these Rapidox measurements and now making proper batch corrections.

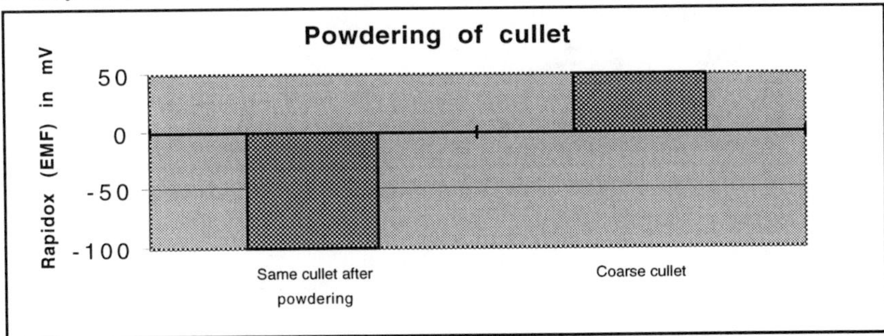

Figure 3 : Coarse cullet becoming more reducing after pulverisation

3.3. Effect of additions of pollutants to cullet

Systematic measurements of the redox state of melts prepared from polluted cullet, with controlled additions of identified organic materials have been performed in the BSN-laboratory using contaminants usually found in recycled cullet. The reducing effects on the glass melt by a single organic contaminant like plastics, paper, sugar, and fats have been proven and quantified by applying the Rapidox analysis.

Some mineral pollutants such as road stones, coal, asphalt appear to be very reducing in the glass melt as well. Surprisingly, a mineral with a strong oxidising effect has been identified in the cullet : calcium sulfate in the shape of white porous pellets (originating from cat litter !!)

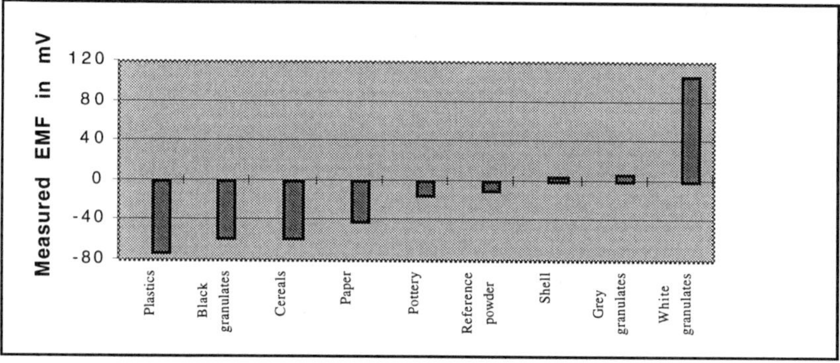

Figure 4 shows the effects of additions of 0.5 weight % of identified pollutants to powdered cullet on the redox state of the glass prepared from these mixtures.

3.4 Redox state of glasses from the production lines

The EMF-values measured by Rapidox in melts of glass products at 1300°C are compared to the calculated redox numbers or colors of these glass products. Figure 5 shows the EMF-values for different glass products from BSN factories.

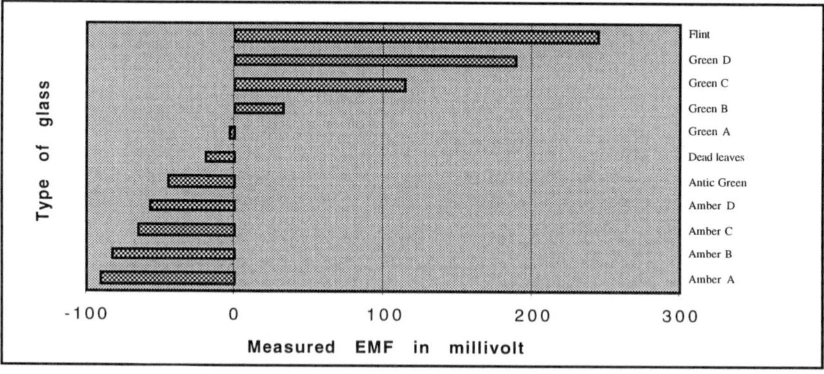

Fig.5- Different glass types currently produced in BSN Emballage

   Advances in Fusion and Processing of Glass II

At any time, the redox state can be checked by sampling parisons or bottles and measuring the redox state using the redox sensor or measuring the internal optical transmission at a wavelength of 1000 nanometer (T1000 =% transmission at this wavelength).

Figure 6. below relates the T1000 values measured for different products to the EMF-values measured by the redox sensor. This implies that the redox state of a batch or glass melt sample measured by Rapidox is very well related to the color of the glass and can be used to correct the batch composition, keeping the redox and color of the glass constant.

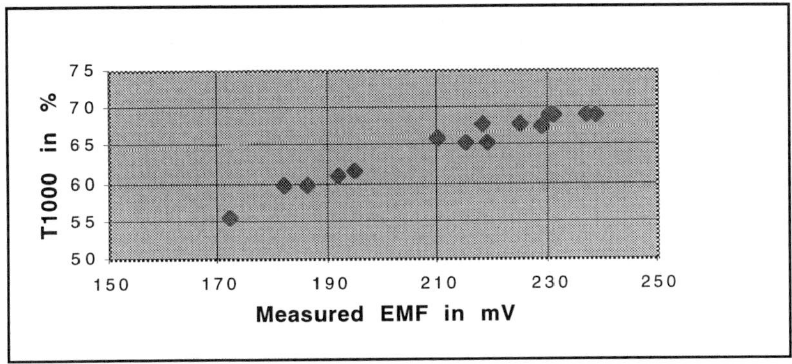

Fig.6- Transmission % at 1000 nanometer and Rapidox measured at 1300°C

3.5 Color changes

In the case of a voluntary color change of the glass, the programme of batch changes can directly be followed by measuring the redox state of small batch melts or the glass products, using the redox sensor.

Figure 7, shows the redox change (expressed as EMF in mV) of the glass during the transition from flint glass to dark amber glass, as a function of time. By this way, the color change can be controlled rapidly and this can shorten the transition time or helps to shorten the transient stage during which often blisters in the products occur.

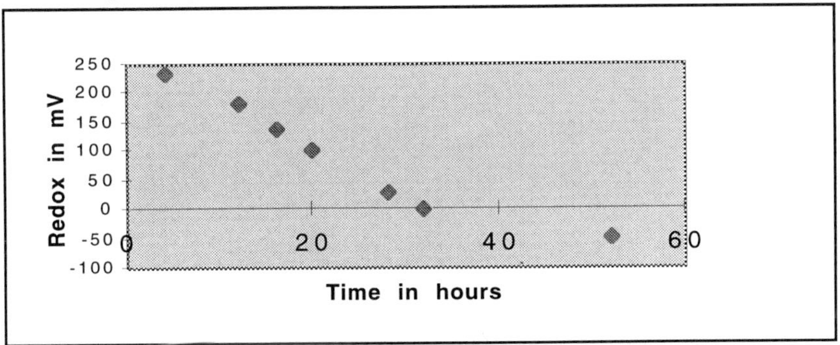

Fig.7 Measured Redox values for a color change Flint to Amber as function of time

4. Future applications of the redox sensor (Rapidox)

The Rapidox equipment, recently in use at BSN, is a very convenient tool dedicated to characterize the redox state of all kind of recycled cullet. This will improve the way the quality of the cullet can be judged and the ability to make fast and proper batch correction. The redox state of the cullet can be measured within 2 hours after sampling. It enables the furnace managers to stabilise redox state which also will reduce the problems of reboil and poor fining. More and more results are obtained, especially in the laboratory, which shows relations between redox state and sulfate fining, effect of the size of coke grains in the batch on the redox state, the redox states sensitive for blister formation etcetera.
The reproducibility and accuracy of the redox state characterisation of cullet melts and batch melts by using the Rapidox system has been proven in the BSN container glass factories.

BSN EMBALLAGE. R&D. March 1997. T. TONTHAT

References
(1). A.J.Faber, R.Beerkens, Redox control during melting of cullet-rich glass batch, Int.Glass Journal (1996) N° 87.
(2). Rapidox is a registered trademark of Heraeus Electro-Nite for the Redox state measuring system. Contact Dr. Jacques PLESSERS. Fax. 011 60 04 01 (Belgium)

Separation of Bands for Ferrous and Hydroxyl Groups in the Spectra of Na_2O-CaO-MgO-SiO_2 Glasses.

E. N. Boulos [1], L. B. Glebov [2]

[1] Glass Division, Ford Motor Company, Dearborn, MI
[2] Center for Research and Education in Optics and Lasers, University of Central Florida, Orlando, FL

ABSTRACT

A number of **Na_2O-CaO-MgO-SiO_2** glasses doped with different concentrations of iron and water were studied. A simple algorithm was developed to describe absorption by water and iron in each glass and to simulate the whole assemblage of spectra for pure and iron doped silicate glasses. Only two parameters, which can be determined from experimental absorption spectra, are necessary for spectra simulation. Absorption spectra of hydroxyl groups and ferrous ions for the spectral region 2900 - 1000 nm were determined by comparing of experimental absorption spectra of glasses melted under different conditions. A new absorption band for ferrous ions was found in the region of 2980 nm. This band gives significant contribution to the absorption coefficient in the region of the water band with maximum at 2850 nm. The spectral shape of hydroxyl group and ferrous ion bands remains stable for all glasses within the accuracy of measurements and calculations (≈ 5 %). These spectral data allow one to study the effect of water on the valence state of the ions in the glass and to control the absorption of glasses which are used for heat screening.

1. INTRODUCTION

The measurement of absorption spectra for determination of concentration of components in solids and liquids is a very common approach. Such important components in oxide glasses as iron (ferrous ion) and water (hydroxyl ion) possess absorption bands in the near IR region. The reliable determination of band locations for these components is a difficult problem because of overlapping of their absorption spectra.

In oxide glasses the ferrous ion (Fe^{2+}) possesses a main absorption band in the near IR region with a maximum near 1100 nm. This band extends to the shorter wavelength visible region [1] and affects glass coloration. There is also a band of ferrous ion at ~2000 nm [2, 3]. There is no evidence of absorption bands for ferrous ion at longer wavelengths. The shortest wavelength absorption band for hydroxyl containing alkali-silicate glasses possesses a maximum at ~2850 nm and is attributed to stretching vibration of **OH** groups associated with some structural elements of the glass network [4]. The absorption bands for **OH** molecular ions, which are strongly bonded with the glass network, are at 3600 nm and 4250 nm [4].

Both ferrous ions and hydroxyl ions absorb in the same spectral region. To our knowledge, there are no data for the bulk multicomponent silicate glasses that describe absorption spectra of these ions in the region between the main bands at 1100 and 2850 nm, and the level of overlapping near these maxima. Hence, the goal of this work is to analyze the overlapped absorption spectra of iron and **OH** groups in **Na$_2$0-CaO-MgO-SiO$_2$** glass in the near IR region, to describe separately the shapes of absorption spectra for both iron and hydroxyl groups, and to describe experimental spectra as their linear combination.

2. EXPERIMENTAL

The glass composition chosen for this work is similar to the standard float glass widely used in the glass industry: (w.%) **14Na$_2$0-9CaO-4MgO-73SiO$_2$**. The approach was to compare the absorption spectra of glasses doped with different concentration of iron. Glasses were made from both regular batch materials, which were melted in conditions that are usually used in the glass industry, and chemicals with transition metals contamination in the range of 1 ppm that were melted in high purity conditions.

For the first method, we used sand, soda ash, limestone and dolomite as the main components of the batch. Part of sodium oxide was introduced as sodium nitrate to produce oxidizing melting conditions. A carbocite (carbon powder) was added to the batch to produce reducing melting conditions. Deionized water (0 - 20 ml) was added to the batch before filling in a crucible. The glasses were melted in 100 cm^3 platinum crucibles in a gas furnace for 6 hours at 1430°C and poured into a graphite mold. The glass mold was annealed at 540°C for 2 hours, and then cooled to room temperature at the rate ~1 °C/min. The concentration of iron in these glasses was ~0.04 wt.% if no iron was added to the batch and was varied up to 0.9 w.% if iron oxide was added to the batch.

For the second preparation method, we used purified chemicals with iron concentration in the range from 1 to 3 ppm. Glass melting technology is described in our previous work [5]. High purity iron oxide was added to the batches to produce samples with concentrations of iron from 0.001 to 0.9 w.%. All of these glasses were melted in 100 cm^3 fused silica crucibles in an electric furnace. To obtain oxidized or reduced glasses, a part of these specimens were then melted with oxygen bubbling through the liquid glass in the furnace or with carbon added to the batch. The concentration of iron in the specimen was ~5 ppm if no iron was added to the batch.

Absorption spectra were measured using double beam spectrophotometers (Perkin-Elmer Lambda 9 and Perkin-Elmer 330) in the spectral region 200 to 3200 nm.

3. RESULTS AND DISCUSSION

a) Experimental spectra.

Absorption spectra of about 30 doped and non-doped glasses that were melted from both natural and high purity materials in oxidized, neutral and reduced conditions, with and without water in the batch were measured. Some examples of their spectra are

                              Advances in Fusion and Processing of Glass II

presented in Fig. 1. One can see a sharp absorption band with maximum near 2850 nm in all spectra. This band is ascribed to **OH** vibrations in multicomponent silicate glasses [4]. It was found that the absorption at 2850 nm varies in glasses that were melted under different conditions and doped with different concentrations of iron. One can also observe a wide absorption band with maximum near 1075 nm. This is the absorption band for ferrous ion [1-3]. This band is observed for all glasses except non-doped high purity ones. Its intensity is changed when concentration of iron and redox conditions of melting are changed. An absorption spectrum of the high purity glass normalized at 2850 nm is presented in Fig. 2, curve 1. No absorption in pure glasses was measured in the region of the maximum of the ferrous band near 1075 nm within the accuracy of the measurement of about 0.005 cm^{-1}. One can see several overlapping bands between the two mentioned bands of water and iron in both doped and non-doped glasses, but it is difficult to ascribe them to any components of the studied glasses.

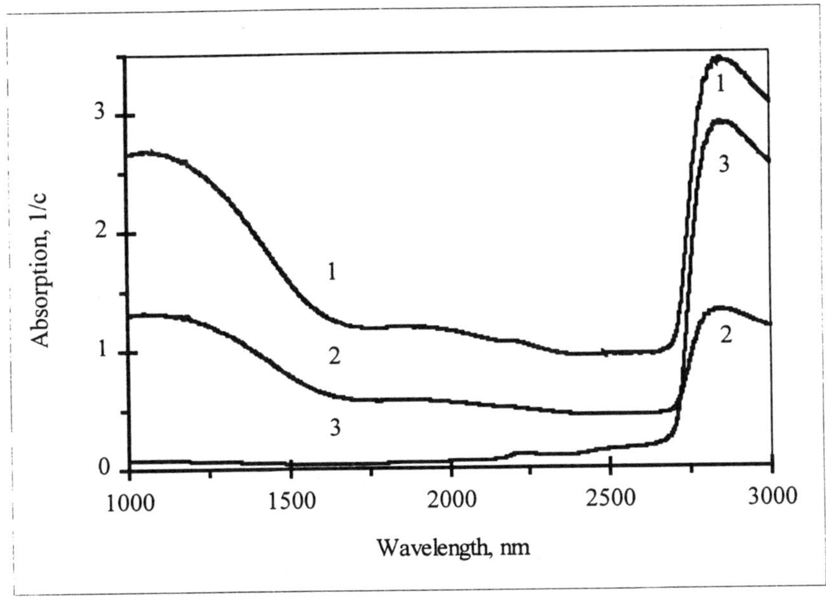

Fig. 1. Absorption spectra of **Na$_2$0-CaO-MgO-SiO$_2$** glass in the near IR region.
1 - glass doped with 0.9 % Fe$_2$O$_3$, melted in a gas furnace in reduced conditions from natural raw materials;
2 - glass doped with 0.5 % **Fe$_2$O$_3$**, melted in an electric furnace in neutral conditions from high purity raw materials;
3 - glass doped with 0.04 % **Fe$_2$O$_3$**, melted in a gas furnace in oxidized conditions from natural raw materials;

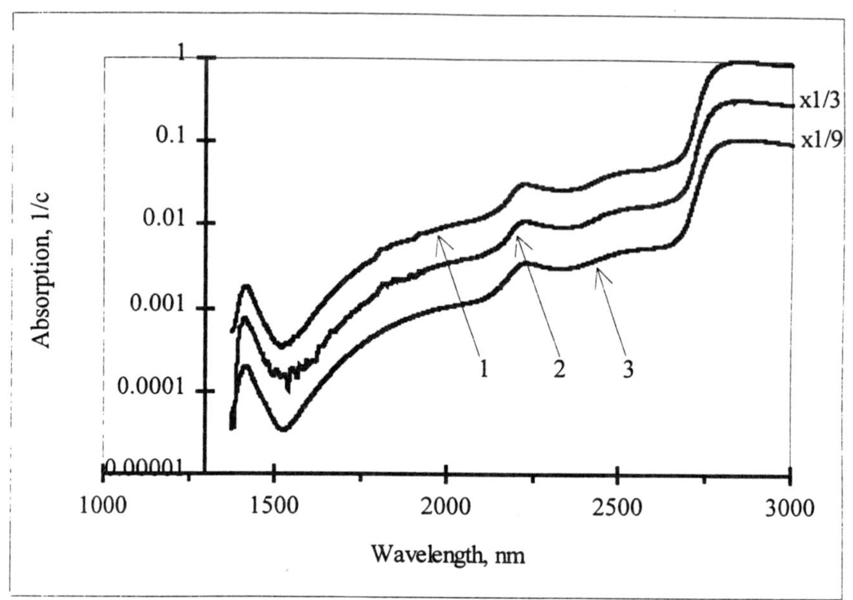

Fig. 2. Normalized at 2860 nm absorption spectra of water
in **Na$_2$0-CaO-MgO-SiO$_2$** glasses.
1 - measured and averaged for a number of non-doped high purity glasses,
2 - calculated with formula (4) from experimental spectra for glasses doped with
0.04 and 0.9% **Fe$_2$O$_3$** melted in gas furnace from neutral raw materials,
3 - calculated as a sum of Gaussian bands to fit experimental data.

b) Extraction of the hydroxyl absorption spectrum

It is necessary to eliminate the absorption bands of iron to determine the spectral
shape of the water absorption bands for the glass. One can assume that there are no
components, except water, in the high purity glasses that have significant absorption
bands in the spectral region from 1000 to 3000 nm. Actually, the intrinsic vibration bands
of the silicate glass matrix are positioned to the long wavelength side of the discussed
region [4, 6]. The main absorption band of ferrous ion with a maximum near 1075 nm has
a specific absorption of ~9 cm^{-1}/wt.% **Fe** [7]. This means that for high purity glass, the
absorption coefficient of ferrous ion in the near IR region should be less than 10^{-3} cm^{-1}
and, consequently, cannot be detected by spectrophotometer in samples with thicknesses
less than 1 cm. It was confirmed in this work, that one could see no detectable absorption
in high purity glasses in the region of 1100 nm. Therefore, one can suppose that the
absorption spectrum presented in Fig. 2, curve 1, is the absorption spectrum of water.

For glasses doped with iron, one can suppose that experimental IR absorption
spectrum of silicate glass doped with iron is the sum of absorption spectra of water and
ferrous ions :

$$A_{rec,k}(\lambda) = c_{w,k}A_w(\lambda) + c_{Fe2,k}A_{Fe2}(\lambda), \qquad (1)$$

where $A_{rec,k}$ is the recorded absorption coefficient of the glass sample #k, A_w and A_{Fe2} are specific absorption coefficients of hydroxyl and ferrous ion, respectively, $c_{w,k}$ and $c_{Fe2,k}$ are the concentrations of hydroxyl and ferrous ions in the glass #k, respectively, and λ is the wavelength. The values of multiphonon absorption of water in spectral region near the maximum of the iron absorption band at 1075 nm are very small and were detected for the high purity fibers only [8]. Therefore, one can write the absorption in the region of the maximum of the ferrous band as:

$$A_{rec,k}(\lambda_1) = c_{Fe2,k}A_{Fe2}(\lambda_1), \qquad (2)$$

where λ_1 =1075 nm. Comparing spectra of two glasses with different concentrations of iron: #k1 and #k2, the ratio of the absorption coefficients at λ_1 (the ferrous ratio) in accordance with (2) is:

$$R_{Fe2} = \frac{A_{rec,k1}(\lambda_1)}{A_{rec,k2}(\lambda_1)} = \frac{c_{Fe2,k1}}{c_{Fe2,k2}}. \qquad (3)$$

Multiplying the equation (1) for the glass #k2 by R_{Fe2} and subtracting it from the equation (1) for the glass #k1, yields:

$$A_w(\lambda) = \frac{A_{rec,k1}(\lambda) - R_{Fe2}A_{rec,k2}(\lambda)}{c_{w,k1} - R_{Fe2}c_{w,k2}}. \qquad (4)$$

One can see that equation (4) presents the absorption spectrum of hydroxyl as a linear combination of experimentally measured absorption spectra. We can obtain an absorption spectrum of hydroxyl in arbitrary units only and not specific water absorption because concentrations of water in equation (4) are unknown. An example of such spectrum that was normalized at 2850 nm is presented in Fig. 2, curve 2. One can see that this spectrum has the same shape as the spectrum of high purity glass (see curve 1). This allows us to conclude that spectra shown in Fig. 2 are the absorption spectra of water in the studied glasses. The averaged spectrum based on the measurements in all investigated glasses was modeled as a sum of Gaussian bands and is presented in Fig. 2, curve 3. RMS deviations of averaged spectrum from all studied experimental spectra do not exceed 5 %.

c) Extraction of ferrous absorption spectrum

It should be noted that concentrations of water and ferrous ions in the formula (1) are unknown, and there is no evidence that the spectral region in the near IR exists where iron does not absorb. Therefore, determination of the spectral shape of hydroxyl ion does not allow one to find the spectral shape of iron in the same manner. Then some other features of absorption spectra should be used to find the absorption spectrum of hydroxyl groups. One can see in Fig. 1 that the short wavelength edge of the hydroxyl band in the region 2750 nm is very sharp in comparison to all of the other fractions of spectrum. This means that the derivative of the experimentally measured absorption coefficient over the wavelength must have an outstanding band in the mentioned region. Fig. 3, curve 3 shows this very sharp band near 2740 nm, which distinguishes from the other bands in all samples independently on the presence of iron and melting conditions. Consequently, this band can be considered as a significant evidence of water absorption participation in the spectrum studied. In accordance with equation (1), the spectral derivative of the ferrous ion absorption can be written in the following form:

$$c_{Fe2,k} \frac{d}{d\lambda} A_{Fe2}(\lambda) = \frac{d}{d\lambda} A_{rec,k}(\lambda) - c_{w,k} \frac{d}{d\lambda} A_w(\lambda) . \qquad (5)$$

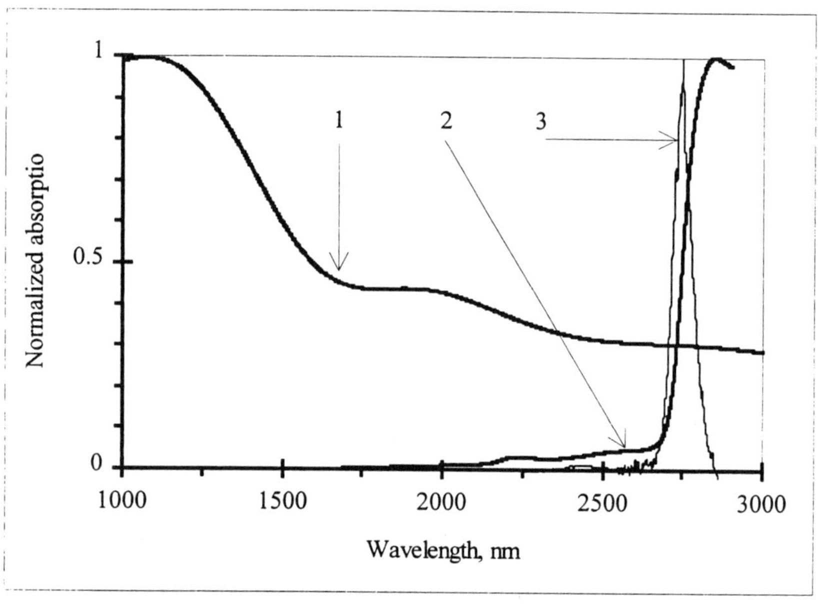

Fig. 3. Normalized absorption spectra of ferrous (curve 1) and hydroxyl (curve 2) ions in **Na₂0-CaO-MgO-SiO₂** glasses and a derivative of hydroxyl absorption spectrum.

This derivative should be equal to zero in the spectral region of 2850 - 2700 nm (see curve 3 in Fig. 3) if the hydroxyl absorption is completely eliminated from the linear combination of recorded and hydroxyl spectra in formula (5). Because absolute values of water concentration and $A_w(\lambda)$ in (5) are not known, we have to use normalized absorption spectrum of hydroxyl $A_{w,N}(\lambda)$. Then one can define a coefficient before

                    Advances in Fusion and Processing of Glass II

$A_{w,N}(\lambda)$ as $R_w A_{rec,k}(\lambda_2)$ where λ_2= 2850 nm is the maximum of the water band. Physical meaning of R_W is clear from the following consideration. The recorded absorption of glass in the maximum of the water band at λ_2 is completely determined by water if R_W=1 or by iron if R_W=0. Then, we can redefine coefficients in (5) and equate it to zero:

$$\frac{d}{d\lambda} A_{rec,k}(\lambda) - R_w A_{rec,k}(\lambda_2)\frac{d}{d\lambda} A_{w,N}(\lambda) = 0. \qquad (6)$$

To solve equation (6) in the entire spectral region between λ_2 and nearest minimum of the derivative of the water absorption spectrum at λ_3 =2700 nm, let us calculate an integral of (6) in this region and solve it for R_W:

$$R_{w,k} = \frac{1 - \dfrac{A_{rec,k}(\lambda_3)}{A_{rec,k}(\lambda_2)}}{1 - A_{w,N}(\lambda_3)}. \qquad (7)$$

Then normalized ferrous absorption spectrum can be derived from (1) with using (6) and (7):

$$A_{Fe2,N}(\lambda) = \frac{A_{rec,k}(\lambda) - R_{w,k} A_{rec,k}(\lambda_2) A_{w,N}}{A_{rec,k}(\lambda_1)}. \qquad (8)$$

These spectra were calculated for all studied glasses and very similar shapes were observed. The averaged absorption spectrum of ferrous ion calculated using formula (8) is presented in Fig. 3, curve 1. One can see that ferrous ions exhibit not only well known absorption bands at 1075 nm and 2000 nm but one more absorption band with maximum near 2980 nm. Usually, this band cannot be observed because of masking by the strong absorption band of water. Ferrous band cannot be confused with the hydroxyl band because it is about ten times wider than the last band. It is important that ferrous ion has a significant absorption in the region of the maximum of the well known absorption band of water and can strongly influence the results of the spectrophotometric water concentration measurements.

d) Reconstruction of experimental spectra

Hence, we determined absorption spectra of water $A_{w,N}(\lambda)$ and ferrous ion $A_{Fe2,N}(\lambda)$ normalized at λ_2=2850 nm and λ_1=1075 nm, respectively. These two spectra should describe the experimental absorption spectra in glasses doped with hydroxyl and ferrous ions. We believe that the spectrum of glass is the sum of ferrous and hydroxyl bands with unknown coefficients because normalized spectra are not specific per unit of concentration. It is necessary to find these coefficients for the simulation of the experimental spectrum. It is clear that the coefficient before the second term of the right side of (1) is $A_{rec,k}(\lambda_1)$ because only ferrous absorption is significant in this region. In this case ferrous absorption at λ_2 is $A_{rec,k}(\lambda_1)A_{Fe2,N}(\lambda_2)$. It should be noted that $A_{Fe2,N}(\lambda_2)=0.3A_{Fe2,N}(\lambda_1)$ for the studied glasses. The rest of the absorption at λ_2 is determined by water: $A_{rec,k}(\lambda_2)-A_{rec,k}(\lambda_1)A_{Fe2,N}(\lambda_2)$. Finally, the total absorption spectrum can be written as:

$$A_{calc,k}(\lambda) = p_{w,k}A_{w,N}(\lambda) + p_{Fe2,k}A_{Fe2,N}(\lambda),$$
$$where \quad p_{w,k} = A_{rec,k}(\lambda_2) - A_{rec,k}(\lambda_1)A_{Fe2,N}(\lambda_2), \quad p_{Fe2,k} = A_{rec,k}(\lambda_1) \qquad (9)$$

One can see that coefficients for the spectrum reconstruction are calculated directly from the experimental data and not fitted. This approach was applied for all experimental spectra measured in this work. It was found that all spectra that were calculated as a linear combination of hydroxyl $A_{w,N}(\lambda)$ and ferrous $A_{Fe2,N}(\lambda)$ ions for oxidized and reduced, doped and undoped glasses coincide with accuracy of few percents with recorded spectra. No spectra with RMS deviation more that 3 - 5 % were obtained among the studied glasses. As a result, all absorption spectra in the region 1000 - 2900 nm of **Na_2O-CaO-MgO-SiO_2** glasses doped with iron up to 0.9 %, melted in gas and electric furnaces, under oxidized and reduced conditions can be described with an accuracy better then 5 % by two coefficients and tabulated spectra of ferrous ion and hydroxyl group in accordance with formula (9). Some distortions of spectral shape of water and iron absorption were obtained in glasses melted under different conditions. These distortions should be a subject of the further study. Meanwhile, within the standard accuracy of measurements of several percent, the approximation proposed in this work can be used for the near IR absorption spectra analysis.

5. CONCLUSIONS

Comparison of experimental absorption spectra in **Na_2O-CaO-MgO-SiO_2** glasses melted under different conditions allows one to obtain normalized absorption spectra of hydroxyl and ferrous ions glasses in the spectral region 3000 - 1000 nm. Absorption spectra of hydroxyl groups and ferrous ions can be separated and extracted from experimental data. There is a new absorption band for ferrous ion with maximum at 2980 nm. This band makes a significant contribution to the absorption coefficient in the region of the water band with maximum at 2850 nm. Linear combination of water and iron absorption describes the whole assemblage of complicated experimental spectra in all studied glasses with accuracy better than of 5 %. Two parameters for calculation can be determined directly from experimental absorption spectra. The spectral shape of hydroxyl and ferrous ions remains stable for all studied glasses within the accuracy of about 5%.

Authors would like to thank L.N. Glebova, M.V. Kryukova, and T.V. Smirnova for assistance in glass melting and I.A. Barskaya and N. Malani for chemical analysis of glasses studied.

Advances in Fusion and Processing of Glass II

REFERENCES

1. W. A. Weyl, *Coloured Glasses*, The Society of Glass Technology, Sheffield, 1951.
2. A. Bishay, A. Kinawi. *Phys. Non-Crystalline Solids, Proc. Int. Conf.*, Delft (1964) 589.
3. R. J. Edwards, A. Paul, R. W. Douglas. Phys. Chem. Glasses, 13 (1972) 131.
4. E. N. Boulos, N. J. Kreidl, *J. Canadian Ceramic Soc.*, 41 (1972) 83.
5. L. B. Glebov, L. B. Popova, M. N. Tolstoi, V. V. Rusan. *Sov. J. Glass Phys. Chem. 2 (1976) 547.*
6. H. Scholze. *Glass: nature, structure, and properties*, NY, Springer, 1991.
7. L. B. Glebov, V. G. Dokuchaev, G. T. Petrovskii, M. N. Tolstoi, *Fiz. Khim. Stekla. (Sov. J. Glass Phys. Chem. In Russian), 10 (1984) 90.*
8. P. Kaiser, et al. *J. Opt. Soc. Amer.*, 63 (1973) 1141.

STRUCTURAL UNITS IN ALKALI BORATE GLASS STUDIED BY MOLECULAR DYNAMICS

Byeongwon Park and Alastair N. Cormack
New York State College of Ceramics at Alfred University
Alfred NY 14802 USA

ABSTRACT

The structure of alkali borate glass is investigated by Molecular Dynamics simulation and is seen to be comprised of well defined grouping of basic BO_3 and BO_4 units as proposed by Krogh-Moe[1]. We used a coordination dependent potential scheme to simulate borate glass. The computer simulated alkali borate glass consists of structural grouping such as BO_3, BO_4, triborate and di-triborate etc. and the fraction of tetrahedrally coordinated boron reached a maximum at 40% addition of Na_2O but is a slightly smaller fraction amount than that deduced from experiment.

I. INTRODUCTION

The alkali borate glass is famous for its anomaly in properties such as thermal expansion coefficients, viscosity, and molar oxygen volume which are not usually observed in boron free glasses. There have been many attempts to develop the structural models of alkali borate glasses. It is found that alkali metal ions can change coordination state of boron from 3 to 4 and/or generate nonbridging oxygen ion. According to Zachariasen's random network model of glass[2], alkali borate glass consists of BO_3 and BO_4 which are randomly connected. However, Krogh-Moe asserted that borate glasses were not only a random network of BO_3 triangles and BO_4 tetrahedra but they also contained well-defined and stable groups as segment of the disordered framework which occur in related crystalline materials[1]. Experimental evidence for the existence of large grouping in borate glasses is provided by NMR[3], IR[4], and density measurements[5].

Extensive NMR measurement by Bray [6] lead to the proposal of a structural model in which superstructural units bigger than basic BO_3 and BO_4 structural units are connected randomly to each other. Feller and coworkers have published a series of density measurements of alkali (Li, Na, K, Rb, Cs, Ag and Tl) borate glasses[7-9]. They developed a semi-empirical model for density which makes use of the structural ideas of Krogh-Moe. They used five different boron configuration in the modeling process and claimed that they were able to yield the relative volumes of the structural groupings over wide range glass compositions. They also argued that it is possible to calculate a packing fraction of the each of the units. Feller et al.[10] used [10]B NMR spectroscopy for an investigation of glass structure and found that the alkali borate glass structure is consistent with the Krogh-Moe structural model. In particular, it was assumed that for 0<R<0.4, the glass is constructed from boroxol, tetraborate and diborate units: for 0.4<R<1.0, tetraborate, diborate, metaborate and loose BO_4 units were used; and for 1.0<R<1.86, metaborate, loose BO_4, pyroborate and orthoborate units were used; where R is metal ion to boron ratio.

Wright et al.[11] have surveyed a wide range of crystalline borates to propose a set of criteria for the formation of vitreous borate networks, taking into account the topological degrees of freedom necessary for conventional glass formation. They concluded that borate crystallography can provide considerable insight into the structure of the corresponding glasses. They developed a number of criteria to discuss the formation of vitreous borates containing superstructural units and used them to further refine a structural model originally proposed by Bray[6].

Molecular Dynamics simulation has been a powerful tool in study of structural and dynamical properties of glasses and there have been numerous simulation studies of borate glass[12-14]. However, they have not given results that matched experiments well while in contrast to silicate glass. This may be due to the complexity of boron-oxygen bonding, which is not only partially covalent, but also exhibits a change in boron coordination depending on its environment. For B_2O_3 glass simulation[14-16], almost standard simulation has failed to reproduce structural units called boroxol ring which are believed to exist on the basis of Raman Spectroscopy[17] and neutron diffraction experiment[18]. Alkali borate glass simulation[19] reproduced the basic structural units such as BO_3 triangle and BO_4 tetrahedron but failed to reproduce larger structural groupings which are believed to exist in the glasses.

We have successfully simulated boric oxide[20] and alkali borates[21] by introducing so called coordination dependent potential, developed by Takada[22]. We were able to reproduce the basic structural units, boroxol rings and connecting BO_3 in boric oxide glass. The same scheme was applied to alkali(Na) borate glass and here we report results of MD simulation focusing on the structural units in alkali borate glasses.

II.EXPERIMENTS

New Improved Interatomic Potential Model

Takada et al. [23,24] have reported two improved interatomic potential models obtained by fitting to ab initio quantum mechanical data. Experimental data of borate crystals show that bond lengths are strongly dependent on the coordination state, so the potential may be explicitly formulated to depend on the instant coordination state during the simulation. However standard MD code does not allow a continuously variable potential with respect to coordination state change because atom labels are not usually allowed to change during the course of a simulation. The conventional MD code was severely modified to implement a coordination dependent potential scheme and successfully applied to boric oxide glass simulation. More details can be found in reference 21, 22 and 24.

A Buckingham type potential is applied to interactions between O-O, B-B and Na-O and a Morse type potential is applied to the B-O interaction. In addition, three body potentials are applied to O-B-O with equilibrium angels of 120° for BO_3 triangles and 109.47° for BO_4 tetraheda. The potential parameters used may be found in reference 21.

Simulation Procedure

The number of atoms, target density and simulation box size are given in table 1. The initial configuration is based on crystalline B_2O_3 and sodium atoms are put into interstitial sites randomly. This initial structure is first heated to a nominal 6000K to disorder it and then cooled down to 3000K , 1500K, 600K and finally to 300K. The total number of timesteps for each temperature is 10,000 and each timestep is 1 fs. The position of the atoms, averaged over the last hundred timesteps, were used to obtain radial distribution function, bond angle distribution, and coordination state.

Advances in Fusion and Processing of Glass II

Graphical visualization was achieved using commercial software; Cerius^2 and InsightII from Molecular Simulation Incorporated.

Table 1. Density values of x Na$_2$O • (1-x)B$_2$O$_3$ glasses. The corresponding number of atoms and box sizes are also included.

Na$_2$O Content, x (mol fraction)	Number of Atoms				Density (g/cm^3)	Cell Size (Å)
	Na	B	O	Total		
0.10	16	162	251	429	2.049	17.101
0.15	27	159	252	438	2.133	17.052
0.20	38	150	244	432	2.209	16.925
0.30	60	140	240	440	2.334	16.858
0.40	80	120	220	438	2.424	16.805
0.50	110	110	220	440	2.479	16.925
0.60	137	80	208	438	2.499	17.069

III. RESULTS AND DISCUSSIONS

Pair Distribution Functions

Pair distribution functions(PDF) for the 0.4 Na$_2$O • 0.6B$_2$O$_3$ system are given in figure 1. The basic features of sodium borate glass were well reproduced. The distance for B-O is 1.35Å; O-O, 2.31; Na-O 2.29 Å; B-B, 2.46Å; B-Na, 3.12Å; Na-Na, 2.73Å.

The coordination states for each atom is calculated for the final hundred configurations and a specific label is assigned based on its coordination state. B$^{[3]}$ is for boron with three fold coordination; B$^{[4]}$ with four fold coordination, NBO is a nonbridging oxygen which has only one boron within the cutoff distance; BO is bridging oxygen which has two boron atoms within cutoff distance, 1.8Å.

The B-O PDF as shown in figure 2, can be deconvoluted into B$^{[3]}$-O and B$^{[4]}$-O based on coordination state of boron. The average distance of B$^{[3]}$-O is 1.35Å and B$^{[4]}$-O is 1.45Å. In addition, the B$^{[3]}$-O PDF is deconvoluted again based on the coordination state of oxygen. The B$^{[3]}$-O PDF is comprised of B$^{[3]}$-NBO with distance 1.31Å and B$^{[3]}$-BO at 1.36Å. The B$^{[3]}$-NBO distance is shorter than B$^{[3]}$-BO, as expected. However, the PDF peak for B$^{[4]}$-NBO is negligible. It means most of NBOs are associated only with trigonally bonded boron, B$^{[3]}$ and they have alkali ions nearby.

Bond Angle Distribution.

Figure 3 shows the bond angle distribution of O-B-O. The curve is comprised of two peaks, one from O-B$^{[3]}$-O where B$^{[3]}$ is trigonally coordinated boron with an angle of 120° and the other from O-B$^{[4]}$-O where B$^{[4]}$ is tetrahedrally coordinated boron with an angle of 109°. As expected, added sodium converts the coordination state of boron from 3 to 4.

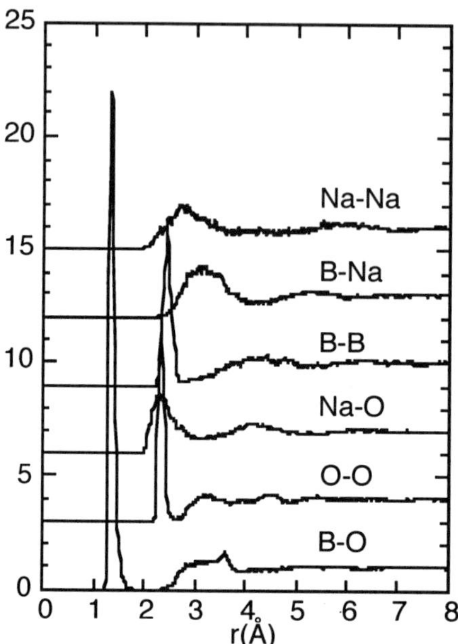

Fig 1. Pair distribution functions of 0.4 Na$_2$O•0.6 B$_2$O$_3$ glass at room temperature.

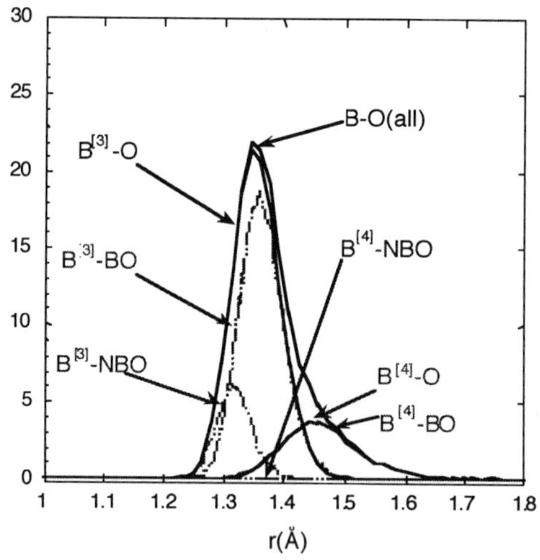

Fig. 2. Deconvolution of B-O pair distribution function of 0.4 Na$_2$O•0.6 B$_2$O$_3$ glass. B$^{[3]}$ denotes trigonally bonded boron, B$^{[4]}$, tetrahedrally bonded boron, NBO nonbridging oxygen, BO, bridging oxygen.

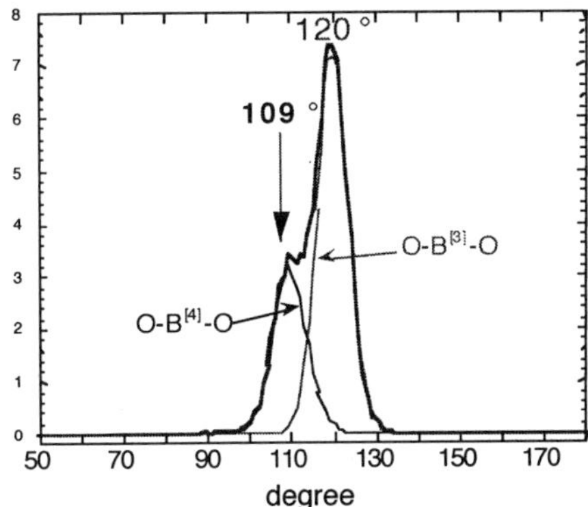

Figure 3. Bond angle distribution for O-B-O of 0.4 $Na_2O \bullet 0.6\ B_2O_3$ glass. The curve is decovoluted into two peaks, O-B[3]-O and O-B[4]-O.

Variation of BO_4 with Added Sodium Oxide

Figure 4 shows the variation of BO_4 tetrahedra as a function of sodium content in soda borate glass. For comparison, the theoretical $x/(1-x)$ curve assuming all the added alkali convert a BO_3 triangle to a BO_4 tetrahedron without generating nonbridging oxygen, the NMR data of Jellison[3] and MD simulation by Verhoef[19] for Cs borate and Li borate are also depicted.

In our current simulated glass, N_4 reached a maximum at 40 mol% sodium oxide addition which matches Jellison's NMR data, although the absolute concentration of N_4 in the simulated glass is smaller. This is due to generation of NBOs from the very first addition of sodium oxide into glass, in contreast to the interpretation of the NMR data.

A number of reasons may be advanced to explain this. First of all, the glass may have high fictive temperature. The amount of BO_4 decreases with increasing temperature. Secondly, cutoff distance used to determine the oxygen coordination state may lead to low concentration of BO_4 tetrahedron. The pure B_2O_3-II crystal structure consists of four-fold coordinated boron and two and three fold oxygen. Therefore if the cutoff distance for determining oxygen coordination state is increased, there is a greater tendency to generate four fold boron. Finally the force constant for O-B-O may not be sufficiently optimized to produce enough BO_4.

IV. CONCLUSIONS

Sodium borate glass was successfully simulated by implementing a coordination dependent potential scheme. The glass consists of not only BO_3 and BO_4 basic structural units but also structural groupings such as boroxol ring, triborate and di-triborate etc. The added sodium converts BO_3 triangles to BO_4 tetrahedra and generates NBO simultaneously. The N_4, fraction of tetrahedrally coordinated boron reaches maximum at 40 mol% Na_2O but the absolute amount is lower than experimental data. This may result from high fictive temperature of the simulated glass, smaller cutoff distance and not fully optimized force constant for O-B-O three body term.

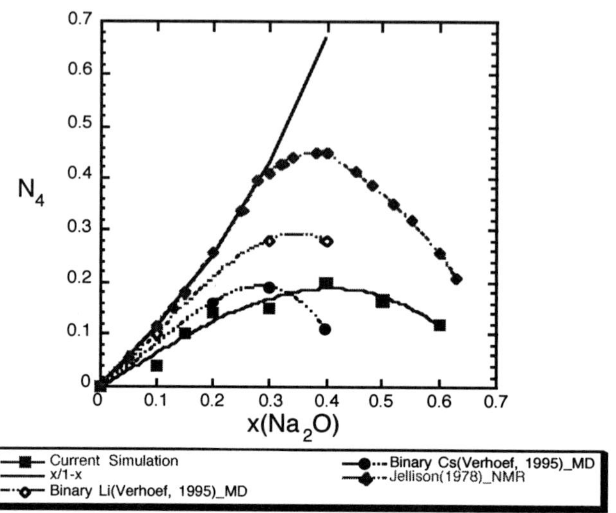

Figure 4. Variation of tetrahedral coordinated boron by added sodium in simulated glass.

REFERENCES

[1] J. Krogh-Moe, "The structure of vitreous and liquid boron oxide,"*J. Non-Cryst. Solids*, **1**, 269-84 (1969).

[2] W. H. Zachariasen, "The atomic arrangement in glass,"*J. Am. Ceram.Soc.*, **54**[10], 3841-51 (1932).

[3] G. E. Jellison, S. A. Feller, and P. J. Bray, "A re-examination of the fraction of 4-coordinated boron atoms in the lithunm borate system,"*Phys. Chem. Glasses*, **19**[3], 52-3 (1978).

[4] E. I. Kamitsos, A. P. Patsis, and G. D. Chryssikos, "Infrared reflectance investigation of alkali diborate glass,"*J. Non-Cryst. Solids*, **152**, 246-57 (1993).

[5] M. Shibata, C. Sanchez, H. Patel, S. Feller, J. Stark, G. Sumcad, and J. Kasper, "The density of lithium borate glasses related to atomic arrangements,"*J. Non-Cryst. Solids*, **85**, 29-41 (1986).

[6] P. J. Bray, "Structural models for borate glasses,"*J. Non-Cryst. Solids*, **75**, 29-36 (1985).

[7] H. C. Lim and S. Feller, "The density of low metal content rubidium, cesium, silver, and thallium borate glasses related to atomic arrangements,"*J. Non-Cryst. Solids*, **94**, 36-44 (1987).

[8] H. P. Lim, A. Karki, S. Feller, J. E. Kasper, and G. Sumcad, "The density of potassium borate glasses related to atomic arrangement,"*J. Non-Cryst. Solids*, **91**, 324-32 (1987).

[9] A. Karki, S. Feller, H. P. Lim, J. Stark, C. Sanchez, and M. Shibata, "The density of sodium-borate glasses related to atomic arrangement,"*J. Non-Cryst. Solids*, **92**, 11-9 (1987).

[10] S. A. Feller, W. J. Dell, and P. J. Bray, "[10]B NMR studies of lithium borate glasses,"*J. Non-Cryst. Solids*, **51**, 21-30 (1982).

        Advances in Fusion and Processing of Glass II

[11]A. C. Wright, N. M. Vedishcheva, and B. A. Shakhmatkin, "Vitreous borate networks containing superstructural units: a challenge to the random network theory,"*J. Non-Cryst. Solids*, **192&193,** 92-7 (1995).

[12]A. N. Cormack and Y. Cao, "Molecular dynamics simulation of silicate glasses,"*Molecular Eng.*, **6,** 183-227 (1996).

[13]B. Vessal, "Simulation studies of silicates and phosphates,"*J. Non-Cryst. Solids*, **177,** 103-24 (1994).

[14]A. H. Verhoef and H. W. d. Hartog, "A molecular dynamics study of B_2O_3 glass using different interaction potential,"*J. Non-Cryst. Solids*, **146,** 267-78 (1992).

[15]T. F. Soules, "A molecular dynamic calculation of the structure of B_2O_3 glass,"*J. Chem. Phys.*, **78**[8], 4032-6 (1980).

[16]W. Soppe, C. v. d. Marel, W. F. v. Gunsteren, and H. W. d. Hartog, "New insights into the structure of B_2O_3 glass,"*J. Non-Cryst. Solids*, **103,** 201-9 (1988).

[17]G. E. Walrafen, S. R. Samanta, and P. N. Krishnan, "Raman investigation of vitreous and molten boric oxide,"*J. Chem. Phys.*, **72**[1], 113-20 (1980).

[18]A. C. Hannon, D. I. Grimley, R. A. Hulme, A. C. Wright, and R. N. Sinclair, "Boroxol groups in vitreous boron oxide: new evidence from neutron diffraction and ineleastic scattering studies,"*J. Non-Cryst. Solids*, **177,** 299-316 (1994).

[19]A. H. Verhoef and H. W. d. Hartog, "Structure and dynamics of alkali borate glasses: a molecular dynamics study,"*J. Non-Cryst. Solids*, **182,** 235-47 (1995).

[20]B. Park and A. N. Cormack, "Molecular dynamics simulation of B_2O_3 glass using coordination dependent potential", in *International Conference on Borate Glasses, Crystals and Melts*, Ed. by A. C. Wright,1996

[21]B. Park and A. N. Cormack, "Molecular dynamics simulation of alkali borate glass using coordination dependent potential", in *Structure and Dynamics of glasses and glass Former*, Ed. by C. A. Angell, T. Egami, J. Kieffer, U. Nienhaus, and K. L. Ngai,Boston,1996

[22]A. Takada, C. R. A. Calow, and G. D. Price, "Computer modelling of B_2O_3: part II. Molecular dynamics simulation of vitreous structures,"*J. Phys.:Condens. Matter*, **7,** 8693-722 (1995).

[23]A. Takada, C. R. A. Catlow, J. S. Lin, G. D. Price, M. H. Lee, V. Milman, and M. C. Payne, "Ab initio total energy pseudopotential calculations for polymorphic B_2O_3 crystals,"*Phys. Rev. B*, **51**[3], 1447-55 (1995).

[24]A. Takada, C. R. A. Catlow, and G. D. Price, "Computer modelling of B_2O_3: part I. new interatomic potential, crystalline phase and predicted polymorphs,"*J. Phys.:Condens. Matter*, **7,** 8659-92 (1995).

HIGH-TEMPERATURE RADIATIVE CONDUCTIVITY OF COMMERCIAL IRON-CONTAINING GLASSES

Oleg A. Prokhorenko and Oleg V. Mazurin
Thermex Company
24/2 Odoevskogo
St.Petersburg, 199155, Russia

ABSTRACT

Results of measurements of high-temperature IR-spectra of six iron-containing float glass melts are presented. The spectra of these melts were divided into four characteristic zones having specific temperature dependencies of absorption coefficients. It is demonstrated in what extent the differences in absorption coefficients of the studied melts within each of these zones lead to the differences in radiative conductivity of melts. It is concluded that for obtaining dependable radiative conductivity data for computer simulation of melting processes one has to measure optical spectra of any new melt of the same type in the wavelength range at least up to 2.7 μm. However absorption coefficients of a molten glass at high temperatures and radiative conductivity for selected type of glasses can be roughly estimated by the use of spectra measured at room temperature.

INTODUCTION

During several last years important steps towards development of a really dependable program for a computer simulation of processes in commercial glass-melting tanks were done [1]. Accordingly, the importance of obtaining correct data for such properties of commercial molten glasses has been greatly increased. Radiative conductivity belongs to such group of properties. It is known that at present the only way to obtain dependable data of this kind is to measure high-temperature absorption spectra of the melts (see for example Ref. [2, 3]).

The authors of the present paper have developed the method of measurements of absorption spectra at high temperatures which gives repeatable results with a reasonable accuracy [4]. By using this method a number of commercial glasses were studied including quite a few float glasses.

It is known that the compositions of some types of float glasses which are produced by different companies are quite similar to each other. This fact leads to a rather obvious question: is it not possible to use a certain mean, the most probable temperature dependence of the radiative conductivity for a group of commercial glasses having similar compositions without necessity to study nearly the same glass produced by each new company (of probably by each new glass-melting furnace of the same company) separately?

The authors of the present paper tried to find an answer to this question by taking as an example a number of iron-containing float glasses with nearly the same iron content (1.55±0.15 wt. %) and with additions of cobalt and selenium.

The objectives of this work were as follows.
- To compare absorption spectra of iron-containing float glasses produced by different companies.
- To compare temperature dependencies of absorption coefficients in several typical wavelength zones of spectra for the same group of glasses.
- To analyze the influence of the difference in the level of absorption within each zone for the studied glasses on the radiative conductivity of the glasses.
- To analyze the influence of the difference in absorption spectra of the studied glasses on the radiative conductivity of these glasses and to compare results with the present level of precision of the measurements for an individual glass composition.

EXPERIMENTAL PROCEDURE

For the measurements of IR absorption spectra of studied glasses two spectrophotometers (SF-2 and SF-3) built by Thermex Company were used. Measurements of IR absorption spectra were performed in the wavelength range from 0.8 to 3.8 μm at temperatures from 20 to 1450°C.

Four-step procedure of determination of temperature dependence of absorption coefficients of glass in near infrared region was applied. This procedure makes it possible to obtain dependable absolute values of absorption coefficients at high temperatures. It consists of the following steps:

Step I. Measurements of the absorption coefficients K_r at room temperature (absorption coefficients are calculated on the natural logarithmic basis).

Step II. Measurements of the changes in absorption coefficients (ΔK_t) caused by the heating of the sample of glass from room temperature to the temperature T_I (it is the temperature corresponding to viscosity of the glass equal to 10^{12}-10^{10} Poise at which the sample of the glass can be measured without deformation). For the studied float soda-lime-silica glasses T_f=600°C is taken. Absolute values of absorption coefficients K_I at T_I are determined by the following formula: $K_I = K_r + \Delta K_I$

Step III. Measurements of the changes in absorption coefficients (ΔK_{x-I}) of molten glass positioned between sapphire windows in the temperature range from T_I up to temperature of reversibility (T_{rev}). The temperature of reversibility is the highest temperature to which one can heat a sample and afterwards cool it down without an appreciable difference between absorption coefficients obtained in the course of heating and cooling. We have found that for regular float glasses T_{rev} is equal to 1150±50°C. Absolute values of K_x at temperatures T_x (lying within the temperature range T_I - T_{rev}) are calculated by using the following equation: $K_x = K_I + \Delta K_{x-I}$. Absolute values of K_{II} corresponding to $T_{II}= T_{rev}-100$ $(K_{II} = K_I + \Delta K_{II-I})$ are also determined.

Step IV. Measurements of optical absorption of molten glass positioned between sapphire windows from T_{II} up to 1450°C. The freshly assembled cell is heated from room temperature to 1450°C with the rate 25 K/min. As the experiments have shown, it is enough to ensure that for the great majority of commercial glasses the time of passing the temperature interval corresponding to the intensive nucleation rates is smaller than the induction time of nucleation. Then the melt is cooled down to T_{II} with the same rate and kept at this temperature for a time needed to reach the constant temperature and to measure absorption spectrum. Then the cycles of heating and cooling are repeated several times. The changes of the absorption coefficients of the melt (ΔK_{y-II}) are registered. Absolute values of K_y at temperatures T_y lying within the temperature range T_{II}-1450°C are calculated by using the following equation: $K_y = K_{II} + \Delta K_{y-II}$

Thus the whole temperature dependence of absorption spectrum is found. Multiple measurements of the glass samples are performed at each of the steps that allows to calculate the confidential limits for absorption coefficient values.

Mean confidential limits of the measured absorption coefficient for fiducial probability 95 % (for the studied group of glasses) were in the range between 0.05 (at 20°C) and 0.15 (at 1450°C).

RESULTS OF MEASUREMENTS

Six float glasses produced by different companies were measured. In this paper the mentioned glasses have numbers from 1 to 6. The absorption spectra of glass No.2 measured at four different temperatures are shown in Fig.1.

Within the whole temperature range of measurements main features of the spectra remain the same. At short wavelengths there is a high enough maximum connected with the absorption caused presumably by bivalent iron. The center of the maximum is positioned at 1.15-1.25 μm. An increase in temperature leads to the decrease and slight broadening of the maximum and to a slight shift of it towards higher wavelength. In the range from ~1.6 to ~2.65 μm a zone

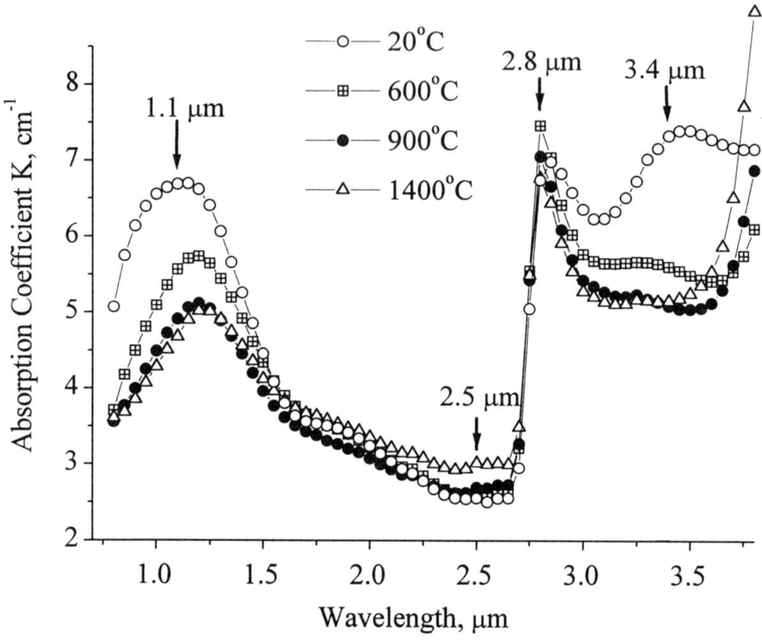

Fig.1. Spectra of glass No. 2 at several temperatures.

of comparatively low absorption exists. At about 2.8 μm there is a high peak resulted from the absorption by water. According to Scholze [5] it is connected with "free" (comparatively weakly bonded with the glass-forming network of a studied glass) hydroxyl groups. The height of the peak decreases with temperature. Scholze supposed that the broad maximum at 3.4-3.6 μm is connected with OH groups being in the immediate vicinity of a nonbridging oxygens which gives rise to the hydrogen bonds between the protons of OH groups and nonbridging oxygens. With increasing temperature the decrease in the height of this band is especially pronounced. At the same time an increase in temperature leads to the shift of the tail of the band connected with vibrations of Si-O bonds to lower wavelengths.

                    Advances in Fusion and Processing of Glass II

Fig.2 presents results of measurements of absorption spectra of all six of studied float glasses at 1400°C. It is seen that the general character of all six spectra are the same although the level of absorption for various glasses is different.

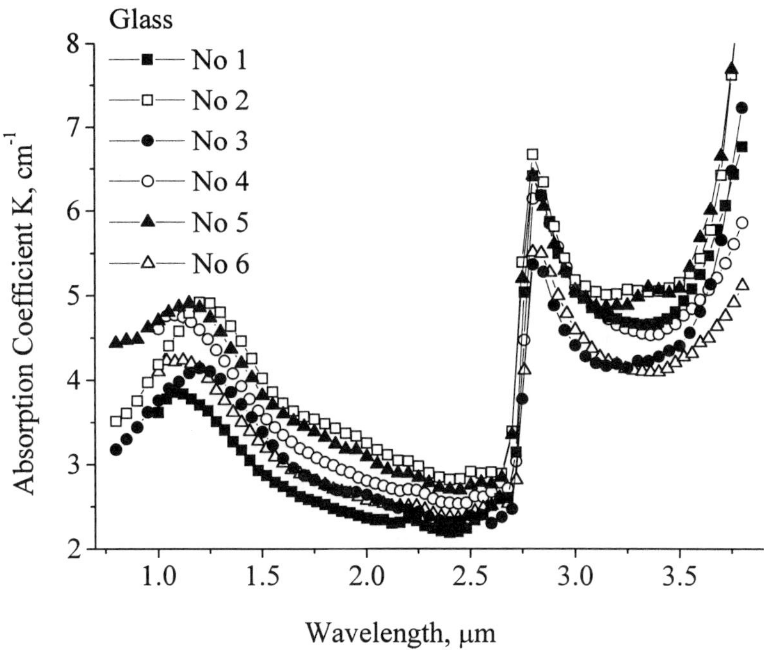

Fig.2. Absorption spectra of six float glasses at 1400°C.

As it follows from Fig.1, the influence of temperature on absorption of various glasses can be analyzed if one presents temperature dependencies of absorption for several characteristic wavelengths. One can select four wavelength zones within which the influence of temperature can be considered as approximately the same. These zones are shown in Fig.3. Zone I covers the band connected with absorption of bivalent iron and corresponds to the range 0.8-1.5 μm. As a characteristic wavelength of this zone 1.1 μm can be taken. Zone II covers the area of the highest transparency of the glass, namely the range from 1.5 to 2.7 μm with characteristic wavelength 2.5 μm. Zone III corresponds to a sharp band connected with the absorption of "free" OH groups (from 2.7 to 3.1 μm with characteristic wavelength 2.8 μm). Zone IV covers both the peak connected with absorption by hydrogen bonds and the short-wave tail of the silicon-oxygen network band (from 3.1 to 3.8 μm with characteristic wavelength 3.4 μm). It is worth nothing that the temperature dependence of absorption coefficient in the narrow right part of zone IV is different from that in the main part of this zone. However, the influence of this small range on the thermal conductivity of melts is insignificant and it is possible to neglect this factor.

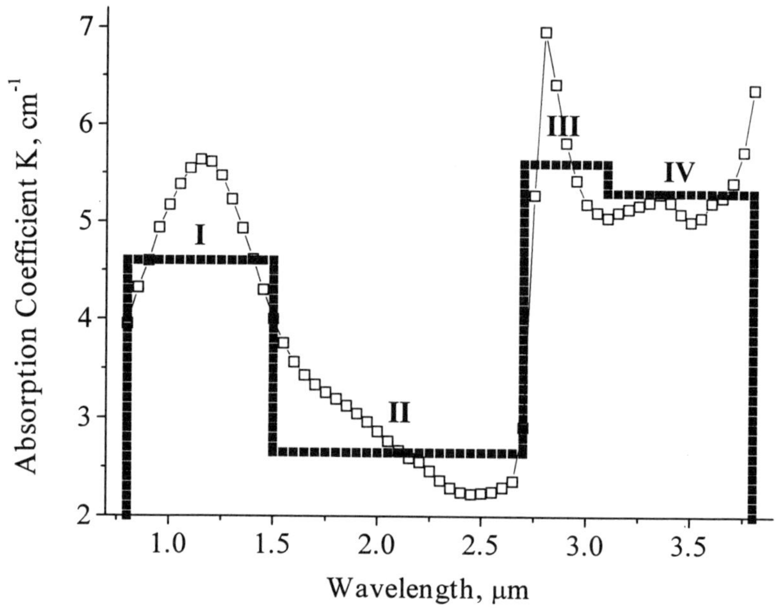

Fig.3. Division of spectra into four zones.

Fig.4 presents temperature dependencies of absorption coefficients for all the studied glasses at the above mentioned characteristic wavelengths. It is seen from the figure that the influence of temperature is quite specific for each of the wavelength in question. The details of the shapes of the dependencies for different glasses at the same wavelength are at variance with each other. However, the general character of all dependencies at one wavelength remains about the same. Thus it could make possible to predict roughly values of absorption coefficients at high temperatures (600-1450°C) by the use of extrapolation of data obtained at room temperature. A precision of such prediction can be found by the evaluation of difference between temperature changes of K in the course of heating from

                         Advances in Fusion and Processing of Glass II

20°C to the high temperatures obtained for different glasses. Influence of uncertainty   of

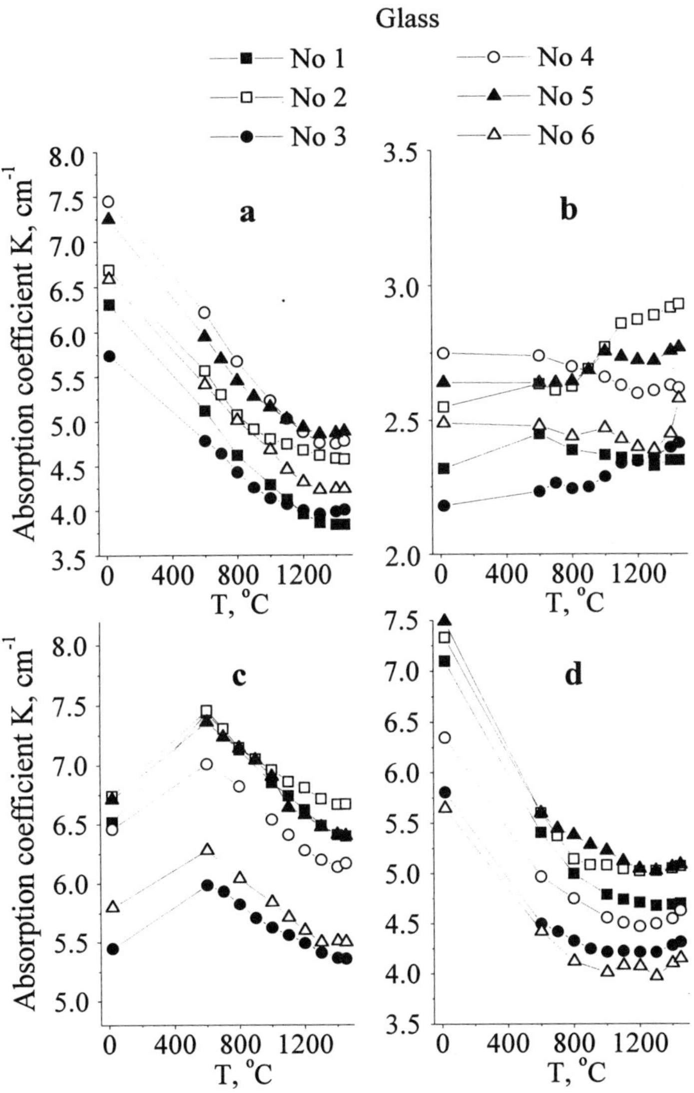

Fig. 4. Temperature dependencies of absorption coefficients for selected
wavelengths: **a)** 1.1 μm; **b)** 2.5 μm; **c)** 2.8 μm; **d)** 3.4 μm

prediction of absorption coefficients on the radiative conductivity will be presented in the following section.

The problems connected with the interpretation of dependencies $K = f(T, \lambda)$ are too complicated to be discussed in the present communication.

DISCUSSION

For practical purposes it is important to know, how the maximal difference between absorption coefficients of all studied glasses in each of the defined above

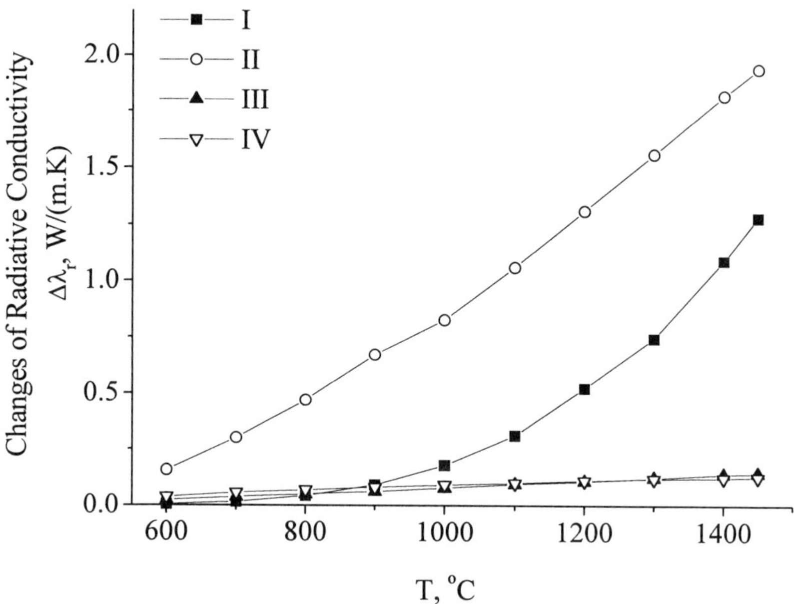

Fig. 5. Changes in radiative conductivity caused by changes of level of absorption coefficient within four zones.

zones could influence the values of radiative conductivity of glass melts at any temperature within the studied temperature range. Such information is presented in Fig.5. It is clearly seen that the corresponding effect for zones III and IV is comparatively very small. It is connected with the high level of absorption in these zones within the whole studied temperature range. It means that if one determines the average spectrum for all studied glasses in these two zones, he could use it for calculations of radiative conductivity of float glasses without necessity to perform measurements of absorption spectra for each new glass. At the same time the scatter of temperature dependencies of absorption coefficients in zones I and II leads to great differences in radiative conductivity values calculated for different glasses. One can see that the character of the discussed dependencies for these two zones are quite different. There are four main factors which should be taken into account in this case. The first factor is the value of the maximal scatter of absorption coefficient. It is considerably greater for zone I than for zone II (cf. Figs 4a and 4b).

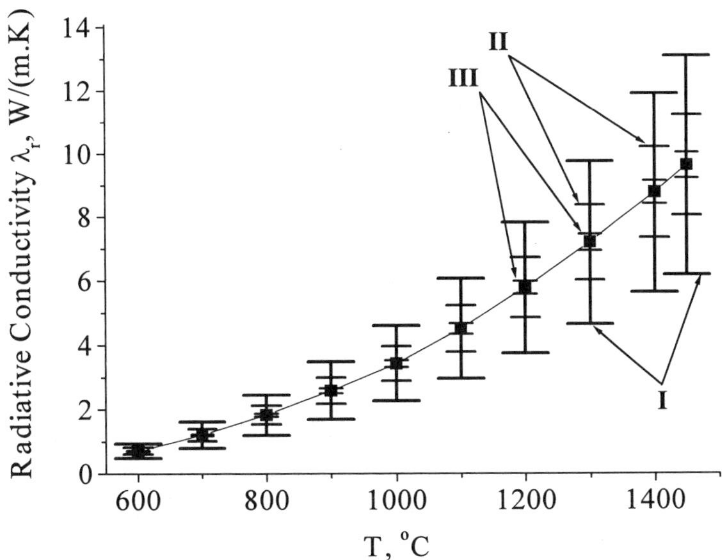

Fig. 6. Distribution of radiative conductivity values for the studied glasses around a mean temperature dependence (I), uncertainty of prediction of radiative conductivity by the use of values measured at room temperature (II) and typical confidential limits of a radiative conductivity measurement (III).

The second factor is the difference of the character of the temperature dependence of K for different glasses. It is much greater for zone II than for zone I. The third factor is the absolute level of absorption in the corresponding zones. The higher the level the smaller the influence of absorption coefficient variations on radiative conductivity values. It is seen that in general the second factor is predominant in our case. The fourth factor is the shift of the maximal intensity of radiation towards the shorter wavelengths with an increase in temperature. The influence of this factor is clearly seen in the figure.

Fig. 6 presents results of the comparison of the possible differences in calculated values of radiative conductivity caused by three different sources: (I) difference of absorption spectra for different float glasses; (II) uncertainty of prediction of the high-temperature radiative conductivity values by the use of spectra measured at room temperature and mean temperature dependence of absorption coefficients; (III) confidential limits of measurements of K for any specific glass. It seems impossible to use mean high-temperature absorption spectra for the calculation of heat-transfer processes in any glass belonging to the group of iron-containing float glasses.

It is possible, however, to estimate very roughly the high-temperature radiative conductivity characteristics basing on the corresponding spectra measured at room temperature and average temperature dependence of K.

CONCLUSION

For iron-containing float glasses produced by various companies, the shapes of near IR-spectra, and the character of temperature dependencies are similar enough. However, the differences in levels of absorption between various glasses could be quite pronounced.

As a result of above mentioned difference, the mean temperature dependence of radiative conductivity for the group of iron-containing glasses could not be used for calculations of temperature profiles in glass melts of the above mentioned group. Before doing such calculations one has to perform measurements of high-temperature spectra for each particular glass of interest. At the same time one can use absorption spectra measured at room temperature and mean temperature dependencies of absorption spectra in the range $20\text{-}1450^\circ C$ for determination of the required characteristics with more or less satisfactory precision.

REFERENCES
[1] R.G.C. Beerkens, Van der T. Heijden and E. Muijsenberg, "Possibilities of Glass Tank Modelling for the Prediction of the Quality of Melting Processes," *Ceram.Eng.Sci.Proc.,* **14** [3-4] 139-160 (1993).

[2] S.V. Tarakanov, V.K. Leko and O.V. Mazurin, "Some Problems of Precise Measurements of Heat Transfer Coefficients in Glass Melts. Part 1. Measurements of Effective Conductivity," *Glastech.Ber.Glass Sci.Technol.,* **68** [10] 301-311 (1995).

[3] J. Endrys, F. Geotti-Bianchini, L. De Riu, "Study of the high-temperature spectral behavoir of container glass," *Glastech.Ber.Glass Sci.Technol.,* **70** [5] 126-136 (1997).

[4] O.A.Prokhorenko and O.V.Mazurin, "Method of High-Temperature Measurements of Radiative Conductivity of Glass-Forming Melts," *Fizika i Khimiya Stekla (English translation: Glass Physics and Chemistry)*. To be published.

[5] H.Scholze, "Der Einbau des Wassers in Glasern: 2. UR-Messungen an Silikatglasern mit systematisch variierter Zusammensetzung und Deutung der OH-Banden in Silikatglasern," *Glastech.Ber.*, **32** [4] 142-152 (1959)

THE TIME-DEPENDENT CHANGE OF RADIATION-INDUCED OPTICAL BAND AS CRITERIA OF GLASS DISORDERING

Alexei Diikov
S.I.Vavilov State Optical Institute,
Babushkina st. 36/1, St.-Petersburg, Russia, 193171
E-mail:diikov@dal.usr.pu.ru

ABSTRACT

The evolution of the spectral shape of the radiation-induced optical band of the deep level defect in amorphous and glassy material was simulated using a computer taking into account microscopic structure of the defect. The model suggests that the energy of local phonon connected with the defect has a wide distribution through glass matrix at short times after the irradiation pulse due to the presence of random local strains in the disordered atomic network of the glass. This distribution determines the observed in experiment average spectral shape of the optical band of the deep level defect. The evolution of the phonon distribution due to reduction of glass disordering leads to evolution of average spectral shape of the optical band, the maximum of which in the case of absorption shifts to the side of high energy. The amount of the shift can be considered as the degree of glass disordering after irradiation.

INTRODUCTION

It is well known that intense radiation (laser, electrons, γ-quanta) cause defect formation in material. In glass these defects create additional absorption and luminescence bands. These radiation-induced optical bands are changed drastically [1, 2] at once after the irradiated pulse on the times of the order 10^{-8} -10^{-3} seconds. The investigations of the dynamics of optical band evolution at short times after the irradiated pulse are of great interest because they give information about the glass structure during irradiation.

In this work evolution of the spectral shape (SSB) of multi-phonon radiation-induced optical band in glass is calculated on the assumption of the strong glass disordering at short times after irradiating pulse. The parameters of SSB is connected with the glass disordering in the terms of the distribution of the energy of local phonon of the radiation-induced defect.

THE MODEL OF EVOLUTION OF SPECTRAL SHAPE OF THE BAND

The presence in material after an irradiated pulse of considerable quasistatic fluctuations of the density and local stresses lead to the wide distribution of the parameters of the adiabatic vibronic term of the radiation-induced deep level defect. This means that the energy of the local phonon connected with this adiabatic vibronic term will have a wide distribution too. To obtain the experimental observed absorption profile of

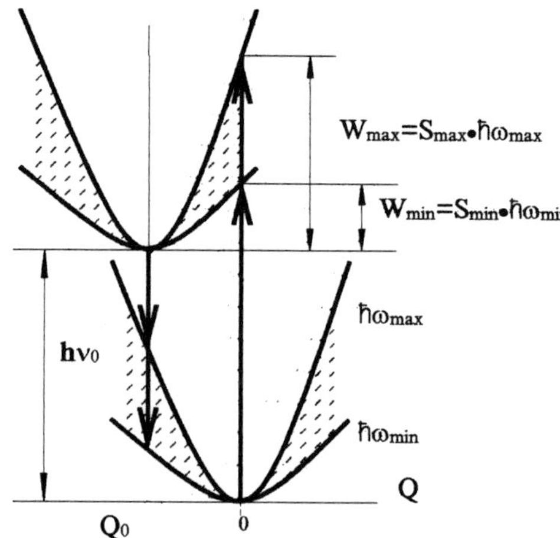

Fig.1. The fluctuations of the polaron energy shift **W**, associated with the fluctuations in the elastic constants of the adiabatic terms of a defect due to structural disorder after an irradiation pulse. The vertical arrows identify the quantum transitions of absorption and photoluminescence. $\hbar\omega_{max}$, $\hbar\omega_{min}$ are the energies of local phonon corresponding to the maximum and minimum fluctuations in the elastic constants; $h\nu_0$ is the energy of zero phonon electronic transition. S is Hyang-Rhys factor; Q_0 is the coordinate polaron shift.

the induced band it is necessary to average SSB of the defect over the distribution of the energies of the local phonon. The averaging procedure alters significantly the SSB compared with the usial Gaussian function [3]. In the course of time the defects which have a small curvature of adiabatic vibronic terms will disintegrate faster, because of defect annealing with the characteristic time $\tau \sim \exp(E_b/kT)$, where E_b is the barrier height which the defect must overcome to disintegrate. This process leads to the reduction of defects with small phonon energies in the phonon distribution and as a result the averaged SSB is changed. The speed and the value of this change will depend upon the glass relaxation after the irradiated pulse and the degree of glass disordering before and during irradiation.

The exactly solvable model of SSB evolution will be discussed below on the basis of the following assumptions:

i) Coordinate polaron shift Q_0 of the defect in the glass matrix [4] is fixed and does not depend on the magnitude of the energy of the phonon (Fig. 1). This assumption simplify calculations and suggests that the change

Advances in Fusion and Processing of Glass II

of the lattice Q_0 in the process of defect annealing is determined only by the defect type and does not depend from the defect environment in the glass matrix. In view of the fixed Q_0 and the linear dependence of the Huang-Rhys factor S [5] on the phonon frequency ($S = (1/(2\hbar)) \cdot m \cdot \omega \cdot (Q_0)^2$; m is the effective mass of a center) S can be represented conveniently in the following form: $S = (\hbar\omega/\hbar\omega_{max}) \cdot S_{max}$, where $\hbar\omega_{max}$ is the maximum magnitude of the energy of the phonon, S_{max} is the Huang-Rhys factor which corresponds to the polaron energy shift for ω_{max}; $W_{max} = S_{max} \cdot \hbar\omega_{max}$ (Fig. 1).

ii) The spectral shape of one defect (without averaging) is taken in the most observed form and described by a Gaussian function (in the limit of strong vibronic coupling) and has the following form [6]:

$$I(h\nu, \hbar\omega) = \frac{1}{[2\pi S \cdot (\hbar\omega^2) \cdot cth(\hbar\omega/2kT)]^{1/2}} \cdot exp(- \frac{(h\nu - h\nu_0 - S\hbar\omega)^2}{2S \cdot (\hbar\omega^2) \cdot cth(\hbar\omega/2kT)}), \quad (1)$$

where $h\nu$ is the energy of absorption; $\hbar\omega$ is phonon energy; $h\nu_0$ is the energy of the zero phonon electronic transition (Fig. 1); S is the Huang-Rhys factor; k is the Boltzmann constant and T is the temperature.

iii) The distribution of energies of local phonon is a beta distribution with the density :

$$P(\hbar\omega) = \{(1/G) \cdot (\hbar\omega - \hbar\omega_{min})^{(\alpha - 1)} \cdot (\hbar\omega_{max} - \hbar\omega)^{(\beta - 1)}, \hbar\omega_{max} > \hbar\omega > \hbar\omega_{min}; 0, \hbar\omega < \hbar\omega_{min},$$
$$\hbar\omega > \hbar\omega_{max}\},$$

$$G = \int_{\hbar\omega_{min}}^{\hbar\omega_{max}} (\hbar\omega - \hbar\omega_{min})^{(\alpha - 1)} \cdot (\hbar\omega_{max} - \hbar\omega)^{(\beta - 1)} \, d(\hbar\omega), \quad (2)$$

where $\hbar\omega_{min}$ and $\hbar\omega_{max}$ are the minimum and maximum energy of the local phonon (Fig.1); α and β are parameters associated with the distribution. The chose of such distribution is determined by presence of finite boundaries of distribution that corresponds to the minimum and maximum phonons.

iiii) The time-dependent change of the defect concentration is described by the annealing kinetics [5]:

$$N(\hbar\omega, t) = N_0 \cdot P(\hbar\omega) \cdot exp(-t/\tau(\hbar\omega)), \quad \tau(\hbar\omega) \sim \tau_0 \cdot exp(E_b(\hbar\omega)/kT),$$
$$(3)$$

$$E_b(\hbar\omega) \approx 2 \cdot W = 2\hbar \cdot (\omega^2/\omega_{max}) \cdot S_{max},$$

where τ_0 is the characteristic life-time of the defect with the minimum phonon energy $\hbar\omega_{min}$, W is the polaron energy shift (Fig.1), N_0 is initial concentration of defects.

Under the assumptions described above we can write the final expression for evolution of average spectral shape of the absorption band I_{av}:

$$I_{av}(h\nu, t)= \int_{\hbar\omega_{min}}^{\hbar\omega_{max}} N(\hbar\omega,t)\cdot I(h\nu, \hbar\omega)\cdot d(\hbar\omega) \qquad (4)$$

THE RESULTS OF CALCULATION AND DISCUSSION

The results of the calculations on the basis of equation (4) are shown in Fig. 2 and 3. In Fig. 2 is shown the distribution of defect concentration $N(t, \hbar\omega)$ as a function of phonon energy $\hbar\omega$ at the four different points in time:

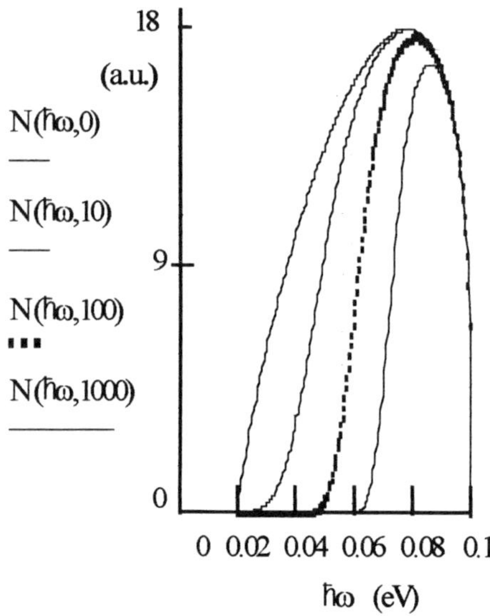

Fig. 2. The time-dependent change of the defect concentration $N(\hbar\omega,t)$ as a function of phonon energy $\hbar\omega$ at the four different points in time $t=0$, $t=10\cdot\tau_0$, $t=100\cdot\tau_0$, $t=1000\cdot\tau_0$, τ_0 is the characteristic life-time of the defect with the minimum phonon energy $\hbar\omega_{min}$ at $t=0$.

$t=0$, $t=10\cdot\tau_0$, $t=100\cdot\tau_0$, $t=1000\cdot\tau_0$ with the typical parameters for defects in vitreous SiO_2: $\hbar\omega_{max}=0.1$ eV, $\hbar\omega_{min}=0.02$ eV, $S_{max}=7$, $\alpha=1.8$, $\beta=1.3$, $N_0=1$, $T=300$ K, $h\nu_0=5.1$ eV ($h\nu_0=5.1$ eV is taken to fit absolute energies of absorption to the band of E' center). In the course of time the distribution evolves from the wide at t=0 to narrow t=1000·τ_0.

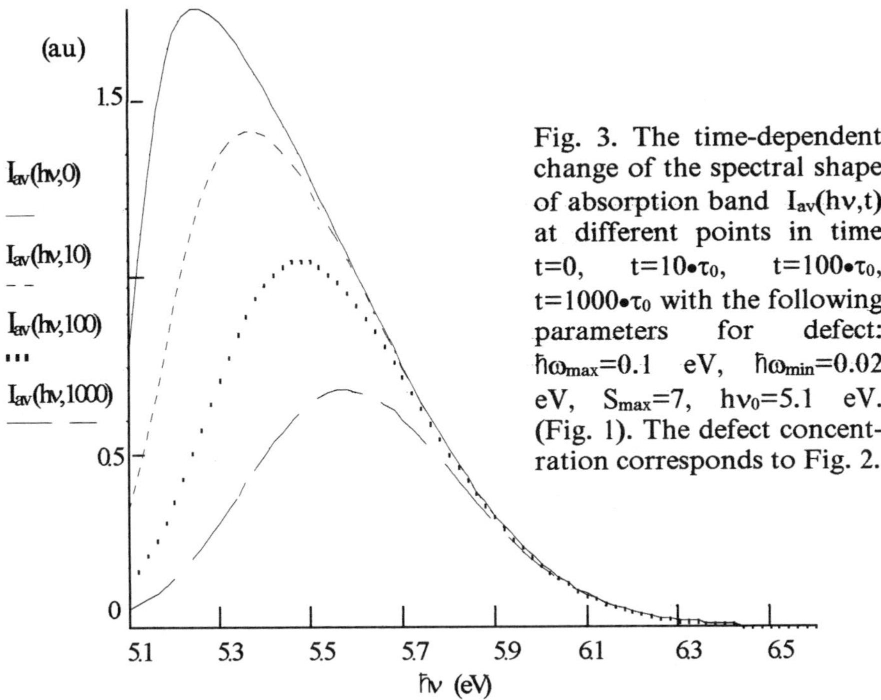

Fig. 3. The time-dependent change of the spectral shape of absorption band $I_{av}(hv,t)$ at different points in time t=0, t=10·τ_0, t=100·τ_0, t=1000·τ_0 with the following parameters for defect: $\hbar\omega_{max}$=0.1 eV, $\hbar\omega_{min}$=0.02 eV, S_{max}=7, hv_0=5.1 eV. (Fig. 1). The defect concentration corresponds to Fig. 2.

Fig. 3 shows the average SSB of absorption calculated using distribution shown in Fig. 2. It is clearly seen that asymmetrical broadening of the SSB (t=0) evolves to the Gaussian line shape (t=1000·τ_0) with the shift of SSB maximum to the side of high energy.

The results of the calculation of the SSB evolution agreed with the experimental results on the evolution of an electron-induced E` absorption band [2]. The estimation showed that the minimum energy of local phonon connected with E' center increases several times with the course of time after an irradiated pulse.

The calculation carried out above shows that the information obtained from the time-dependent change of radiation-induced optical bands can be used to estimate the distribution of energies of local phonon and as a consequence the degree of glass disordering.

CONCLUSIONS

In this paper is described the evolution of the radiation-induced optical band of a deep level defect after influence of radiation. The glass disordering is connected with the distribution of the energy of local phonon of radiation-induced center which comes from considerable fluctuations in elastic constants of the disordered material. The spectral shape of the optical band of absorption is calculated by averaging the spectral shape of the band of one center by distribution of the energy of local phonon connected with this center. The results of calculations show that the averaged spectral shape of the band at short times after the irradiated pulse is assymetrically broadened and its maximum in the course of time shifts in the side of high energy. The suggested model allow to relate the drastic change of radiation-induced optical bands seen at short times (10^{-8} -10^{-3} seconds) after the irradiated pulse with the large fluctuations of local strains of the disordered structure.

REFERENCES

[1] J. Stathis and M. Kastner, "Time resolved photoluminescence in amorphous silicon dioxide", Physical Review B, **35**, [6] 2972-2979 (1987).

[2] C. Itoh, T. Suzuki, N. Itoh, "Luminescence and defect formation in undensified and densified amorphous SiO_2", Physical Review B, **41**, [6] 3794-3799 (1990).

[3] V. Mashkov, A. Diikov, "New mechanism of inhomogeneous broadening of optical spectra of deep-level defects in insulating glasses", Soviet Physics. Solid State (Fizika tverdogo tela), **34**, [11] 1773-1775 (1992).

[4] N. Mott and E. Davis, "Phonons and Polarons"; pp. 80-114 in Electron Processes in Non-Crystalline Materials, 2nd ed. Clarendon Press, 1979.

[5] J. Bourgoin and M. Lannoo, "Optical properties"; pp. 105-138 and "Annealing of Defects"; pp. 266-289 in Point Defects in Semiconductors II, Experimental Aspects. Edited by Manuel Cardona, Springer Series in Solid-State Sciences 35, Springer-Verlag, 1983.

[6] J. Markham, "Interaction of normal modes with electron traps", Reviews of Modern Physics, **31**, [4] 956-989 (1959).

DETERMINATION OF THE SKIN VISCOSITY OF GLASS VIA FIBER ELONGATION

William C. LaCourse and Po Shen
NYS College of Ceramics
Alfred University
Alfred, NY 14802

ABSTRACT

The use of fiber elongation techniques for studying the formation and resulting properties of thin, high viscosity surface layers is described. The governing equations indicate that the existence of layers on the order of 10 nm thickness may be studied. The technique has been applied to studies of the process of gas phase dealkalization using SO_2, SO_3, and $C_2H_4F_2$ by observing elongation during treatments. Results on elongation of fibers with a previously dealkalized surface layer are also provided.

INTRODUCTION

The composition of a glass surface may bear little resemblance to that of the bulk, a fact that has important implications for glass properties such as chemical durability and surface tension which depend primarily on surface interactions. Furthermore, properties not generally considered surface dependent can be strongly influenced by the surface composition. Primary amongst these is the flow behavior of glass during forming, where surface, or "skin" viscosity is critical. Viscosity differences between surfaces and bulk during forming can result from compositional gradients, caused by diffusion controlled dealkalization and volatilization. They generally exist over 10 - 1000 nm. In the present paper we show that a simple fiber elongation test can be used to study existing high surface viscosity layers and the process by which they are formed during high temperature dealkalization treatments.

The elongation rate of a uniform fiber of radius r, in which the elongation is uniform over a length L is given by

$$\varepsilon = \frac{F_t\ L}{3\pi r^2\ \eta} \tag{1}$$

where F_t is the total force acting on the fiber and η is the viscosity. F_t is the sum of three terms

$$F_t = F_f + F_w - F_\sigma \tag{2}$$

where F_f is due to the weight of the fiber, F_w is any added weight and F_σ is the force due to the surface tension, σ. $(F_\sigma = \pi\ r\ \sigma)$ F_f and F_w cause elongation while F_σ causes the fiber to shrink. Therefore, from eq. 1 and 2, when $F_f + F_w = F_\sigma$ the sample will neither elongate or contract. This fact has been used to determine surface tension in various atmospheres.[1]

Consider a "composite" fiber of overall radius r_2 and with a surface layer of thickness $r_2 - r_1$, as shown in Figure 1. Viscosity profiles for two different types of surface layers are also provided. Fig. 1b shows the profile expected when the surface layer is of constant composition. Figure 1c shows a profile resembling those expected from dealkalization or ion-exchange processes. Since each segment of the fiber is constrained to elongate at the same rate, the observed elongation rate will be

$$\varepsilon = \frac{F_t\ L}{3\pi r_2{}^2\ \eta_{eff}} \qquad (3)$$

where η_{eff} is the effective viscosity of the composite. For the profile of Fig. 1b

$$\eta_{eff} = \frac{\eta_1\ \pi\ r_1{}^2\ +\ \eta_2\ \pi(r_2{}^2 - r_1{}^2)}{\pi r_2{}^2} \qquad (4)$$

Fig. 2 illustrates the effect of layer (skin) thickness for values of η_2/η_1 greater than 1. Extremely thin layers can have a strong influence on the observed elongation, when the skin viscosity is higher than the bulk. Thin surface layers with viscosity's less than the bulk are not easily detected. Present calculations assume η_1 is 10^7 Pa·sec. The glass density and surface tension are assumed to be 2.5 g/cm^3 and 0.3 N/m respectively. It is further assumed that the density and surface tensions are the same for the bulk and surface layer. For present purposes $F_w = 0$ so that the fiber elongates under its own weight. By suitably choosing the fiber diameter one can observe the effects of surface layers over a wide range of thicknesses. Figure 3 shows the minimum skin thickness required to observe a 1 0% reduction in elongation rate as a function of overall fiber

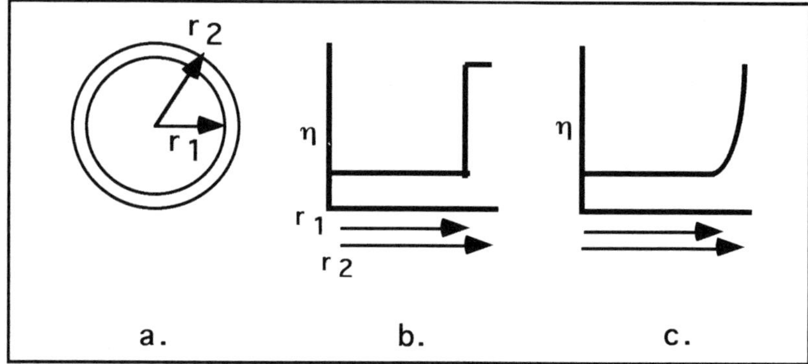

Figure 1. Viscosity profiles in "composite" fibers.

Advances in Fusion and Processing of Glass II

diameter, for viscosity ratios from 10 to 1000. With $\eta_2/\eta_1 = 1000$ for a 0.5 mm diameter fiber the required skin thickness is less than about 0.05 microns.

When the skin forms due to diffusion controlled processes one expects a continuous variation in viscosity. (Fig. 1c) Here the effective viscosity is

$$\pi\, r_2{}^2\, \eta_{eff} = \pi\, r_1{}^2\, \eta_1 + 2\pi \int_{r_1}^{r_2} \eta(r)r\; dr \tag{5}$$

where $\eta(r)$ describes the variation of viscosity along the fiber radius from r_1 to the surface. $\eta(r)$ depends on the compositional profile and its dependence on viscosity. Neither may be known and it therefore difficult to apply the formula to real cases. The overall effect of a variable skin viscosity is to reduce the effectiveness of a given layer thickness in changing the measured viscosity of the composite. However, calculations show that very thin layers can still be observed.

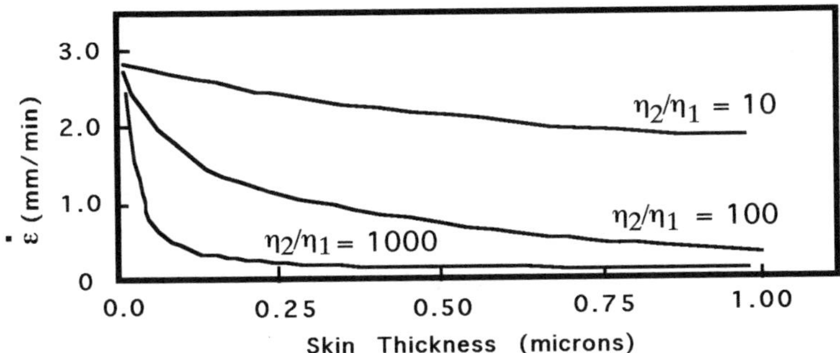

Figure 2. Elongation rate of a composite fiber as a function of skin thickness and viscosity.

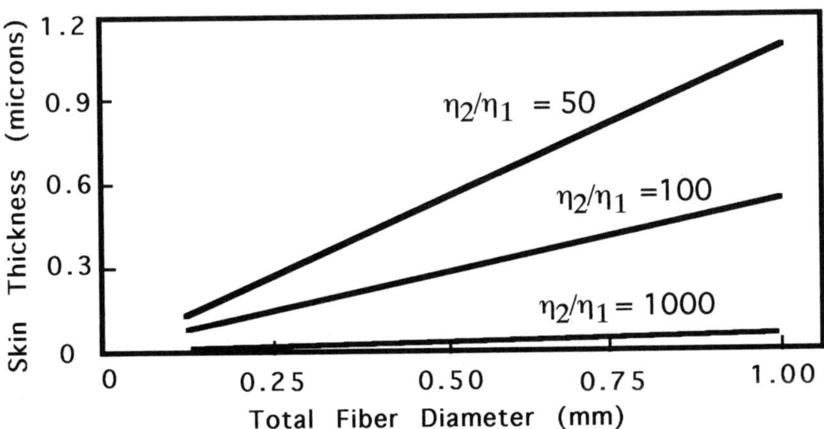

Figure 3. Skin thickness required to reduce effective viscosity by 10 %.

EXPERIMENTAL

A simple test of the model was performed with a composite fiber formed by placing a 3 mm diameter rod of a soda-lime silicate glass inside a tube (3 mm i.d.) of borosilicate, fusing the two surfaces, then pulling a fiber. η_1 and η_2, were determined from experiments on fibers pulled from the soda-lime and borosilicate glasses respectively, Calculated viscosity's, assuming a step gradient, of the composite fiber were in excellent agreement with experiment over the viscosity range encompassing the softening point. We have also used fiber elongation to study several surface treatments important in glass processing or use. These include SO_2, SO_3, HCl, and $C_2H_4F_2$ "dealkalization" treatments used in the glass industry to improve chemical durability (containers) or to provide a lubricious coating (Na_2SO_4) on float surfaces. Treatments are carried out by exposing the hot glass surface to an atmosphere containing the "active gas" above the annealing temperatures for times ranging from a 1 minute to about 30 minutes, depending on application. Some studies were carried out on glasses which had been pre-treated, then analyzed. In other cases elongation was monitored while the treatments were being carried out.

The treatments all produce a "high-silica", (high viscosity) surface due to the following "dealkalization" reactions.[2,3,4]

$$(Si - O^-)R^+ \; + \; HCl \; (or \; HF \; from \; C_2H_4F_2) \; \leftrightarrow \; (Si - OH)glass \; + \; NaCl(surf) \tag{6}$$

$$2(Si - O^-)R^+ \; + \; SO_2 + \tfrac{1}{2}O_2 \; + \; H_2O \; \leftrightarrow \; 2(Si - OH)glass \; + \; Na_2SO_4(surf) \tag{7}$$

$$2(Si - O^-)R^+ \; + \; SO_3 \; + \; H_2O \; \leftrightarrow \; 2(Si - OH)glass \; + \; Na_2SO_4(surf) \tag{8}$$

While alkali (R^+) are the primary diffusing species, divalent ions such as Ca^{++} may also be removed. Note also that if treatments are carried out at high temperatures, R^+ may also be removed by simultaneous diffusion of Na^+ and O^{-2}. A critical, probably simultaneous, reaction occurs at high temperature in which the Si-OH groups react to form Si-O-Si linkages via

$$2(Si\text{-}OH) \; \leftrightarrow \; Si\text{-}O\text{-}Si + H_2O \tag{9}$$

It is this reaction that finally yields the SiO_2 rich, high viscosity surface layer.

RESULTS

Results of a few experiments which demonstrate the technique are summarized here. Figure 4 illustrates the effect of prior SO_3 treatments (600 ^oC for 15 minutes or 2 hours) on the elongation rate of a soda-lime silicate glass at 740 ^oC. These treatments are known to create a dealkalized, silica rich surface layer. Present results show that the layer has a significantly higher viscosity that the original glass. Calculations assuming a "step" gradient and a 0.5 micron skin thickness indicate that the 2 hour dealkalization treatment creates a surface layer with a viscosity more than 2800 times that of the bulk. The increased elongation rate at longer times is due to diffusion of alkali into the dealkalized surface layer.

Figure 5 illustrates the relative effectiveness of SO_3, $C_2H_4F_2$, and a combination of the two gasses. The treatment times was 1 minute at 600 ^oC, and

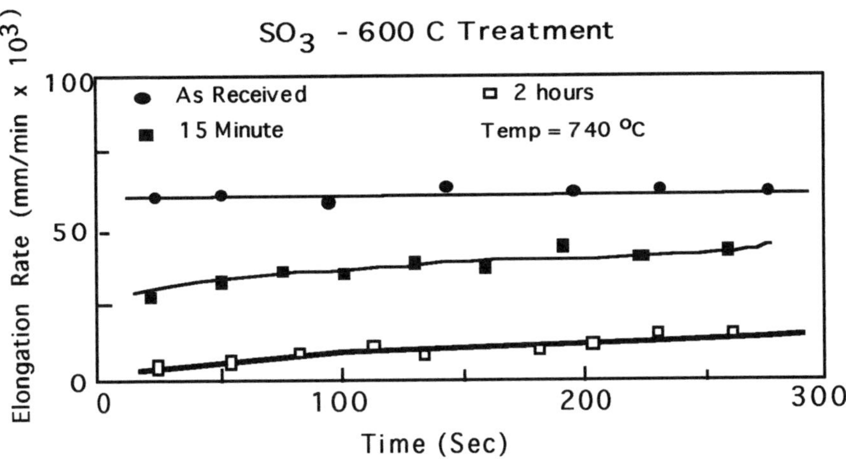

Figure 4. Effect of SO$_3$ treatment on elongation rate at 740 ^{o}C.

the elongation tests were carried out at 700 ^{o}C. The initial contraction is observed in most treated fibers and is due to a relaxation of the surface layer as residual H_2O is removed via equation 9. The results indicate that treatments with a combination of $C_2H_4F_2$ and SO_3 are more effective than treatments with only one gas. This is in agreement with experiments carried out by Ryder et al.[5]

The technique can also be used to investigate changes in surface viscosity and surface tension during exposure to various gasses. Figure 6, illustrates the effect of moist air on elongation of soda-lime silicate glass. The test begins with dry air in the atmosphere, and initially, the elongation rate decreases, presumably due to the removal of hydroxyl from the surface and interior of the fiber. This has two effects. First, the surface tension increases, and secondly, if SiOH groups are removed from a thin surface layer, the surface viscosity increases. When moist air (about 25% H_2O) is then allowed into the system the elongation rate increases by about a factor of 10. This large effect cannot be accounted for by surface tension alone, and suggests a rehydration of a thin surface layer.

Observation of the elongation of fibers exposed to SO_3 + H_2O at high temperature provides important information regarding the processes occurring during treatments. (Figure 6) The fiber first begins to shrink, going quickly through a maximum rate of shrinkage, and eventually stopping. When the H_2O is removed from the environment the fiber almost instantaneously begins to elongate but the rate decreases rapidly until the fiber is essentially stable The following processes are thought to control.

a. The initial shrinkage is due to the dealkalization reaction (eq. 8 for SO_3). The removal of Na^+ and incorporation of Si-OH groups leads to a reduction in molar volume of the surface.

b. The fact that the material continues to contract indicates that the surface layer viscosity remains low, due to the incorporation of H_2O via a high Si-OH content. The continued contraction can be due either to a very high surface tension (unlikely), to the continued dealkalization process, or to a combination of both. Additional experiments have shown that the rate of contraction is related to the concentration of water in the atmosphere. When no water is present the sample contracts very slowly then stops.   Other experiments also show that treatments with $SO_2 + H_2O$ are less rapid, in agreement with other experimental results.

    c. Elongation of the specimen after the removal of moisture from the atmosphere indicates that the dealkalization process is stopped, or slowed dramatically. Importantly, it also shows that the surface layer is of a low viscosity to start, but increases with time and that the surface due to the reaction described in eq. 9. The surface does not reach an "equilibrium" structure/viscosity for some time after the treatment is stopped. Most commercially treated surfaces would therefore be at some non-equilibrium structure with the residual surface water content depending on the exact conditions of treatment.

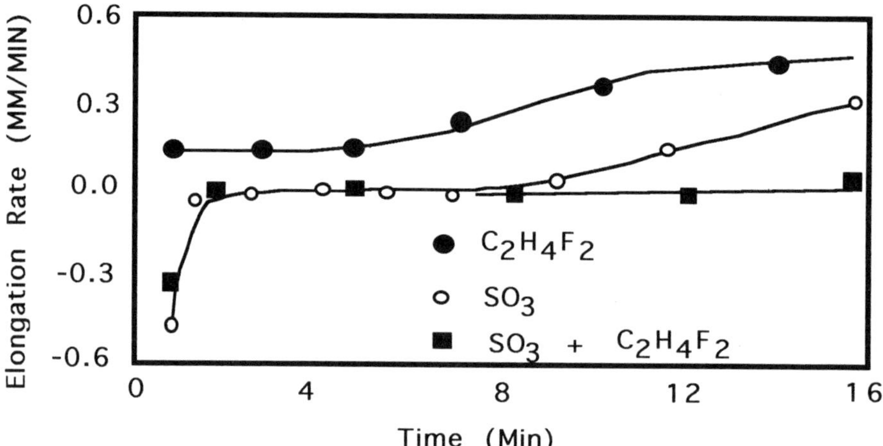

Figure 5. Elongation rate at 720 ºC for glass treated with various "acid gasses" at 600 ºC for 1 minute.

SUMMARY

1. Fiber elongation can be used to obtain information regarding the viscosity of thin surface layers on a constant viscosity core. Skin thicknesses of less than 0.01 microns are easily detected if the viscosity is sufficiently high.

2. Real time investigations in which elongation is monitored during exposure to acidic gas at temperatures near the softening point, yields information regarding the relative efficiency of various dealkalization treatments, the mechanisms of such treatments and the state of the glass surface after treatment.

3. Removal of water from glass surfaces at high temperature creates a high viscosity skin. The process is reversible and the viscosity decreases as the surface reacts with atmospheric moisture.

Advances in Fusion and Processing of Glass II

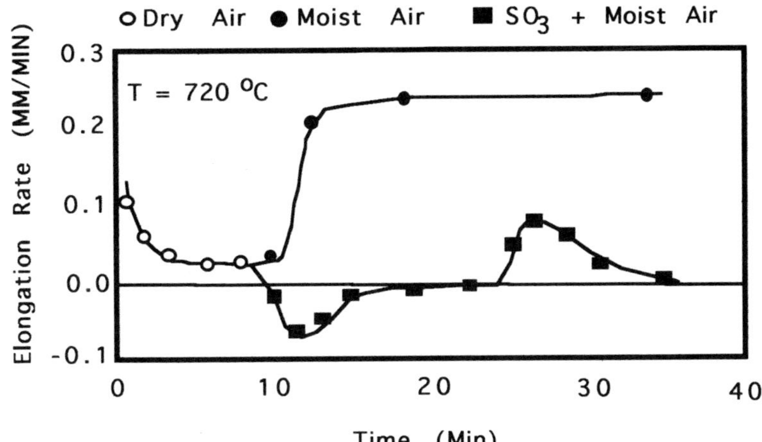

Figure 6. Elongation during exposure to various atmospheres. In the bottom curve the treatment is stopped after approximately 25 minutes, after which the fiber begins to elongate. (See text)

REFERENCES

[1] M.M. Parikh, "Effect of atmosphere on the surface of glass", J. Am. Ceram. Soc., **4**1(1), 18-22 (1958).

[2] R. W. Douglas and J.O. Isard, "The action of water and of sulfur dioxide on glass surfaces" J. Soc. Glass Tech. **33** 289-335 (1949).

[3] H.A. Schaeffer, M. Stengel and J. Mecha, "Dealkalization of glass surfaces utilizing HCl gas" J. Non-Cryst. Solids **80** 400-04 (1986).

[4] R. K. Brow and W. C. LaCourse, "Fluorine treatments of soda-lime silicate glass surfaces" J. Am. Ceram. Soc. **66** C123-5 (1983)

[5] R.J. Ryder, W. J. Poad and Carlo G. Pantano, "Improved Internal Treatments for Glass Containers" J. Can. Ceram. Soc., **5 1** 21-27 (1982)

Advances in Materials for Glass Making

IN SITU OBSERVATION OF THE DISSOLUTION OF REFRACTORY INCLUSIONS IN GLASS

C.H. Yoon, W.C. LaCourse, and J.O. Byun
NYS College of Ceramics at Alfred University
Alfred, NY 14802

ABSTRACT

The dissolution rate and bubble nucleation tendencies of zircon, Al_2O_3, ZrO_2, and AZS refractory in glass were measured utilizing hot stage microscopy in the temperature range from 1150°C to 1400°C. It was found that melting temperature, melting time, and glass composition strongly affect to the dissolution of refractory inclusions in glass.

INTRODUCTION

Dissolution mechanism and rate of dissolution are important for the understanding of refractory corrosion, ceramic-to-metal sealing, ceramic coating, and the rate of conversion of batch to glass. This conversion rate is governed by the dissolution rate of the most refractory particles or inclusions. Sometimes these batch components or impurities in raw materials may survive in the form of stones or cords. Refractory particles may also provide sites for heterogeneous nucleation of gas bubbles[1].

The dissolution kinetics and mechanisms of refractories have been studied for a long time[2-4]. In an effort to understand how dissolution of refractory particles influences the final properties of glass, it was necessary to study how fast refractory particles can be dissolve into glass. Hot-stage microscopy provides a convenient tool for studying batch melting and bubble generation of glasses. The present paper reports on the use of hot stage microscopy study of refractory inclusions in which the effect of temperature, melting time, and glass composition was evaluated.

EXPERIMENTAL PROCEDURE

Commercial SLS glass and A and B glaze frits were employed. The compositions of glasses are shown in Table 1. For the dissolution tests, Al_2O_3, Zircon, ZrO_2, and AZS refractory particles were chosen and sieved through ASTM standard sieve No. 100 remaining on sieve No. 200. For the best results, spherical shaped particles were selected utilizing optical microscopy. Each glass composition was prepared using Pt crucible and glass frit of each composition was remelted on the Pt crucible (D 7.5mm X H 3mm) for dissolution studies. The Pt crucible was placed in hot stage microscope and a single refractory particle was inserted into glass. When desired temperature was attained, the melt was held at that temperature and real-time images were taken through a CCD camera and recorded using an s-video recorder. After these real time images were imported into a computer, the area occupied by a single particle at certain time was calculated using NIH image software. This result was normalized in order to compensate the particle size and surface area differences. For each set of dissolution tests, at least duplicate runs made and the average numbers were plotted against the melting time and temperature. The effect of dissolution on bubble growth and movement, was studied in a similar manner.

Table 1. Composition of glasses used. (mol%)

Glass	SiO_2	Na_2O	CaO	PbO	MgO	Al_2O_3
SLS	72.1	13.4	7.4		5.9	1.2
A	70	15	15			
B	70	15	10	5		

RESULTS

Figure 1 shows the dissolution of Al_2O_3 as a function of time and temperature. In case of SLS glass, an initial rapid dissolution rate followed by a slower linear portion were observed at 1250°C and 1350°C. It may be explained by the build up of a high viscosity product layer around Al_2O_3 inclusion. Due to the density differences and convection currents, the thickness of the product layer becomes constant and then dissolution rate also becomes constant. Reed et al. reported similar results[5]. At higher temperatures, this short initial rapid dissolution disappeared and the full curve exhibited a fairly constant slope above 1350°C. Similar trends were also observed in A and B glasses. Generally, dissolution rate can be determined by two different processes: the rate of reaction at the interface and diffusion of solute. Dissolution rates were calculated from the slope of the major liner portion by the least-square method. This is shown in Table 2. As shown in Figure 1 and Table 2, the dissolution rate of Al_2O_3 depends on the glass composition and temperature/viscosity.

Figure 2 shows the dissolution of zircon and ZrO_2 as a function of time and temperature. It is well known that zircon can decomposed at 1676°C ± 7°C. The decomposition temperature also strongly depends on the glass composition as has been shown by Kato et al.[6] In the present study, as shown in Figure 2 and Table 2, the dissolution rate of zircon in SLS glass is much faster than the dissolution rate of ZrO_2. In the case of zircon, the dissolution rate is accelerated by zircon decomposition. From current results, it is possible to assume that when zircon is decomposed, SiO_2 from zircon dissolves into surrounding glass as does some of the precipitated ZrO_2.

Advances in Fusion and Processing of Glass II

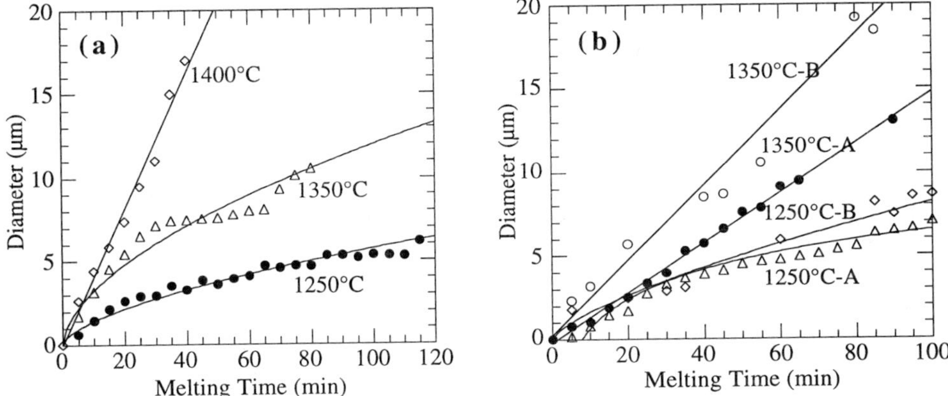

Figure 1. Dissolution of Al_2O_3 in (a) SLS glass and (b) glass A and B as a function of time and temperature.

Table 2. Dissolution rate of various inclusions in SLS glass.

(unit: 10^{-7} cm/s)

Inclusion	Glass	1150°C	1250°C	1350°C	1400°C
Al_2O_3	SLS	-	0.710	2.518	5.970
	A	-	1.127	2.495	-
	B	-	1.374	3.628	-
ZrO_2	SLS	-	0.322	1.068	3.055
Zircon	SLS	1.459	2.330	6.341	-
AZS[*]	SLS	-	1.078	1.511	4.058

*Composition of AZS Refractory(wt%): Al_2O_3 49.7, ZrO_2 33.0, SiO_2 15.3, and other oxides 2.0

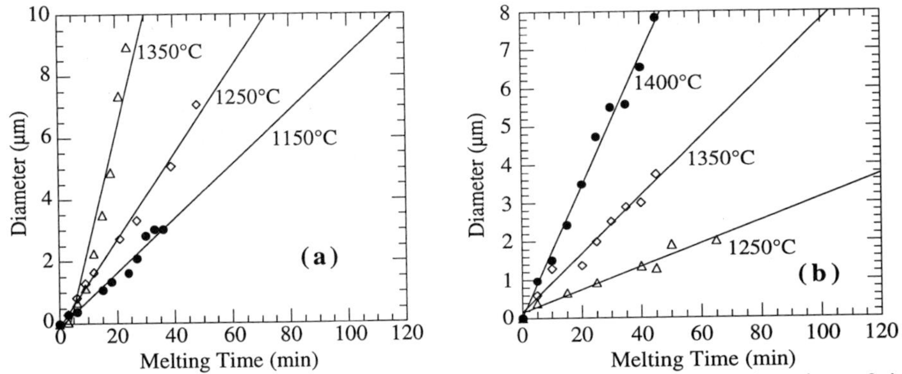

Figure 2. Dissolution of (a) zircon and (b) ZrO_2 in SLS glass as a function of time and temperature.

Dissolution rate can be controlled by two different processes: the rate of reaction at the interface and diffusion of solute. Diffusion process can be expressed by the Noyes-Nernst equation;

$$n/A = k(C_o - C_s) = (D/d)(C_o - C_s) \qquad (1)$$

where C_o, C_s = the concentration in the bulk glass and saturated glass, respectively, k = the rate constant (=D/d), n/A = the rate of dissolution (g/cm^2 s), D = the effective diffusion coefficient, and d = the effective product layer thickness. The constant, k, is obtained from equation (1). In this calculation, the solubility limit of Al_2O_3 and ZrO_2 were obtained from reference 7. The apparent activation energy of the rate constant, E_k, can be obtained from the slope in Arrhenius plot.

$$E_k = -R(dlnk / d(1/T)) \qquad (2)$$

where, R = gas constant, T = temperature in K. The results are shown in Table 3.

Table 3. Calculated dissolution activation energy of inclusions in SLS glass.

Inclusion	E_k (kcal / mol)
Al_2O_3	65
ZrO_2	83
Zircon	29

The dissolution of AZS refractory particle in SLS glass was shown in Figure 3. AS shown in Table 2, the dissolution rate of AZS in SLS glass is higher than the rate for ZrO_2 and lower than the rate for Al_2O_3. From these results, we can assume that as Al_2O_3 from AZS particle dissolves into surrounding glass, it may form Al_2O_3-rich diffusion layer. This layer may slow down the dissolution of ZrO_2.

Figure 4 shows the photomicrographs of bubble growth near refractory inclusions in SLS glass at 1250°C. As a result of frit softening, entrapped air leads to the formation of bubbles at the interface of zircon and glass (see Figure 4(a)). Small bubbles at the interface are growing until they arrive to maximum size to rise to the surface (see Figure 4(b)). In case of Al_2O_3 inclusions, Small bubbles formed at the interface of Al_2O_3 (see Figure 4(c)) and these bubbles easily moved to the top of glass melt after a certain time period (see Figure 4(d)). Detailed studies of bubble generation have not been attempted here. However, previous results suggest that the local compositional changes in the vicinity of the dissolving particles can lower the solubility of various gasses.

CONCLUSION

The dissolution rate of refractory inclusions in glass can be measured utilizing hot stage microscopy at the temperature range from 1150°C to 1400°C. Dissolution of refractory inclusions in glass depends on the melting temperature, melting time, and glass composition. The order of dissolution rate of refractory

Advances in Fusion and Processing of Glass II

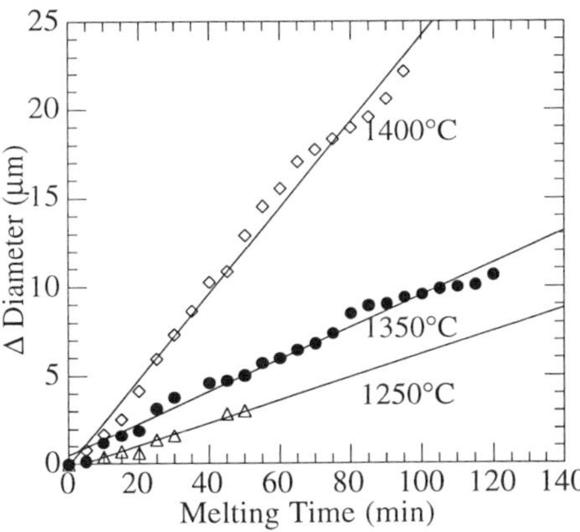

Figure 3. Dissolution of AZS refractory particle in SLS glass as a function of time and temperature.

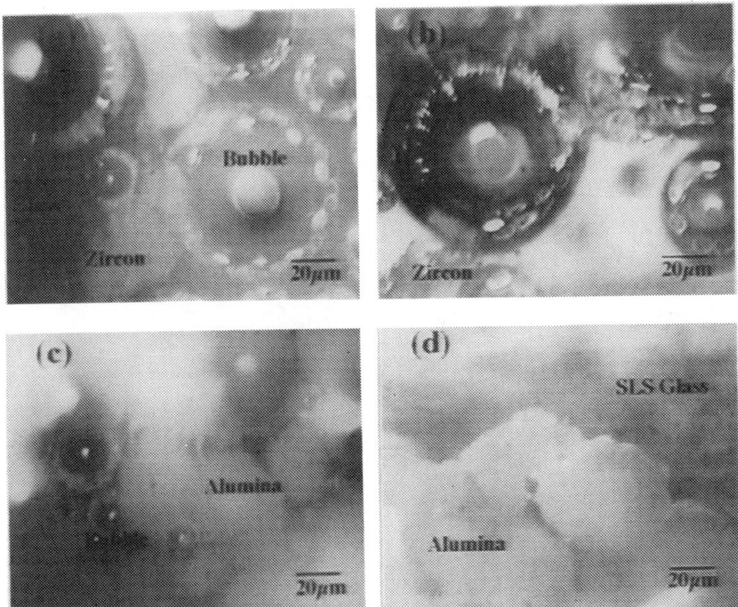

Figure 4. Photomicrographs of bubble growth near refractory inclusions in SLS glass at 1250°C. (a) zircon for 0 min., (b) zircon for 20 min., (c) Al_2O_3 for 0 min., and (d) Al_2O_3 for 20 min.

inclusions in SLS glass is Zircon > Al_2O_3 > AZS > ZrO_2. Refractory inclusions in SLS glass can be the source of heterogeneous bubble nucleation site.

ACKNOWLEDGMENTS

The authors wish to thank the NSF Industry-University Center for Glass Research Center and the Whitewares Research Center of the Center for Advanced Ceramic Technology at Alfred university for support of this study.

REFERENCE

[1] L.Nemec and J. Zluticky, "Possibilities of Photographic Recording of Some Processes Occurring in Molten Glass," *Sklar Keram.*, **29**[12], 353-58 (1979).

[2] A.R. Cooper, Jr. and W.D. Kingery, "Dissolution in Ceramic Systems: I, Molecular Diffusion, Natural Convection, and Forced Convection Studies of Sapphire Dissolution in Calcium Aluminum Silicate," *J.Am. Ceram. Soc.*, **47**[1] 37-43 (1964).

[3] T.S. Busby and J. Eccles, "A study of the Solution of Single Crystals of Corundum in Molten Glass," *Glass Technol.*, **5**[3] 115-23 (1964).

[4] R. Hutchins III, "Dissolution Kinetics in Viscous Systems Where Diffusion and Free Convection are Rate Controlling," *Glass Technol.*, **7**[2] 42-53 (1966).

[5] L. Reed and L.R. Barrett, "The Slagging of Refractories:II. The Kinetics of Corrosion," *Trans. Brit. Ceram. Soc.*, **63** 509-34 (1964).

[6] K. Kato and N. Araki, "The Corrosion of Zircon and Zirconia Refractories by Molten Glasses," *J. Non-Cryst. Solids*, **80** 681-87 (1986).

[7] L.J. Manfredo and R.N. McNally, "Solubility of Refractory Oxides in Soda-Lime-Glass," *J. Am. Ceram. Soc.*, **67**[8] C-155-C-158 (1984).

Advances in Industrial Gas Supply for Glass Manufacturing

Susan M. Sattan and Dante P. Bonaquist
Praxair, Inc.
175 East Park Drive
Tonawanda, NY 14150-7891

ABSTRACT

Oxygen, nitrogen and hydrogen are important materials used in the manufacturing of glass. Glass furnace throughput and efficiency are enhanced by conversion to oxy-fuel combustion while NOx emissions are reduced. Nitrogen and hydrogen are critical for maintaining the quality of glass from float facilities. This paper reviews recent advances in gas supply systems relevant to the glass industry. We cover improvements to cryogenic and vacuum pressure swing adsorption technologies that have contributed to capital cost reductions and enhanced energy efficiency of the corresponding gas supply system. The development of systems co-producing oxygen and nitrogen in ratios required by float facilities is described. Simplifications to cryogenic plants eliminating the need for oxygen compressors to deliver product at pressures typical for glass furnace use are reviewed. A small on-site hydrogen generator having the potential to replace liquid hydrogen as the preferred supply method for small volume consumers is introduced.

INTRODUCTION

The use of industrial gases by the glass industry has enabled increased productivity of manufacturing systems and improved product quality. For example, the float glass industry uses a mixture of hydrogen and nitrogen as a protective atmosphere to maintain the molten tin bath thereby improving the quality of the glass produced. Melting productivity is enhanced with oxygen. Relatively small amounts of oxygen can be used in oxy-boost systems where supplemental burners are installed in existing furnaces. In addition to the increased throughput, NOx emissions are also reduced. A complete conversion to oxy-fuel combustion can lead to even greater improvements in both productivity and emissions. It is clear from these examples that industrial gases qualify as key materials for glass making.

COMBUSTION GRADE OXYGEN SUPPLY

Commercial oxygen purity is typically set by the needs of large volume end users. Since systems for the supply of industrial gases were originally developed around the high oxygen purity needs of the steel and chemical industries, large scale cryogenic systems making more than 1.6 kg/s (150 tons per day) of oxygen will generally be capable of producing oxygen at 99.5% or better. Depending upon the specific application, a higher oxygen purity can be desirable. For example, certain chemical oxidation processes specify higher purity oxygen to minimize the loss of reactants and products due to purging of inert components. Figure 1 shows the concentration of nitrogen and argon impurities as a function of oxygen purity for oxygen produced from a cryogenic system. The argon content of the oxygen product remains approximately constant until the nitrogen impurity almost completely disappears. Oxygen and argon are difficult to separate from each other compared to oxygen and nitrogen. As a result, more complex plants with increased energy consumption are needed to make oxygen at a purity greater than 96%.

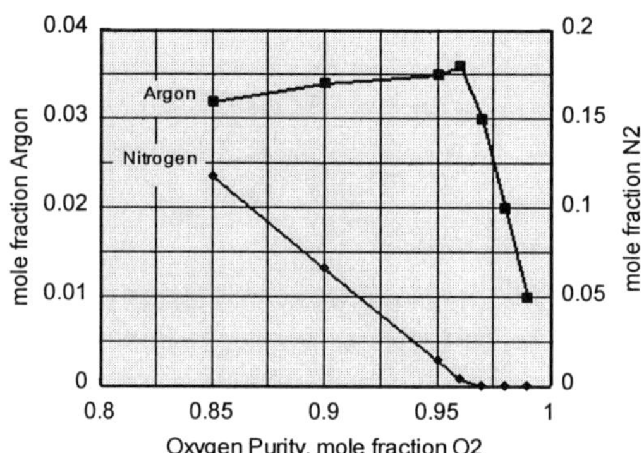

Figure 1. Distribution of Impurities in Cryogenically Produced Oxygen

Oxy-fuel combustion has the advantage of a reduction in the quantity of nitrogen moving through the heat recovery system and furnace. Fuel efficiency, throughput and emissions are all improved by conversion to oxy-fuel combustion. For the glass industry, oxygen purity between 90 and 95%, referred to as "low purity oxygen", is optimum in terms of benefits measured against cost. For this purity level required, the vacuum pressure-swing adsorption (VPSA) unit may be used. The distribution of the impurities in the oxygen from this system are shown in Figure 2. The VPSA system is a method of separating the components of air by adsorption. A purity of 90% oxygen can be obtained from the VPSA with the major impurities of argon (4.5%) and nitrogen (5.5%). For higher oxygen purities (up to 95%) the argon impurity remains constant and the nitrogen impurity decreases.

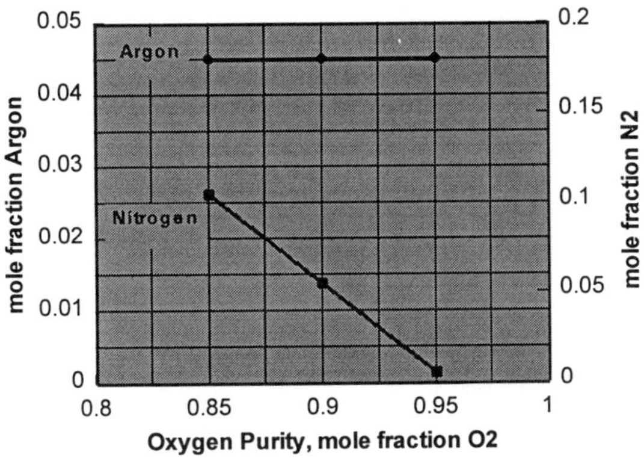

Figure 2. Distribution of Impurities in VPSA Produced Oxygen

Oxygen volumes needed by the glass industry will range from 0.1 to 5.2 kg/s (10 to 500 tons per day). Supply systems based on cryogenic distillation of air and vacuum pressure-swing adsorption (VPSA) are both applicable in the 90 to 95% purity range as shown in Figure 3. Cryogenic systems are generally favored based on the cost of delivered oxygen at capacities greater than about 1.6 kg/s (150 tons per day). Figure 4 shows that the cost of oxygen from cryogenic systems rises dramatically above about 95%, while continuing to drift downward for oxygen purity below 90%. The cost of oxygen from VPSA systems is insensitive to oxygen purity in the 90 to 95% purity level.

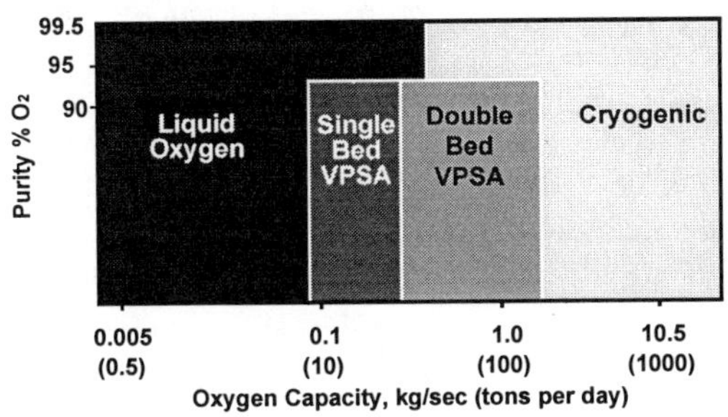

Figure 3. Modes of Supply for Oxygen

Advances in Fusion and Processing of Glass II

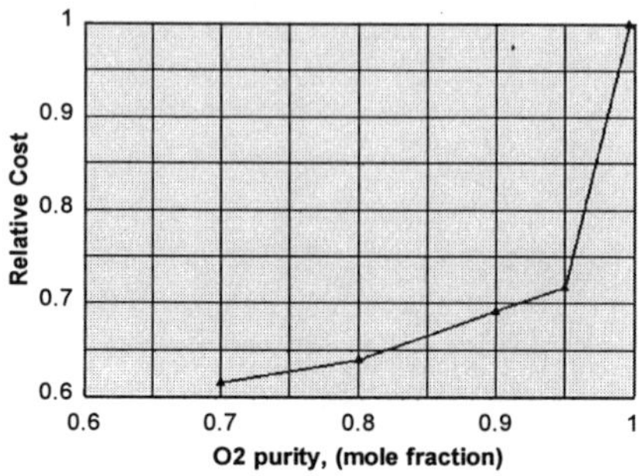

Figure 4. Relative Cost of Contained O_2, Capital + Power

CRYOGENIC DISTILLATION

The majority of atmospheric (nitrogen, oxygen and argon) industrial gases consumed today are obtained from plants employing cryogenic distillation. These plants are often located on the customer's site or near-by so that the gas is supplied through a dedicated pipeline. This scenario of a dedicated cryogenic plant has allowed the supply system to be tailored to the exact customer specifications. The industrial gas industry refers to such systems as "on-site" plants.

The heart of a basic cryogenic plant for oxygen production is the double distillation column. As shown in Figure 5, the double column consists of low and high pressure distillation columns that are thermally linked. This linkage is provided by the main condenser that simultaneously acts as the reboiler of the low pressure column and the condenser of the high pressure column. The low pressure column is also known as the upper column since it is often placed above the high pressure column. For the same reason, the high pressure column is also known as the lower column.

Ambient air is filtered as it is drawn into the inlet of the primary air compressor. For plants in the 2.1 to 5.2 kg/s (200 to 500 ton per day) capacity range, the feed air compressor consists of 2 to 4 discrete stages of compression with cooling of the air between stages and following the last stage to remove the heat of compression. The high pressure feed air, now at about 6.2×10^5 Pa (75 psig), enters a purification system know as a prepurifier. This unit removes most of the contaminants from the air by selective adsorption. Water and carbon dioxide are removed in the prepurifier because they would freeze and cause solid plugs in the cryogenic heat exchangers and distillation system. Hydrocarbons heavier than propane are also removed by the prepurifier to preclude their concentration in liquid oxygen within the low pressure distillation column.

                    Advances in Fusion and Processing of Glass II

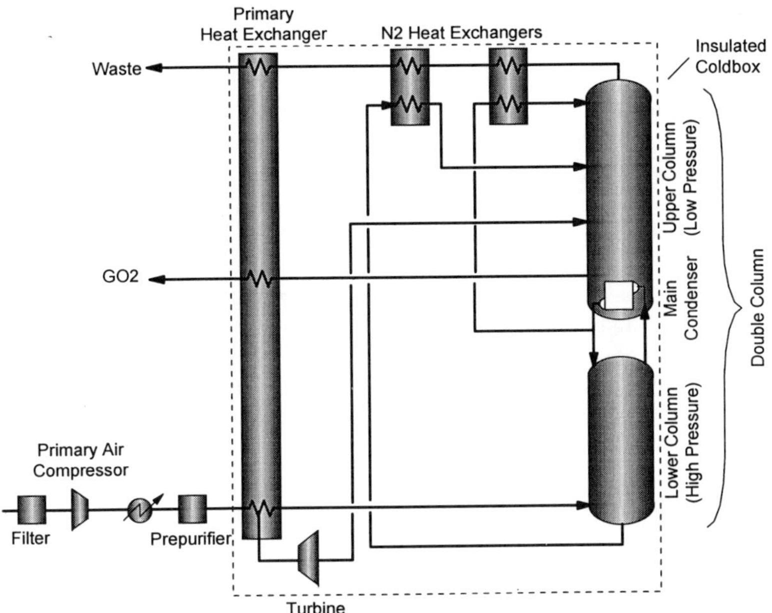

Figure 5. Typical Cryogenic Oxygen Plant

The clean, dry high pressure air then passes through a heat exchanger where it is cooled to cryogenic temperatures against warming product streams. A portion of the air is withdrawn from the heat exchanger and expanded in a turbine to provide the required refrigeration due to system heatleak, thermal inefficiencies, and the withdrawal of liquid products. The main air stream is fed to the high pressure column where an initial separation takes place. The high pressure column fractionates the feed air to generate a relatively pure overhead nitrogen vapor and an impure oxygen rich bottom liquid. The oxygen rich liquid withdrawn from the bottom of the high pressure column is subcooled against colder waste gas (composed mainly of nitrogen) and fed to the low pressure column where its oxygen concentration is increased. The overhead nitrogen vapor of the high pressure column is condensed against boiling liquid oxygen in the base of the low pressure column. The resulting nitrogen liquid is used to reflux both the high and low pressure columns. The low pressure column completes the separation of its feed streams into the product oxygen and nitrogen. For low purity (90 to 95%) oxygen, argon will tend to concentrate in the product oxygen. Oxygen vapor is withdrawn from the bottom of the low pressure column and warmed to near ambient conditions in the primary heat exchanger. The nitrogen vapor is removed from the top of the low pressure column and is similarly warmed in the heat exchangers to recover the refrigeration from the low temperature stream.

The boiling points of nitrogen and oxygen are such that in a distillation column, the nitrogen will concentrate in the rising vapor while the oxygen will concentrate in the descending liquid. As stated earlier, argon will tend to remain with the oxygen for low purity oxygen production. The operating pressure of the low pressure column is set to permit the nitrogen vapor to flow out of the plant without the need for any type of blower. The operating pressure of the high pressure column is established by the thermal link provided by the main condenser. Typical main condenser temperature driving force is about 2 degrees Celsius. Therefore, the pressure at the top of the high pressure column is the dew point pressure of the overhead nitrogen vapor at a temperature 2 degrees Celsius above that of the boiling liquid oxygen in the bottom of the low pressure column. The temperature of the boiling liquid oxygen will vary slightly depending upon its purity thereby providing a relationship between oxygen purity and plant power consumption. For example, reducing the oxygen purity from 95% to 90% reduces the high pressure column pressure and therefore the feed air pressure to give about a 5% reduction in plant energy consumption.

ADVANCES IN CRYOGENIC SUPPLY SYSTEMS

There have been significant improvements over the past 10 years in cryogenic air separation technology and components used in the assembly of gas supply systems. Combined with the reduced purity requirements of the glass industry, these advances lead to lowering the cost of oxygen supplied by cryogenic means by 30% or more during that period.

The production of low purity oxygen for the glass industry permits the utilization of energy efficient distillation methods. Low purity oxygen production does not require the separation of oxygen from argon. In the separation process, the argon merely behaves like additional oxygen; so the separation is between an oxygen-argon mixture and nitrogen. The double column system previously shown can be modified to distribute low pressure column boiling energy over a range of temperatures reducing the energy consumption for the production of 90% oxygen by as much as 25% compared to 99.5% oxygen. Figure 6 is one example of a class of low purity oxygen plants that supply low pressure column boiling energy discretely at two temperature levels. The effect of this is to reduce the pressure ratio between the low and high pressure column. Since the low pressure column operates at a constant pressure just above ambient, the reduction in pressure ratio translates to a reduction in feed air pressure and therefore a reduction in plant energy consumption. The reduced pressure ratio can lead to lower capital cost as well, since fewer air compressor stages are needed to efficiently compress the air. For cryogenic plants used to supply oxygen to the glass industry requiring capacities from 1.6 to 5.2 kg/s (150 to 500 tons per day), energy contributes between 30 and 50% of the total cost of oxygen (not including the cost of liquid storage and vaporization for backup needs). The total impact of energy efficiency improvements can be as much as 12.5% of the total cost of oxygen.

Advances in Fusion and Processing of Glass II

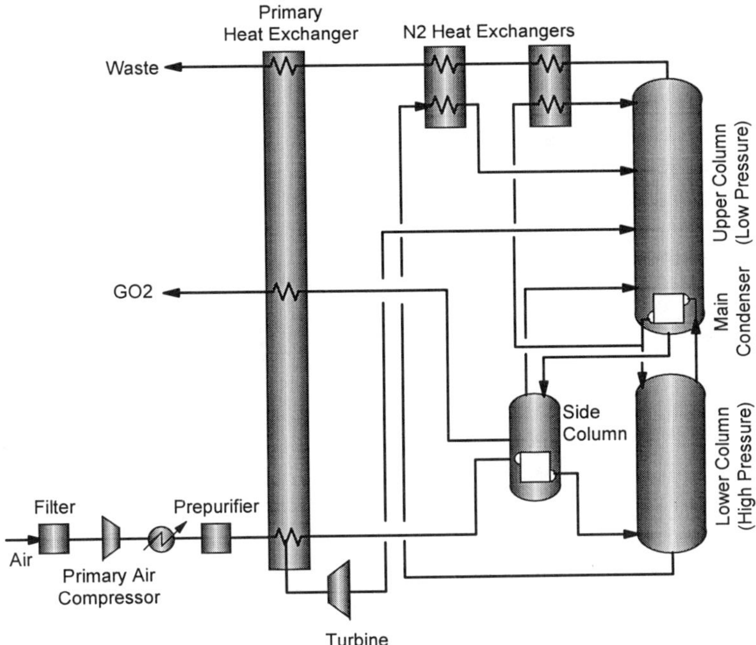

Figure 6. Improved Cryogenic Oxygen Plant

Further improvements have come in the form of high efficiency low cost expanders that are needed by the cryogenic plant to overcome refrigeration losses and to permit the production of liquid oxygen for backup. At 2.6 kg/s (250 tons per day), about 1.0×10^5 J/s (100 kW) of refrigeration at -123.15 degrees Celsius (150K) is required to supply the total refrigeration needs of the plant. Recent advances in expander design have lead to increased efficiency and reduced capital cost. The efficiency gains result in improved capability for liquid production and high recovery of oxygen contained in the feed air. Today's low purity plants are capable of producing at least 5% of the oxygen as liquid without compromising the cost and efficiency of the plant for gas production. Greater quantities of liquid, including some high purity liquid, may be produced by resorting to more complex process schemes. Expanders are commonly coupled to single stage air boosters that further enhance the performance of the plant by recovering the work available from the expander, using it to increase the expansion ratio. Figure 7 is an example of this practice made economical by low cost expanders and boosters developed by the air separation industry.

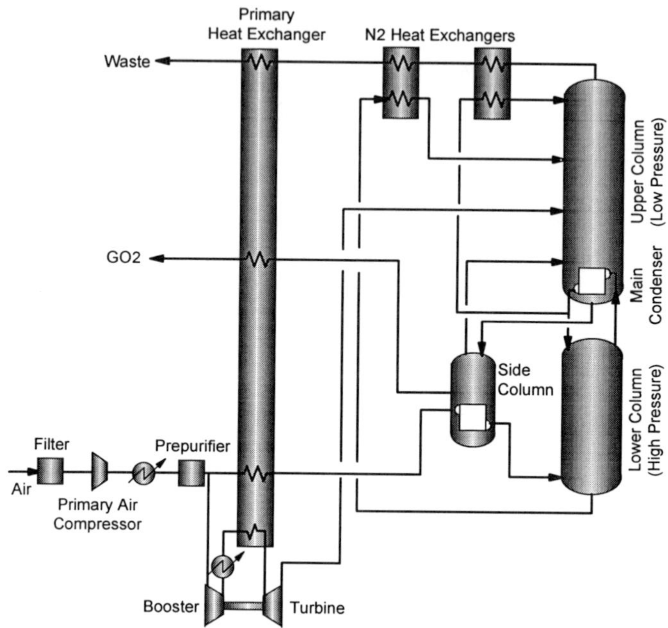

Figure 7. Improved Cryogenic Oxygen Plant with Turbine-Booster

Perhaps one of the most significant improvements in cryogenic plant technology has come with the replacement of distillation trays, know as sieve trays, with structured packing. Structured packing is comprised of corrugated metal sheets that are bundled together to provide the surface area necessary for vapor-liquid contact within the distillation column (see Figure 8). The pressure drop within the column obtained with structured packing is as little as 10% of that for sieve trays. Structured packing is usually installed only in the low pressure column where the pressure drop improvements are leveraged by a factor of nearly 3 across the main condenser. This translates into an additional 5% energy savings. The low pressure drop of structured packing enables process arrangements that are particularly advantageous for supply of low purity oxygen and high purity nitrogen from the same plant for float glass facilities. In addition to these benefits, structured packing allows the low pressure distillation column to operate over a very wide capacity range (as much as 3 to 1) while reducing the cost of the plant. The cost of columns based on structured packing is now equal to or lower than their sieve tray counterparts.

Advances in Fusion and Processing of Glass II

Figure 8. Structured Packing

Recent advances in process control and optimization technology have been applied to cryogenic oxygen supply systems to provide a number of benefits. Model Predictive Control (MPC) technology uses a dynamic model of the process to predict its response to changes in adjustable process variables. MPC is now used to control cryogenic supply systems where there is a need to make rapid changes in plant capacity and product mix. Although these situations are rare in glass manufacturing, the industrial gas industry's knowledge of the dynamic response of cryogenic systems has developed to the point where very tight product purity specifications, including those with upper and lower limits, are easily satisfied to maintain the product at a constant purity level. Real time optimization and supervisory control systems are used to maintain peak performance of the cryogenic oxygen plant.

Capital costs of cryogenic oxygen plants have seen significant reduction due to the use of standard components and flexible modular design concepts. As a result, it is possible to tailor the characteristics of a cryogenic supply system to the needs of a specific customer without incurring the costs related to custom design, fabrication, construction and commissioning.

VPSA SYSTEM

Pressure swing adsorption for oxygen production involves the use of synthetic zeolites (molecular sieves) to adsorb nitrogen from a pressurized air stream passing over a fixed adsorbent bed. The zeolites are complex crystalline inorganic polymers that are made up of a framework of AlO_4 and SiO_4 tetrahedrals. These are linked together by the sharing of the oxygen ions. The linkage constructs a structure with channels and metal cations. The nitrogen and other contaminants in the air are adsorbed based on their electrical attraction to the cations that are within the zeolite framework.

VPSA systems operate with feed pressures typically above atmospheric pressure during the adsorption step. De-sorption takes place under a vacuum. Vacuum de-sorption increases the effective nitrogen adsorption capacity of a given molecular sieve. The basic VPSA process involves passing air over an adsorbent bed until that bed is saturated with nitrogen. At this point, the bed is removed from service and regenerated. To accomplish this, the pressure is lowered, and the nitrogen de-sorbed. The regenerated bed is purged with a small quantity of the product oxygen, re-pressurized and returned to service. Cycle time, from the beginning of the adsorption step through regeneration to the point where the bed is ready to begin a new adsorption step, is about 1 minute. VPSA systems have become the supply technology of choice for low purity oxygen at capacities less than 1.6 kg/s (150 ton per day) because of their low capital and operating cost. The past 5 years have seen a steady increase in the upper limit of VPSA capacity.

The number of adsorption beds used in VPSA systems have ranged from 1 to 4 depending on the molecular sieve and the specific details of the regeneration step. Two bed systems are commonly used for supply of low purity oxygen to the glass industry. Figure 9 is a schematic of a typical 2 bed VPSA. The system consists of a suction filter, air blower, switching valves, two adsorbent beds, vacuum blower and oxygen compressor. The use of two beds allows for continuous operation where one bed is in the adsorption step and the other is in the regeneration step. An oxygen receiver or surge tank is used so that the switching of adsorbent beds will not effect product pressure or flow variations.

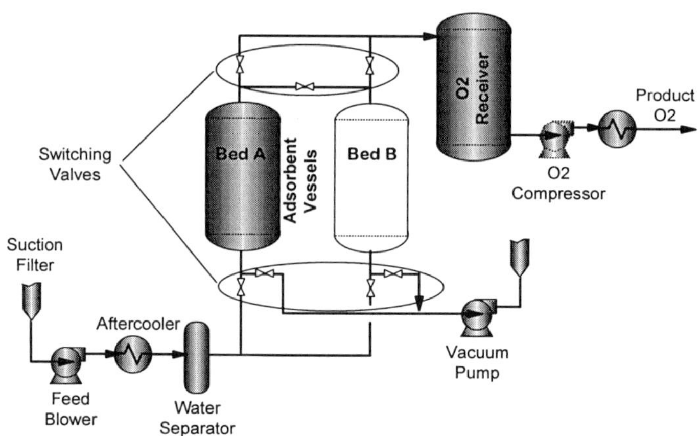

Figure 9. VPSA Oxygen Supply System

Air is compressed to just over atmospheric pressure by the feed blower and cooled to near ambient temperature in the aftercooler. Following removal of liquid water, the air then passes through switching valves to one of two adsorbent beds, Bed A, where water vapor, carbon dioxide, hydrocarbons and nitrogen are adsorbed (see Figure 10). Oxygen product at 90 to 95% purity leaves the adsorption bed through a series of switching valves entering the oxygen compressor for delivery to the customer. As Bed A approaches saturation, the feed air is switched to Bed B that was previously regenerated. Now Bed A undergoes regeneration having its pressure reduced to sub-atmospheric pressures by the vacuum blower. The nitrogen, hydrocarbons, carbon dioxide and water

                                   Advances in Fusion and Processing of Glass II

vapor de-sorb at the lower pressure as shown in Figure 11. At the end of the evacuation period, a small amount of product oxygen is used to sweep the remaining de-sorbed components from the void spaces around the molecular sieve particles within the bed prior to re-pressurization with air to begin the next adsorption step. The control system in the VPSA controls the purity to the desired level (between 90 and 95%) with the constant impurity of about 4.5% argon and the balance nitrogen.

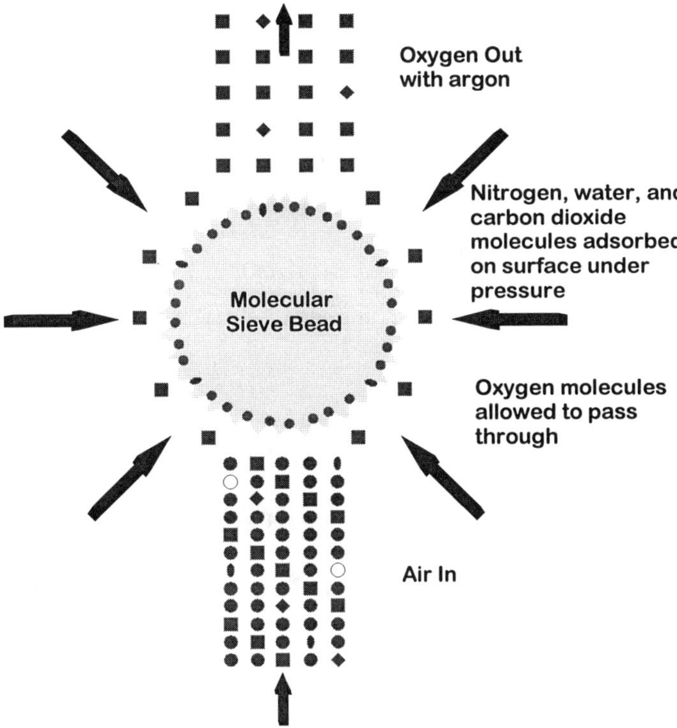

Figure 10. Adsorption

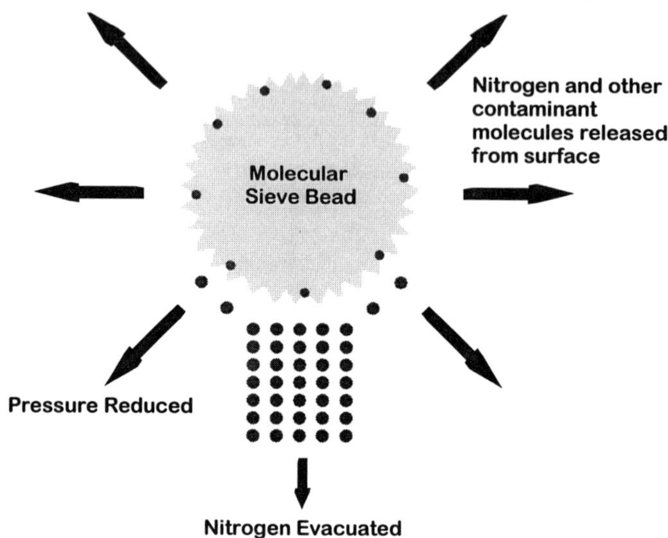

Figure 11. De-sorption

ADVANCES IN VPSA SUPPLY SYSTEMS

Key improvements in adsorbents, process design, systems design, component packaging, and control have combined to achieve a steady reduction in the cost of oxygen from these supply systems while continuing to enhance the reliability. Increased nitrogen selectivity and capacity of the adsorbents continue to be the primary drivers for overall VPSA supply system improvements. The development of adsorbents with higher selectivity increases the recovery of product oxygen from the feed air stream by allowing a larger fraction of the oxygen to pass through the bed during the adsorption step. Ideally, the adsorbent material should have a high capacity for nitrogen and also a high selectivity for the nitrogen over the oxygen. Figure 12 illustrates that the improved adsorbent material has a significant increase in the capacity for nitrogen with only a small increase in the capacity for the oxygen. Increased adsorbent capacity allows the use of less adsorbent material for a certain capacity supply system. Both the adsorbent material and the vessels in which it is contained contribute significantly to the capital cost of a VPSA oxygen supply system. By reducing the amount of adsorbent used, the power used by the air compressor and vacuum pump is decreased.

Advances in Fusion and Processing of Glass II

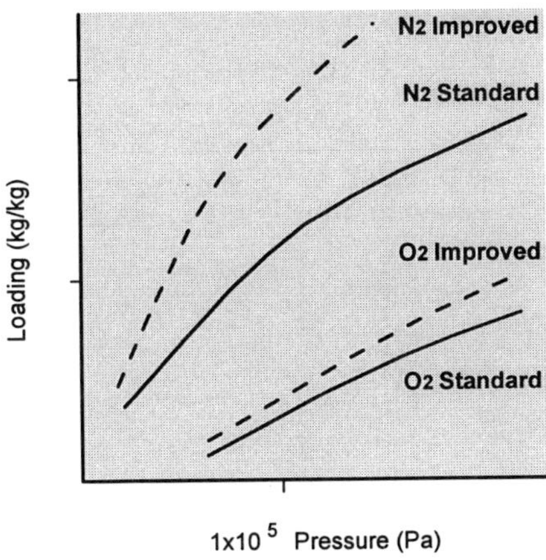

Figure 12. O_2/N_2 Adsorption Isotherms

Optimization has been performed in the area of process cycle development to reduce the capital cost and power consumption. Based on a specific adsorbent, the temperature, adsorption pressure, regeneration pressure, cycle time and cycle steps must be designed to result in a lower cost. Although the temperature for the system is generally near ambient, the pressure required must be optimized. The selection of the cycle time reflects directly on the amount of adsorbent required. For example, if the cycle time is reduced, the amount of adsorbent required is similarly reduced, but the process is not as efficient. Figure 13 demonstrates the optimization of the process cycle.

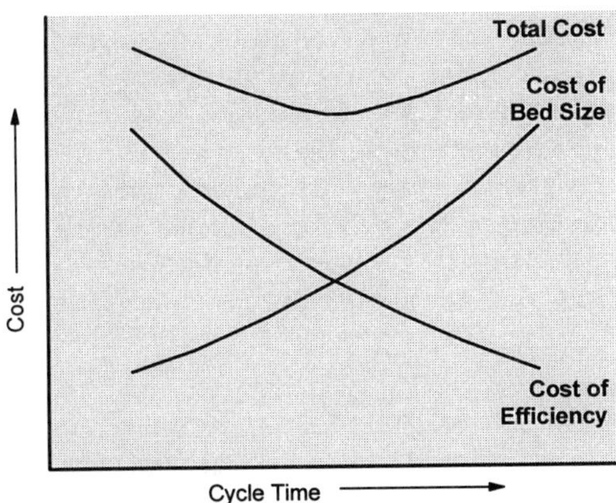

Figure 13. Cycle Time Selection

Advances in Fusion and Processing of Glass II

369

The design of the system itself has been another area where improvements have reduced the cost of oxygen supplied. Typically, the trade-off is between an incremental cost and process efficiency. By increasing the number of beds and cycle steps, the efficiency is increased but the capital cost also increases because of the additional components: valves, vessels, and piping. Figure 14 graphs the total cost as a function of the number of beds required. The optimum, based on the current technology, is a two bed system.

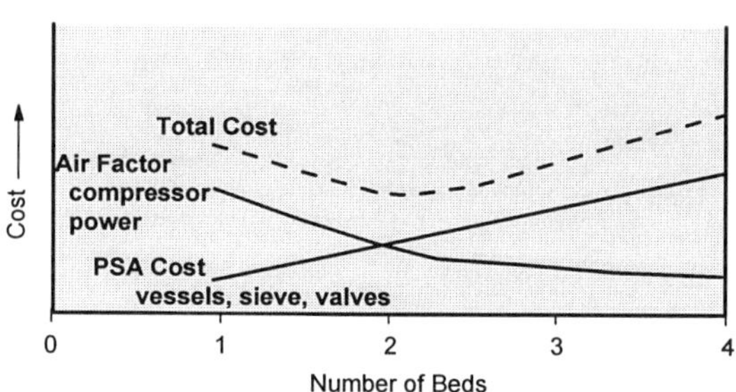

Figure 14. PSA System Optimization

The design of the adsorbent bed itself has been the subject of ongoing developments. Sophisticated techniques for modeling of fluid flow have reduced the void volume within the beds and their supports. These models are applied to the design of the adsorbent bed and their supports to minimize purge and blowdown losses. This work has contributed to capacity and efficiency improvements.

Significant cost reductions have been realized through the development of modular VPSA product lines. These product lines consist of predesigned elements for feed compression and evacuation, separation, product compression and backup that can be linked together to meet the needs of a variety of customers. The packages are constructed into truck shippable skids to maximize the amount of shop assembly and to minimize the labor performed in the field. This approach reduces the overall cost by reducing the costly engineering and field installation by using more efficient shop assembly. In addition, the project schedule can be shortened. As a result, a particular class of customers, such as the glass industry, receive the benefits of improvements in technology that are motivated by a broader base of customers.

Advances in controls have also helped to reduce the cost of oxygen to the customer. For example, load tracking can be implemented in a system where the product demand fluctuates. By turning down the VPSA when the demand is low, the power consumption is decreased and the operating cost is therefore reduced. This flexibility can be very beneficial to customers with varying use rates.

In addition, remote operation is often used with the VPSA system. This allows the unit to be started or stopped without the need for a customer service technician at the

                    Advances in Fusion and Processing of Glass II

production site. Key operating parameters can also be continuously monitored through the remote communication to ensure that the plant is performing optimally.

LOW AND MODERATE OXYGEN PRESSURE DELIVERY SYSTEMS

The low purity oxygen used by the glass industry is needed over a range of pressures from 1.4×10^5 to 4.5×10^5 Pa (5 to 50 psig). VPSA supply systems utilize an oxygen compressor or oxygen blower to meet these needs. Cryogenic oxygen plants can be designed to deliver the oxygen at an elevated pressure without a separate oxygen compressor. To accomplish this, low pressure oxygen liquid rather than vapor is withdrawn from the bottom of the low pressure column (see Figure 15). The pressure of the liquid is increased using a low cost, reliable cryogenic pump. The liquid is then vaporized at the elevated pressure by condensing an air stream. The pressure of the condensing air stream, established by the oxygen boiling pressure, is somewhat greater than the pressure of the feed air. The flow rate of the higher pressure air stream is approximately 30% of that of the feed air stream. It is obtained by further compressing a fraction of the feed air. The compression stage used for this purpose is usually integrated with the primary air compressor for the moderate oxygen boiling pressures needed by the glass industry. Thus, the installation of an entirely separate compressor to pressurize the product oxygen is avoided. Capital cost savings and efficiency improvements are realized with this arrangement.

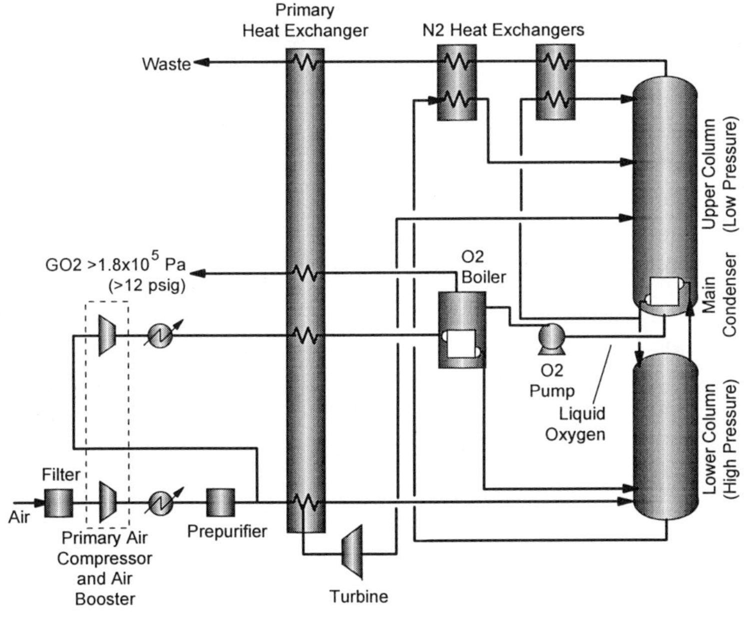

Figure 15. Cryogenic Oxygen Plant for Moderate O2 Pressures

With a delivery pressure less than 1.8×10^5 Pa (12 psig), it is possible to avoid the use of a pump and additional compression stages, taking advantage of the elevation of the low pressure column to pressurize the oxygen product. Figure 16 shows an oxygen plant where liquid oxygen is withdrawn from the base of the low pressure column. This liquid oxygen is then vaporized against partially condensing feed air in a heat exchanger (known as a product boiler) that is at a lower elevation within the oxygen plant coldbox. Recall that the low pressure column is stacked above the high pressure column. The hydrostatic head available from this elevation difference is sufficient to pressurize the oxygen product to as much as 1.8×10^5 Pa (12 psig). The feed air does not need compression above its normal pressure to partially condense against the boiling oxygen in the product boiler. Praxair built a 5.2 kg/s (500 ton per day) oxygen plant for a large oxy-fuel glass furnace conversion project based on this process in 1993. The plant is extremely reliable, cost effective and efficient. Reliability and low capital cost stem from its simplicity. It is energy efficient since the oxygen is pressurized without further compression of either the oxygen or the feed air. The same plant is capable of producing high purity nitrogen (up to 25% of the oxygen capacity) at a pressure of 5.2×10^5 Pa (60 psig) directly from the coldbox.

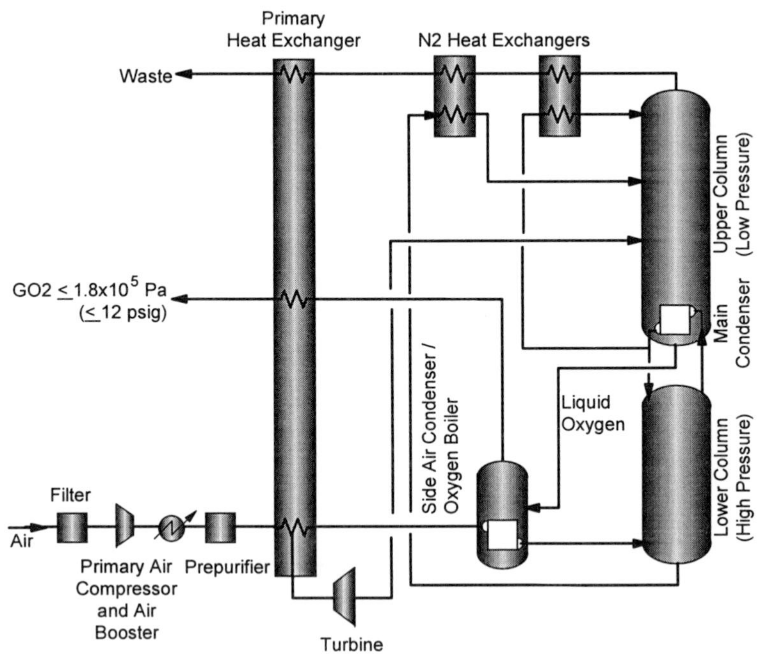

Figure 16. Cryogenic Oxygen Plant for O$_2$ Pressures $\leq$ 1.8x10^5 Pa (12 psig)

INDUSTRIAL GAS SUPPLY FOR FLOAT GLASS FACILITIES

The float glass industry is unique in the field of glass because it requires the supply of oxygen, nitrogen and hydrogen. A cryogenic system can cost-effectively meet the needs of the high purity nitrogen and the low purity oxygen. Both of these products can be delivered at the use pressures to the customer without the use of separate product compressors.

A low purity cryogenic oxygen plant may be designed to meet the high purity nitrogen requirements of a float glass facility with only an incremental increase in capital and energy consumption over the base plant. A nitrogen to oxygen flow ratio up to 0.8 can be met without the use of external nitrogen or oxygen compressors. Higher flow ratios will require either oxygen or nitrogen compression. Figure 17 illustrates how the high purity nitrogen can be removed directly from the higher pressure column and used as product. If the use pressure of the nitrogen is slightly less than that of the operating pressure of the high pressure column, the nitrogen stream can be throttled to the lower pressure.

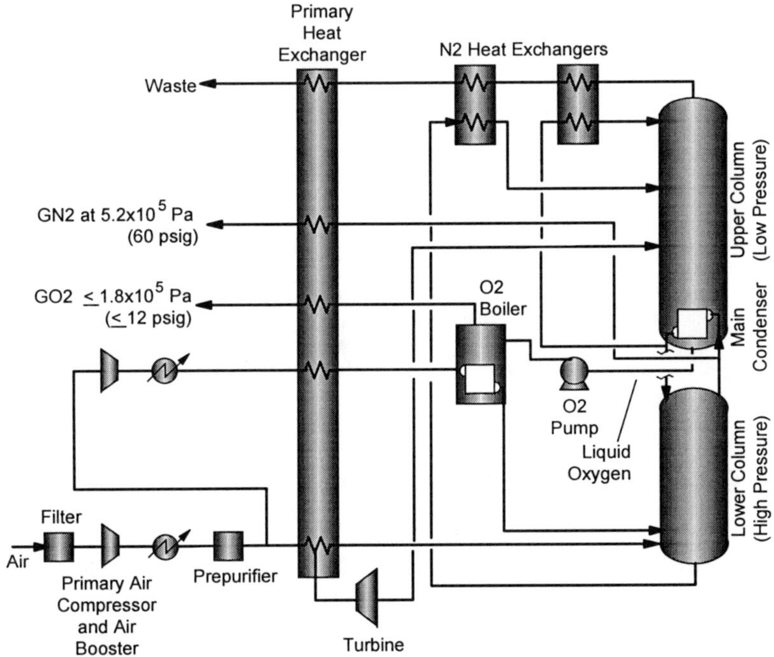

Figure 17. Cryogenic Oxygen Plant for Moderate O_2 Pressures with N_2 at 5.2x10^5 Pa (60 psig)

In a cryogenic system, the majority of the costs associated with the production of nitrogen are also included in the production of oxygen. The incremental cost of purifying the nitrogen is less than the cost of installing and operating a separate nitrogen plant. By taking advantage of the co-production of the oxygen and nitrogen, a significant cost savings can be achieved.

RELIABILITY AND BACKUP SYSTEMS

The uninterrupted supply of oxygen and nitrogen is crucial to glass manufacturing, as it is in many industries. The high reliability of on-site production systems is augmented by a backup system that vaporizes liquid to maintain a constant flow of gases. Overall availability of on-site plants is greater than 99% with most unplanned outages being of a short duration. When planned outages for scheduled maintenance are included in this figure, the reliability is about 98%. Figure 18 gives the typical outage duration for cryogenic oxygen plants. A large fraction of outages are caused by faults in the electric utility provider's system, merely requiring the restart of the industrial gas on-site production system. A restart is performed by automatic control systems. Customer service technicians, located at a remote operations center, will often monitor the status of plants being restarted.

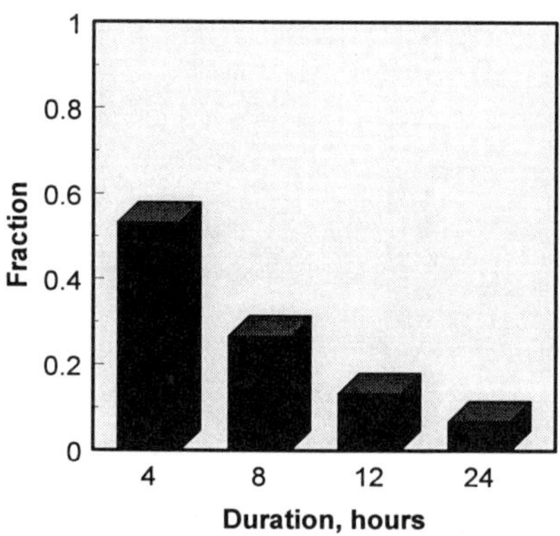

Figure 18. Distribution of Unplanned Outage Duration — On-site Cryogenic Systems

Backup systems for both cryogenic and VPSA plants are similar. The primary difference is that cryogenic oxygen plants are capable of producing some liquid oxygen to refill or maintain inventory within storage tanks. The key elements of the backup system are shown in Figure 19. The backup system typically operates without the need for utilities. It is triggered automatically by several means, including a slight reduction in supply system pressure. The backup system is also utilized to meet the customer's peak gas demands when they exceed the capacity of the on-site production system.

                          Advances in Fusion and Processing of Glass II

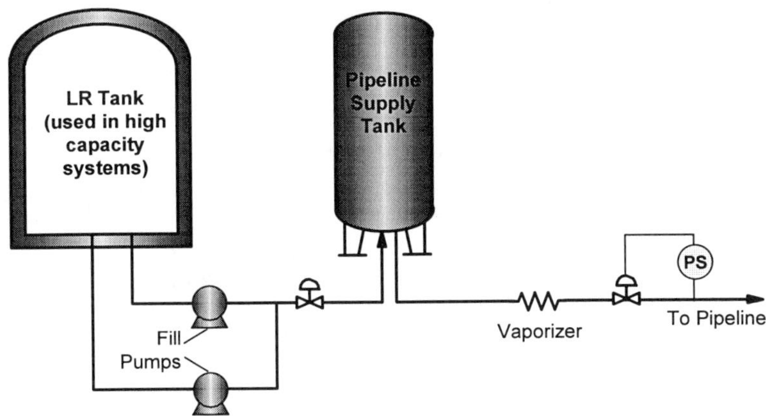

Figure 19. Typical Backup System

The backup system is best viewed as an extension of a vast industrial gas production and distribution infrastructure. The inventory of liquid throughout this network is constantly monitored. The failure of an on-site production facility triggers the reallocation of liquid production and distribution resources to maintain backup inventory while a repair is made.

ON-SITE SUPPLY OF HYDROGEN

The float glass industry currently uses molten tin in the glass making process. Nitrogen and hydrogen are mixed to create a protective atmosphere around the tin and the glass. Nitrogen is used as an inert gas around the tin and glass interface. Hydrogen reduces any oxides of tin that do form and scavenges any oxygen that may enter the float tank. Previously, the only economical hydrogen supply sources for the quantities used by the float glass industry were liquid hydrogen and compressed gas in tube trailers. Transportation of the liquid or compressed gas is a major component of hydrogen cost. Compared to atmospheric gases, the number of facilities producing merchant hydrogen is small.

Recently, Praxair, Inc., using low pressure natural gas reforming technology developed by International Fuel Cell (IFC), has created a line of small on-site hydrogen generators. These generators are designed to produce from 0.0447 to 0.223 m^3/s (6,000 to 30,000 SCFH) of hydrogen in increments of 0.0447 m^3/s (6,000 SCFH) at pressures of up to 1.4x10^6 Pa (190 psig). The purity of the product is up to 99.999% hydrogen. These units may offer flexibility and cost advantages to the float industry. Figure 20 is a schematic of the Praxair hydrogen generator.

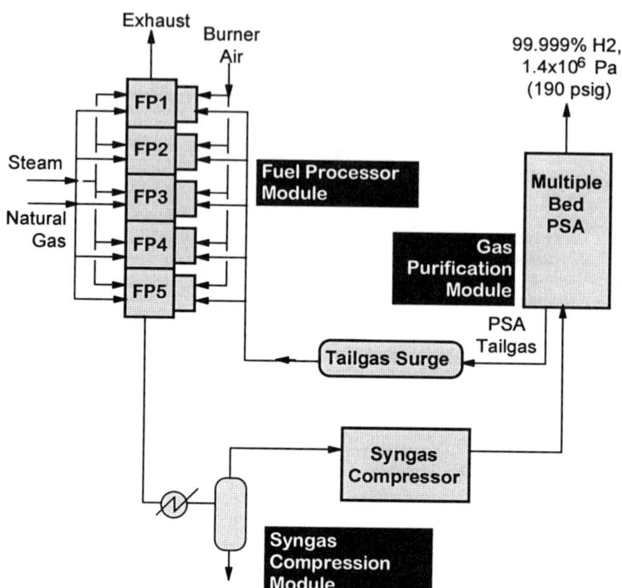

Figure 20. Hydrogen Generation System Block Flow Diagram

A fuel process or module, with one to five IFC reformers, is combined with gas compression and purification modules to provide a complete hydrogen supply system. The gas purification is done using a multiple bed PSA system. Natural gas, water and electric power are the only utilities required by the system. Liquid backup is added to guarantee uninterrupted supply of hydrogen.

SUMMARY

Advances in the area of industrial gas supply have resulted in a lower cost to the customer while at the same time increasing reliability. The area of cryogenic technology has seen several improvements in the distillation column arrangement. Examples of some of the improvements include reduction in energy requirements by taking advantage of the lower purity oxygen requirements of the glass industry and eliminating process compression equipment while still providing oxygen at the desired pressure. Improvements in the distillation equipment such as the use of structured packing in place of sieve trays in the column have also helped to reduce the cost.

The VPSA unit for the supply of low purity oxygen has also experienced significant process improvements while also achieving reductions in cost. There has been research and development in the areas of adsorbent technology, process design, system design, and component packaging. Improvements in each of these areas has contributed to the reduction in the price of the oxygen supplied.

   Advances in Fusion and Processing of Glass II

The overall availability of industrial gas plants has been increased to the current level of over 99%, with most unplanned outages being of a short duration. Improvements have been made in this area by monitoring the plants continuously and providing the required backup liquid in an efficient backup system.

The float glass industry requires nitrogen and hydrogen in their process, in addition to oxygen. By taking advantage of cryogenic technology, the nitrogen can be supplied from the oxygen distillation unit at a cost substantially below that of a separate nitrogen supply system. The development of on-site hydrogen supply enables the supply of hydrogen to sites where transportation costs are prohibitive.

Low purity oxygen supply is a relatively new market for the industrial gas industry. New technology and process arrangements are continuously being developed and implemented into the low purity oxygen supply systems. Future gas supply systems should see improved efficiency and reliability that will result in a lower cost for the glass customer.

REFERENCES:

1. Campbell, M. J., D. A. Lagree and J. Smolarek, "Advances in Oxygen Production by Pressure Swing Adsorption," presented at AIChE Spring National Meeting, Houston, TX, (April 7-11, 1991).

2. Michael, Keith P., "Industrial Gas: Surveying Onsite Supply Options," Chemical Engineering, (January 1997).

3. Snyder, W. J., F. N. Steigman and A. Tasca, "Oxygen Firing and Glass Melters," Glass Machinery Plants & Accessories, (March 1995).

4. Sacks, R.E., and O'Connor, D. "The Use of Structured Packing in Cryogenic Air Separation Plants," Proceedings of Kryogenika '96, The Fourth International Conference - Praha Czech Republic, (1996).

5. Thorogood, R.M., "Developments in Air Separation," Gas Separation & Purification, Vol 5 (1991).

6. Bonaquist, D. P., "Air Boiling Cryogenic Rectification System with Staged Feed Air Condensation," U. S. Patent 5,611,219, (March 18, 1997).

7. Halvorson, T.G., Victor, R.A. and Farris, P.J., "On-Site Hydrogen Generator for Vehicle Refueling Application," presented at World Car Conference, Riverside, CA, (January 21, 1997).

8. Schroeder, R.W., Campbell, M.J., Lagree, D.A. and Smolarek, J., "On-Site Oxygen Supply for Oxy-Fuel-Fired Glass Furnaces," presented at the Glass Problems Conference, Ohio State University, Columbus, OH, (October 1992).

9. Sinicropi, M., "Two-Bed VPSA System," presented at Praxair Technology Center, Tonawanda, NY, (August 5, 1993).

HIGH TEMPERATURE DEFORMATION OF AN AZS REFRACTORY[1]

A. A. Wereszczak J. Heide, and T. P. Kirkland
High Temperature Materials Laboratory
Oak Ridge National Laboratory
Oak Ridge, TN 37831-6069

G. V. Srinivasan
Advanced Materials and Process Engineering Laboratory
University of British Columbia
Vancouver, BC V6T 1Z4

S. M. Winder
UK Software Services, Inc.
Grand Island, NY 14072

ABSTRACT

The tensile and compressive deformation behavior of a commercially-available, fusion-cast alumina-zirconia-silicate (AZS) refractory was characterized at several temperatures between 700-1200°C (1290 - 2190°F) in ambient air. The compressive and tensile stresses were continuously measured as functions of strain, displacement or strain rate, temperature, and specimen position (in reference to where they were machined from the originally cast block). The maximum tensile and compressive stresses (i.e., flow strengths) of the AZS refractory were described using an empirical power-law model and an empirical exponential model. Within 95% confidence, both models satisfactorily represented the flow strength as a function of strain rate and temperature. AZS material 6 in. from the base of the cast block was found to be more resistant to deformation than material 44 in. from the base of the block, presumably due to lower glassy phase content at the 6 in. position. Additionally, the AZS refractory was found to be more resistant to deformation in compression than in tension for equivalent strain rates and temperature. Lastly, the elastic modulus of this material dropped by approximately 50% between 25 and 700°C ($\approx$ 80 GPa to $\approx$ 40 GPa), and continued to decrease through 1200°C to approximately 10% of its 25°C value.

I. INTRODUCTION

There is incentive to minimize or eliminate crack formation in cast AZS blocks during their solidification. Studies have been performed to fundamentally model the cooling processes [1] and crack formation in AZS refractories using heat flow and stress generation algorithms [2-3]. These models have provided important insight into the solidification and crack formation processes; however, the development of an all-encompassing model has been hindered by the lack of an engineering database of the AZS's high temperature mechanical behavior. To understand crack formation during cooling, one must understand the temperature profiles and the correspondingly produced thermal strains, along with the subsequently induced stresses. For experimental testing to appropriately mimic this behavior, high temperature tests need to be performed under strain-control, and *not* under stress- or load-control as they are for conventional creep tests.

The intent of the present study was to mechanically characterize the high temperature *flow strength* of an AZS refractory *in strain-control* as a function of temperature, strain rate (which

---

1   Research sponsored by the U. S. Department of Energy, Assistant Secretary for Energy Efficiency and Renewable Energy, Office of Transportation Technologies, as part of the High Temperature Materials Laboratory User Program under Contract DE-AC05-96OR22464, managed by Lockheed Martin Energy Research Corporation.

would be effectively proportional to a cooling rate), and relative position within the cast block. This dependent parameter is important because, deterministically speaking, if a thermomechanical stress which forms during cooling exceeds the material's flow strength, then crack initiation will occur. Regarding the independent parameter of "position", the microstructure of the AZS block was inhomogeneous due to the large cast size of the block (66 in. height with 18 in. x 12 in. cross-section) and large thermal gradients that existed during solidification. Subsequently, gradients in composition and glassy phase content resulted, so the mechanical performance of the AZS refractory as a function of position was sought to understand the mechanical effects of this inhomogeneity.

II. EXPERIMENTAL PROCEDURES
IIA. Material and Testing

The tensile and compressive flow strength of a commercially-available AZS refractory were characterized at several temperatures between 700-1200°C (1290-2190°F) in ambient air. The manufacturer's reported composition for this AZS refractory showed it to contain 39.5% ZrO_2, 45.8% Al_2O_3, and 12.9% SiO_2. Tensile tests were performed in strain control, while strain-control testing in compression was unsuccessful and not used because the test machine's relatively high compliance acted to inhibit stable control. Consequently, compression tests were performed in displacement control (after calculating what the associated strain rate was likely to be). A button-head cylindrical specimen geometry (with a nominal dimension of 0.5 in. [12.7 mm] diameter) was used for the tensile tests, and a cylindrically shaped specimen (0.5 in. [12.7 mm] diameter x 1.5 in. [38.1 mm] height) was used for the compression tests. A high temperature contacting extensometer with a 25 mm gage length was used to control and/or measure the strain. For both test types, the compressive and tensile stress response was continuously measured as a function of strain, displacement or strain rate, and temperature. The identified maximum tensile or compressive stress was designated as the *flow strength*. A schematic of a generated stress-strain curve is illustrated in Fig. 1. Additionally, when linear stress-strain behavior was observed, the elastic modulus was calculated. Specimens were machined out of a cast block as a function of position (height), with specimens coming from either a 6 in. (152 mm) or 44 in. (1117 mm) height from the block's base. Originally, specimen position within the 12 in. (305 mm) x 18 in. (457 mm) cross-section was also considered, but results did not show any consistent trend, so the consideration of position within this plane was omitted as an independent

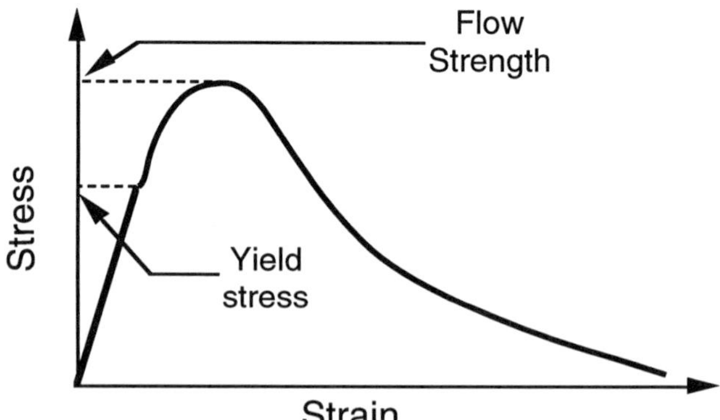

Fig. 1.   Schematic of a generated stress-strain curve, and the definition of *flow strength*. The elastic modulus was determined only for those stresses less than the yield stress.

                        Advances in Fusion and Processing of Glass II

parameter in the present study. A substantially greater number of compression specimens than tension specimens were tested, so most of the shown flow strength results were from compression tests.

IIB. Analysis

The applicability of two empirical models to describe the flow strength was examined. The steady-state creep or strain rate ($d\varepsilon/dt$ or $\dot{\varepsilon}$) is commonly related to the stress and temperature using an empirical Arrhenius power law or Norton-Bailey [4] creep equation:

$$d\varepsilon/dt = \dot{\varepsilon} = A\ \sigma^n \exp[\ \frac{-Q_{NB}}{RT}\ ], \tag{1}$$

where A is a pre-exponential constant, σ is the stress, n is the stress exponent, Q_{NB} is the Norton-Bailey (NB) activation energy, R is the gas constant, and T is absolute temperature. In the present study though, strain rate was an independent parameter, and the maximum stress, or flow strength (σ_{Flow}), was determined. Rearranging Eq. 1 and solving for this strength yields:

$$\sigma_{Flow} = B\ \dot{\varepsilon}^{1/n} \exp[\ \frac{Q_{NB}}{nRT}\ ], \tag{2}$$

where B is the new pre-exponential constant.

The second model acknowledges the possibility that the σ_{Flow} is exponentially-related to the strain rate according to:

$$\sigma_{Flow} = C \exp[\ D\ \dot{\varepsilon}\ ] \exp[\ \frac{Q_E}{RT}\ ], \tag{3}$$

where C and D are constants, and Q_E is the exponential activation energy. This model is a simplification of the hyperbolic sine function, which is sometimes used to represent creep data [5]. For "large" stresses the hyperbolic sine function simplifies into the exponential form represented by Eq. 3.

Multilinear regression was performed using Eqs. 2-3 to determine the coefficients for specimens machined from the 6 in. or 44 in. height. By performing the analysis in this manner, it was assumed that the same dominant (or flow-strength-controlling) deformation mechanism was active at all temperatures and stresses. The validity of this assumption was assessed by the goodness-of-fit of this equation to the experimental data and the "reasonableness" of the obtained values of the strain rate exponent and the activation energies. To compress all the data onto one graph for each model, temperature-compensated analysis was used.

III. RESULTS & DISCUSSION

The temperature-compensated compressive flow strength as a function of strain rate using the NB model is shown in Fig. 2. 95% confidence estimates for the regressed functions (using Eq. 2) are shown. By inputting a strain rate and temperature into the shown functions for the deformation behavior, one can determine the compressive flow strength. The results show that for an equivalent strain rate and temperature that the AZS material machined from a 6 in. height is conclusively more resistant to compressive deformation than AZS material machined from a 44 in. height. Although not shown, Q_{NB} for the tensile flow strengths was equal to 550 kJ/mol, and n was 5.1. The n value is consistent with deformation accommodation through cavity formation [6]. The activation energy (Q_{NB}) is consistent with the activation energy of silicate glass viscous flow in glass-ceramic materials [7-8].

Advances in Fusion and Processing of Glass II   381

The temperature-compensated flow strength as a function of strain rate using the exponential model is shown in Fig. 3. 95% confidence estimates for the regressed functions (using Eq. 3) are shown. The exponential model can consider a non-constant strain rate exponent, as evidenced by the fit at the fastest test strain rate. Similarly to Fig. 2, the AZS is conclusively shown to be more resistant to deformation near the cast block's base. The correlation of the fits with the data using the two models are satisfactory, and one is not conclusively better than the other.

A couple practical matters regarding Figs. 2 and 3 are noteworthy. Firstly, note that the activation energies in Figs. 2 and 3 are not the same even though the same data are being regressed. This does not insinuate that the analyses are erroneous, it simply reflects the fact that Q_{NB} and Q_E were mathematically determined using different models and do not strictly represent the same entity. Secondly, the AZS manufacturer provided microstructural information that showed that the ZrO_2 content was greater in the block at the 6 in. position than in the 44 in. position, and that the secondary glassy phase volume fraction was less in the 6 in. position than in the 44 in. position; this information is consistent with the greater deformation resistance of the 6 in. AZS material.

For equivalent magnitudes of strain rate and temperature, the flow strength of the AZS material was found to be 10-12 times larger in compression than in tension at 900°C and 7-9 times larger at 1000°C. Analogous anisotropy with creep resistance has been observed in polycrystalline ceramics containing an amorphous secondary phase [9-10]. If one approaches Figs. 2-4 deterministically, then one may use them to predict the onset of crack formation. For example, if a tensile stress in the AZS block located near the 44 in. position is greater in magnitude than the tensile flow strength, then the surrounding material may not be able to relax it quickly enough, and a crack may initiate. Information such as this may be used to develop more mature models to predict crack formation in these AZS blocks during their solidification.

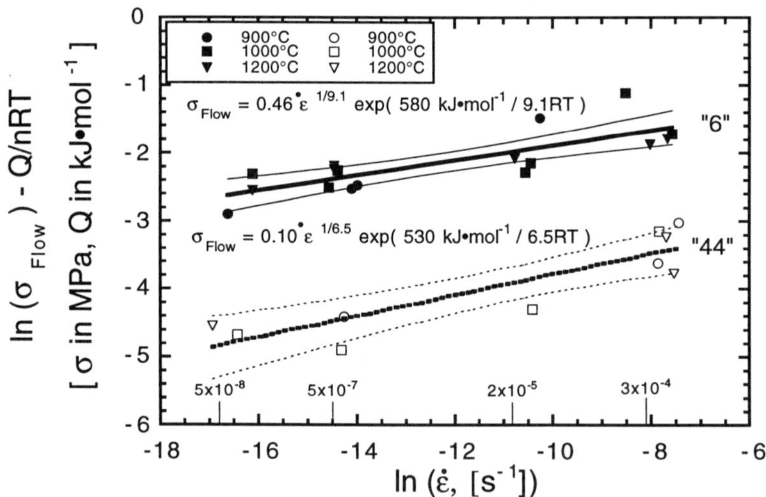

Fig. 2.   Temperature-compensated compressive flow strength as a function of strain rate using the modified Norton-Bailey model (Eq. 2) for AZS material machined either 6 in. or 44 in. from the cast block's base.

    Advances in Fusion and Processing of Glass II

The elastic modulus of the AZS dropped by approximately 50% between 25 and 700°C ($\approx$ 80 GPa to $\approx$ 40 GPa), and continued to decrease through 1200°C to approximately 10% of its 25°C value, as shown in Fig. 4. The prevalence of linearity in this AZS material essentially ceases at or near 1200°C as several of the specimens tested at 1000 and 1200°C exhibited no linearity at all, so an elastic modulus was not defined. Dilatometry showed that the AZS material had an effective glass transition temperature of 755°C, so the reduction in effective elastic modulus at temperatures equal to and above 700°C is plausible.

IV. CONCLUSIONS

An empirical power-law model and an exponential model may be used to satisfactorily represent the flow strength of an AZS refractory as a function of strain rate and temperature. With 95% confidence, AZS material 6 in. from the base of the cast block was found to be more resistant to deformation than AZS material 44 in. from the base of the fusion cast block. Additionally, the AZS refractory was found to be more resistant to deformation in compression than in tension for equivalent strain rates and temperature. Lastly, the elastic modulus of this material decreased by approximately 50% between 25 and 700°C ($\approx$ 80 GPa to $\approx$ 40 GPa), and continued to decrease through 1200°C to approximately 10% of its 25°C value.

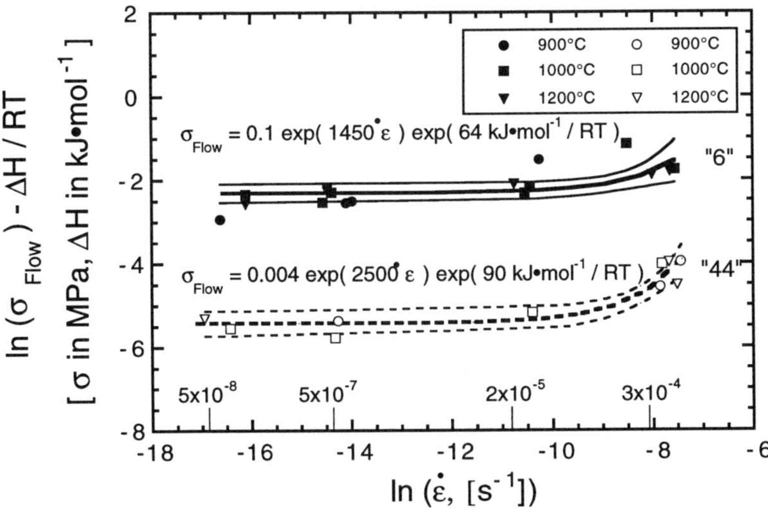

Fig. 3.   Temperature-compensated compressive flow strength as a function of strain rate using the exponential model (Eq. 3) for AZS material machined either 6 in. or 44 in. from the cast block's base.

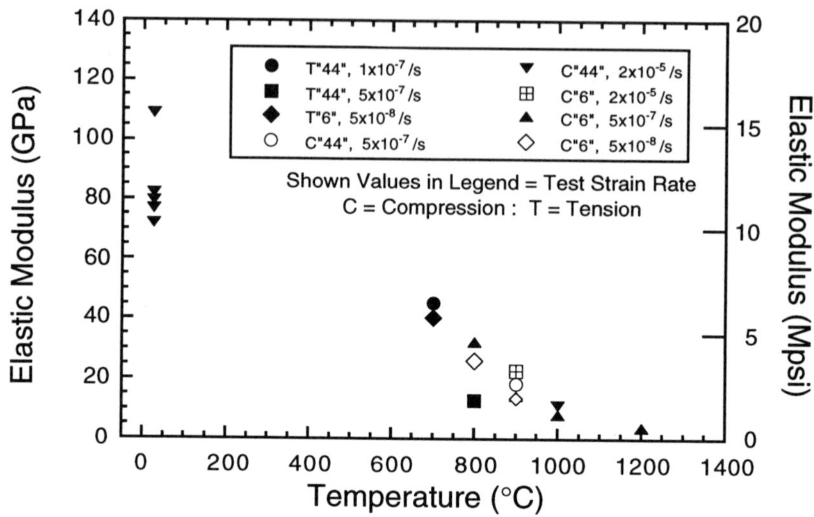

Fig. 4.   The AZS's elastic modulus as a function of temperature.

V.   ACKNOWLEDGMENTS

The authors wish to thank Drs. K. Breder and C. -H. Hsueh for reviewing the manuscript and for their helpful comments.

VI.   REFERENCES

[1]   V. Gottardi, A. Trotta, G. Michelotto, and S. Bauer, *Glass Tech.*, **21** 120-124 (1980).
[2]   S. L. Cockroft, J. K. Brimacombe, D. G. Walrod, and T. A. Myles, *J. Am. Cer. Soc.*, **77** 1505-1511 (1994).
[3]   S. L. Cockroft, J. K. Brimacombe, D. G. Walrod, and T. A. Myles, *J. Am. Cer. Soc.*, **77** 1512-1521 (1994).
[4]   F. H. Norton, <u>The Creep of Steel at High Temperature</u>, McGraw Hill, New York, 1929.
[5]   C. M. Sellars and W. J. McG. Tegart, *Mem. Sci. Rev. Metall.*, **63** 731-746 (1966).
[6]   W. R. Cannon and T. G. Langdon, *J. Mat. Sci.*, **18** 1-50 (1983).
[7]   R. Morrell and K. H. G. Ashbee, *J. Mat. Sci.*, **8** 1253-1270 (1973).
[8]   R. Raj and C. K. Chyung, *Acta Metall.*, **29** 159-166 (1981).
[9]   W. E. Luecke and S. M. Wiederhorn, *Key Engineering Materials*, **89-91** 587-592 (1994).
[10]   A. A. Wereszczak, T. P. Kirkland, H. -T. Lin, M. K. Ferber, C. -W. Li, and J. A. Goldacker, In press, *Cer. Engrg. Sci. Proc.*, **18** (1997).

    Advances in Fusion and Processing of Glass II

CORROSION OF REFRACTORIES IN A DC-PLASMA ARC FURNACE PROCESSING MIXED WASTES

S. K. Sundaram[a]
Pacific Northwest National Laboratory[b]
P. O. Box 999
Richland WA 99352

ABSTRACT

The corrosion of six different refractory candidate materials was studied using a modified ASTM C621 method for a DC-plasma arc furnace processing mixed waste. Two sets of static tests, with and without carbon addition, were completed. The radius loss after the test was reported as corrosion. Representative refractory-glass interface microstructure and chemistry were studied using scanning electron microscopy (SEM) and line energy dispersive spectrometry (EDS), respectively. Monofrax K3 (without carbon) was the least corroding material followed by Ruby. In the presence of carbon, Monofrax K3 was reduced and the reduced species dissolved into the glass melt, enhancing the corrosion. The most corroding refractory, ZS1500, decomposed by the attack of the glass melt and the silica phase dissolved into the glass melt, resulting in higher corrosion. These results suggest that preferential corrosion was the basic mechanism of corrosion of Monofrax K3 and Ruby. This left a corrosion-resistant chromium-rich layer at the interface that reduced further refractory corrosion.

---

[a] Member, The American Ceramic Society

[b] Pacific Northwest National Laboratory is operated for the U. S. Department of Energy by Battelle under Contract DE-AC06-76RLO 1830.

INTRODUCTION

Plasma technology has a wide range of applications for treating a variety of waste materials. The DC-plasma arc furnace is a robust, high-temperature thermal treatment system being tested to treat mixed wastes as a part of a project supported by the U.S. Department of Energy's (DOE's) Mixed Waste Focus Area (MWFA). In this system, a stable DC arc is created by applying a potential between a graphite electrode and the graphite hearth of the furnace. The thermal energy produced by this arc creates and maintains a molten bath of material (glass/slag and/or metal) in the furnace hearth. Waste materials fed into the system are melted into the bath. Organics are pyrolyzed at the high operating temperature and may be destroyed in the plenum or in a suitable afterburner. Oxide materials, including many of the hazardous and radioactive species, are immobilized in the durable glass/slag phase, while metals separate out of the melt and are removed as a second phase. A major advantage of this technology is its potential application to waste streams containing a wide range of materials such as debris, trash, metals, and soil.

Although the DC plasma arc technology is well demonstrated commercially for melting and processing of various metals in metallurgical industries, several technology deficiencies have been identified by the MWFA program for its application to the treatment of radioactive wastes[c]. Refractory corrosion is one of the barriers to implementing plasma technology for the treatment of radioactive/mixed waste. The high temperature (melt temperatures in excess of 1500 °C), the variability in melt chemistry, and the operating conditions create a hostile environment for the materials in contact with or containing the melt. A preliminary evaluation[d] of refractory corrosion in plasma/torch furnaces indicated significant gaps in the materials knowledge between the data necessary to advance and develop plasma technology and the data available. Bridging these gaps offers significant economic benefits in terms of longer furnace life, lower repair or replacement cost, lower shutdown time, higher productivity, greater security, sand saving time, labor, and potentially life by avoiding catastrophic failure of materials in service.

---

[c] Mixed Waste Focus Area Technology Development Requirements Document (TDRD), 1997, *Radionuclide Partitioning,* INEL/EXT-97-00304

[d] S. K. Sundaram, "Corrosion of Slag and Metal Contact Materials in Plasma Arc/Torch Furnaces," MWFA, TTP No. RL3-6-MW51, March 1997.

Advances in Fusion and Processing of Glass II

An integrated corrosion evaluation of refractories has been undertaken under the MWFA program. Both laboratory crucible tests as well as bench-scale tests have been used in determining the corrosion of selected refractory candidate materials. In the bench-scale tests, the effects of process variables on the corrosion have been determined. The data from these tests have been compared to the data from the laboratory crucible tests.

This paper summarizes the results of laboratory crucible tests. The background data available in the literature were reviewed, the chemistry and microstructure of the representative refractory-glass interfaces were studied, and the corrosion mechanisms of the refractories proposed.

BACKGROUND

Refractory scoping tests were performed at the Diagnostic Instrumentation and Analysis Laboratory (DIAL) at Mississippi State University in March, 1995[1]. A 250 kW plasma torch attached to a gimbal mount to allow for material consisted of briquettes with a composition reflecting a "typical" low-level mixed waste heterogeneous debris waste of composition. Thirty-three different materials, from groups of oxides, nitrides, carbides, borides, fused cast refractories, and composites were tested. Two tests, a multiple cycle test and a continuous operation test, were conducted to assess the performance of refractory materials. A graphite crucible was used to contain the test materials. Because many of the samples were odd-shaped, the walls and bottom of the crucible was notched to accommodate the samples and minimize sample movement during the tests. Since these were scoping tests, no efforts were made to account for the effects of cross-contamination. After the tests, only the very high chromium refractory and aluminum nitride with 0.25 wt.% yttria survived. Oxide samples, silicon carbide samples, and fused cast refractories showed evidence of thermal shock. Severe corrosion was observed at the melt line of all the refractories tested. Corrosion in samples near the bottom of the crucible was much less than near the surface of the melt. The results were qualitative and inconclusive.

There have been attempts to evaluate refractories during or after processing wastes as a part of furnace operation. Scientific Applications International Corporation (SAIC) performed preliminary refractory testing[2] by constructing hearths of the sample materials. Tests were conducted in two different systems: the 1.2-MW torch proof-of-principle (POP) system and a 200-kW torch system located at SAIC's Science and Technology Applications Research (STAR) Center. The systems were operated on batch mode with melt temperatures typically in the range of 1600-1700°C. The materials tested were Ruby®, Korundal®, Harklase®, Nocacon®, Harbide® brick and castable, and Nuline R20® (all from Harbison-Walker), and Monofrax K3® from Monofrax Inc. Under the acidic slag (basicity of 0.38) conditions, there were no significant differences between the performance of any of the oxide refractories. In the case of basic slag (basicity of 2.5), the Monofrax K3 performed the best, followed by (in decreasing order) Ruby, Harklase, Korundal, and Zircon. The Monofrax K3 was susceptible to thermal shock. The Ruby brick performed well and was less susceptible to thermal shock. Non-oxide materials such as silicon carbide and the Nuline refractory (containing 20% carbon) were significantly inferior in performance to the oxide materials under all slag conditions; a definite crucible size effect on refractory performance, dramatic decrease in refractory wear, and an increase in volume of the crucible were observed.

The limited (mostly qualitative) corrosion data available from the furnace tests have been evaluated and are summarized in Table I. No systematic effort has been reported in the literature. The major refractory corrosion issues and their impact on the DC-plasma arc technology are as follows:

- The corrosion rates of the refractories being considered/used in the waste melts have not been determined; therefore, the life of the refractories, which dictates the life of the furnace, can not be predicted. The general rule of thumb for glass-contact refractories is that the corrosion rate doubles for each 50-100 °C increase in the interface temperature[3]. Lower corrosion rate means less materials used, lower idle time, higher productivity, lower repair cost, lower operating cost, and less maintenance. Additional cost saving on time and labor comes from avoiding the processing problems caused by seriously corroding refractories that pollute the melt and cause these problems (e.g., high viscosity, segregation, lower product quality).

- The refractory failure modes (usually corrosion-related) have not been determined. At the present state of knowledge on materials of construction for plasma technology system for waste treatment, prediction of failure of refractories is not possible. Testing helps to identify and predict corrosion-related failures. This translates to a significant cost advantage in terms of accident avoidance, contamination prevention, and greater safety.

- The corrosion mechanism has not been established. Understanding the corrosion mechanism will help in identifying and choosing the best suited

                      Advances in Fusion and Processing of Glass II

refractory for different locations of the plasma furnaces for processing of mixed wastes. This will also make it possible to predict refractory behavior for changes in waste composition and type, system operational changes, and fluctuations.

- The effects of processing conditions (temperature, atmosphere (redox), and chemistry of the wastes) on refractory corrosion have previously been unknown. Fluctuations in temperature, redox, and chemistry of the waste provide harsh conditions for the refractories. An increase in temperature enhances corrosion. Iron oxide and other transition metal oxides seriously change redox. The halides and sulfates are known to attack the refractories locally and cause catastrophic corrosion failure of materials in contact. Investigation of these effects help to avoid conditions that cause serious refractory failure during furnace operation, which can be expected to significantly save costs and provide greater worker safety.

EXPERIMENTAL

Waste and glass compositions

Many of the DOE mixed and radioactive wastes must be treated to destroy or immobilize hazardous constituents. A large number of waste streams identified for treatment reside at the Idaho National Engineering Environmental Laboratory (INEEL) and the Savannah River Site (SRS). A significant portion of the waste at SRS is specifically designated for vitrification (60% of the mixed transuranic waste MTRU). The surrogate wastes studied in this effort support the mixed waste treatment activities at the INEEL and the SRS.

The heterogeneous nature of mixed waste streams will have a significant impact on many aspects of arc furnace operation and performance. The waste form and/or composition may affect partitioning behavior of key elements, product durability, system component durability, and operability of product removal systems, as well as overall system operability. Table II presents the composition of the SRS debris surrogate used in this testing program. This formulation mimics the feed material supplied to the Diagnostic Instrumentation and Analysis Laboratory (DIAL) of Mississippi State University by the Savannah River Technology Center[4]. In addition to the base composition given below, hazardous metals, radionuclides, and radionuclide surrogates are added at 1000 ppm.

Table I. Refractories Corrosion in Plasma Technology Testing

Plasma Technology	Refractories/Evaluation
Transferred-Arc/South Africa	Serious dissolution of MgO crucible observed.
Plasma Energy Corp. (PEC) 150-kW Plasma Tilt Furnace, INEEL	Corrosion observed. No detail.
DC-plasma arc furnace Mark I & II, PNNL, MIT & EPI	Graphite hearth.
Plasma Calcination, 1500-kW DC arc torch operated in blown-arc mode, WSTC and WHC	Incomplete coupon test followed by visual evaluation & mass balance. Alumina-silica (90-RAM-PC) corroded. Ruby (alumina-chromia) significant corrosion in the Cupola shaft.
Plasma Arc Furnace, US Bureau of Mines	Complete dissolution of MgO hearth observed.
Plastic Arc Centrifugal Treatment (PACT system, Retech, Inc., CA; EPA Site & INEEL	Hearth refractory corrosion (Proprietary). No reported testing/evaluation.
Plasma Hearth Process (PHP), SAIC 1.2 MW torch hearth Proof-of-Principle (POP) system, 200-kW torch system at SAIC's Science and Technology Applications Research (STAR) Center	Best performance among the materials tested: Ruby (Harbison-Walker) Monofrax K3 (Monofrax Inc.)

   Advances in Fusion and Processing of Glass II

Table II. SRS Surrogate Debris Waste Recipe for FY-97 DC Arc Testing

Material	Wt.%
Carbon Steel	14.05
Perlite	14.05
PVC	14.05
Soda/Lime Glass	14.05
Wood	14.05
Alumina	10.18
Activated Carbon	6.31
Silica	5.60
Portland Cement	5.60
Aluminum	2.04
Total	100.00

The debris wastes required additives for producing a low-viscosity glass at 1300 to 1500°C. Soil was chosen as a glass-forming additive because it is inexpensive and easily available. This criterion ensures that the glass can be poured from a continuously drained furnace. The oxide composition of SRS (L area) soil in wt % is $Al_2O_3 = 3.5$, $CaO < 0.1$, $Fe_2O_3 = 1.0$, $K_2O < 0.2$, $MgO = 0.1$, $Na_2O = 0.9$, $SiO_2 = 86.0$, $TiO_2 = 0.4$, and minor constituents = 8.1. Lime was added with each soil type to a given waste to reduce the melting temperature of the final glass. The oxides of sodium and potassium typically reduce glass melting temperatures at lower weight percentages compared to lime; however, these oxides are extremely volatile at high temperatures and would vaporize out of the glass.

Refractory candidate materials

Based on the data from melting studies and the bench-scale non-radioactive processing tests completed using SRS soils, the baseline feed of 40% SRS debris and 60% SRS soil was selected for further investigation. All the bench-scale arc furnace coupon tests at PNNL used this baseline feed composition. The targeted glass composition representing baseline composition used for bench-scale tests are presented in Table III.

Table III. Targeted Glass Composition

Oxides	Wt. %
Al_2O_3	12.80
BaO	0.06
CaO	24.38
CdO	0.15
Fe_2O_3	4.00
K_2O	0.98
MgO	1.94
MnO	0.01
Na_2O	5.27
P_2O_5	0.02
SiO_2	50.13
SO_3	0.09
TiO_2	0.17
ZnO	0.02
Total	100.00

On the basis of the limited data available in the literature and the expertise at PNNL on design and operation of DC-plasma arc and other furnaces[5,6,7], the potential candidate refractory materials were chosen. The candidate materials and manufacturers are listed in Table IV.

Table IV. Refractory Candidate Materials for Corrosion Testing

Refractory	Manufacturer
Monofrax K3® (Fused Cast)	Monofrax Inc.
Magnesia-Carbon (Nuline R20®)	Harbison-Walker
Chrome-Magnesia (C104®)	Corhart Refractories
Dense Zircon (ZS1500®)	Corhart Refractories
Alumina-Chrome (Ruby®)	Harbison-Walker
Aluminum Nitride (AlN) (with yttria as sintering aid)	Advanced Refractory Technology (ART)

Chrome and chrome-containing refractories (Monofrax K3, C-104, and Ruby) were chosen for their corrosion resistance to common silicate glasses and slags. Nuline R20 was selected for its extensive use in steel industries. Monofrax K3, Ruby, ZS1500, and AlN performed well in tests conducted at DIAL. Therefore, dense zircon and aluminum nitride (with yttria) were also tested.

Corrosion testing

Processing of mixed wastes involved separation of metal and slag. The ASTM C621-84 (1989)[8] test was modified to incorporate this aspect of mixed-waste processing. The crucible tests were done in ambient air atmosphere. But the atmosphere in a DC-plasma arc furnace was generally reducing (CO/CO$_2$ and H$_2$/H$_2$O). The crucible tests were included in the present evaluation to test the refractory in highly oxidizing atmosphere.

The slag-metal melt was contained in an AZS (alumina-zirconia-silica) crucible. The test glass melt was prepared in the laboratory using conventional chemicals (oxides, carbonates). The refractory coupons (0.5 in diameter in the case of cylindrical samples and about 0.5″ × 0.5″ in the case of square cross-section) were immersed in the melt and allowed to corrode for three days at 1450-1550°C. After the completion of the test, the coupons were removed from the melt, cooled and sliced at glass line and half-down locations for corrosion measurement. The loss of diameter was determined and radius loss was reported as the corrosion.

Two sets of samples were tested. A total of nine tests were completed. The test conditions were as follows:

Set 1	Six different refractory coupons (Monofrax K3, Nuline R20, Ruby, C104, ZS1500, and aluminum nitride); 100 g of glass + 100 g of steel shot; 1550°C; 48 h
Set 2	Three refractory coupons (Monofrax K3, Ruby, and ZS1500); 100 g glass + 150 g of steel shot + 12.5 g of activated carbon; 1450°C; 48 h

RESULTS AND DISCUSSION

Corrosion data

The results of the tests are summarized in Table V. In the absence of carbon, Monofrax K3 is the least corroding refractory followed by Ruby and ZS 1500. Carbon addition increased corrosion of Monofrax K3 significantly. The Monofrax K3 coupon corroded so severely that no measurements could be made on the corroded samples. In the presence of carbon, Ruby had lower corrosion than ZS 1500. Generally, half-down corrosion was less than glass line corrosion, as expected, except in the case of ZS 1500 (without carbon). This is attributed to reaction of the refractory with the molten glass, formation of a reaction layer and flaking away into the glass melt.

Table V. Summary of Laboratory Corrosion Test Results

Refractory	Location	Corrosion (mm) Without Carbon	Corrosion (mm) With Carbon
Monofrax K3	Glass Line	1.48 ± 0.30	Completely corroded
	Half-Down	0.82 ± 0.34	Completely Corroded
Ruby	Glass Line	2.85 ± 0.45	1.18 ± 0.42
	Half-Down	2.07 ± 0.35	1.07 ± 0.29
ZS 1500	Glass Line	2.69 ± 0.18	1.61 ± 0.11
	Half-Down	3.13 ± 0.17	1.42 ± 0.15

Three of the refractories (Nuline R20 and AlN expectedly and C104 unexpectedly) failed in the first set of laboratory tests. Therefore, they were not tested in the second set. In the case of Nuline R20 (Magnesia-Carbon), carbon oxidized off by oxygen in and above the melt and magnesia dissolved into the melt. Similarly, AlN was oxidized off and alumina dissolved into the glass melt. In the case of C104 (Chrome-Magnesia), magnesia dissolved into the glass. The remaining porous skeleton collapsed into the melt.

Refractory-glass interfaces

Monofrax K3 (without carbon)-glass interfacial features of the sample at the half-down position are shown in Figure 1 (Left-Glass; Right-Refractory). A distinct chrome-rich reaction layer (about 600 microns thick) was seen at the refractory-glass interface. This data was also supported by line EDS data (not shown) across the line marked on the figure.

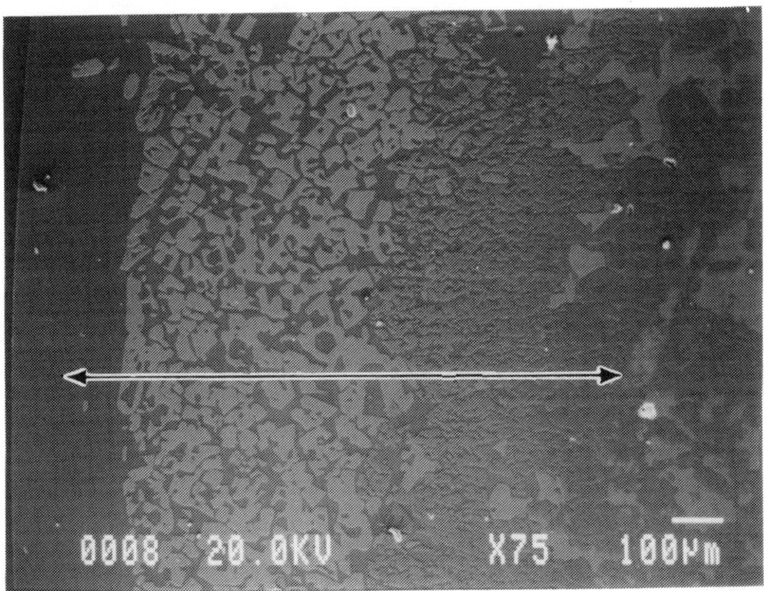

Figure 1. Monofrax K3(without carbon) - Glass Interface
(1550°C; 48 h; Half-down position; Left-Glass; Right-Refractory)

In the presence of carbon, no chrome-rich interfacial layer was observed as shown in Figure 2 (Left-Refractory; Right-Glass). The carbon creates a reducing atmosphere, which in turn reduces oxides of iron and chromium in the refractory. As these species continue dissolving into the glass melt, fresh Monofrax K3 surface is exposed to the melt for further corrosion. The line EDS data (not shown) indicated disconnected Cr-rich as well as Fe-rich regions in the vicinity of the refractory-glass interface.

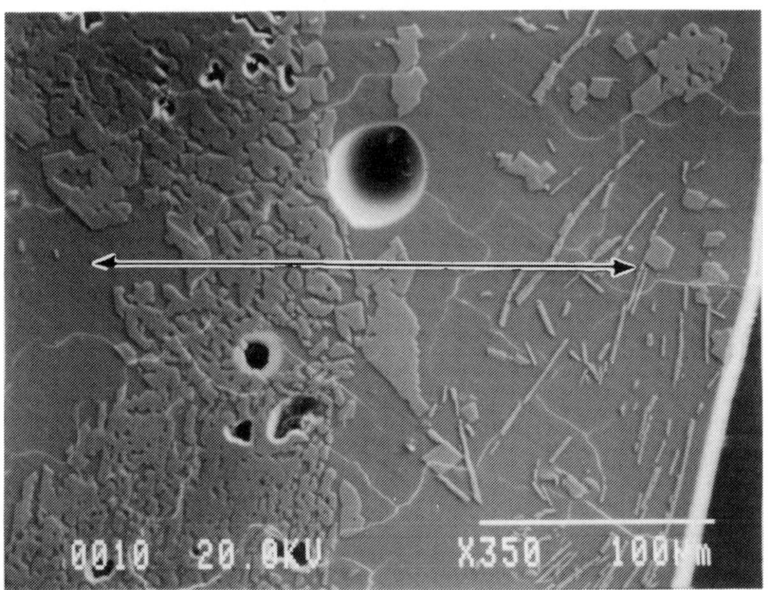

Figure 2. Monofrax K3(with carbon) - Glass Interface
(1450°C; 48 h; Half-down position; Left-Refractory; Right-Glass)

Figure 3 shows the interfacial features of Ruby refractory (Left-Glass; Right-Refractory).   Line EDS (not shown) indicated a short chromium-rich and aluminum-deficient region (about 400 microns thick) at the interface. The most corroding refractory in the laboratory tests, ZS1500, showed severe attack by the glass melt as shown in Figure 4 (Left-Glass; Right-Refractory). A cluster of fine particles were observed at the interfacial region. The line EDS data (not shown) across the interface that the particles were zirconia and silica particles.

Corrosion mechanisms

Preferential corrosion of specific phases is the basic corrosion mechanism in the case of Monofrax K3 (without carbon) and Ruby. In both the cases, a chromium-rich interfacial layer clearly supports this mechanism. As the phases of lower corrosion resistance (spinel and alumina) dissolve into the glass melt, the chromium-rich layer is formed. This corrosion-resistant layer acts as a barrier for further reaction between the glass and refractory, thus protecting the refractory from further corrosion. Similar observations have also been made with Monofrax

K3 on service in a slurry-fed melter [9]. In the presence of carbon, enhanced corrosion of Monofrax K3 is caused by the reduction of the refractory and dissolution into the glass melt. This was supported by data from the bench-scale tests data (not shown).

Decomposition of the zircon refractory followed by the dissolution of SiO_2 into the glass melt is the cause of the severe corrosion of ZS1500. Though ZrO_2 is a highly corrosion-resistant phase[10], these particles are separated by the dissolving silica phase. This is consistent with the higher corrosion observed in this refractory.

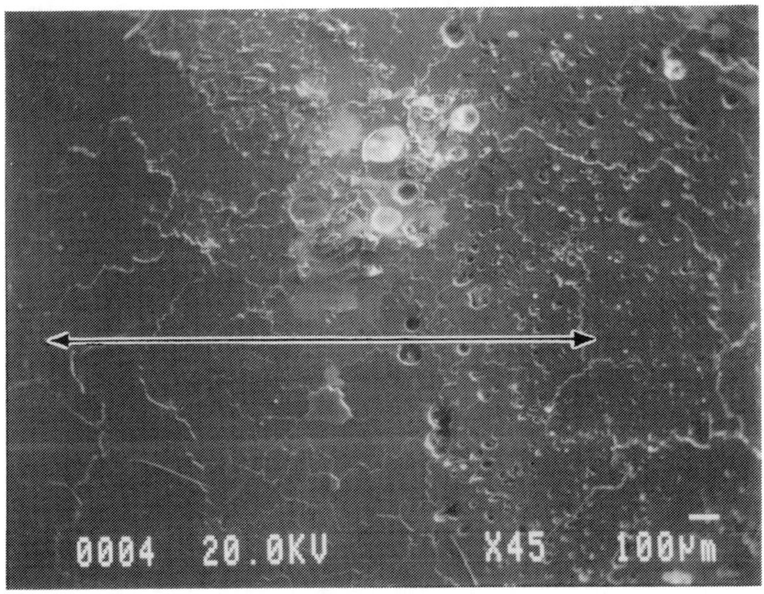

Figure 3. Ruby - Glass Interface
(1550°C; 48 h; Half-down position; Left-Glass; Right-Refractory)

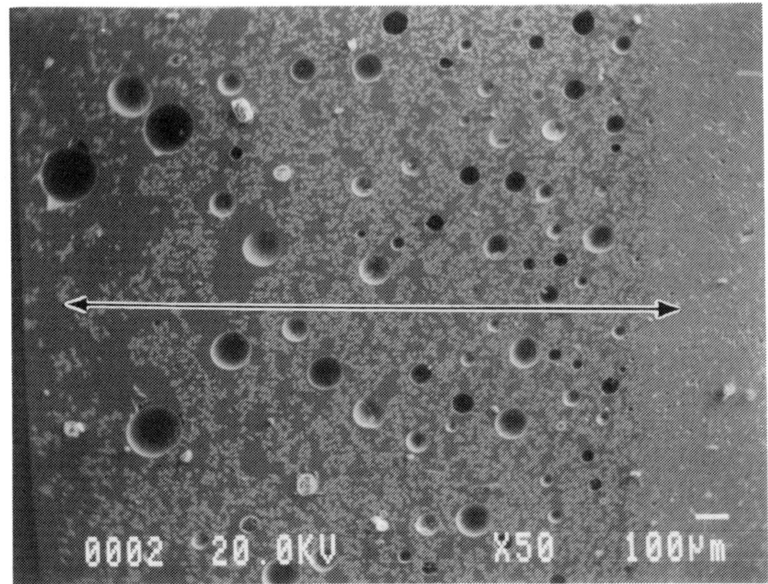

Figure 4. Zircon - Glass Interface
(1550°C; 48 h; Half-down position; Left-Glass; Right-Refractory)

CONCLUSIONS

The following conclusions were made based on the data generated in the present evaluation:

- Monofrax K3 was the least corroding refractory (without carbon) followed by Ruby.

- Monofrax K3 was found unsuitable for the processing organics-containing feeds as in the present application. Carbon addition to the glass enhanced the corrosion significantly due to reduction of Fe and Cr oxides of the refractory and dissolution into the glass melt.

- Corrosion mechanisms: Monofrax K3 formed a chromium-rich interface due to preferential corrosion of the other phases. This layer reduced further corrosion of the refractory. Ruby also preferentially corroded Al_2O_3 away,

      Advances in Fusion and Processing of Glass II

leaving a chromium-rich layer at the refractory-glass interface. ZS1500 decomposed into ZrO_2 and SiO_2 on reacting with the glass melt followed by dissolution of SiO_2 into the glass melt. This weakens the structural integrity of the refractory resulting in enhanced corrosion.

ACKNOWLEDGMENTS

The author acknowledges the support of David A. Lamar and William F. Bonner of PNNL. The study and preparation of this paper was supported by the Mixed Waste Focus Area of the DOE Office of Science and Technology.

REFERENCES

[1] J. C. Marra, J. W. Congdon, A. L. Kielpinski, R. F. Schumacher, A. A. Ramsey, J. Etheridge, and R. Kirkland, "Corrosion Assessment of Refractory Materials for Waste Vitrification", Proceedings of the Corrosion of Materials by Molten Glass Symposium, Ceramic Transactions, **78**, pp. 289-293 (1996).

[2] G. L. Leatherman, and J. A. Batdorf, "Crucible Materials for the Plasma Treatment of Radioactive Waste", Proceedings of the Corrosion of Materials by Molten Glass Symposium, Ceramic Transactions, **78**, pp. 289-293 (1996).

[3] W. Trier in Glass Furnaces (Design, Construction, and Operation), Translation by K. L. Loewenstein, Society of Glass Technology, Sheffield, UK, pp.60-75 (1987).

[4] R. A. Green, B. Nail, P. Norton, P-R. Jang, J. P. Singh, W. Okhuysen, F-Y. Yueh, H. Zhang, J. Andol, B. Kirkland, J. C. Marra, R. F. Schumacher, and J. Congdon, "Refractory Lifetime Testing for Westinghouse Savannah River Company," Diagnostic Instrumentation and Analysis Laboratory at Mississippi State University, March (1995).

[5] G. L. Smith, S. K. Sundaram, and P. A. Smith, "Glass-Refractory Material Interactions", PVTD-C95-02.03M, August (1995).

[6] C. J. Freeman, D. A. Lamar, and S. K. Sundaram, "Evaluation of High-Temperature Electrode and Refractory Materials for Application in Hanford High-Level Waste Vitrification", PVTD-T3C-95-139, September (1995).

[7] Proceedings of Materials for Vitrification Systems Workshop, Chairman: K. J. Imrich, Savannah River Technology Center, October 1-2 (1996).

[8] ASTM C621-84 (Reapproved 1989), "Standard Test Method for Isothermal Corrosion Resistance of Refractories to Molten Glass," American Society for Testing and Materials (1993).

[9] S. M. Barnes, G. J. Sevigny, and R. W. Goles, "Corrosion Experience with a Slurry-Fed Ceramic Melter," Proceedings of the Treatment and Handling of Radioactive Wastes, American Nuclear Society, pp. 203-207 (1982).

[10] T. S. Busby, "Progress in Refractory Usage in the Glass Industry," Glass Technology, **28** (1), pp. 30-37 (1987).

COMPRESSIVE CREEP BEHAVIOR OF FUSION-CAST ALUMINA REFRACTORIES[1]

A. A. Wereszczak and T. P. Kirkland
High Temperature Materials Laboratory
Oak Ridge National Laboratory
Oak Ridge, TN 37831-6069

G. A. Pecoraro and R. A. New
Glass Technology Center
PPG Industries, Inc.
Pittsburgh, PA 15238-0472

ABSTRACT

The compressive creep responses of commercially-available 100%β and 50%α - 50%β alumina fusion-cast refractories were measured in ambient air at 1400°C (2550°F), 1510°C (2750°F), and 1593°C (2900°F), and at static compressive stresses between 0.17 to 1.03 MPa (25 to 150 psi). These results are compared with the creep response of a commercial silica refractory. The compressive creep results at equivalent stresses showed that the 50%α - 50%β alumina was more creep resistant than the 100%β alumina, while the conventional silica was the most creep resistant of the three. A calculated creep rate - stress exponent equaling unity for all three materials was consistent with diffusion being the rate-controlling mechanism. The determined creep activation energy of 130 kcal/mol for the 50%α - 50%β alumina was consistent with literature values. The obtained low-value of 36 kcal/mol for the 100%β alumina appeared to be a consequence of surface-microstructural-changes which occurred during testing, and which subsequently affected the measurement of creep strain at the 1510°C (and especially 1593°C). An activation energy of 1000 kJ/mol (240 kcal/mol) was determined for the conventional silica, and its relatively high value is believed to be a consequence of a larger volume fraction of a lower-viscosity secondary phase present in the refractory at 1593°C compared to 1510°C.

I.   INTRODUCTION

Alumina fusion-cast refractories are attractive candidates for crowns in oxy-fuel fired glass furnaces and furnace modules because of their greater chemical inertness in this environment over that of conventional silica. A higher soda vapor concentration in oxy-fuel fired furnaces results in possible melting and vapor-refractory attack of conventional silica. However, fusion-cast alumina refractories are more resistant to soda vapor attack than silica and also exhibit low glassy phase exudation, which minimizes the potential for viscous glassy and stone defects in the manufactured glass.

A disadvantage of the fusion-cast aluminas is their relatively high density, which results in a heavier furnace crown (compared to a silica crown for instance). The cross-sectional area of an alumina crown cannot be smaller than the silica crown because of structural design considerations. Thus, owing to the density comparisons, an alumina crown would be 2.5 times heavier than a conventional silica crown. Due to this higher weight, creep or subsidence becomes a concern in regards to the possible lessening of service life and dimensional stability. Since an absence of published creep data for these fusion-cast alumina materials exists, interest existed in the present

---

1   Research sponsored by the U. S. Department of Energy, Assistant Secretary for Energy Efficiency and Renewable Energy, Office of Transportation Technologies, as part of the High Temperature Materials Laboratory User Program under Contract DE-AC05-96OR22464, managed by Lockheed Martin Energy Research Corporation.

study to generate engineering creep data for candidate fusion-cast aluminas which: (1) would help assess their creep performances, and (2) provide information for the systematic structural design of glass furnace crowns made from them.

II.   EXPERIMENTAL PROCEDURES
IIA.   Materials Description and Testing Procedures

Two commercially-available brands of fusion-cast alumina and a commercially-available silica refractory brand were tested. One brand of alumina (93.3 wt%) was reported by the vendor to contain 100% of the β phase, and the second brand of alumina (94.5%) was reported to have a 50/50 blend of the α and β phases. The silica had a vendor-reported content of 96.1% silica.

Creep tests were performed in ambient air using an electromechanical test machine in load control. The load train consisted of concentrically aligned α-SiC push rods which were surrounded by a resistance-heated clamshell furnace. High purity (99.5%) alumina disks ($\approx$ 3 mm thick) were inserted between the specimen ends and the push rods to prevent corrosion of the SiC. All specimens were preloaded to approximately 0.03 MPa ($\approx$ 4 psi) during furnace heatup. The specimens were then soaked at temperature for approximately 15-20 hours prior to the initiation of creep testing to allow achievement of thermal equilibrium of the furnace, load train, and specimen.

Specimens from both the fusion-cast aluminas were tested at 1400°C (2550°F), 1510°C (2750°F), and 1593°C (2900°F). Silica specimens were tested at 1510 and 1593°C. All specimens were loaded sequentially to at least 0.17 and 0.34 MPa (25 and 75 psi), which are representative stresses of service. Many of the specimens were loaded to additional stresses. A minimum of 50 hours of creep data at each stress was sought.

IIB.   High Temperature Extensometry Measurements

A high temperature contacting extensometer was used to continuously measure specimen contraction due to the compressive load, and a schematic of the test apparatus is shown in Fig. 1. Sapphire or α-SiC contacting rods were used, and the resolution and gage length of this extensometer was approximately 1 μm (0.00004 inches) at temperature and 40.00 mm (1.57 inches), respectively.

The employed extensometer system and the load-train-cooling system was an outcome of much time and effort to achieve a stable signal. This was necessary in order to sample the relatively small amounts of creep strain that these three materials exhibited. A laser extensometer (a transmitter and a receiver diametrically opposed on the exterior of the furnace) was initially used in

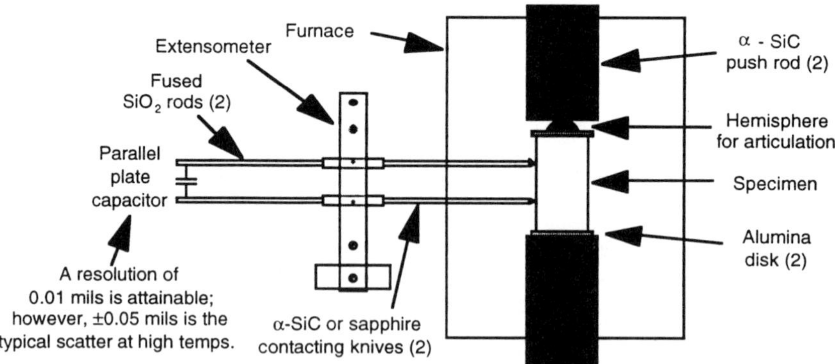

Fig. 1.   Schematic of the contacting extensometer, load train, and furnace arrangement used for the compressive creep measurements.

conjunction with specimen flags, but at 1510°C and 1593°C, the lack of a stable signal was unacceptable. The style of the contacting extensometer shown in Fig. 1 is frequently used by the ORNL authors with other high temperature testing of structural ceramics, so its usage was favored after the laser extensometer system was deemed unsatisfactory. However, additional issues needed to be addressed to achieve the most stable signal possible, even with the contacting extensometer. The laboratory-supplied cooling water to the push-rod couplers had a temperature variability that manifested itself in a cyclic displacement reading that was in phase with the cooling-water temperature cycle. A closed-loop, self-dedicated water-chiller was installed into the cooling system, and the laboratory-supplied water was no longer used; this eliminated the cyclic strain variations due to water-temperature changes. Lastly, room temperature fluctuations (e.g., a door being opened to the laboratory) also affected the strain reading. The whole extensometer (outside of the furnace) was then enveloped by a plastic box that had laboratory-supplied, pressurized air connected to it. This kept the extensometer (the stand, the SiO_2 rods, and the capacitor) thermally stable, which consequently resulted in a stable extensometer signal. Through these efforts, an extensometer stability of ± 0.001 mm (± 0.05 mils) was achieved at temperature.

IIC. Data Reduction & Engineering Creep Parameters

The steady-state compressive creep rate ($\dot{\varepsilon}_{s.s.}$) was related to the applied compressive stress and temperature using an Arrhenius power law or the Norton-Bailey [1] creep equation:

$$\dot{\varepsilon}_{s.s.} = d\varepsilon/dt_{s.s.} = A\,\sigma^n\,\exp(-Q/RT), \qquad [1]$$

where A is a pre-exponential constant, σ is the applied stress, n is the creep rate - stress exponent, Q is the activation energy, R is the gas constant, and T is the absolute temperature. Multilinear regression was performed to determine the constants A, n, and Q for each of the three refractories investigated. By performing the analysis in this manner, it is assumed that the same dominant (or rate-controlling) creep mechanism was active at all temperatures and stresses. The validity of this assumption is assessed by the goodness-of-fit of this equation to the experimental data and the reasonableness of the obtained values of n and Q.

III. RESULTS & DISCUSSION

An example of the relative differences in the creep resistances of the 100%β alumina, the 50%α - 50%β alumina, and the conventional silica are shown in Fig. 2. Typically for a given temperature and stress, the conventional silica was the most creep resistant refractory of the three, followed in turn by the 50%α - 50%β alumina, and the 100%β alumina.

The surface microstructure of the 100%β alumina changed at 1510 and 1593°C affecting the fidelity of the surface-measurement of specimen contraction with the contacting extensometer. Scanning electron microscopy of the 100%β alumina tested at 1510°C (and especially at 1593°C) showed that the surface grains of the as-received 100%β alumina were depleted of sodium as a consequence of testing at these temperatures, and likely transformed the surface grains to the α alumina phase. This change occurred in surface grains inwards of 1 mm into the bulk of the specimens, as shown in the micrographs in Fig. 3. After tens of hours of creep testing at 1510°C, the extensometer measurement became erratic due to the change in surface microstructure, so the testing was ceased. The change in surface microstructure at 1593°C occurred so rapidly that contact extensometry was unsuitable, and no creep data were generated at this temperature.

The compressive creep measurements made on the 50%α - 50%β alumina and the conventional silica were less problematic. The 50%α - 50%β alumina did show some analogous sodium depletion effects at 1593°C as the 100%β alumina did, but to a lesser degree. The conventional silica crept so much slower than the two fusion-cast aluminas that longer durations of testing were required in order to determine a confident estimate of the creep rate.

The creep rates as a function of compressive stress are shown for the 100%β alumina, the 50%α - 50%β alumina, and the conventional silica in Figs. 4-6, respectively. The values for the coefficients in Eq. 1 are shown in Figs. 4-6 as well. The residuals of the regression analysis of Eq. 1 showed that the correlation was quite good for all three refractories. All three materials showed a stress exponent equivalent to unity, which was an indication that the rate-controlling mechanism of creep was diffusion [2]. The obtained low-value of 150 kJ/mol (36 kcal/mol) for the 100%β alumina appeared to be a consequence of the surface-changes which occurred during testing and affected the measurement of creep strain at the 1510°C (and the subsequent effect it had on the multilinear regression result). The determined creep activation energy of 550 kJ/mol (130 kcal/mol) for the 50%α - 50%β alumina was consistent with literature values for the creep of aluminas [3-4]. An activation energy of 1000 kJ/mol (240 kcal/mol) was determined for the conventional silica, and its relatively high value is believed to be a consequence of a larger volume fraction (≈ 50% more) of secondary phase liquid present in the refractory at 1593°C compared to 1510°C [5]. Supporting this, the viscosity of many calcium silicate glasses decreases several orders of magnitude above 1500°C [6]. Unfortunately, no literature values for the creep activation energy for conventional silica refractory (or fusion-cast alumina for that matter) were located to compare with. The relative creep resistances as function of temperature are shown in Fig. 7. Like the differences in creep resistances evident in Fig. 3, the relationships in Fig. 7 for the three refractories show that for a given temperature and stress the conventional silica was the most creep resistant, followed next by the 50%α - 50%β alumina, and lastly the 100%β alumina.

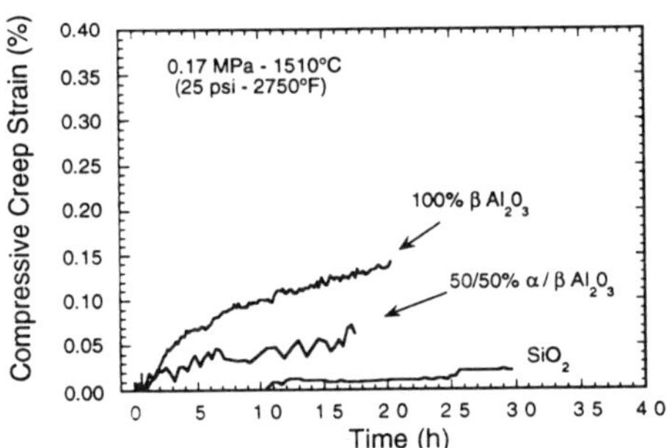

Fig. 2.   Comparison of relative creep resistances of the three refractories tested.

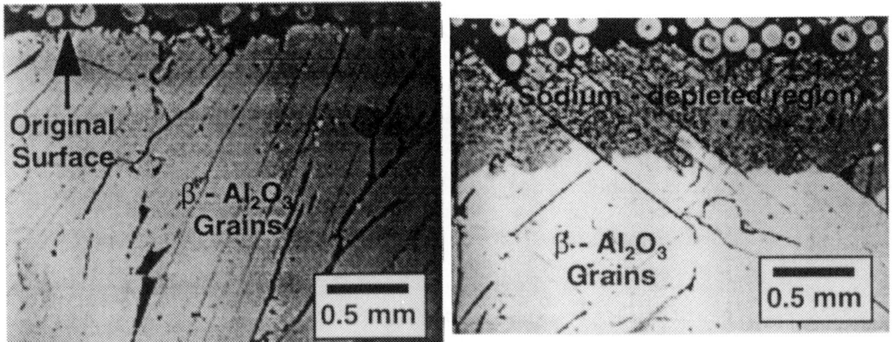

Fig. 3.   The as-received surface microstructure of the 100%β alumina (left) changed as a consequence of testing at 1510 and 1593°C (right). The dark phase at the tops of both photos is mounting epoxy. Secondary electron SEM images shown.

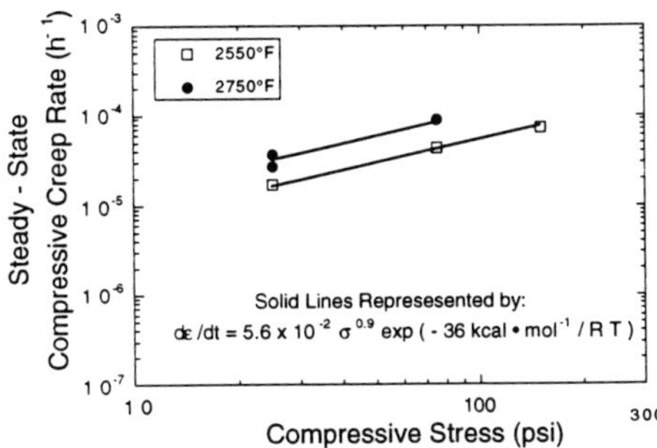

Fig. 4.   Creep rate as a function of stress for the 100% β alumina.

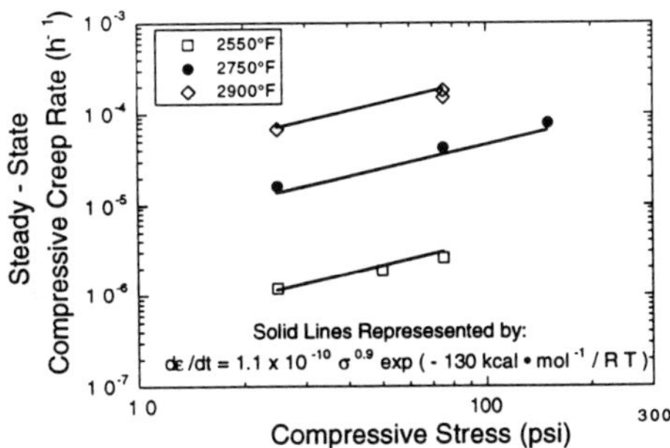

Fig. 5.   Creep rate as a function of stress for the 50% α / 50% β alumina.

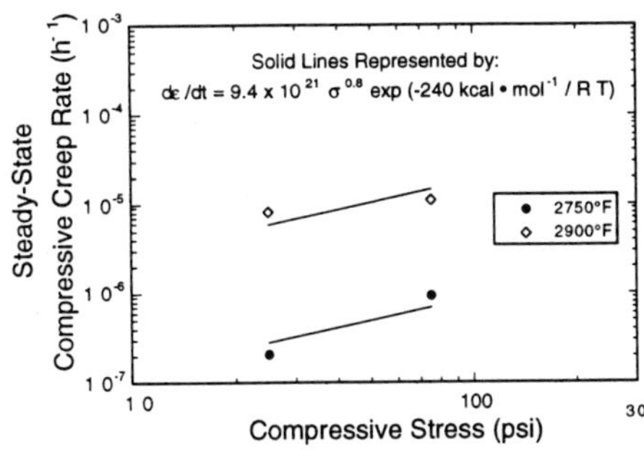

Fig. 6.   Creep rate as a function of stress for the silica.

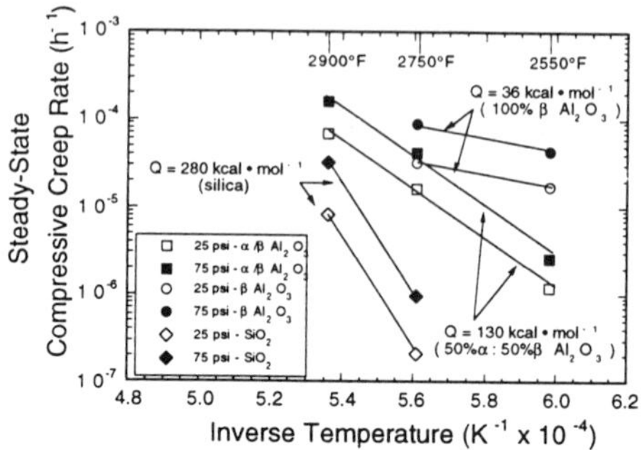

Fig. 7.   Comparison of creep rate as a function of temperature for the three refractories.

IV.   CONCLUSIONS

The compressive creep results at equivalent stresses showed that the 50%α - 50%β alumina was more creep resistant than the 100%β alumina, while neither were as creep resistant as the conventional silica. A calculated creep rate - stress exponent equivalent to unity for all three materials was consistent with diffusion being the rate-controlling mechanism. The determined creep activation energy of 550 kJ/mol (130 kcal/mol) for the 50%α - 50%β alumina was consistent with the literature, while the obtained low-value of 150 kJ/mol (36 kcal/mol) for the 100%β alumina appeared to be a consequence of surface-microstructural-changes which occurred during testing, which subsequently affected the measurement of creep strain at the 1510°C (and especially 1593°C). An activation energy of 1000 kJ/mol (240 kcal/mol) was determined for the conventional silica, and its relatively high value is believed to be a consequence of a larger volume fraction of lower-viscosity liquid present in the refractory at 1593°C (i.e., a greater ease of viscous flow) compared to 1510°C.

V.   ACKNOWLEDGMENTS

The authors wish to thank Drs. E. Y. Sun and T. R. Watkins for reviewing the manuscript and for their helpful comments, and Dr. Roger L. Scriven, Director of Research, PPG, for permission to publish the results.

VI.   REFERENCES

[1]   F. H. Norton, <u>The Creep of Steel at High Temperature</u>, McGraw Hill, New York, 1929.
[2]   W. R. Cannon and T. G. Langdon, "Review Creep of Ceramics," *J. Mat. Sci.*, **18** 1-50 (1983).
[3]   S. I. Warshaw and F. H. Norton, "Deformation Behavior of Polycrystalline Aluminum Oxide," *J. Am. Cer. Soc.*, 45 479-486 (1962).
[4]   V. S. Bakunov, "High-Temperature Creep of Refractory Ceramics. Kinetics and the Effect of Experimental Conditions," *Refractories*, **35** 177-183 (1994).
[5]   <u>Modern Refractory Practice, 4th Ed.</u>, Harbison-Walker Refractories Company, The William Feathers Company, Cleveland, OH, 1961.
[6]   <u>Handbook of Glass Properties</u>, N. P. Bansal and R. H. Doremus, Academic Press, Inc., New York, 1986.

COMPRESSIVE CREEP RESISTANCE OF MgO REFRACTORIES AT TEMPERATURES $\geq$ 1400°C (2550°F)[1]

A. A. Wereszczak and T. P. Kirkland
High Temperature Materials Laboratory
Oak Ridge National Laboratory
Oak Ridge, TN 37831-6069

W. F. Curtis
Glass Technology Center
PPG Industries, Inc.
Pittsburgh, PA 15238-0472

ABSTRACT

Compressive creep tests on one foreign and four domestic commercially-available brands of MgO refractories were conducted at temperatures $\geq$ 1400°C (2550°F) in ambient air at compressive stresses of 0.10, 0.20, and 0.30 MPa (14.5, 29.0, and 43.5 psi). All five MgO brands had a MgO content in excess of 96% with $CaO/SiO2$ ratios equal to or greater than 1.9, and firing temperatures in excess of 1535°C (2800°F). For each material, the compressive creep behavior was determined as a function of applied compressive stress and temperature. Creep resistance was found in some instances to vary significantly among the five brands with some significant contraction even occurring at a negligible stress at 1550°C. The values of the determined creep-stress exponents for these brands suggests that their compressive creep is dominated (or rate-controlled) by a diffusion mechanism, while the values of the determined activation energies suggest that creep in these MgO materials is accommodated by grain boundary sliding through viscous deformation of the silicate grain-boundary phase.

I. INTRODUCTION

MgO refractories are widely used in soda-lime glass furnace regenerators because of their good corrosion resistance to the alkaline environment. However, the regenerator checker packing and crown applications are susceptible to creep deformation during service because they are subjected to compressive stresses at high temperatures. The use of a creep-resistant refractory material is a requirement for optimum service life and performance in these applications.

There are several domestic and foreign vendors of MgO refractories, each claiming the suitability and/or superiority of their brand for utilization as regenerator checker packing and crown applications. However, it is difficult for the end user to make equitable and confident comparisons of the high temperature deformation behavior (i.e., creep or subsidence) among the brands because of differences in standards and test methods employed by the respective vendors. In an attempt to rate brand performances, an experimental study to mechanically characterize, assess, and compare the creep resistance of five commercially available MgO refractories was performed.

II. EXPERIMENTAL PROCEDURES

Five commercially-available brands of MgO were tested. The vendors reported that all five contained greater than 96% MgO, with CaO/SiO_2 ratios equal to or greater than 1.9, and were fired

---

1   Research sponsored by the U. S. Department of Energy, Assistant Secretary for Energy Efficiency and Renewable Energy, Office of Transportation Technologies, as part of the High Temperature Materials Laboratory User Program under Contract DE-AC05-96OR22464, managed by Lockheed Martin Energy Research Corporation.

at temperatures in excess of 1535°C (2800°F). Cylindrical specimens (nominal dimensions: 38 mm diameter x 76 mm length) were core drilled, and the ends were ground parallel to within 0.013 mm. Three specimens of each brand were prepared.

Creep tests were performed in ambient air using an electromechanical test machine in load control. The load train consisted of concentrically aligned α-SiC push rods which were surrounded by a resistance-heated clamshell furnace. High purity (99.5%) alumina disks ($\approx$ 3 mm thick) were inserted between the specimen ends and the push rods to prevent corrosion of the SiC. A high-temperature contacting extensometer was used to continuously measure specimen contraction due to the compressive load. The extensometer's contacting rods were made from sapphire, and did not react or fuse to the MgO specimens during any of the creep tests. The resolution of this extensometer was approximately 1 μm (0.00004 inches) at temperature, and it had a gauge length of 40.00 mm (1.57 inches). All specimens were preloaded to approximately 0.03 MPa ($\approx$ 4 psi) during furnace heatup. The specimens were then soaked at temperature for approximately 15-20 hours prior to the initiation of creep testing.

A total of 15 specimens were tested, and the text matrix is shown in Table I. One specimen from each brand was tested at 1400°C (2550°F), 1475°C (2685°F), and 1550°C (2820°F). All specimens were loaded sequentially at three stresses: 0.1, 0.2, and 0.3 MPa (14.5, 29.0, and 43.5 psi) for approximately 75 hours at each stress. The repeatability in the creep performances of like-brand-specimens is unknown because only one specimen per condition was examined; the reader should exercise caution and recognize that *some* variability may exist in the high temperature mechanical performances of these materials. Part way through the completion of the test matrix, it was observed that many of the specimens exhibited significant contraction *during the heatup and soak steps*, so for the remainder of the tests, this specimen contraction (compressive strain) was measured continuously with the contacting extensometer and computer-aided data acquisition.

The steady-state compressive creep rate ($d\varepsilon/dt_{s.s.}$) was related to the applied compressive stress and temperature using an Arrhenius power law or Norton-Bailey [1] creep equation:

$$d\varepsilon/dt_{s.s.} = A\,\sigma^n \exp(-Q/RT), \qquad [1]$$

where A is a constant, σ is the applied stress, n is the stress exponent, Q is the activation energy, R is the gas constant, and T is absolute temperature. Multilinear regression was performed to determine the constants A, n, and Q for each brand of MgO refractory. By performing the analysis in this manner, it is implicitly implied that the same dominant (or rate-controlling) creep mechanism is active at all temperatures and stresses. The validity of this assumption is assessed by the goodness-of-fit of this equation to the experimental data and the reasonableness of the obtained values for n and Q.

III.   RESULTS & DISCUSSION
IIIA.   Creep Performance

An example of a typical creep curve is shown in Fig. 1. Note the contraction during the first 19.8 hours even at the low applied stress. This contraction may be due to the continuation of sintering and/or microstructural rearrangement occurring in the system. All the refractories exhibited this effect to some degree, and the amounts for each are summarized in Table II. For a superstructure that is several feet in size, this low-load contraction may be significant and perhaps needs to be included in the superstructure's design. The contraction effects were most dramatic in materials B, C, and D, while materials A and E showed the least amount of contraction.

After the 15-20 hour soak, the specimens were compressively loaded to 0.1 MPa for approximately 75 hours, followed in turn by loading to 0.2 and 0.3 MPa for equivalent durations. The creep strain curves for each stress typically showed a primary region followed by a long duration of slowed-transient or steady-state creep. An increasing creep rate (or tertiary creep region) was never observed in any of the tests. The steady-state compressive creep rates of all five MgO refractories for all three stresses were determined, and examples of these are shown in Fig. 1.

After the steady state creep rates were determined, their values were related (via multilinear regression) to the applied compressive stress and test temperature using the Arrhenius power law equation shown in Eq. 1. The natural logarithm of both sides of Eq. 1 was taken, and then commercial spreadsheet software was used to determine the coefficients (lnA, n, and Q/R) of the regressed linear equation for all five MgO refractories. An example of this is illustrated in Fig. 2, and the values of the parameters A, n, and Q are listed in Table III for all five brands.

Using the parameters in Table III, the relative creep resistances among the five MgO refractories may be compared. As an arbitrary example, the steady-state creep rates at 0.2 MPa and 1500°C (2730°F) would be: 1.7×10^{-5}/h for material A; 3.6×10^{-5}/h for material B; 4.8×10^{-5}/h for material C; 6.2×10^{-5}/h for material D; and 2.9×10^{-5}/h for material E. Material A was the most creep resistant MgO material, followed in order by materials E, B, C, and D.

The reasons why some of the MgO materials were more creep resistant than others have not been fully understood as of this writing. This indicates the need to correlate microstructure and chemistry to creep performance, which will serve as the focus of continued future work. To design

Table. I. Test matrix (stresses and temperatures) for the compressive creep tests.

MgO Refractory ID	Stresses (MPa) at 1400°C or 2550°F	Stresses (MPa) at 1475°C or 2685°F	Stresses (MPa) at 1550°C or 2820°F
A	0.1, 0.2, and 0.3	0.1, 0.2, and 0.3	0.03, 0.1, 0.2, and 0.3
B	0.1, 0.2, and 0.3	0.1, 0.2, and 0.3	0.03, 0.1, 0.2, and 0.3
C	0.1, 0.2, and 0.3	0.03, 0.1, 0.2, and 0.3	0.03, 0.1, 0.2, and 0.3
D	0.1, 0.2, and 0.3	0.03, 0.1, 0.2, and 0.3	0.03, 0.1, 0.2, and 0.3
E	0.1, 0.2, and 0.3	0.03, 0.1, 0.2, and 0.3	0.03, 0.1, 0.2, and 0.3

Note 1: The amount of sintering-induced-contraction was only measured with the 0.03 MPa stress. A creep rate for this stress was not determined.

Note 2: Stress conversions: 0.03 MPa = 4 psi; 0.1 MPa = 14.5 psi; 0.2 MPa = 29.0 psi; and 0.3 MPa = 43.5 psi.

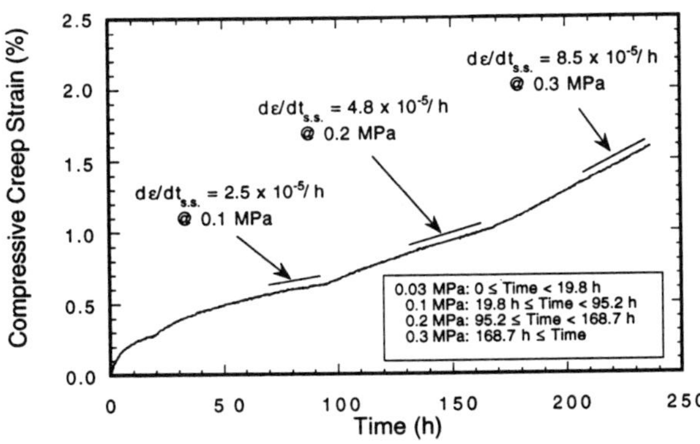

Fig. 1. Example of a typical creep history. The shown strain-time profile was generated with material "B" at 1550°C.

a creep-resistant MgO refractory, theoretically, one would want to maximize the MgO content and have a high CaO/SiO_2 ratio. Additionally, the lowest fraction of secondary phase present, in general, should result in the best creep resistance. In polycrystalline ceramics, it is recognized that an amorphous secondary phase dictates much of the bulk material's deformation behavior above its glass transition temperature. Thus as the amount of secondary phase is reduced, the bulk creep resistance tends to increase as a consequence. With MgO refractories, a higher CaO/SiO_2 ratio results in a higher liquidus temperature for this binary system [2], so the refractory should be more resistant to creep, all else being equal. Other physical factors that can impact creep resistance are the volume fraction of secondary glassy phase, the amount of porosity, and the grain size. With polycrystalline ceramics, it has been observed that creep rate is inversely proportional to the square

Table. II. Percent of contraction of MgO refractory prior
to the application of the first test compressive stress[§]

MgO Refractory ID	1400°C or 2550°F (hours[*])	1475°C or 2685°F (hours[*])	1550°C or 2820°F (hours[*])
A	DNM	DNM	0.02 ± 0.01[††] (13)
B	DNM	0.2 ± 0.05[†] (20)	0.27 ± 0.01[††] (20)
C	DNM	0.2 ± 0.05[†] (20)	0.38 ± 0.01[††] (18)
D	DNM	0.31 ± 0.01[††] (17)	0.35 ± 0.01[††] (18)
E	DNM	0.02 ± 0.01[††] (15)	0.04 ± 0.01[††] (18)

§ = A "low" compressive stress was actually applied (≈ 4 psi) during this test segment; it was used to maintain a slight preload in the specimen and the testframe's load train.

* = Duration for the amount of shown contraction.

DNM = Did Not Measure. The contraction was not experimentally measured for this condition; *however, this does not necessarily mean that no contraction occurred.*

† = This accuracy reflects that the contraction was estimated from a chart recorder.

†† = This accuracy reflects that the contraction was measured with the contacting extensometer.

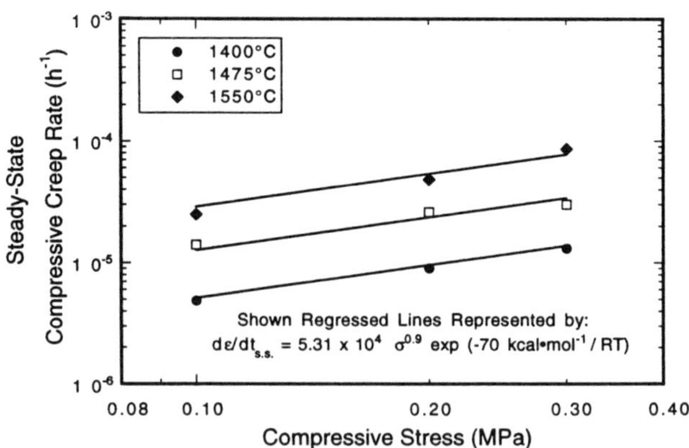

Fig. 2. Example of the creep data (creep rate as a function of stress) regressed with Eq. 1. The shown strain-time profile was generated with material "B".

                                    Advances in Fusion and Processing of Glass II

of grain size diameter for Nabarro-Herring creep, or to the cube of diameter for Coble creep. As a consequence, a larger grain size diameter will tend to decrease the creep rate as a consequence. Other factors that can impact some of the physical aspects of MgO refractories and also have a negative impact on their creep resistance are the presence of certain chemical impurities. Levels of boron as low as 500 ppm in the grain boundaries have been shown to significantly lessen the creep resistance of MgO refractories [3]. Another impurity that could contribute to the poor creep resistance is iron. It has been reported [4] that larger iron contents have been responsible for faster creep rates in other MgO refractories, so this too could be a factor.

IIIB. Stress Exponents, Activation Energies, and Mechanisms

The determined stress exponents for materials A, B, and E were equivalent to unity, which is indicative of the dominance (or rate-controlling) of a diffusion or Coble creep mechanism [5]. The low values (i.e., they are lower than the expected minimum value of 1) of the stress exponents for materials C and D appear peculiar; however, their low values are believed to be: (1) a direct consequence of the contraction effects that were represented in Table II and (2) an artifact of the assumption that a single deformation mechanism was active at all temperatures and stresses. Such low values for other MgO refractories have been reported in the literature [6]. The activity of the contraction mechanism was likely related to the time-at-temperature, and perhaps was active for many tens of hours before it asymptotically stopped. Materials C and D (and B to a lesser extent) strongly exhibited this contraction effect as evidenced by the large amounts they exhibited as shown in Table II. The measured creep rates for the 0.1 MPa loadings of materials C and D appeared to be relatively fast (compared to the subsequent measured creep rates at 0.2 and 0.3 MPa). This observation was attributed to the likelihood that both: (1) creep due to the 0.1 MPa stress was occurring and (2) contraction due to continued sintering and/or microstructural rearrangement was still active. For materials C and D, it was evident that the contraction effects were still active even at the 0.2 MPa (i.e., after 100 h at temperature) loadings. The consequence of these two mechanisms being active on the multilinear regression was that the analysis yielded an apparently low-valued stress exponent. Reiterating, the utilization of the multilinear regression assumes that a single rate-controlling mechanism is dominant for all temperatures and stresses: the observation that two mechanisms are active during early test times suggests that the single-acting-mechanism assumption is invalid for materials C and D.

The range of activation energies (50 -70 kcal/mol) for materials A, B, and E is consistent with activation energy values reported for other calcium-silicate MgO refractories in the literature [7-11], and is equivalent to the activation energy for viscous flow in a silicate melt [12]. Similar to the anomalously-low stress exponents for materials C and D, the activation energies shown in Table III for these same materials reiterates the invalidation of the single-acting-mechanism assumption for the usage of Eq. 1.

Table. III. Values of power law multilinear regression fit

$$[\ d\varepsilon/dt_{s.s.} = A\,\sigma^n \exp(-Q/RT) \ : \ d\varepsilon/dt_{s.s.} \text{ in units of } h^{-1} \ ]$$

MgO Refractory ID	Pre-Exponential Constant "A"	Creep Rate Stress Exponent "n"	Activation Energy "Q"
A	184	1.2	210 kJ/mol = 50 kcal/mol
B	5.31×10^4	0.9	290 kJ/mol = 69 kcal/mol
C	5.44×10^{-4}	0.5	24 kJ/mol = 5.7 kcal/mol
D	1.53×10^{-4}	0.1	11 kJ/mol = 2.6 kcal/mol
E	190	0.9	210 kJ/mol = 50 kcal/mol

IV. CONCLUSIONS

The creep resistances were found to vary significantly among five commercially-available MgO refractories. The most-creep resistant MgO refractory crept approximately four times slower than the least creep resistant MgO refractory at the same stress. The variation in creep performance among the five brands was likely due to mutual differences in a host of physical and chemical factors which will be investigated and discussed at a later date. Three of the five MgO brands exhibited contraction effects as low as 1475°C (2685°F), and for two they lasted long enough to affect the interpretation of their creep analysis. The values of the determined creep-stress exponents for these five brands suggests that their compressive creep is dominated (or rate-controlled) by a diffusion mechanism, while the values of the determined activation energies suggest that creep is accommodated by grain boundary sliding through viscous deformation of the calcium silicate grain boundary phase.

V. ACKNOWLEDGMENTS

The authors wish to thank Drs. H. -T. Lin and K. C. Liu for reviewing the manuscript and for their helpful comments, and Dr. Roger L. Scriven, Director of Research, PPG, for permission to publish the results.

VI. REFERENCES

[1]   F. H. Norton, <u>The Creep of Steel at High Temperature</u>, McGraw Hill, New York, 1929.

[2]   C. G. Bergeron and S. H. Risbud, CaO-SiO$_2$ Binary Phase Diagram, <u>Introduction to Phase Equilibria in Ceramics</u>, The American Ceramic Society, Columbus, OH, 1984.

[3]   G. E. Shaffer, <u>Compressive Creep of Magnesia Refractories</u>, M. S. Thesis, The Pennsylvania State University, State College, PA, 1973.

[4]   R. W. Evans, P. J. Scharning, and B. Wilshire, "Factors Affecting the Creep Strength of Magnesia Refractories," *Trans. J. Br. Ceram. Soc.*, **84** 108-110 (1985).

[5]   W. R. Cannon and T. G. Langdon, "Review Creep of Ceramics," *J. Mat. Sci.*, **18** 1-50 (1983).

[6]   S. V. Gilbert, "Creep Equation Model Used to Study Checker Subsidence," *Glass Ind.*, September, 14-21, 24 (1985).

[7]   P. J. Dixon-Stubbs and B. Wilshire, "Factors Affecting the Deformation Behaviour During Creep of Magnesia Refractory Raw Materials," *Trans. J. Br. Ceram. Soc.*, **80** 180-185 (1981).

[8]   R. W. Evans, P. J. Scharning, and B. Wilshire, "Creep of CaO/MgO Refractories," *J. Mat. Sci.*, **20** 4163-4168 (1985).

[9]   R. F. Krause, Jr., "Compressive Strength and Creep Behavior of a Magnesium Chromite Refractory," *Ceram. Engrg. Sci. Proc.*, **7** 220-228 (1986).

[10] P. J. Dixon-Stubbs and B. Wilshire, "High Temperature Creep Behaviour of a Fired Magnesia Refractory," *Trans. J. Br. Ceram. Soc.*, **80** 180-185 (1981).

[11] T. Vasilos, F. B. Mitchell, and R. M. Spriggs, "Creep of Polycrystalline Magnesia," *Science of Sintering*, **18** 65-67 (1986).

[12] P. Richet, "Viscosity and Configurational Entropy of Silicate Melts," *Geochim. Cosmochim. Acta.*, **48** 471-483 (1984).

SUPERSTRUCTURE REFRACTORY SELECTION FOR OXYFUEL MELTING OF LEAD-ALKALI-SILICATE GLASS

J Kynik (St. George Crystal Inc., Jeannette, PA.)
SM Winder (UK Software Services Inc., Grand Island, NY.)
KR Selkregg (Monofrax Inc., Falconer, NY.)

ABSTRACT

St. George Crystal converted from an all-electric *cold top* 'lead crystal' glassware production process to oxyfuel melting in June 1996. This presentation will describe the reasons for moving to oxyfuel combustion, and the superstructure refractory selection processes involved. A report on the current state of the furnace and refractories after ~13 months of oxyfuel operation at ~1500°C will also be given.

INTRODUCTION

St. George Crystal, Ltd. is a small company, located in Jeannette PA., specializing in machine manufacture of 24% 'lead crystal' glassware products. The firm employs about 265 people and adds a significant contribution to the Jeannette community and economy. St. George produces giftware (including candlesticks, votives, candy dishes, hurricanes, etc.), decorative lamp products, stemware etc.

Prior to the mid 1996 oxyfuel rebuild, St. George manufactured from a hexagonal electric furnace with capacity to melt 24 metric tons of crystal glass per day. The furnace was powered through molybdenum electrodes and was state of the art in 1981. The furnace configuration had 4 forehearths to deliver molten gobs to 3 press machines and one blowing machine. The electric furnace output was somewhat disappointing for the following reasons:
* difficult furnace operation and crystal color control,
* high molybdenum electrode consumption,
* high percentage of solid inclusion defects, and
* excessive environmental waste production and personnel safety issues.

The majority of the furnace performance dissatisfaction was attributable to incompatibility of the crystal glass composition and the electrode material. Molybdenum consumption was on the order of 160" of 2" diameter electrode rod per week. The rapid loss of electrode material contributed to a stability problem in maintaining good operating conditions. The electrode advancement process damaged refractory materials and contributed to producing solid inclusions in the glass (stones). The advancement process also contributed to premature electrode holder failure, necessitating expensive replacement of holder assemblies ($5,000 per incident x 24 electrode locations = $120,000).

An effect of electrode corrosion was the reduction of PbO to metallic lead. This created a downward drilling problem in the melter bottom, promoting stone defects in the glass bath. The molten lead corroded bottom joints as it drained downwards with unpredictable frequency and duration. In addition, the drainage could be somewhat violent in nature, creating safety issues to employees working under the melter. The metallic lead product is defined to be a hazardous waste, and must be disposed of in accordance with regulations pertaining to hazardous waste transport. St. George was generating over 26 metric tons of solid lead metal drippings per year, creating a strong incentive to decrease this problem.

The high molybdenum content in the glass resulted in a cloudy or grayish hue to the 'crystal' glass product. It also reduced iron (even at low levels) - which accentuated a greenish tint in the glass. Visually St. George 'crystal' did not exhibit the clarity or sparkle of competitor's product, and the market perceived the glassware to be inferior in appearance. In 1994, the mission was to determine the next melting process for St. George to advance into the 21st century. The new process needed to eliminate the 15% - 20% glass quality defective rate, and give manufacturing additional production capacity.

Non-electric melters are relatively inexpensive to construct and operate. Oxyfuel furnace technology is a U.S. development, and was considered best available control technology (BACT) by The Department of Environmental Resources (DER) and other regulatory agencies. Preliminary discussions with DER staff on the prospect of an oxyfuel melter at St. George were very well received. This set the stage for embarking on a radical departure from prior operations. The requirements of the project were as follows:
* meet longevity needs of a minimum of 6 yrs continuous operation,
* produce less than a 1% glass quality related defects, and
* install a robust and forgiving process.

THE REFRACTORY SELECTION PROCESS
Particular attention was given to selection of superstructure refractories, since industry operating experience with oxyfuel melters had demonstrated excessive corrosion of furnace crowns and generation of liquid corrosion products on vertical walls[1].

The St. George melt composition contains 60.5% SiO_2, 24.2% PbO, 9.3% K_2O, 4.2% Na_2O, minor amounts of Sb_2O_3 and ZnO, and impurities at low concentrations. Thermodynamic modeling was used to ascertain the major corrosive vapor phase components expected in the atmosphere of the proposed oxyfuel furnace, and results are presented as Figure 1. This melt composition is chemically similar to TV funnel glass in the identity of major volatile species produced above it.

Figure 1. Major Corrosive Vapor Species In The St. George Oxyfuel Melter Atmosphere (Calculated)

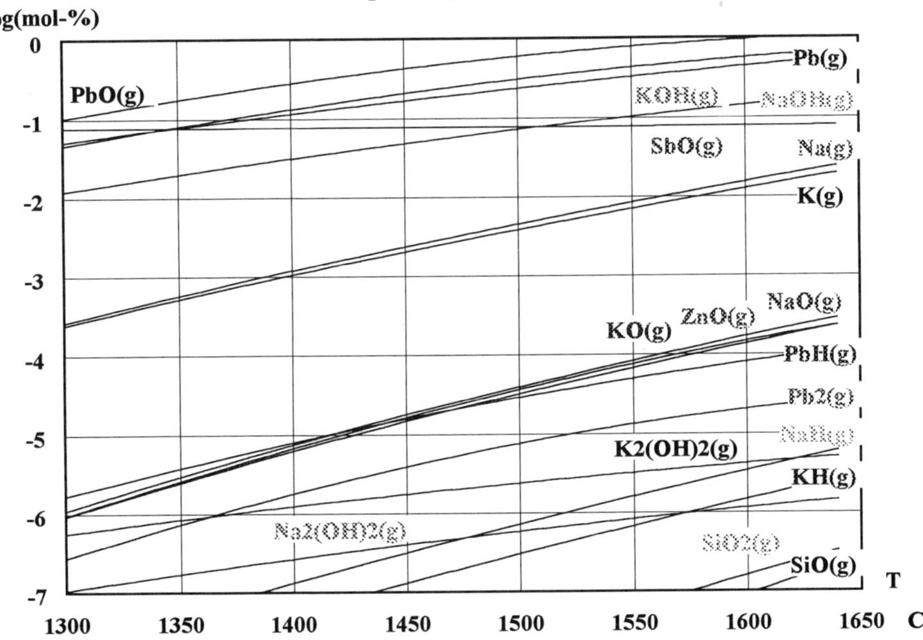

A series of bonded and fused-cast refractory blocks were exposed to vapors above a St. George glass melt in the Monofrax Superstructure Refractories Laboratory Corrosion Test. After 50 hrs exposure under oxyfuel conditions, samples were analyzed using SEM/EDS in order to determine the degree and mode of any refractory degradation.

Bonded silica and bonded mullite refractories were penetrated to considerable depth by a large volume of glassy phase (liquid at test temperature). Further investigation showed that liquid at the surface and in sub-surface layers of these refractories contained high concentrations of alkali, and a PbO content which decreased as a function of depth into the refractory.

The proposed corrosion mechanism for these bonded silica-based refractories involved corrosive vapors penetrating open porosity and dissolving in existing (essentially alkali-free) grain boundary phases, thereby decreasing melting temperatures and expanding the liquid volume present. Alkali and lead ions in these silicate boundary phases then triggered dissolution of the refractories crystalline structure. In the case of silica, the reaction product was entirely liquid phase silicate and in the case of mullite, the reaction products were liquid phase alkali-alumino-silicate and crystalline corundum (Al_2O_3). In either case, corrosion of the refractory occurred by penetration of alkali deep into the bulk of the refractory. Presence of PbO decreased silicate boundary phase viscosity close to the refractory surfaces, thereby accelerating corrosion kinetics.

Fused-cast AZS refractory was also attacked by species originating in the furnace atmosphere, but to a lesser extent than the more porous bonded products. Again the actively corroding species were alkalis, which dissolved in the glassy silicate matrix of this refractory and diffused into the bulk of the material. This triggered dissolution of the crystalline corundum phase, thereby expanding the volume of liquid present, and leaving only zirconia crystals in the solid state. PbO concentrations decayed rapidly as a function of depth into the refractory, but at the surface acted to decrease liquid silicate viscosity. The entire surface of the refractory was found to be evolving towards an end state of lead-alkali alumino-silicate liquid containing relatively insoluble zirconia crystals. This reaction product would represent a large defect potential to the glass bath.

Monofrax M fused-cast αβ-alumina refractory surface converted from a mixture of α-Al_2O_3 and β-$Na_2O.9Al_2O_3$ grains, to layer containing only α-Al_2O_3 crystals - see Figure 2(a). The mechanism of conversion appeared to involve dissolution of the β-$Na_2O.9Al_2O_3$ crystals in the small amount (<1.5%) of calcium silicate boundary phase present in this refractory, and re-precipitation of α-Al_2O_3 solid. The dissolved Na_2O apparently diffused along grain boundaries to reach the refractory surface and exit, as NaOH, into the furnace atmosphere. Previous work has shown that considerably greater presence of alkali hydroxide in the furnace atmosphere can prevent this $\beta \rightarrow \alpha$ conversion, or even promote $\alpha \rightarrow \beta$ conversion[2].

Monofrax H is a β-$Na_2O.11Al_2O_3$ refractory containing only ~0.2% SiO_2. In the furnace, the $\beta \rightarrow \alpha$ transformation occurred by direct loss of NaOH vapor from the crystals to the environment, without a solution / re-precipitation mechanism, and this resulted in a different α-Al$_2$O$_3$ grain morphology at the refractory surface. Monofrax M and H fused-cast alumina refractories did not react with the oxyfuel glass-melting environment, except in losing NaOH from their surface layers. Unlike silicate-based refractories, M and H did not generate large liquid volumes when exposed to oxyfuel glass-melting atmosphere, and defect potential was therefore minimized.

Figure 2. Monofrax M $\alpha\beta$-Alumina Superstructure Refractory Surface Converts To α-Alumina In St George Oxyfuel Glass-Melting Atmosphere. *Lighter Grains Are α-Al$_2$O$_3$, Darker Gray Crystalline Matrix Is β-Na$_2$O.9Al$_2$O$_3$.*

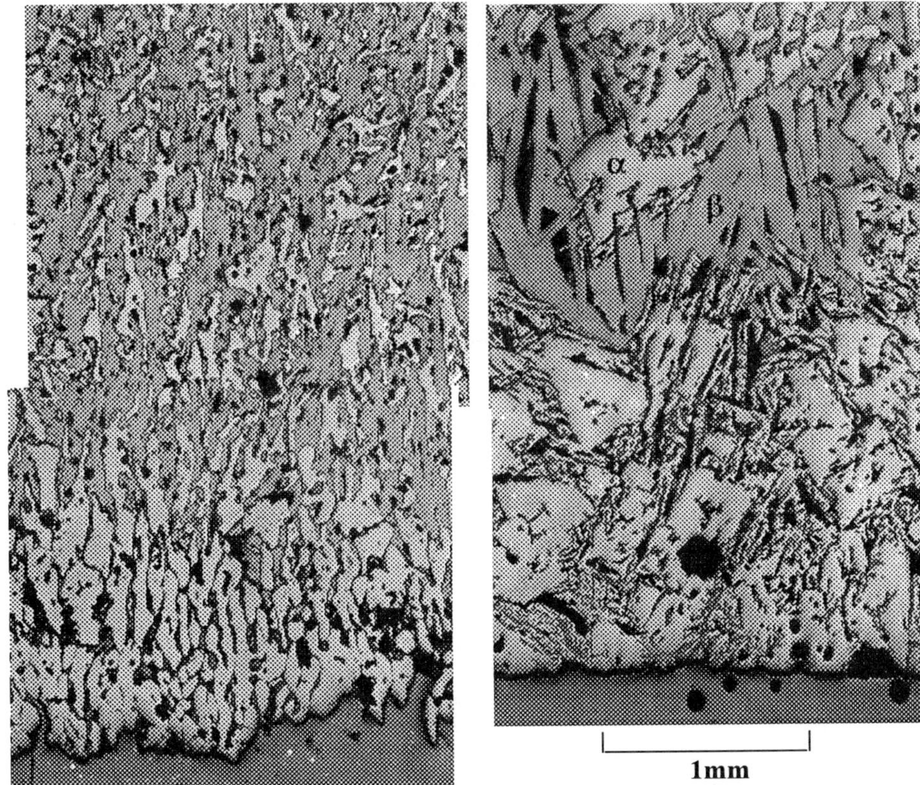

Figure 2(a). Exposure In Monofrax Laboratory Test	**Figure 2(b). Exposure In St George Oxyfuel Melter**

These results led to selection of Monofrax M fused-cast αβ-alumina refractory for the St. George oxyfuel melter crown. Monofrax CS3 fused-cast AZS was chosen as breastwall, frontwall and backwall refractory since AZS had been proven by application in various TV funnel glass melters to that time.

CURRENT CONDITION OF SUPERSTRUCTURE REFRACTORIES

The St. George melter was inspected by video camera in May 1997 in order to allow investigation of the state of the refractories. The observations made then are still applicable at the time of writing of this paper, after 13 months operation at ~1500°C.

The Monofrax M (Regular Cast) refractory crown appeared to be in excellent condition, completely dry with no signs of liquid formation at any place. All joints were unaltered, although occasional blocks exhibited a slightly spalled corner or edge - which occurred during crown construction. Since Monofrax M is considerably stronger than Monofrax H, it is felt that the choice of the β-alumina product in this application may have resulted in considerably more edge chipping.

The AZS superstructure showed evidence of corrosion, with a wet surface and some run-down.

Frontwall / crown and backwall / crown joints were seen to be open, with evidence of liquid phase running down out of the joints. This run-down may be the result of reaction between vapor phases from the furnace atmosphere and the insulation above the crown.

In order to learn more about conditions in the operating melter, a core of Monofrax M refractory was suspended down a thermocouple hole, close to the crown centerline above the batch pile, and exposed to the oxyfuel environment for ~14 days of operation. The exposed refractory was investigated using SEM/EDS and exhibited surface conversion to α-alumina, as experienced in the laboratory corrosion test, see figure 2(b). There was no sign of any batch dust contamination on this core.

Lessons learned from inspecting the crown and ~13 months of operation suggest some changes to superstructure refractory application that might allow further improvement in any future rebuilds. The Monofrax M refractory crown is stable in this environment, but the AZS superstructure is exhibiting signs of corrosion. It is felt that an entire fused-cast alumina superstructure would decrease defect potential. Application of a cement seal on the crown cold face offers the possibility of decreasing any degradation of the crown insulation.

SUMMARY
Performance targets were set as follows:
* 5.5 - 6.0 million BTU's/ ton energy input,
* 6.6 - 20.0 lbs/hr baghouse dust accumulation,
* <1% glass defective rate,
* pull requirements of 35 metric tons per day, and
* very stable furnace operation, melting costs to be less than electric.

Current performance is as follows:
* 5.5 - 6.5 million BTU's/ ton energy input depending on tonnage,
* 8.5 - 10.0 lbs/hr dust collector accumulation,
* substantially improved glass quality, optimization will improve output,
* achieved 32 ton per day pull (8 tons greater than the old melter), and
* melting costs better than all electric furnace.

Monofrax M fused-cast $\alpha\beta$-alumina refractory has been found to offer chemical
and physical stability in this oxyfuel glass-melter, but AZS superstructure
refractory shows signs of corrosion and defect generating liquid run-down.
Selection of fused-cast alumina for the entire superstructure may offer lower defect
potential in any future rebuild.

There is possible evidence for corrosion of insulation materials as a source of
liquid run-down. A layer of cement between the crown and insulation pack may be
a useful strategy to delay degradation of insulation material in future rebuilds.

This oxyfuel conversion project is considered a substantial success and yielded
the following benefits:-
* 30% capacity increase,
* 20% reduction in defective product (yielding more usable capacity),and
* positive environmental and personnel exposure outcomes.

REFERENCES

[1].SM Winder and A Brach, "Refractory Selection For Oxyfuel Glass-Melting
Furnace Superstructure", Presented at The HVG Colloquium, Mainz, Germany,
(13th March 1997). To be published.
[2]. SM Winder and JR Mackintosh, "Testing Oxyfuel Furnace Crown Materials",
Glasteknisk Tidskrift, **1** [52] 20 (1997).

Surface Engineering/Coating

LASERS IN GLASS SURFACE PROCESSING

Rudolf Weissmann
University of Erlangen-Nuernberg
Institute of Materials Science, Glass and Ceramics
Martensstr. 5
D-91058 Erlangen, Germany

ABSTRACT

Lasers are very suitable tools for processing of glasses like cutting, polishing, enamelling or structuring. Due to the short wavelength of excimer laser, micro-structuring of glass and ceramic surfaces is possible. Examples are presented like marking of glasses or microline structuring of float glass surfaces. In comparison to flame and furnace processing the CO_2 laser is far superior, because it allows treating glass very precisely. The laser radiation induces a very fast heating of glass within a small surface layer, other regions of the glass body are not influenced. Due to this controlled heating, thermal cutting of hollow and flat glasses, polishing and enamelling of glass surfaces are possible.

INTRODUCTION

In recent years lasers have found a wide field of applications in industrial manufacturing. Especially cutting and welding with high power of CO_2 lasers has been successfully incorporated in production lines of the metal industry. For surface engineering of glasses, laser working is a relative new tool [1-3]. Common methods are based on mechanical and chemical interactions like grinding, polishing, engraving, etching, etc. (Table I). Laser treatment belongs to the group of physical processes. The advantages are flexibility, punctiform tool geometry, avoidance of tool wear and mechanical load onto the glass. Especially the last point is very important. Because glass is a very brittle material. critical tension stresses have to be avoided during the engineering process.

Table I. Processing of glasses and glass surfaces

Mechanical	Chemical	Physical
Grinding	Etching	Flame polishing
Polishing	Polishing	Thermal strengthening
Cutting	Ion exchange	Ion bombardment
Drilling		
Sawing		***Laser processing***
Sandblasting		

There are further differences between glass and metals leading to the different behaviour. The heat conduction of glass is two orders lower than that of metals. The conduction heat transport is therefore much slower in glass. The result is a steeper temperature gradient that induces high thermal stresses in glass. Another difference between glass and metals is the viscoelastic behaviour of the glasses. Above a characteristic temperature, the transformation temperature, the glass changes from the elastic to viscoelastic state. This means that the glass begins to flow. The characteristic property is the viscosity that strongly depends on temperature. This viscoelastic behaviour makes cutting and welding processes more difficult. On the other hand this behaviour can be used in a favourable way for surface engineering of glasses. By suitable choice of laser properties and process parameters the following surface treatments of glasses are possible:

- Polishing of rough surfaces
- Surface melting of enamels
- Engraving of surfaces
- Cutting of glasses
- Soldering and welding

Also for new glass surface technologies the laser technology can find many applications:

- Laser enhanced CVD
- Surface crystallisation and phase separation
- Surface diffusion
- Microstructuring for integrated optics

The present paper summarises the important features of industrial lasers, shows examples for glass processing, and discusses possible practical applications in the glass industry.

INDUSTRIAL LASER SYSTEMS

EXCIMER LASER

Excimer lasers are high-power lasers with short pulses (10 to 50 ns), low repetition rates (20 to 1000Hz), and a pulse energy ranging from 0.1 to 2 J. The laser-active medium is composed of an inert gas and a halogenide [3,4]. The emitted ultraviolet radiation ranges from 193 (ArF laser) to 351nm (XeF laser). Due to these short wavelengths, the excimer lasers induce photochemical reaction at the glass surfaces. Without the detour of thermal decomposition, chemical bonds are cracked by ionisation and dissociation. Another advantage of the excimer laser is the uniform beam intensity profile. Therefore the laser is suited for large-area material treatment without focusing the laser beam.

The pulse duration of the excimer laser is in the nanosecond range. This is advantageous, because the energy, applied within a locally well defined region, is almost completely used for material removal. Only a negligible part flows as heat into the surrounding material. The consequences are rather low temperature gradients and uncritical stresses. Using focused beams, very high energy densities can be achieved, so that non-linear effects result. The short wavelength allows a focusing down to submicrometer features. Due to these reasons excimer lasers are predominantly used for surface modification and contactless marking. The short wavelength offers the possibility of application in the area of microstructuring of surfaces in the micrometer range. Further characteristics for the excimer laser are a larger rectangular beam cross section and a homogeneous intensity distribution

CO_2 LASER

The CO_2 laser is one of the most important laser systems for material processing, especially for metals. There is a relatively high efficiency using the continuous - wave (cw) irradiation mode of this laser. CO_2 laser radiation with a wavelength of 10.6 µm is strongly absorbed in glass due to the Si - O vibrational band at $\lambda = 9$ to 11µm. The absorption depth is very small, in the range of 1 - 2 µm, so that the CO_2 laser radiation can be considered as a surface heat source [6,10]. From the surface of the glass the energy is transfered deeper into glass by heat conduction and radiation. The strong absorption on the glass surface leads to steep temperature gradients in the glass because the heat conductivity is rather low. These temperature gradients can induce undesired critical thermal stresses in the glass. To avoid cracking and fracture the glass specimen must be preheated to a temperature just beneath the glass transformation temperature T_g. The following laser heating is now in the viscoelastic regime where thermal stress relaxation is possible. For practical work it is necessary to control the surface temperature by pyrometric temperature measurement.

APPLICATIONS OF EXCIMER LASER

MATERIALS ABLATION

By exceeding a material-specific energy density threshold, material removal starts. This ablation process is accompanied by formation of a plasma, whose intensity, size, shape and spectral composition depends on the working parameters and the material itself. Ablation threshold, ablation rate, and surface morphology are determined by the physical and chemical properties of the material and the laser parameters, like energy density, number of applied pulses, pulse repetition rate, and size of the irradiated area.

Table II points out clearly that fairly transparent glasses have higher ablation rates than absorbing crystalline materials. Surface defects, like grain boundaries, microcracks, pores, etc. support the absorption of laser radiation in ceramics, so that the ablation starts with the first pulse. In glasses, the ablation process needs several starting pulses depending on the energy density. During these first pulses the absorption increases due to the creation of internal absorption centres, e. g., colour centres [7].

Table II. Maximal absorption rates of different glasses and ceramics

Material		ablation rates (μm/pulse)
glass	fused silica	3
	borosilicate glass	3
	float glass	2.5
	athermal glass	08
	lead glass	0.4
ceramics	Al_2O_3	0.2
	ZrO_2	0.23
	SiC	0.28
	Si_3N_4	0.19

Concerning the ablation rates two groups can be distinguished. The transmitting glasses, fused silica and borosilicate glass, exhibit high ablation thresholds and ablation rates that decrease with increasing energy density. The morphology of the structures is rough break-outs destroy the edges. Most other glasses have ablation rates which increase rather strongly at the threshold and increase more slightly at higher energy densities. The morphology shows thermal effects such as surface melting and viscous flow.

Because precise microstructuring of most glasses is difficult, if the ablation rate is large, it seems better to use excimer lasers with shorter wavelengths. The radiation of these lasers will be absorbed stronger.

APPLICATIONS [4]

Examples for contactless marking of glass surfaces with a few pulses and with a single mask are shown in figures 2 and 3. The letters "F, L, E" (Forschungsverbund Lasertechnologie Erlangen) on float glass are rather large and well-separated by straight edges and clearly defined corners. The other letters on lead glass are much smaller, but still easy to distinguish and readable so that the code numbers can be printed easily in glass without disturbing the macroscopic appearance.

Another application may be the microstructuring of optical layers on the base of SiO_2. Figures 4a and b give an principal idea of the production of plane waveguides, which require dimensions of a few micrometers. Line structures, which satisfies these demands, have been obtained successfully in float glass. The width of the lines presented is about 20 μm, and structures down to 5 μm have recently been realised.

Figure 1. Micrograph of the marking of float glass

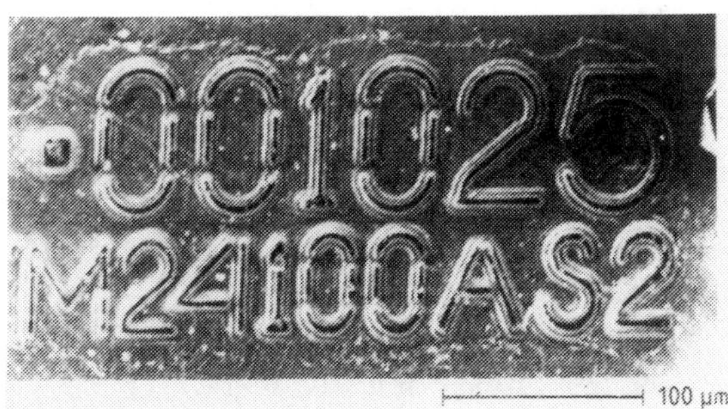

Figure 2. Micrograph of the marking of lead glass

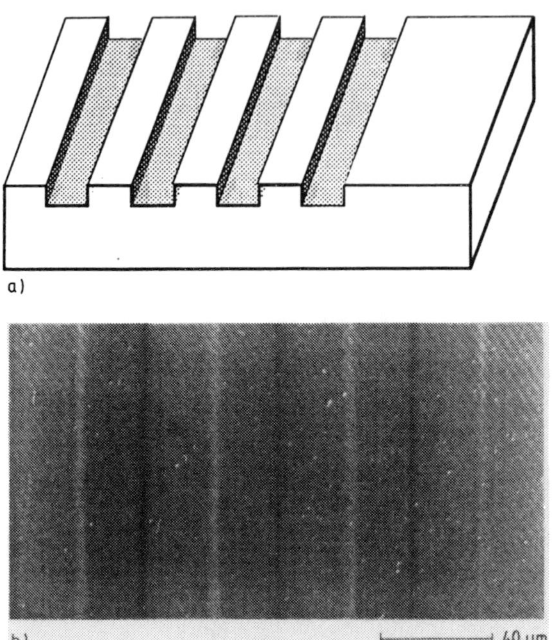

Figure 3. Microstructure of float glass. a) schematic, b) micrograph

APPLICATION OF CO_2 LASERS

CUTTING OF HOLLOW GLASSES BY THERMAL STRESSES

The aim of cutting glass with a CO_2 laser is to generate locally and temporally limited tensile stresses at the glass surface by inducing a very high temperature gradient. The resulting thermal stress is approximately:

$$\sigma = \alpha E \Delta T$$

where α is the thermal expansion, E is the elasticity modulus of the glass and ΔT is the temperature difference. The higher the temperature difference, the higher the resulting tensions.

The great advance of laser heating cutting compared with other methods like cutting by gas flame heating or separation with cutting tools is the excellent straightness of the cut. To achieve cutting there is no slitting necessary and there is no need to grind the resulting edge, meaning no grinding waste. Smoothing the edge can be done by gas flame or even defocused laser beam directly after cutting and preheating. A focused CO_2 laser beam with a power density of 5000 W/cm² is able to cut a rotating hollow glass (radius is 8 cm, revolutions per minute are 450) after 20 rotations.

     Advances in Fusion and Processing of Glass II

Cracking of glass happens during the heating period with laser radiation. The reason for this are bending stresses due to the thermal expansion of the tube within the heated area. Figure 4 illustrates this effect.

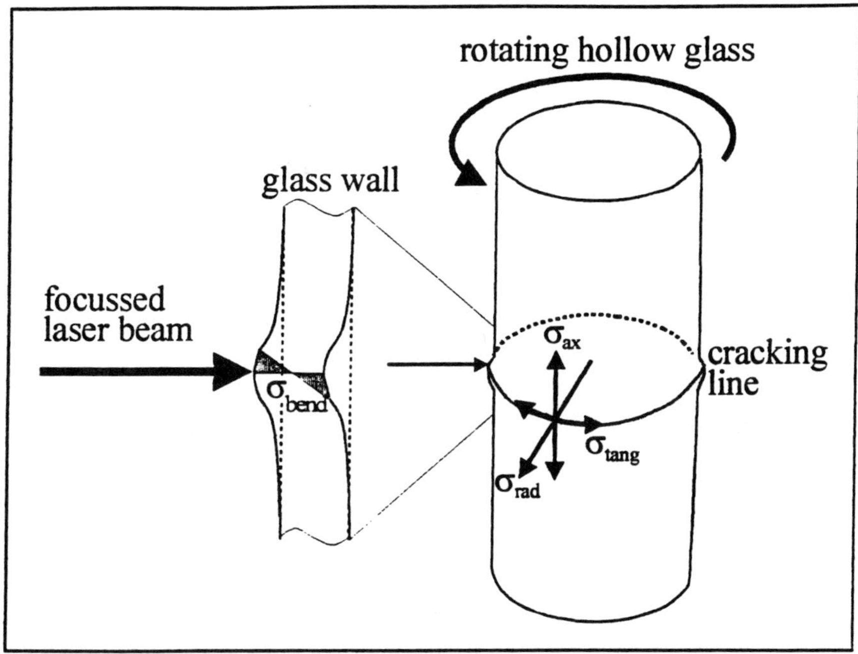

Figure 4. Stresses in a rotating hollow glass during laser irradiation

CUTTING OF GLASS PLATES

Traditionally plane glass is cut by producing a straight crack with a diamond tool or a hard metal wheel and then finally broken. This technique is simulated by CO_2 laser treatment by the following way. A specially configured CO_2 laser beam induces a thermally stressed region within the glass surface and this area is quickly cooled. The resulting tension stresses produce a sharp well-defined straight crack along the path of the laser spot and permit an easy separation of the glass plate. By using a fine microscopic initializing mark, the exact place where the crack is to start is determined. The laser power, the beam spot geometry and the cooling system are adjusted according to the type and the thickness of the material and customer requirements. The optimal cutting speed can be determined. The great advantage of this method compared with hard metal wheel cutting is that the cutt edge shows no microcracks, which increases significantly the edge breaking strength of the glass plates. This method is also best suited for cutting of glass with an approximate thermal expansion coefficient from 3.3 to 15 10^{-6} K^{-1} and material thickness from 0.05 mm to 10 mm.

POLISHING OF GLASS SURFACES [5]

Three methods for polishing a rough glass surface are commonly used. First is the fire polishing technique with a gas burner, second is acid polishing, third is mechanical polishing. All have disadvantages: During fire polishing the glass can be deformed or, for lead-crystal glass, undesired metallic lead segregations can result. Acid polishing and mechanical polishing produce a lot of waste according to the glass composition, even lead-, barium-, fluorine-, arsenic-, and (or) antimony-containing waste.

Laser polishing is similar to fire polishing, but the higher power of the laser beam achieve the heating of the glass surface much faster. If the surface temperature in the laser-irradiated area is high enough, the glass viscosity is lowered locally at the surface. By adjusting the temperature, the surface is polished homogeneously. If the energy of the laser beam is too low, the glass surface does not get hot enough, and no smoothing results. Too high energy leads to very high temperatures and evaporations and bubbles in the glass surface appear. For glasses with a high thermal expansion coefficient, meaning a poor thermal shock resistance, preheating near to T_g is inevitable. After the laser polishing the resulting stresses decrease during tempered cooling.

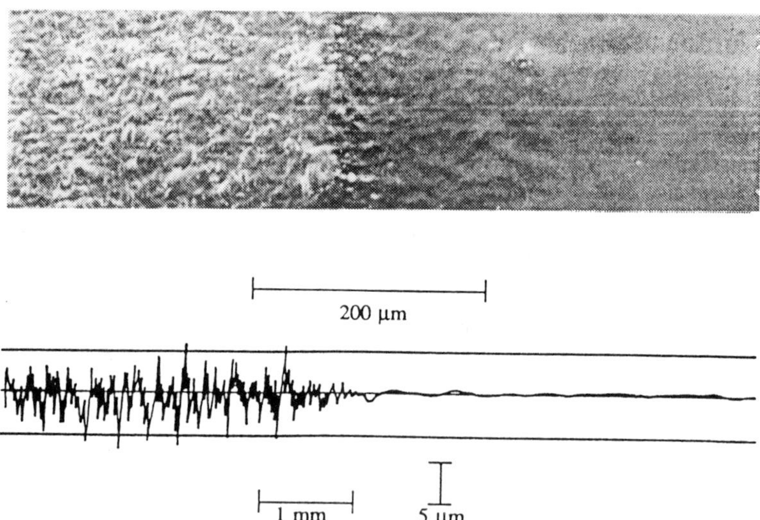

Figure 5. Surface and roughness of ground lead crystal-glass unpolished and laser polished

In figure 5 the success of laser polishing of a lead-silicate glass surface is demonstrated. The Scanning Electron Microscope (SEM) photograph shows that the previously rough surface appears perfectly smooth. At the bottom of figure 5, the reduction of roughness by laser polishing is demonstrated. The surface properties are equal to acid-or fire-polished glasses.

Concentration differences between irradiated and untreated glass are not measurable with Energy Dispersive Spectral analyses (EDS) and Secondary Ion Mass Spectroscopy. Therefore no detectable amounts of PbO evaporate during the laser treatment, and no environmental pollution occurs. But not only lead-crystal glass can be polished with a laser. Rough surfaces, sharp edges and joins at soda-lime-silica, borosilicate and even silica glass can be smoothed with laser, too.

ENAMELLING OF GLASS [8,9]

Glass enamels are used for flat glass decoration like windshields and architectural glass. Also decoration of hollow glass products like drinking glasses, tumblers or bottles is done with glass enamels, similar to china enamelling. In contrast to china decoration, the temperatures for firing glass enamels must not be so high. Only 630 °C for flat glass and 590 °C for hollow glass is tolerable. Higher temperature causes deformations to the glass body. Therefore glass enamels have, in a low-melting glass flux, a high PbO amount. These enamels have a poor chemical and erosive resistance, meaning a lack in dishwashing durability.

Enamel firing with laser needs preheating and tempered cooling, too. The firing technique with CO_2 laser does not need a low-melting enamel any longer. The ingredients of enamels can be changed arbitrarily, and even high-melting enamels used for china decoration can be fired on glasses. For of CO_2 laser radiation heating only the surface properties are decisive. So firing takes place at temperatures higher than 1000 °C, whilst other parts of the glass remain at the preheating temperature. Only if the firing process with laser lasts too long or the glass is very thin does deformation happen as the heat gets into the depth. Figure 6 shows temperature increase by decrease in feed rate for different power densities of the laser beam. The curves in figure 6 were measured for lead-free china enamels on float glass, the radius of the circular laser beam was 1.5 mm. Further details are several regions with different melting behaviors of the enamel. In region I the temperature for firing is not high enough, melting of enamel is uncompleted, and a rough and matte surface remains. The best firing results are in region II, whilst the enamel is overheated in region III, meaning bubbles and destruction of color.

Although the firing process lasts only seconds, there is the danger of deformation of the ground glass. For a 2 mm thick ground glass the optimum firing extends over a wide range, but a 1 mm thick float glass allows only very short firing times. Here region II is very small (hatched area).

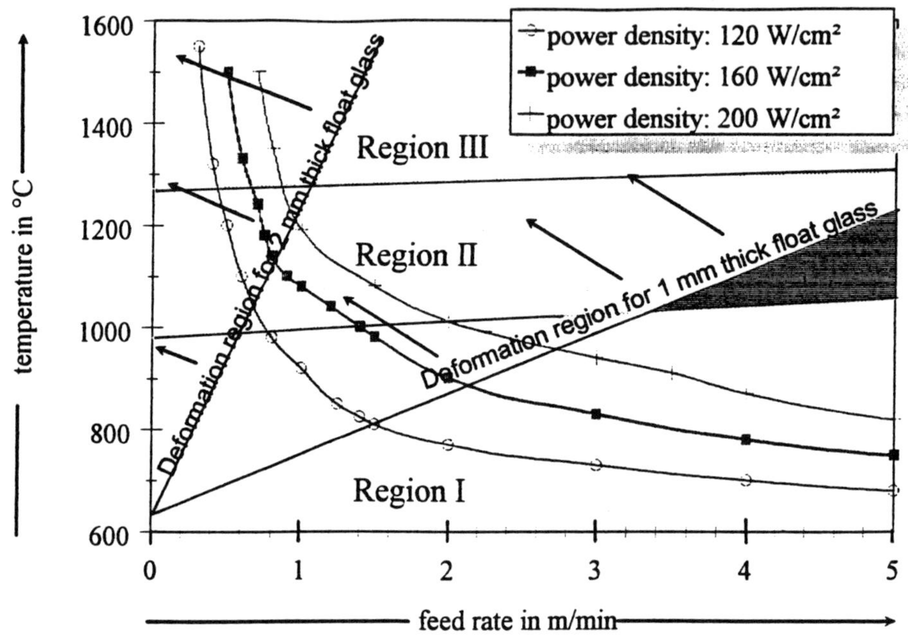

Figure 6. Temperature development for different feed rates and power densities for a lead-free enamel for fast firing china.

With a CO_2 laser it is feasible to fire various enamels to various ground glasses. Most important for laser firing is a colour pigment that endures high temperatures. The power distribution of the laser beam should be very homogenous across its diameter, otherwise an inhomogeneouly melted enamel results.

CONCLUSION
The demonstrated examples show that lasers are a very good tool for glass processing like marking, microstructuring, cutting, polishing or enamelling. The procedures are non-contacting and very exact. The results are mostly better than the conventional methods or at least equal. For working on very large areas, laser radiation is not a suitable tool, but treating small regions is very effective. Non-

 Advances in Fusion and Processing of Glass II

destructive working with laser beams on glasses, especially surface modification, is possible by reducing the resulting stresses. Therefore the glass must be preheated and slowly cooled. Cutting glass with CO_2 laser by thermal stresses makes use of the high tensions the laser beam induces. But there are still many more possibilities for employing laser technique to glass manufacturing.

REFERENCES
[1] G. K. Chui, "Laser Cutting of Hot Glass" , *Ceramic Bulletin*, **54**, 514-518 (1975)
[2] P. Urbanek, "Laser Decoration of Glass" , *Journal of Non-Crystaline Solids*, **38+39**, 891-895 (1980)
[3] V. Pfeifer, "Excimer laser - a novel tool for material fine and microprocessing of glasses", *Glastechn. Ber. Glass Sci. Technol.* **70**, 113-118 (1997)
[4] C. Buerhop, N. Lutz, R. Weißmann and G. Tomandl, "Surface treatment of glass and ceramics using XeCl excimer laser radiation", *Glastechn. Ber.* **66**, 61-67 (1993)
[5] A. Geith, C. Buerhop, R. Jaschek, R. Weißmann, "Polishing of lead crystal glass using cw- CO_2 - laser" , *in B.L. Mordike, editor, 4 th European Conference on Laser Treatment of Materials (ECLAT 92)*, 521-526 (1992)
[6] A. Helebrandt, C. Buerhop, R. Weißmann, "Mathematical modelling of temperature distribution during CO_2 -Laser-irradiation of glass", *Glass Technology*, **34**, 154-158 (1993)
[7] C. Buerhop, "Glasbearbeitung mit Hochleitungslasern", *Universität Erlangen- Nürnberg, Thesis*, Erlangen 1994.
[8] C. Buerhop, R. Weißmann, "Enamelling of Glass by CO_2 - laser treatment", *Glastechnische Ber. Glass Sci .Techn.*, **68**, 147-152 (1995)
[9] K. Hahn, C. Buerhop, R. Weißmann, "Firing PbO-free glass enamels using the cw- CO_2 - laser", Glastechn. Ber. Glass Sci. Techn., **96**, 1-6 (1996)
[10] C. Buerhop, R. Weißmann, "Temperature development of glass during CO_2 - laser irradiation. Part 1. Measurement and calculation", *Glass Technology*, **37**, 69-73 (1996)

CORRELATION BETWEEN ZETA POTENTIAL OF GLASS SURFACES AND MICROINDENTATION DATA IN AQUEOUS ENVIRONMENTS

Jacqueline C. Rolf, William C. LaCourse, James R. Varner, New York State College of Ceramics at Alfred University, Two Pine St., Alfred, NY 14802

Abstract

This study focused on the relationship between zeta potential of the glass surface and its microhardness in aqueous environments for silicate glasses of different compositions. The tests were performed on vitreous silica, soda-lime-silica microscope slides, and an aluminosilicate glass remelted from a glass-ceramic. The points of zero charge were found to be between pH 1.5 to 2 for vitreous silica and the soda-lime-silicate glass, and between pH 2.5 and 3 for the aluminosilicate glass. Maxima in microhardness were expected for all three glasses at the pH of the iso-electric points of the different surfaces. For vitreous silica and soda-lime-silica glass, the expected was observed; but for the aluminosilicate glass, two separate maxima in hardness were measured. An explanation for this observation is offered by consideration of the different surface groups present on the surfaces of the three glasses.

1. INTRODUCTION

The extreme sensitivity of the glass surface to compositional and environmental influences offers an interesting and challenging field for research and development. Especially for surface processing, which continuously increases in importance for product improvements, an understanding of the processes taking place on surfaces is essential. Therefore, in recent years, more emphasis has been placed on the study of surface chemistry of glass and reactions of surface groups.

The effect of surface groups and interactions with environments is of particular importance in grinding and polishing operations. Westwood et. al.[1] found a maximum in drilling rate of soda-lime-silica glass drilled with diamond-studded hemispherical-headed bits, and a maximum in pendulum hardness at the iso-electric point (i.e.p.) of the glass surface in n-alcohols and n-alkanes. They argued that environmentally induced diffusion of sodium ions in the glass could be responsible for the observed effect. It has been shown that the relationship between the i.e.p.s of both the glass surface and of the polishing agent strongly influences the removal rate in polishing[2]. In a study of loose abrasive microgrinding, Golini and Jacobs[3] found that at a certain particle size and grinding pressure, the grinding mode changes from brittle to ductile material removal at the i.e.p. of the glass surface in n-alcohols.

Macmillan et. al.[4] observed a maximum in hardness occurring at the i.e.p. of soda-lime-silica glass in a series of n-alcohols. He stated that this phenomenon should be reproducible in any environment which creates a zero zeta potential on the glass surface. The same relationship has been found by Keulen[5] on a glass containing MgO, which shifts the i.e.p. of the glass to a more alkaline pH value with respect to the i.e.p. of soda-lime-silica glass. His finding strongly supports the theory of Macmillan et. al. In a study of hardness and frictional behavior of glasses Cott (now Rolf) and LaCourse[6] found supporting evidence for the validity of the theory as well. The objective of this study is to investigate the influence of aqueous environments on the surface hardness of glasses of different compositions. The goal is to gain information about the universal validity of the theory suggested by Macmillan et. al.

2.   EXPERIMENTAL

2.1   Zeta Potential Measurements:

The zeta potentials of soda-lime-silica microscope slides, vitreous silica microscope slides and aluminosilicate glass were measured in aqueous solutions of different pH values using a Pen Kem Model 501 Laser Zee Meter. The testing solutions were titrated immediately prior to the experiments from 1/2N HCl and 15N NH_4OH. The aluminosilicate glass was remelted from a commercial glass ceramic.

The glasses were etched in a mixture of 2% HF and 2% H_2SO_4 for 5 minutes, rinsed with distilled water, rinsed with ethanol, and dried in a desiccator. After drying, the glasses were ground in an alumina mortar and passed through a 325-mesh sieve. Immediately prior to filling the solutions into the Laser Zee Meter, the glass powder was mixed into the solution, and the pH of the suspension was measured again.

The principle of operation of the Laser Zee Meter is based on the measurement of the electrophoretic velocity of small particles suspended in a solution under the influence of an electric field caused by high voltage.

2.2   Recording Microindentation Measurements:

The microhardnesses of the same glasses in the same aqueous solutions were determined using a maximum load of 0.5 N using a recording microindenter*. The load was applied with a speed of 0.1 μm/s and held for 10 s under full load before unloading. For each glass composition 12 indentations were made under each condition and the average and standard deviation were determined. The indentations were repeated for randomly selected test conditions

---

* developed by NYSCC at Alfred University

to ensure reproducability. Prior to the tests, the glass slides were etched for 5 minutes in 2% HF-2% H_2SO_4, rinsed with distilled water, rinsed with ethanol, and dried in a desiccator. After placing the slide on a three ball ring on the instrument, a large drop of test solution was placed on the surface of the specimen with a syringe. Twelve indentations were made for each glass/solution combination. The hardness values LVH, L_2VH, and LVH_2 were calculated for each indentation, and their averages determined. LVH is the Vickers hardness at the maximum load; L_2VH and LVH_2 are load-independent hardness values determined by two different mathematical treatments of the load and penetration data obtained during loading. Details can be found in Ref. 7 and 8.

3.   RESULTS AND DISCUSSION

3.1   Zeta Potential

The iso-electric points of vitreous silica and soda-lime-silica were determined to be in the same region of pH; between a pH of 1.5 and 2. The i.e.p. of the aluminosilicate glass was measured at a pH between 2.5 and 3. The i.e.p. of alumina is located at a pH of around 8, which implies that the surface of an alumina particle is positively charged at pH values below 8. The surfaces of silica particles, on the other hand, are negatively charged at pH values above 2. During the zeta potential measurement, the net charge of the surface is determined. The addition of alumina to a silicate glass will create alumina surface sites which will lead to a shift of the i.e.p. to more alkaline pH values because of the location of the i.e.p. of alumina.

3.2   Recording Microindentation

The results of the recording microindentation tests are shown in Figures 1, 2, and 3 for soda-lime-silica, vitreous silica, and the aluminosilicate glass respectively. In agreement with the observations in a previous study by Macmillan et. al.[1], a small local maximum in hardness was found in the vicinity of the i.e.p. for both vitreous silica and soda-lime-silica glass.

All three hardness values calculated from recording microindentation curves are graphed, because the trend in the data is of interest, and if the trend is real, it should be detectable in all three values. The absolute hardness values are much lower than the reported hardness values for these glasses due to the design of the recording microindenter. Since only samples tested with the same device are compared, the deviation from reported values is admissible.

For soda-lime-silica, the trend is not as clear as for vitreous silica. The LVH_2 value is surface sensitive and load independent. Since this study is concerned with the surface effect, the LVH_2 value is the most interesting one. For soda-lime-silica, this value does not show a distinct local maximum, but a steady trend to higher values with increasing alkalinity of the solution. Considering the corrosion mechanisms of glasses, a lower hardness would be expected for solutions of higher alkalinity, because the alkaline solution is assisting the destruction of the network by breaking up bridging oxygen bonds, which should lower the resistance of the surface to the penetration of an indenter. The presence of 2 wt.% Al_2O_3 can explain the higher hardness in the alkaline region for soda-lime-silica, since Al_2O_3 improves the chemical durability of the silicate network in alkaline environments[5] if the ratio of Al_2O_3/Na_2O is smaller than 1, but the effect seems to be more pronounced for the vitreous silica, which cannot be explained by this mechanism.

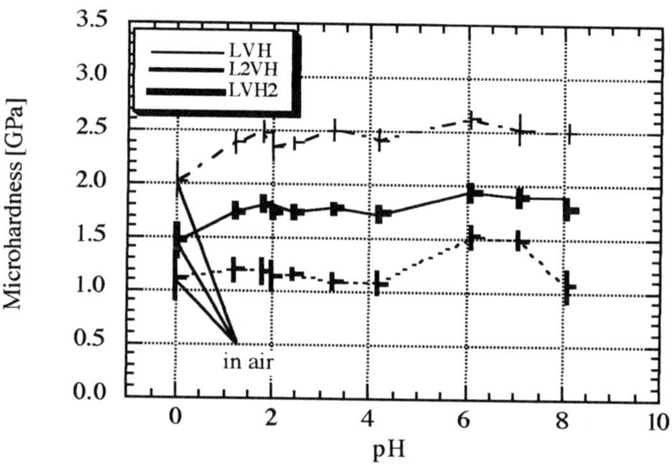

Figure 1:   pH dependence of the microhardness of soda-lime-silica glass. The error of the measurement lies between 5 to 8%.

According to these results, the theory of Macmillan[1] et. al., and the results of Keulen[5], a local maximum was expected in the vicinity of the i.e.p. for the aluminosilicate glass as well. As Figure 3 shows, the expectation was not met, but two local maxima were observed instead. Interestingly, the first of the maxima is found at a pH of about 1.3, which is near the i.e.p. of silica, and the second maxima is located at a pH of 7, which is near the i.e.p. of alumina. A possible explanation is the presence of two distinct surface groups, i.e. the ones formed by the silica and the ones formed by the alumina in the glass. It appears that, in the acidic region, the surface groups of silica dominate the response of the material, and in the alkaline region, the surface groups of alumina are dominating. Since only one maximum in hardness at the i.e.p. of glasses was observed by other researchers[5] and in this work as well for soda-lime-silica microscope slides, which contain 2 wt.% alumina, it is proposed that the phenomenon is concentration dependent. If the surface hardness of a series of soda-lime glasses containing increasing amounts of alumina were to be tested, a transition from one

   Advances in Fusion and Processing of Glass II

maximum in hardness at the i.e.p. of the glass to two distinct maxima in the vicinity of the i.e.p. of silica and alumina respectively could be expected to be found.

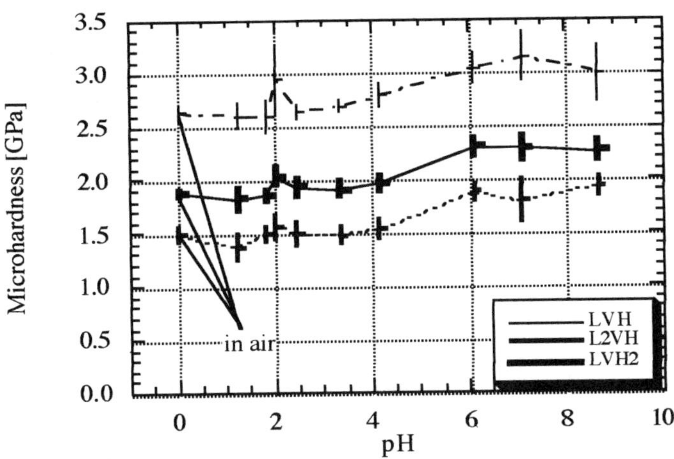

Figure 2: pH dependence of the microhardness of vitreous silica

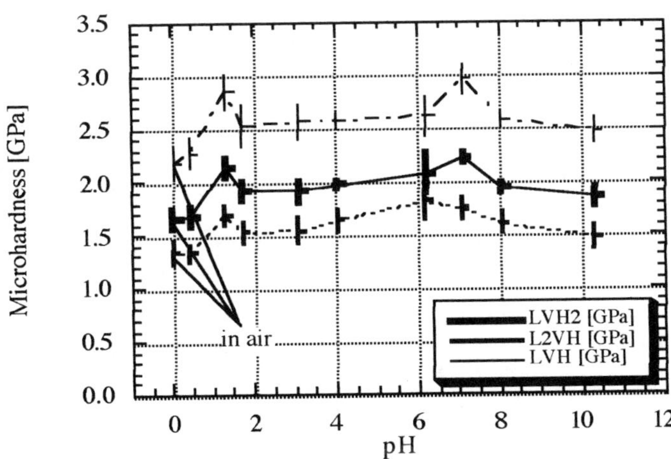

Figure 3: pH dependence of the microhardness of the alimino-silicate glass

4.   CONCLUSION

Recording microindentation tests have shown that a maximum near the i.e.p. of vitreous silica occurs in aqueous environments. The conclusion that this should be a universal property of glass surfaces was shown not to apply to all glasses. An aluminosilicate glass showed two distinct maxima in hardness, one near the i.e.p. of silica, the second one in the vicinity of the i.e.p. of alumina. It appears that a glass can have regions in which certain surface groups are dominating the material's response, depending on the composition of the glass. The absence of the maximum in hardness at the i.e.p. for soda-lime-silica is not fully understood. A possible explanation is the complexity of the surface chemistry in its influence on surface properties. The main problem in fully explaining the data is the non-availability of surface analytical techniques for glasses under the influence of aqueous environments. Further studies investigating the influence of glass composition on the surface properties in aqueous environments could reveal intriguing details about the behavior of surface groups of glasses and open new avenues for surface processing.

ACKNOWLEDGMENTS

The authors would like to thank the NSF-Industry-University Center for Glass Research at Alfred University for providing the funding for this project.

REFERENCES

[1] A. R. C. Westwood, G. H. Pharr, R. M. Latanison, "Adsorption-Sensitive Machining Behaviour of Glass," in *Amorphous Materials*, Wiley, London (1972) pp. 533-543

[2] M. J. Cumbo, "Chemo-Mechanical Interactions in Optical Polishing," *Ph.D. Thesis*, Institute of Optics, College of Engineering and Applied Science, Univ. of Rochester, NY (1993)

[3] D. Golini, S. D. Jacobs, "Physics of Loose Abrasive Microgrinding," *J. Appl. Opt.*, **30** [19] 2761-2777 (1991)

[4] N. H. Macmillan, R. D. Huntington, A. R. C. Westwood, "Chemomechanical Control of Sliding Friction Behavior in Non-metals," *J. Mater. Sci.*, **9** 697-706 (1974)

[5] N. M. Keulen, "Effect of Water Environment on Deformation Behavior of Soda-Lime Silicate Glass," in *Fundamentals of Glass Science and Technology* 1993, Proceedings of the Second Conference of the European Society of Glass Science and Technology, Venice, Italy (1993) pp. 157-162

[6] J. Cott, W. C. LaCourse, S. Jenkins, "Effect of Water on the Frictive Behavior of Glasses," in *Fundamentals of Glass Sci. and Technol.*, 3rd Conference European Soc. of Glass Sci. and Technol., Würzburg, Germany (1995) pp. 289-294

[7] F. Fröhlich, P. Grau, and W. Grellman, "Performance and Analysis of Recording Microhardness Tests," *Phys. Status Solidi*, A, **42**, 79-89 (1977)

[8] H. J. Weiss, " On Deriving Vickers Hardness from Penetration Depth," *Phys. Status Solidi*, A, **99** [2] 491-499 (1987)

CHARACTERIZATION OF SILICATE GLASS SURFACES

Amy B. Jedlicka, Alexis G. Clare
Alfred University
Alfred, NY 14802

Surface modifications, such as sterilization procedures, cleaning and polishing affect how a material behaves when exposed to a specific environment. Contact angle analysis indicates changes in surface energy with surface treatment. Chemical durability studies are performed on a set of silicate glasses to determine the nature species present when exposed to a 37°C aqueous environment.

INTRODUCTION

The introduction of a foreign material into a biological environment, whether intentional (e.g., biomedical applications) or accidental (e.g., inhalation of mineral fibers) results in the development of an interface between the material and the surrounding environment. The nature of the material surface is   critical in determining the subsequent reactions occurring at such an interface. The material surface represents a discontinuity in chemical composition and bonding, hence the surface properties of a given material differ from the bulk properties. A material surface is the termination of an extended three-dimensional structure, and has unsatisfied bonds associated with it. At the surface of the material, the fields of force associated with each atom cannot simply disappear, but will extend past the physical boundary of the solid [1]. This 'force field extension' results in an increase in energy -- the surface energy. This parameter 'surface energy' is not an intrinsic property of the material, but can be altered by various surface treatments. Seemingly innocuous treatments, commonly employed in biomaterial preparation, include sterilization through gas plasma treatment, autoclaving and cleaning [2,3]. These treatments induce great variation in the material's surface energy, which in turn affects the material behavior in a given environment. The post-formation polishing step can (while not changing surface energy) affect the surface adhesion of a material. Studies have shown that cells adhere more strongly to rougher surfaces [4].

The material surface itself may be modified by its surrounding environment (e.g., air, water, tissue). Examining the changes induced within the material as a result of the milieu is important in determining chemical durability. Although silicate glasses are generally believed to be relatively chemically inert, the complex glass / surrounding environment interaction can significantly alter the glass surface chemistry. Bioactive silicate-based glasses demonstrate a wide range of surface modifications caused by interactions with an aqueous environment [5]. Chemical durability studies identify the changes occurring due to the exposure of a specific glass to a certain environment.

MATERIALS AND METHODS

Glass System

The silicate glass system studied is shown schematically in Figure 1. These glasses were chosen to allow fundamental compositional effects on chemical durability to be systematically determined. Pure silica[δ] provides a simple glass composition, through which theories of material / environment interaction can be proposed. Sodium trisilicate demonstrates the effect of addition of a network modifier, Na. An alkaline earth, Ca, is added next to generate soda-lime-silicate[ψ] glass, such as found in a common window glass composition. The two Bioglass® compositions[γ] allow determination of the effect of phosphate additions (45S5), plus calcium fluoride additions (45S5.4F). The two SLABS glasses[β], commercially available fiberglass compositions, include effects of calcia, borate, and magnesia. The SLABS 1 (753D) glass has a higher percentage of silica, and less borate, calcia and magnesia than the SLABS 2 (901) glass.

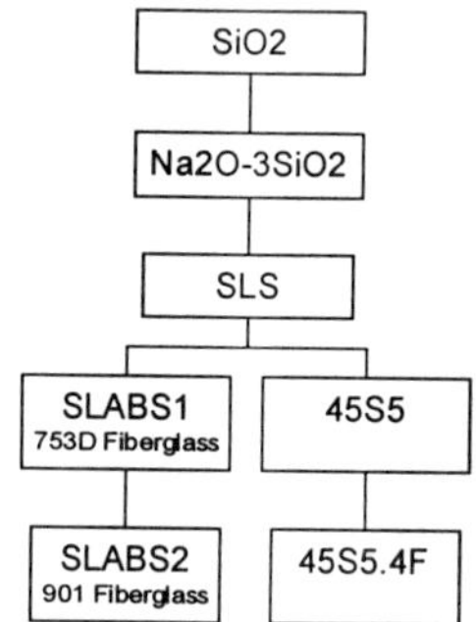

Figure 1. Tree diagram of silicate system studied.

Preparation

The glass samples were formed and/or machined to attain a size of roughly 1 cm^2 in area, with 2 mm thickness. The samples were then polished to a one micron finish, and modified using several different treatments. A roughness study, conducted on soda-lime-silicate glass, involved abrading the surface with 240, 320, 400, and 600 grit SiC paper. The effect of two sterilization treatments -- autoclaving, and gas plasma exposure was tested on soda-lime-silicate glass. Autoclaving involves exposing the samples to steam at temperatures of 124 - 127 degrees C, and pressures of 17 - 19 psi. The gas plasma treatment involved subjecting the samples for 5 minutes to a 100 Watt RF plasma (air), under vacuum. This gas plasma treatment was intentionally severe in an effort to simulate a surface formed under fast-cooling conditions.

Characterization

Surface energy comparisons were drawn through contact angle measurements using the VCA2500XE[π], a video contact angle analysis system, which interfaces a computer with a video camera. In this procedure, a pure, homologous liquid was placed on the glass substrate in a sessile drop configuration. The computer generates a drop profile, and calculates the tangent at the solid - liquid interface.

Roughness parameters were measured using a Tencor® P-10 Surface Profiler[ξ]. Scans, 300 microns in length were taken, with a stylus force of 5 milligrams, and a scanning speed of 2

[δ] Quartz Scientific, Inc., Fairport Harbor, OH

[ψ] Corning No. 2947, Corning, NY

[γ] Donated by U.S. Biomaterials, Alachua, FL

[β] Donated by Johns Manville, Littleton, CO

[π] AST Products, Billerica, MA

[ξ] Tencor, Santa Clara, CA

Advances in Fusion and Processing of Glass II

microns per second. Roughness parameters, such as root-mean-squared roughness, average roughness, maximum roughness, and mean peak height were calculated by the Tencor software program.

Chemical durability studies were performed by soaking the glass samples in distilled water at 37 degrees C. A flint glass container was used as the storage vessel. The ratio of fluid volume to sample surface area was 5.3 cm^{-1}. Solution analysis was performed through pH measurements and inductively coupled plasma emission spectroscopy.

RESULTS AND DISCUSSION

Surface Roughness

It has been suggested that surface roughness affects the measured contact angle according to the equation:

$$\cos \theta_{rough} = r \cos \theta_{smooth}$$

where r is a roughness factor [6-9]. This equation suggests that for a contact angle of less than 90 degrees, the contact angle is decreased by an increase in roughness. This may be explained by examining the sessile drop configuration, and the surface tensions associated with the solid, liquid, and vapor. For a wetting liquid (i.e., $\theta < 90°$), the interfacial energy between the solid and the liquid is less than the interfacial energy between the solid and the vapor, thus the total energy of the system is decreased by the drop spreading. An equilibrium contact angle is reached because the system energy is increased by the increase in liquid-vapor surface area. With a rough surface, the amount of solid surface in contact with the liquid is effectively greater. Thus, as the liquid spreads, there is a greater net energy decrease, and a rough surface is wetted more rapidly. This trend is observed in Figure 2., for a wetting liquid, water, on a soda-lime-silicate surface. These observations correlate well with Wenzel's theory of degree of roughness causing a decrease in measured contact angle [10].

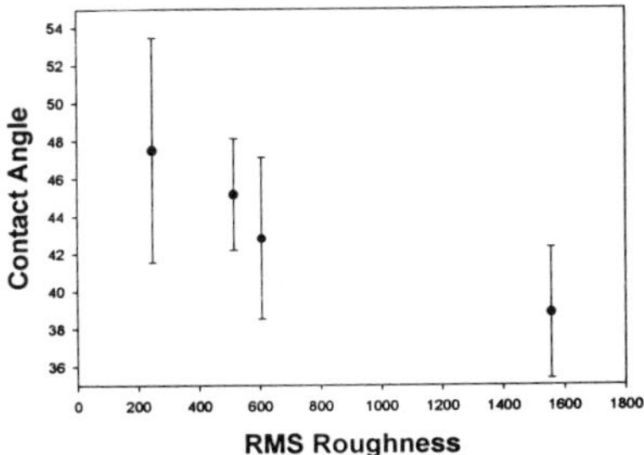

Figure 2. Variability of contact angle with root-mean-squared surface roughness.

The same argument applies for a water-repellent surface, making a rough surface more strongly water repellent. The effect of surface roughness, therefore, is to augment the wetting properties of the solid. This does not mean that surface energy is varied, only that more (or less) liquid is able to wet the surface. For an aqueous based cell culture medium, exposed to a material with a contact angle less than ninety degrees, the wettability is increased. The increased wetting surface area may increase the number of cell attachment sites, increasing cell adherence. The degree of roughness may also provide 'footholds' for the cells to attach to, further increasing cell adhesion.

The root-mean-squared roughness parameter in fig. 2 is calculated through the two-dimensional scan generated by the profilometer. The RMS parameter is defined per the ANSI/ASME standard as

$$\left[\frac{1}{L}\int_0^L y^2 dx\right]^{1/2}$$

where L is the sampling length, and y is the ordinate height of the curve of the profile, relative to the centerline. Figure 3. illustrates a typical two-dimensional profilometry scan.

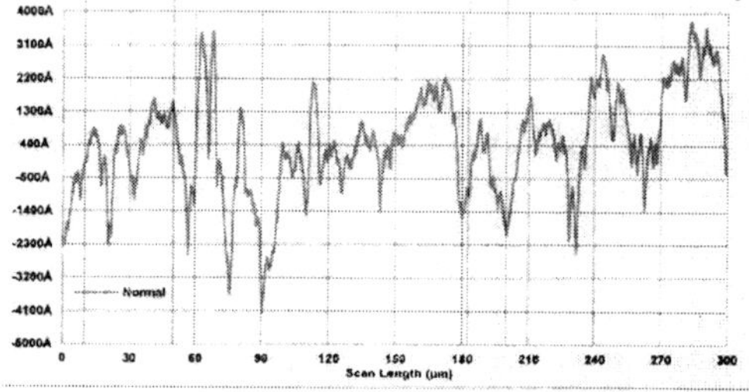

Figure 3. Typical two-dimensional profilometry scan of soda-lime-silicate glass, abraded with 240 grit SiC paper.

Sterilization

The effect of the sterilization procedures (gas plasma treatment (GPT), and autoclaving (Auto)) is shown for soda-lime-silicate glass in Figure 4. (AR = As received) Autoclaving the glass samples causes the surface energy to increase (exhibited by a decrease in contact angle), due to the high temperatures involved in this sterilization process. The gas plasma treatment greatly increases the surface energy of the material. The exposure of the material to the RF plasma leads to desorption of impurities from the surface, due to low energy ion and electron bombardment, thereby exposing "dangling" or unsatisfied bonds [11]. Therefore the wetting liquid is more attracted to the glass surface, and the measured contact angle is lower. Autoclaving the sample after gas plasma exposure (Combo) allows some of these dangling bonds to be satisfied due to the presence of water in the autoclaving process. Figure 5. demonstrates the reversibility of these processes. After exposure to ambient air for 24 hours, the measured contact angle greatly decreases, as compared to storage in a desiccator.

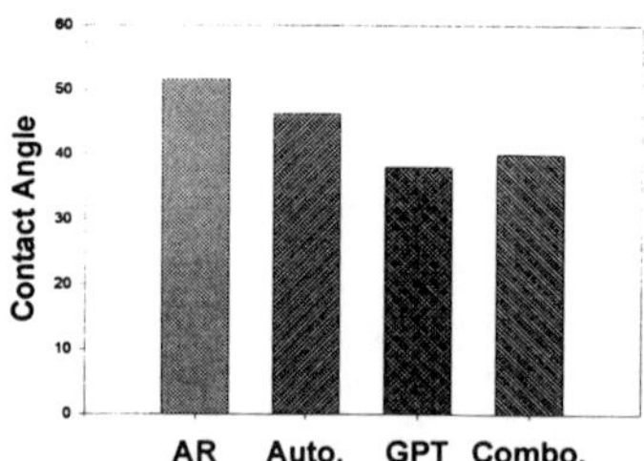

Figure 4. Contact angle dependence on sterilization treatment performed on soda-lime-silicate glass.

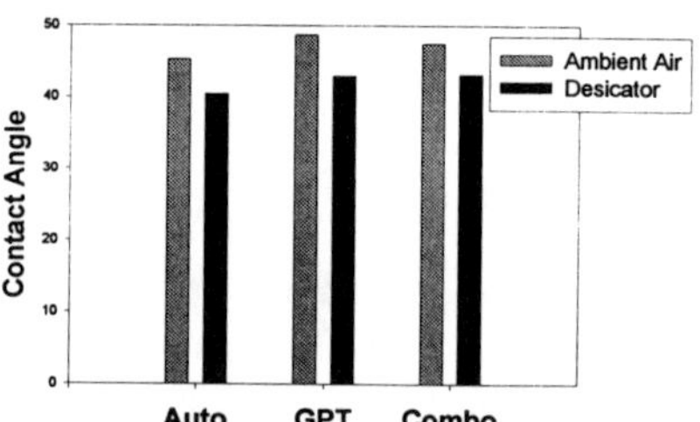

Figure 5. Effect of atmospheric exposure on contact angle of sterilized soda-lime-silicate glass.

Advances in Fusion and Processing of Glass II

Chemical Durability

The behavior of proposed biomaterials in biological environments must be well documented before a material may even be considered for implant use. To form a basis for this type of study, a set of silicate glasses (including two known biocompatible glasses and two commercially used fiberglass compositions) was exposed to distilled water at 37°C for various times. The pH of the resulting 'soaking' solutions was measured, and inductively coupled plasma emission spectroscopy was performed to determine ions present in solution. Figure 6. is a measure of the pH as a function of soak time for five different glasses (SLABS showed no definite trend). The sodium trisilicate glass is a non-durable glass, and the pH increases rapidly due to the exchange of H^+ for Na^+. The soda-lime-silicate glass is extremely durable -- the addition of calcium to soda-silicate glass serves to block sodium diffusion and "tighten" the glass structure due to the stronger ionic field of the small, doubly charged ion. Pure silica is also quite durable, but shows a slight decrease in pH. This decrease is due to the formation of surface silanols (SiOH) and subsequent generation of silicic acid. The fluorine containing bioglass (45S5.4F) is more durable than the 45S5 glass (no fluorine present), supporting Hench and Andersson's theory that the presence of fluoride reduces the rate of dissolution [5].

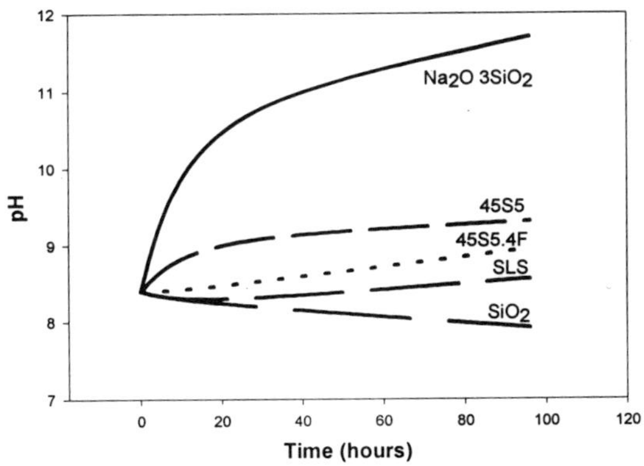

Figure 6. Effect of soak time on pH for various glass compositions (fl. vol./ SA = 5.3 cm^{-1}).

The concentrations of Si, Na, and Ca in the soak solutions of these glasses are shown in Figure 7. (Note: Ionic concentrations for the sodium trisilicate glass could not be measured past 4 days of soak time, due to saturation of the detector.) The durabilities of the five glasses follow the same trend as that seen by the pH measurements. The ionic concentrations for the commercial fiberglass compositions (SLABS 1 and 2) are shown in Figure 8. It appears as though the SLABS 2 glass is less durable than the SLABS 1 glass, however, upon extended scrutiny of the data, this conclusion may be unfounded. The low ionic concentrations require further experiments to elucidate any *in vitro* differences between the two fiberglass compositions.

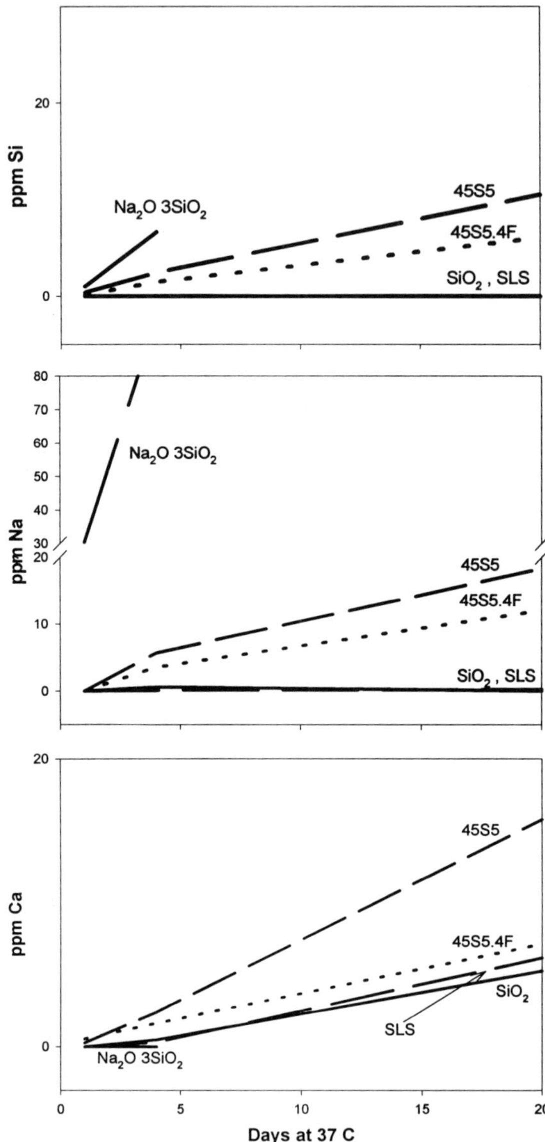

Figure 7. Chemical durability of five silicate glasses.

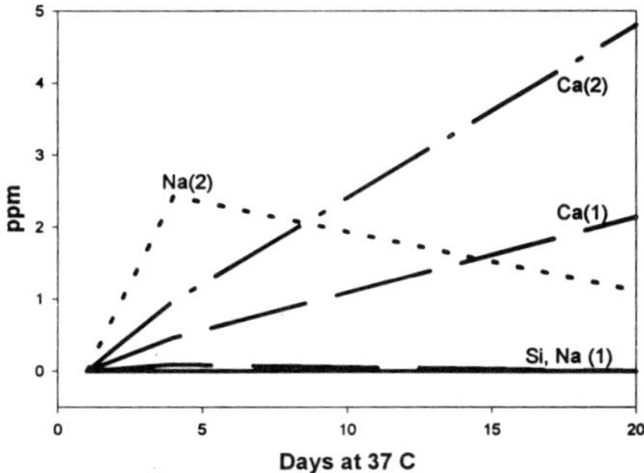

Figure 8. Chemical durability of SLABS glasses (1) and (2).

CONCLUSIONS

Seemingly innocuous sterilization procedures influence material surface properties. Surface characterization is necessary to identify the nature of a material surface prior to its exposure to a biological environment. Surface topography (roughness) causes the measured contact angle to differ and has been cited to affect cell adhesion. Chemical durability studies of the silicate glass system indicate concentrations and types of ions that would be present in a biological system similarly exposed to these glasses.

REFERENCES

1. S.J. Gregg, *The Surface Chemistry of Solids*, 2nd ed., Reinhold Publishing Corporation, New York, 1961.
2. U.M. Gross, C. Muller-Mai, and C. Voigt, "Ceravital® Bioactive Glass-Ceramics," in *An Introduction to Bioceramics*, Advanced Series in Ceramics, Vol. 1, Edited by L.L. Hench and J. Wilson, World Scientific Publishing Co., Singapore, 1993.
3. R.E. Baier, A.E. Meyer, C.K. Akers, J.R. Natiella, M. Meenaghan, and J.M. Carter, "Degradative Effects of Conventional Steam Sterilization on Biomaterial Surfaces," *Biomaterials*, **3** [10] 241-5 (1982).
4. R. S. Delguerra et al., "Optimization of the Interaction between Ethylene-vinyl Alcohol Copolymers and Human Endothelial Cells," *Journal of Materials Science: Materials in Medicine*, **7** [1] 8-12 (1996).
5. L.L. Hench and O. Andersson, "Bioactive Glasses," in *An Introduction to Bioceramics*, Advanced Series in Ceramics, Vol. 1, Edited by L.L. Hench and J. Wilson, World Scientific Publishing Co., Singapore, 1993.
6. A.W. Adamson, *Physical Chemistry of Surfaces*, 2nd Ed., Interscience Publishers, New York (1967).
7. R.N. Wenzel, "Resistance of Solid Surfaces to Wetting by Water," *Industrial and Engineering Chemistry* **28** [8] 988-94 (1936).
8. A.B.D. Cassie, "Contact Angles," *Discussions of the Faraday Society*, **3** 11-16, (1948).
9. R. Shuttleworth and G.L.J. Bailey, "The Spreading of a Liquid Over a Rough Surface," *Discussions of the Faraday Society*, **3** 16-22, (1948).
10. N.R. Morrow and F.G. McCaffery, "Displacement Studies in Uniformly Wetted Porous Media," in *Wetting Spreading and Adhesion*, Edited by J.F. Padday, Academic Press, London, 1978.
11. B. Chapman, *Glow Discharge Processes: Sputtering and Plasma Etching*, John Wiley and Sons, New York, 1980.

DRIFT SPECTRAL CHARACTERIZATION OF VARIOUS CARBOXYLATE CONTAINING ORGANIC COATINGS ON GLASSES

Donghun Lee and R.A. Condrate, Sr.
New York State College of Ceramics
Alfred University
Alfred, NY 14802

ABSTRACT

Organic compounds such as oleic acid, polymethyl methacrylate (PMMA) and polyacrylic acid (PAA) were coated on slide and silica glass powders. The resulting interaction mechanism was determined on the basis of the leaching properties of the glass, and the DRIFT spectra resulting from the interaction of each ion on the glass with coated organics. Soluble components such as sodium and calcium play important roles in the interaction properties of the organics with glass surfaces. The effect of solvent is also discussed.

INTRODUCTION

The interactions of organics with glass surfaces are important for lubricant coatings on glass bottles [1], glass-organic hybrid materials [2], glass-polymer composites [3], and glass surface modifications such as organic protective coatings, anti-dust, and water-resistant coatings [4]. Surfaces of glass fibers have been modified with organics to improve chemical resistance and optical properties [5]. Critical properties such as the durability of the resulting coated glass, etc. are intrinsically related to the structural interaction between the coating material and the glass surface. FTIR spectroscopy is a very useful technique for understanding those kinds of interactions between glasses and organics [6]. In this study, we have investigated the structural interactions of coated surfactants and polymers containing carboxylate groups on the surfaces of the glasses with different compositions, using the DRIFT spectral technique. Such organic compounds can bond on the surfaces of the glasses either as hydrogen-bonded species or by forming metal complexes involving bonds between their carboxylate groups and leachable ions from the glass such as sodium, calcium, and magnesium, etc., or metal ions such as aluminum ions which are more strongly covalently bonded to the silicate network than the other metal ions. These surface species will determine the lubricating or hydrophobic characteristics associated with the resulting glass materials.

EXPERIMENTAL PROCEDURES

Organics such as oleic acid, PMMA and PAA were coated(adsorbed) on slide and silica glass powders. The interaction of the organics and glass was investigated by FTIR spectroscopy. The slide glass includes ions such as sodium, calcium, magnesium and aluminum. After adding alumina to the re-melted slide glass, the organics were also coated on this glass to study the effects of alumina addition on the interaction of the organics. Other slide glass powders were coated with TiO_2 by sol-gel, and then oleic acid and PMMA were coated on them. To investigate roles of soluble ions in the slide glass on the organic coating, the glass powders were washed before and after coating of the organics. For comparisons, the organic compounds were also coated on crystalline silica powder, CaO, MgO, soda-silica glass powder and alumina powder. The effects of solvents on the coating of oleic acid were investigated using ethyl alcohol and benzene.

RESULTS and DISCUSSION

1. Oleic Acid Coating

Figure 1 shows that when oleic acid is coated on the slide glass from ethyl alcohol or benzene, Na^+ Ca^{2+} Mg^{2+} and Al^{3+} ions strongly interact with oleic acid as well as the Si-OH groups. The symmetric C-O stretching mode can be analyzed on the basis of bond formation between cation and carboxylate group [7]. Carboxylate (COOH) groups in oleic acid interact with Si-OH groups through hydrogen bonds. However, $Na,^+$ Ca^{2+} Mg^{2+} and Al^{3+} ions also form complexed bonds (COO-M) with carboxylate groups [8].

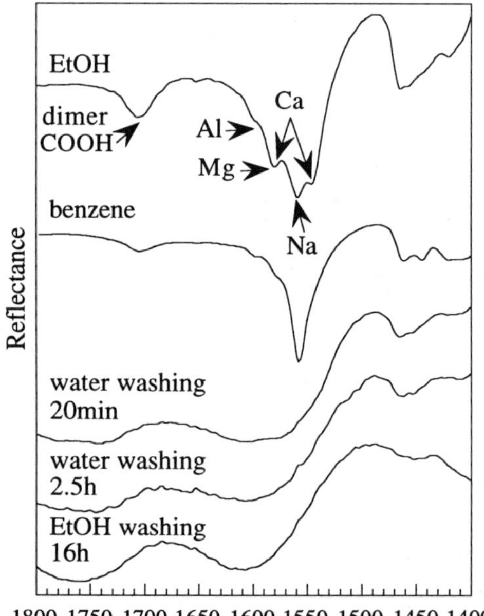

Figure 1. IR spectra of SLS glasses coated with oleic acid from ethyl alcohol or benzene, and then washed with water or ethyl alcohol.

When benzene is used as solvent, IR spectra show a stronger interaction of Na^+ ions than Ca^{2+} and Mg^{2+} ions with the carboxylate group of oleic acid. However, when ethyl alcohol is used, interactions between oleic acid and Ca^{2+} and Mg^{2+} ions increase. It seems that the polarity of the solvent affects the coating of the oleic acid on the slide glass. Ethyl alcohol is more polar than benzene. Therefore, ethyl

                                    Advances in Fusion and Processing of Glass II

alcohol extracts more Ca^{2+} and Mg^{2+} with Na^+ ions from the glass powder to the solvent. The concentration of Na^+ ion will be lowered on the surface. In the case of benzene, the amounts of the extracted ions are very small. Consequently, the coating depends on the relative amounts of ions present on the powder surface, with respect to the original compositions. In the case of this study, the slide glass contains much more Na^+ than $Ca,^{2+}$ Mg^{2+} or Al^{3+} ions The IR bands due to the complexed bonds between Na^+, Ca^{2+} or Mg^{2+} and carboxylate anions disappear after washing the coated slide glass powders with water and ethyl alcohol for different times. Only the band for the COO-Al bond remains. This result indicates that the Na,- Ca- and Mg-oleic acid salts are soluble, and consequently such salt molecules are detached from the glass surface.

The COO-Al bond at ca. 1600 cm^{-1} is relatively strong, and Al^{3+} ions are not soluble in those solvents. It will not be easily removed from the glass surface. Therefore, we added alumina to the slide glass powder via re-melting, and then oleic acid was coated on the resulting powder from benzene. The intensity of the IR band at ca. 1600 cm^{-1} due to the COO-Al complex increased, indicating that increased concentration of aluminum ion stabilized the coated oleic acid by the formation of strong chelate bonds with Al^{3+} ions. When TiO_2 was coated on the slide glass powder, its IR spectrum indicated COO-Ti bonds in the resulting oleate coated powder.

Fig. 2 illustrates the IR spectrum measured for the glass powders coated with oleic acid and then aged in air for 2 months. After aging the powders, the relative intensity of the band at 1706 cm^{-1} corresponding to COOH groups significantly reduces, as compared to the band related to the metal complexed bonds.

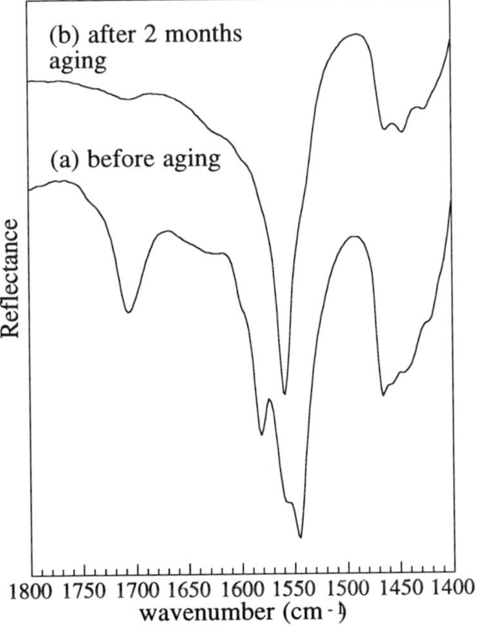

Figure 2. IR spectra of the SLS glasses coated with oleic acid from ethyl alcohol, (a) before aging, and (b) after aging for 2 months in air

Also, the band intensity associated with the COO-Na complex increases at 1558 cm⁻¹, while the bands related with the COO-Ca chelate bonds are observed as weak shoulders. Changes for the band at 1599 cm⁻¹ due to the COO-Al chelate bond are not discernible. These spectra illustrate that aging in air causes Na^+ ions to diffuse from the bulk SLS glass onto the surface, resulting in the formation of more COO-Na bonds. The Na^+ ions extracted by water molecules in air on to the surface interact with COOH groups of the coated oleic acid molecules to form Na-oleate complexes.

2. PMMA Coating

Figure 3 indicates that when PMMA is coated on the slide glass from benzene, the resulting IR spectra show relatively small bands due to the COO-Na bond, as compared to oleic acid coating. This is because it is more difficult to replace CH_3 groups of the $COOCH_3$ groups by Na^+ ions than to replace the H^+ ions of the COOH groups. When alumina is added to the slide glass, the IR band due to COO-Al bonds at 1600 cm⁻¹ increases. An Al^{3+} ion possesses higher polarity than a Na^+ ion, and consequently Al^{3+} ions replace more easily CH_3 groups than Na^+ ions. These results indicate that increasing the amount of Al^{3+} ions in the glasses is one of the methods to improve the adhesion of the organics on the glasses

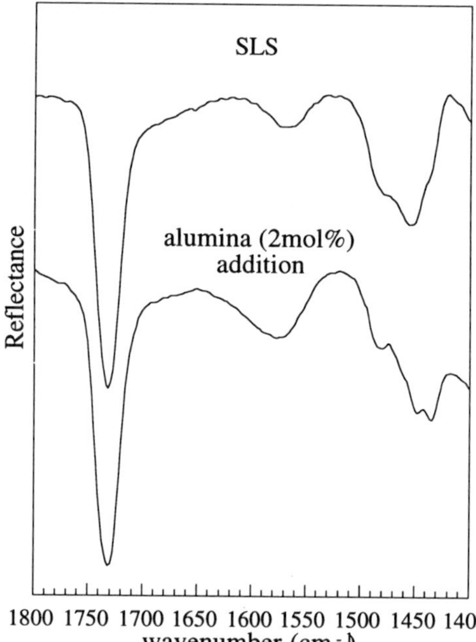

Figure 3. IR spectra of SLS glasses coated with PMMA, before and after addition of alumina.

3. PAA Coating

Polyacrylic acid is a poly-electrolyte material which means that the dissociation constant of its COOH groups depends on the pH of water. Therefore, this study on the adsorption of the polymer was carried out at different pH's in water. In this study, the applied pH's were 2, 5 and 11. The investigated substrates were slide and silica glass powders. When PAA was coated on slide glass powder, more COOH groups dissociated at the same pH as compared to silica glass powder. This result is due to the fact that ions such as Na^+, Ca^{2+}, Mg^{2+}, and Al^{3+} ions as well as OH^- ions in water help dissociate COOH groups. In the case of silica glass, the IR band due to COOH groups was strong even at a pH of 11. COOH groups of pure PAA completely dissociate above a pH of 8. This indicates that the hydrogen bonding between COOH and Si-OH prevents the dissociation of COOH groups of PAA.

PAA can be either coiled or stretched depending on the dissociation of its carboxylate groups and the metal ion contents. Such configurations can be determined by studying the IR stretching modes associated with COOH groups for the various adsorbed polyacrylate species [7]. PAA chains are stretched out when the COOH groups are ionized to COO^- ions because of the electrostatic repulsion between COO^- ions, or coiled when the concentration of metal ions is increased because the metal ions neutralize the negative charge of the COO^- ions. The configuration of the coated PAA on the surfaces is more complicated in this study due to bonding of PAA to metal ions on the surfaces or dissoluted metal ions from the surface.

CONCLUSIONS

Metal complex species form when organic compounds with carboxylate groups are coated on the surfaces of soda-lime-silicate glass. We have found that the formation of various metal complex species is dependent on nature of the interacting materials and the treatment conditions. The interaction mechanism of the organic compounds closely relates to the presence of metal ions possessing high mobility in the glass as well as to the bonding strength of the organic compounds with the adsorption sites on the glass surfaces.

REFERENCES

1. J. Muker and A.S. Sanyal, "A Mössbauer Spectroscopy Study of the Nature of the Bond Between Glass, Tin and the Organic Molecule in hot and Cold End Coating," *Phys. Chem. Glass*, **23** [3] 79-82 (1982).

2. A. Hinsch, "Organic Fluorescent Dyes in Organically Modified Al_2O_3-SiO_2 or TiO_2-SiO_2 Coatings with Varying Polymethylmethacrylate Content," *J. Non-Crystal. Solids*, **147-148** 478-482 (1992).

3. E.J.A. Pope, M. Asami and J.D. Mackenzie, "Transparent Silica Gel-PMMA Composites," *J. Mater. Res.*, **4** [4] 1018-1026 (1989).

4. T.C. Nason and T.-M. Lu, "Thermoplastic and Spectroscopic Properties of Amorphous Fluoropolymer Thin Films," *Thin Solid Films*, **239** [1] 27-30 (1994).

5. L. Klinger and J.R. Griffith, "Fluoropolymer Barriers to Stress Corrosion in Optical Fibers," *J. Mater. Res.*, **2** [6] 876-883 (1987).

6. E. Nishio, N. Ikuta, H. Okabayashi and R.W. Hannan, "Fourier Transform Near-Infrared Attenuated Total Reflectance spectroscopy of Silane Coupling Agent on Glass IRE," *Appl. Spec.*, **44** [4] 614-617 (1990).

7. D.H. Lee and R.A. Condrate, Sr. "Infrared Spectral Investigation of Polyacrylate Adsorption on Alumina," *J. Mater. Sci.*, **31** [2] 471-478 (1996).

8. K. Nakamoto, *Infrared and Raman spectra of Inorganic and Coordination Compounds*, Wiley Interscience, New York 1978.

ACKNOWLEDGEMENT

The authors would like to acknowledge the generous support from the Industrial/University Center for Glass Research at Alfred University.

Advances in Glass Finishing

DETERMINISTIC MANUFACTURING OF PRECISION GLASS OPTICS USING MAGNETORHEOLOGICAL FINISHING (MRF)

Stephen D. Jacobs
Center for Optics Manufacturing and Laboratory for Laser Energetics
University of Rochester
250 East River Road
Rochester, NY 14623

ABSTRACT

Finish polishing of optics with magnetic media has evolved extensively over the past decade. The newest process is called magnetorheological finishing (MRF). In MRF, a magnetic field is used to stiffen a magnetorheological (MR) fluid consisting of micron size particles of iron and cerium oxide in water. When a part is brought into contact with a moving ribbon of this stiffened fluid, a polishing spot is created on its surface. The MR fluid polishing "spot" is effectively an abrasive-charged, subaperture lap which automatically conforms to the local shape of the part. Finishing is accomplished by mounting the part on a rotating spindle and sweeping it through the stationary polishing "spot". The process is deterministic. A computer program can be derived to generate both a dwell time schedule for the MRF machine and a prediction of finished surface shape, using the characteristics of the stable polishing "spot" and information regarding initial surface conditions (depth of subsurface damage, surface microroughness, and surface figure errors). The predictions match the outcome. In this paper we describe the MRF process, including properties of the MR fluid, machine configurations, software for finishing, and experiments to finish a variety of surface shapes (spherical, flat, and aspheric) on the various materials of interest to optics manufacturing.

INTRODUCTION

The concept of MRF is shown in Figure 1. A suspension of micron size magnetic particles and polishing abrasives is contained in a vessel, or trough. Rotation of the trough delivers the suspension to the surface of a spindle-mounted workpiece. With the application of a dc magnetic field in the vicinity of the workpiece, the suspension stiffens and is dragged through the gap formed by the part and the bottom of the trough. A zone of high shear stress forms over a "spot" that contacts and conforms to the workpiece, causing material removal from the workpiece surface. Some unique features of this process are the controllable and conformal nature of the lap, the constant replenishment of the polishing zone with fresh suspension, and the continual removal of glass particles and heat generated in the polishing process.

A workpiece is polished out by sweeping its surface through the zone of high shear stress. Dwell time determines the amount of material that is removed. Part of the illustration in Figure 1 gives a cut-away view which shows the finishing process for a spherical surface. Lens center is polished with the spindle normal to the bottom of the trough. Rotation of the spindle about the lens center of curvature causes annular regions of increasing diameter to come into the zone of high

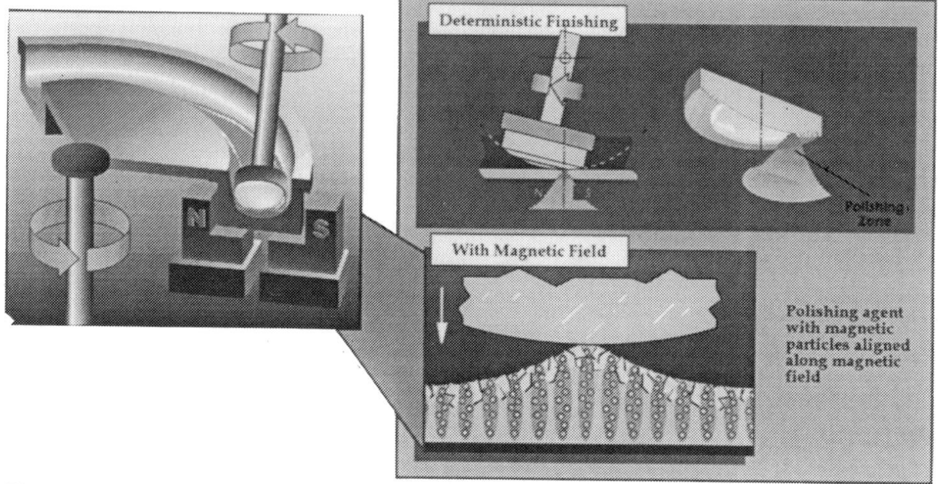

Figure 1 - The MRF concept: a workpiece is immersed directly into the magnetorheological suspension for processing. A part is polished out by sweeping it through the polishing zone, consisting of non-magnetic abrasive particles supported by the field-stiffened magnetic particle "lap".

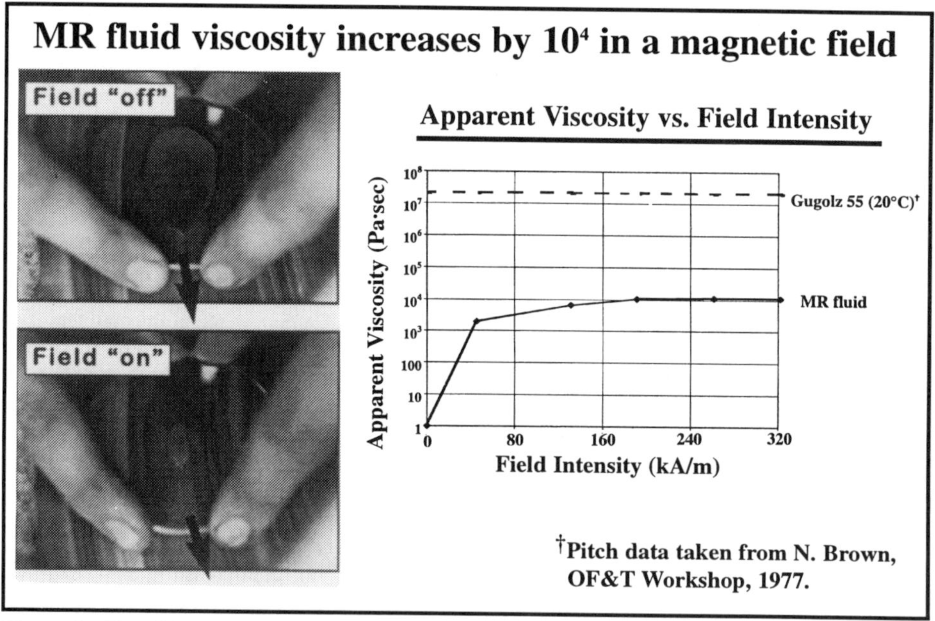

Figure 2 - The ribbon of MR fluid stiffens with the application of a magnetic field. Apparent viscosity (shear rate, 8/sec) increases from 0.5 to 10 000 Pa·sec.

shear stress for polishing. Deterministic finishing to remove subsurface damage, smooth the part surface, and correct figure errors is done with a machine program which drives spindle motion at predetermined velocities through both positive and negative angles. Spherical or aspheric surfaces can be finished in 10 minutes to an hour with the same machine set-up using customized machine programs.

A critical part of the MRF process is the polishing "spot". Its shape is determined by the zone of contact between the MR fluid and the part surface. This depends upon many parameters including the MR fluid composition, the way the MR fluid is mechanically shaped before it passes into the field, the local shape of the part surface being finished, the depth of penetration of the part surface into the MR fluid, the magnetic field intensity distribution in the contact zone, and the physical properties of the part itself. Figure 2 shows that, in the absence of a magnetic field, there is no well defined contact zone. The fluid exhibits an apparent viscosity comparable to that of honey. With the application of a magnetic field, a well-defined contact zone is established as the apparent viscosity of the MR fluid increases by four orders of magnitude. Figure 3 shows a contact zone and the resulting region of material removal for a flat glass surface (Fig. 3a) or a convex glass surface (Fig. 3b), each submerged ~1 mm into a ~2 mm thick ribbon of MR fluid. Removal is largest for those regions of the part which penetrate deepest into the ribbon where the magnetic field intensity is maximum. Because these parts are not mounted on a rotating spindle, material removed is in the form of a trench (flat-Fig. 3a) or the letter "D" tipped on its side (convex-Fig. 3b).

The magnitude of material removed in the polishing spot is used in two ways. Peak removal rates in μm/min give an indication of polishing efficiency for different materials when studied under identical fixed conditions. As is often the case for other processes, hard materials polish more slowly than soft ones, but with MRF these rates are easily quantified. Volumetric removal rates are necessary for determining the machine program and finishing time required to process a specific part geometry. One could consider polishing out the flat in Fig. 3a using only translational motion without part rotation. This strategy might be useful in the context of Si wafer and grating planarization. The favored method for finishing curved surfaces with a spot like that shown in Fig. 3b is to consider annular zones of removal for the part rotating on the spindle, as shown schematically in Fig. 1. This approach is discussed later for spherical and aspheric optics.

The role of the nonmagnet-

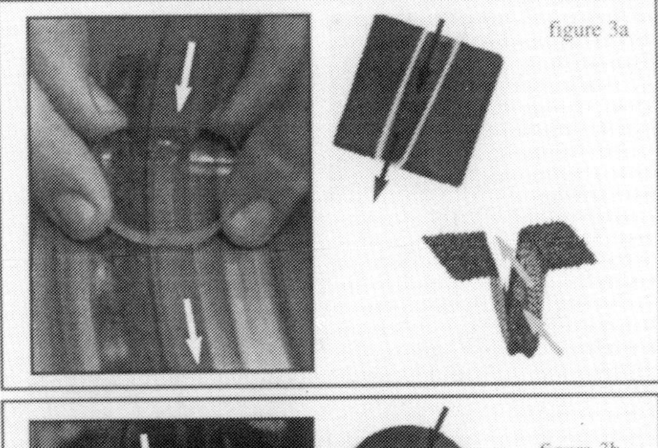

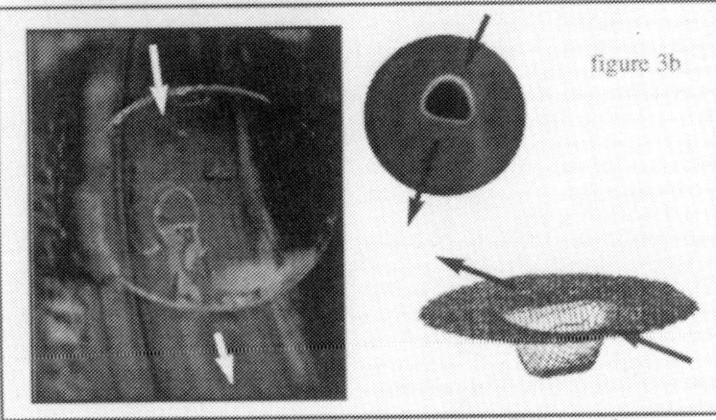

Figure 3 - Polishing spots taken with MRF on glass parts whose surfaces are a) flat, and b) convex. Interferometic views of the zones of removal are also given.

ic polishing abrasives is complex and not completely understood. [Theoretical treatments of the MR fluid under flow in a magnetic field, and speculation on the material removal mechanism are given in references 1 and 2]. Within milliseconds of passing over the electromagnet poles, the magnetic particles form chains along gradients in the magnetic field intensity. This is shown schematically in the expanded box in Fig. 1. Photographs taken during experiments (see Figure 4) confirm that the nonmagnetic polishing abrasives are expelled from the initially homogeneous MR fluid mixture to the surface as the ribbon enters the magnetic field. Such segregation is probably advantageous for finishing , because the abrasive is forced against the part; but this is not good for process stability. Rehomogenization of the MR fluid by aggressive mixing in an external recirculating system is necessary to insure spot removal function stability during a process run and under prescribed set-up conditions from day to day. This is a key to determinism in MRF.

Figure 4 - Behavior of the MR fluid in the polishing zone. Aluminum oxide abrasive rises to the surface of the MR fluid ribbon upon entering the region of high magnetic field.

MAGNETORHEOLOGICAL FLUID

The MR fluid consists of micron size magnetic particles of carbonyl iron (CI) in water, with small concentrations of stabilizers added to retard oxidation of the iron. Although this formulation will polish glass, removal rates are accelerated with the addition of nonmagnetic polishing abrasives. The most commonly used polishing abrasive is cerium oxide, and a "standard" formulation that has been used in many of the experiments reported here consists of (given in vol. %) 36/CI, 55/water, 6/cerium oxide and 3/stabilizers. This fluid exhibits an apparent viscosity of ~0.5 Pa·sec outside of a magnetic field, making it relatively easy to pump to and from an MRF machine. In a magnetic field exceeding 160-240 kA/m (2-3 kG), the apparent viscosity of this MR fluid exceeds ~10 000 Pa·sec (@ a shear rate of 8/sec - see Fig. 2), making it stiff enough to support loads for finishing.

To prepare the MR fluid for experiments, the solid and liquid components are weighed out separately, blended together, and vigorously shaken. Batches in quantities of ~1L are prepared in this way. Once mixed, batches must be constantly agitated to prevent sedimentation. When not being used in a machine, fluids are stored in plastic bottles on a roller mill. During use a batch of MR fluid is placed into a storage reservoir of a fluid delivery system and pumped to / from the

MRF machine. The purpose of this delivery system is to stabilize the MR fluid against temperature changes, sedimentation, and evaporative losses, all of which tend to alter the apparent viscosity.

The delivery system maintains the MR fluid at a nominal apparent viscosity of 0.5 ± 0.05 Pa·sec over the entire work day in several ways. As shown in Figure 5, heat generated in the MR fluid by the finishing process and by flow through the delivery and

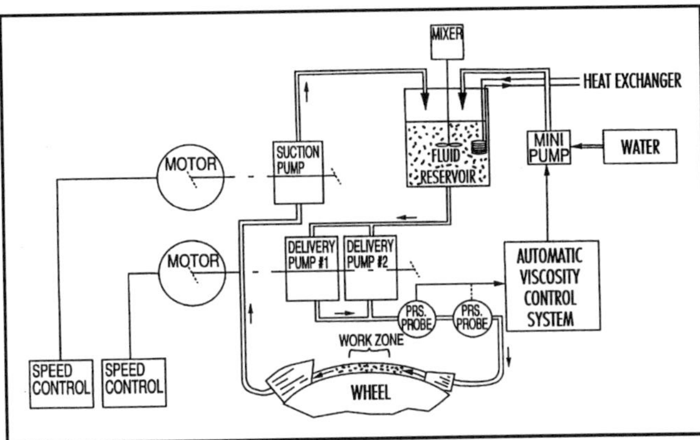

Figure 5 - Schematic diagram of an MRF fluid delivery system. MR fluid is circulated between a reservoir and an MRF machine by peristaltic pumps. Properties are monitored and maintained at pre-determined levels to clamp apparent viscosity at ~0.5 Pa·sec.

pickup tubes is removed through a heat exchanger located in a fluid reservoir. Water lost through evaporation is replaced automatically by a feedback system which senses pressure changes in the delivery tube, and then controls the drip rate of makeup water into the reservoir. Sedimentation of the magnetic particles is prevented by aggressive mixing in the fluid reservoir. This also serves to restore the homogeneous nature of the fluid after field induced segregation in the machine. Peristaltic pumps reliably circulate the MR fluid at flow rates which are adjustable up to 13 L/min. Many MR fluid formulations are stable and may be used and reused over several days. Figure 6 shows initial size histograms for the CI (top) and cerium oxide (bottom) particles[3]. Median particle size for both materials is ~4 µm. The SEM of the MR fluid shown on the right side of Fig. 6.

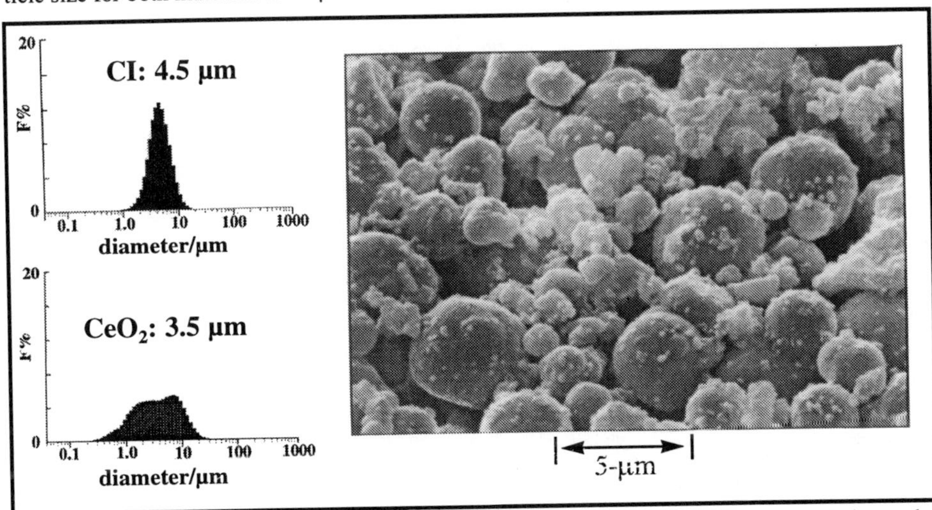

Figure 6 - Histograms showing initial particle sizes of solids in MR fluid. The SEM shows the solids after one week of use.

was taken after a week of use. The round CI particles are essentially unaltered from their initial median particle size, but the irregularly shaped cerium oxide particles have been broken down to smaller sizes by constant milling.

MRF MACHINE PLATFORMS

In MRF, the workpiece is positioned above a moving surface which supports and carries a shaped ribbon of MR fluid into the polishing zone. The dc field from an electromagnet, located just below the carrier surface and centered under the workpiece, stiffens the ribbon before it contacts the part. Two machine configurations are shown in Figure 7. The research testbed[4] in Fig. 7a consists of a rotating aluminum plate with a shallow trough along its rim. Fig. 7b shows the vertical wheel configuration used in the new prototype MRF machine[5]. In both machines, finishing occurs as the rotating, spindle mounted workpiece is swept through the stiffened ribbon. Each has its advantages and limitations.

Figure 7a - Horizontal trough MRF machine.   7b - Vertical wheel MRF machine.

With the horizontal trough configuration it is possible to conduct screening experiments without the fluid delivery system. A small (~50 ml) quantity of MR fluid is circulated in the open trough with two fixed ribbon shapers. One shaper forces MR fluid dispersed by polishing to trough center; the second establishes a small mixing pool and regenerates the ribbon before passing into the polishing zone. New compositions can be evaluated quickly for removal efficiency on a given material, before the onset of sedimentation or evaporative losses. Nonaqueous MR fluids and MR fluids with exceedingly high apparent viscosities can also be tested. Finally, the DC electromagnet can be rotated between two orthogonal orientations, allowing for studies with magnetic field lines approximately parallel or perpendicular to the direction of ribbon motion. Disadvantages of the horizontal trough are its inability to process concave surfaces or parts larger than 50 mm in diameter.

A new MRF machine has been designed, manufactured, and tested which overcomes the geometric constraints of a horizontal trough. As shown in Fig. 7b, this machine consists of a vertical wheel whose rim defines the carrier surface. By fabricating the wheel rim as an arc of a convex sphere, convex, flat, or concave parts can be finished without changing the setup. An MR fluid delivery nozzle and collector are shown on the left and right of the wheel, respectively. The shape of the delivery nozzle orifice helps to define the shape of the MR fluid ribbon incident on the rim. The nozzle and collector are connected to the MR fluid delivery system described earlier.

MATERIAL REMOVAL RATES

By freezing all process parameters, it is possible to use MRF in a study of removal rates for a variety of materials of interest to optics manufacturing. Each part used in work reported here was in the form of a convex spherical plug, 40 mm in diameter with a 70 mm radius of curvature. Standard MR fluid @ 0.5 Pa·sec apparent viscosity was recirculated in the horizontal trough, rotating at 20 rpm (~0.5 m/sec). The ribbon height was ~2 mm with a part/trough gap of ~1 mm.

Removal rates are obtained from spots as illustrated in Figure 8. With the part mounted on a stationary (nonrotating) spindle, a program is used which moves the surface into the field-stiffened ribbon for a dwell time of from ~5-10 sec. The contact angle is fixed at some value between 0° and 5°. Peak and volumetric removal rates are calculated from interferometric data obtained on a Zygo MarkIVxp phase shifting interferometer [6]

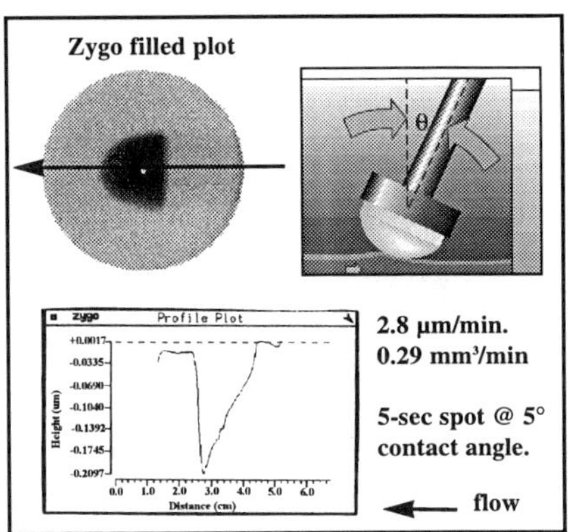

Figure 8 - Removal rates are derived from polishing spots. Interferometry is used to determine the peak and volumetric removal for a part which is positioned without spindle rotation in the MR fluid ribbon at a ~ 5° angle for ~ 5 sec.

after correcting for the initial surface shape. It is important to measure removal rates for surfaces that are initially well polished out (rms surface roughness <50 Å). This avoids the reporting of erroneously high rates of removal which are seen for spots taken on ground surfaces. Usually, three spots are generated over a period of time, and the average result is reported. Data for fused silica suggest that results are good to ±4-5%.

The standard MR fluid composition is effective at removing mass from a variety of optical glasses, phosphate laser glasses (Kigre Q89 and Hoya LHG 8), single and polycrystalline materials, plastics (CR 39) and metals. Figure 9 shows that peak removal rates vary from ~ 8-12 μm/min for lead and phosphate glasses, down to 0.03 μm/min for single crystal sapphire. Rates are seen to increase with decreasing hardness. [For plotting purposes in Fig. 9, the Knoop hardness values for the metals were estimated from a series of plots in Ref. 7]. For the optical glasses in particular, we have found a strong, positive, linear correlation between removal rate and a set of material mechanical properties [8], $E^{5/4}/K_cH_k^2$ (E, Young's modulus; K_c fracture toughness; H_k, knoop hardness).

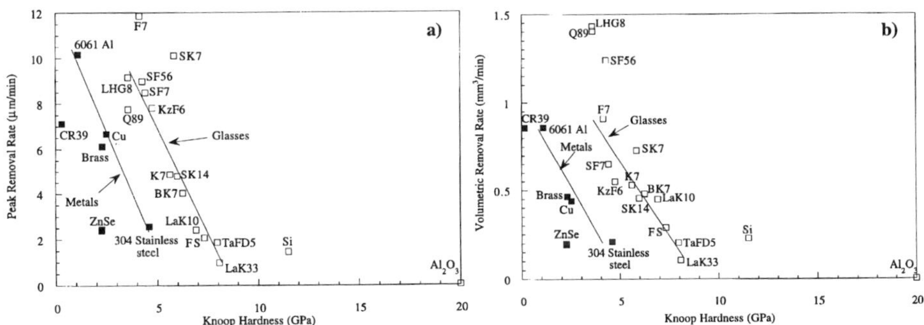

Figure 9 - Rates of a) peak and b) volumetric material removal for a variety of materials under fixed processing conditions described in the text.

The ability of MRF to remove mass from a surface is necessary for eliminating subsurface damage and accomplishing figure correction, but this is not sufficient. MRF must also smooth away surface microroughness. This is done exceedingly well for glass. Final rms surface microroughness, for a variety of optical glasses is ~10 Å. The smoothing process is sensitive to the initial condition of the glass surface and the glass hardness. If the initial rms surface microroughness is less than a few hundred Å, smoothing occurs in 5 - 10 minutes. More time is required for rougher surfaces on harder glasses. More work needs to be done with finer polishing abrasives to determine why rms surface microroughness levels off at ~ 10 Å.

FINISHING SPHERICAL SURFACES

A computer code generates the machine control programs for MRF. Inputs to this code are the material removal spot and the initial surface figure. Outputs are the machine program and a prediction of the final surface figure. There are three operating modes: dc removal, figure correction, or a combination of the two.

An interferogram of the removal spot is acquired and loaded into the code. Alternatively, a previously recorded and stored spot profile may be called up for use from a database. The initial shape of the surface to be finished may consist of another interferogram showing initial deviation from a best fit sphere, or for an aspheric surface the input can be a surface profile obtained with a stylus instrument. The code runs on a PC. Using a series of complex algorithms, the code deconvolves the removal function with the initial surface shape error to derive an operating program for the spindle arm angular controller on the MRF machine. The code specifies angles and velocities for the controller, the number of sweeps required between positive and negative angles, and the total estimated processing time. Finally, the code gives a prediction for figure expected from the process cycle. In what follows, we give several examples which illustrate the capabilities of the code and the MRF process.

Convex Spherical Parts from Fused Silica

One of several convex fused silica parts (40 mm diameter, 58 mm radius of curvature), generated with deterministic ring tool processes [9] on an Opticam®SX [10] at the COM, was polished in three cycles to illustrate dc removal, figure correction, and surface smoothing. Results are given in Table I. In the first 32 minute finishing cycle, 3 μm of material was uniformly removed from the part surface. The surface rms microroughness was reduced from 40 Å to 8 Å. Symmetric surface wavefront error was held to an increase of 0.1 μm for 3 μm of material removed. [The MRF machine configurations in current operation do not permit for the efficient removal of any asymmetric surface features.] The second cycle (discussed more thoroughly below) brought p-v figure error down from 0.42 μm to 0.14 μm. This was accomplished in 6 minutes with the selective removal of ~0.7 μm of material. A third cycle was implemented which removed an addition-

Table I - Summary of results for MR Finishing of a convex fused silica part in 3 process cycles

Cycle		amount removed μm	duration minutes	figure*error μm p-v	areal** roughness Å rms
	initial	---	---	0.31	40
#1	dc removal/smoothing	3.0	32	0.42	8
#2	figure correction	0.7	6	0.14	7
#3	dc removal/figure correction	3.0	42	0.09	8

*symmetric   **Zygo Maxim® 3D Optical profiler, 0.25 mm², unfiltered [6]

Advances in Fusion and Processing of Glass II

al 3 µm of material while further reducing symmetric p-v figure error to 0.09 µm. Areal roughness remained at 8 Å rms.

A portion of the user interface for cycle #2 is shown in Figure 10. Interferograms for the initial, predicted, and actual surface figure errors are shown at the top of the figure. Below each interferogram is a line scan (radial section) depicting the symmetric wavefront error compared to a best fit sphere. This cycle removed a hole at the center of the surface, in substantial agreement with the prediction.

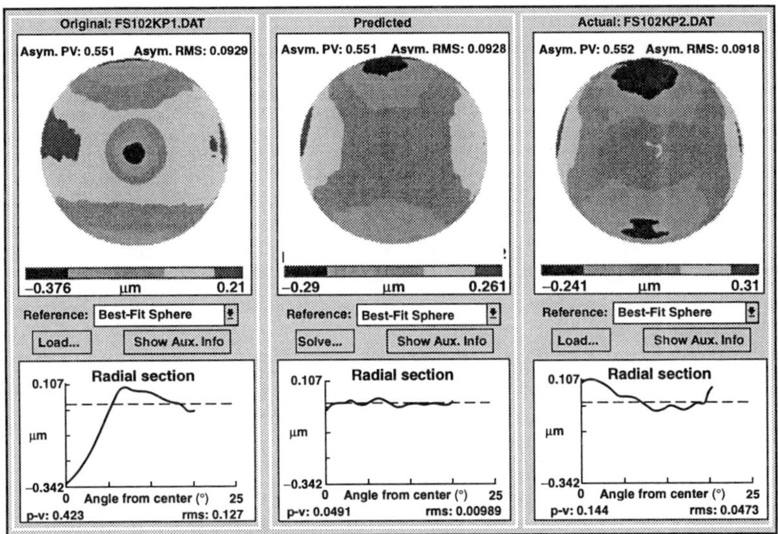

Figure 10 - Initial, predicted, and actual results for a ~6 minute figure correction cycle on a 40 mm diameter convex fused silica part. The central hole was removed in substantial agreement with the code prediction.

Convex Spherical Parts from SK7, BK7, and LaF2 Glasses

Other aspects of the interaction between the code and the machine program are illustrated by the example given in Figure 11. Here, a convex SK7 lens surface (40 mm diameter, 58 mm radius of curvature) is processed in a figure correction cycle. The left column shows a (symmetric) radial section of the initial surface, indicating a hill or bump in the part relative to the best fit sphere. Directly below is a (symmetric) radial removal contour, calculated by the code, for correcting the figure error. It indicates that approximately 0.3 µm of material must be removed, primarily at the center of the part, in order to reduce the p-v wavefront error (0.334 µm) to that shown in the prediction at the top of the middle column (0.052 µm). The machine control program required to perform this operation is shown graphically at the bottom of the middle column. The curve indicates the angular velocities (1 mrf = 0.01° / sec) to be programmed into the MRF machine spindle controller for one sweep of the part through the ribbon. Information is provided at the bottom of the right column on the duration of the correction run (1.2 min) and the number of angular sweeps (2 scans for this example). Actual results from this cycle are shown at the top of the right column. The figure correction achieved (0.07 µm) agrees well with the prediction, both in amplitude and shape.

Recent results for finishing experiments on 70 mm radius of curvature, 40 mm diameter, spherical optical glass parts with significantly different initial figure errors are shown in Figure 12.

The LaF2 glass part in this figure was generated with deterministic ring tool micro-grinding on the COM Opticam® SM. It had an exceedingly low initial microroughness value of 25 Å rms (Zygo NewView® 100, 20x Mirau [6]), with a p-v figure error of 1 λ (90% clear aperture, ref.: best fit sphere). This error was in the form of a central peak. The part was finished on the horizontal trough MRF machine in 15 minutes (2 runs). A total of 0.7 µm of material was selectively removed. The rms micror-

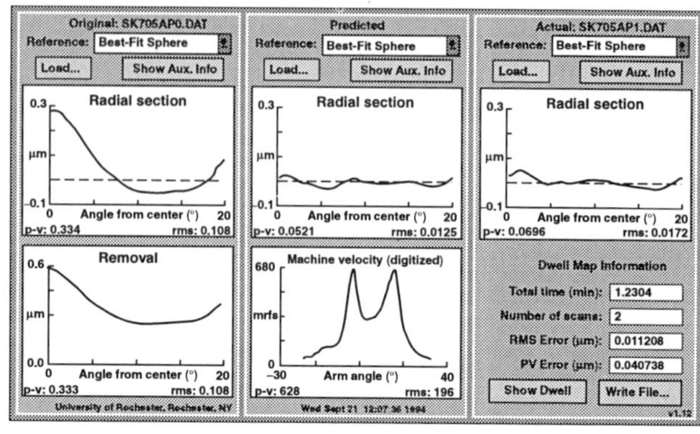

Figure 11 - Details of code for spherical figure correction of an SK7 lens. An MRF control program is generated along with the prediction for surface shape errors. After a 1.2 minute running cycle, the actual p-v figure error of 0.07 µm shows excellent agreement with the prediction of 0.05 µm, both in amplitude and form.

oughness and p-v figure error were reduced to 10 Å and λ/4, respectively. Fig. 12 also shows the results from an experiment on a BK7 part which was generated on the COM Opticam® MicroSX [10] and finished on the vertical wheel MRF machine. The central hole from the microgrinding stage was eliminated in 10 minutes (2 runs) with the programmed removal of 1.4 µm of material. Finished rms microroughness and p-v wavefront quality were 10 Å and λ/7, respectively.

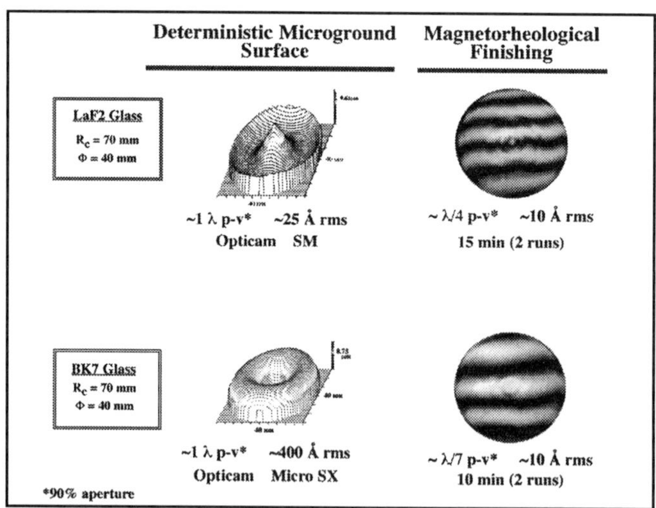

Figure 12 Finishing experiments for widely varying initial surface conditions. The LaF2 glass lens was finished on the horizontal trough MRF machine and the BK7 glass lens was processed on the vertical wheel MRF machine.

    Advances in Fusion and Processing of Glass II

POLISHING OF A PARABOLA INTO A GROUND SPHERICAL SURFACE

A full size lap is difficult to use on an asphere because of the constantly changing local curvature. Subaperture polishing is effective on aspheres since the pad sees only small curvature changes. Because it constitutes a subaperture polishing tool, MRF is uniquely suited to finishing aspheric optical components. The MRF polishing spot conforms to the local shape of the optical surface, thus eliminating pad/workpiece mismatch.

When an optical design calls for an aspheric surface which deviates less than 10 µm from a sphere, it is possible to deterministically generate the sphere and use MRF to produce the finished part.

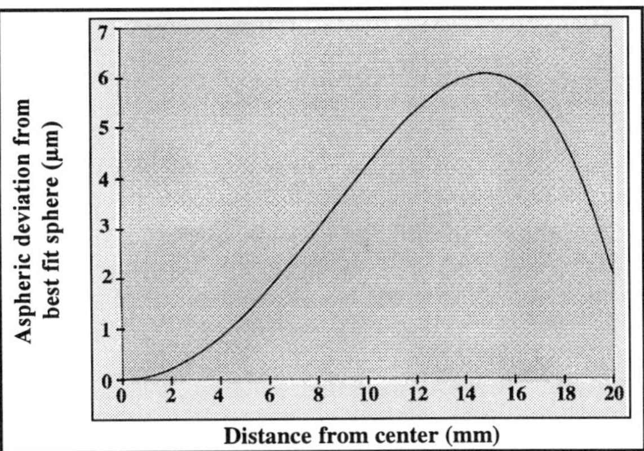

Figure 13 - Parabolic shape required for a 40 mm diameter surface which deviates at most 6 µm from a sphere.

Figure 13 shows the shape required for a parabola whose maximum deviation from a 100 mm base radius sphere is 6µm. Using the prototype machine, three MRF figure correction runs were made in under 40 minutes of machine time on an SF7 glass part. Roughness (Zygo NewView® 100, 20x Mirau[6]) was reduced from > 2100 Å (as ground on the Opticam Micro®SX [10]) to 16 Å rms. The p-v figure error was reduced to ~0.15 µm as measured by Taylor Hobson Form Talysurf [11] stylus profilometry (see Figure 14).

Iteration	Avg. material removed (µm)	Cycle time (min)	Surface rms Å	p-v figure error (µm)
Initial Surface	--	--	2128	5.47
1 (fc)	3.4	22	69	0.88
2 (fc)	1.25	10	27	0.26
3 (fc)	0.5	4	16	0.15

Figure 14 - MRF of the SF7 glass parabola from a ground spherical surface in under 40 minutes. Profilometry shows the resulting figure error to be less than 0.15 µm

SUMMARY

This paper describes a manufacturing technology for optics called magnetorheological finishing (MRF). Fundamental to this technology is an environmentally safe, aqueous suspension of magnetic particles and polishing abrasives, whose viscosity is increased by orders of magnitude in a magnetic field. The stiffened suspension acts as a subaperture "spot" lap which conforms to and polishes out the surface of a workpiece immersed in it. In many different experiments on two different machine configurations, MRF has shown an excellent capability for smoothing ground glass surfaces, correcting figure errors and eliminating subsurface damage. We have demonstrated that, with machine control programs generated by a computer code, both spheres and aspheres can be finished with the same machine set-up, from a variety of optical glasses. The flexibility of MRF, its lack of dedicated tooling combined with a straightforward set of operating protocols, make it ideal as an enabling technology for rapid prototyping of precision optics.

ACKNOWLEDGEMENTS

The author acknowledges the dedication and hard work expended over many years by the following colleagues: Mark Atwood, Paul Dumas, Hugh Edwards, Judah Feingold, Edward Fess, Greg Forbes, Birgit Gillman, Don Golini, Gennady Gorodkin, Sergei Gorodkin, Leslie Gregg, William Kordonski, Vladimir Kordonski, Arne Lindquist, Lowell Mintz, Harvey Pollicove, Igor Prokhorov, Jeff Ruckman, Aric Shorey, and Fuqian Yang. Support for this work is provided by Byelocorp Scientific, Inc., the U.S. Army Materiel Command, and the Defense Advanced Research Projects Agency (DARPA).

REFERENCES

1. W. I. Kordonski and S. D. Jacobs, "Magnetorheological Finishing", *Int. J. Mod. Phys.* **B10**, 2837-2848 (1996).

2. D. Golini, S. D. Jacobs, W. I. Kordonski, and P. Dumas, "Precision Optics Fabrication using Magnetorheological Finishing", to be published in *SPIE CR67: Advanced Materials for Optics and Precision Structures*. Edited by Ealey, Paquin, and Parsonage, San Diego, CA, 26 July 1997.

3. Horiba LA900 Particle Size Analyzer, Horiba Instruments Inc., Irvine, CA 92714.

4. S. D. Jacobs, D. Golini, Y. Hsu, B. E. Puchebner, D. Strafford, Wm. I. Kordonski, I. V. Prokhorov, E. Fess, D. Pietrowski, and V. W. Kordonski, "Magnetorheological finishing: a deterministic process for optics manufacturing", pp. 372-282, in *SPIE 2576: International Conference on Optical Fabrication and Testing*. Edited by Toshia Kasai, Tokyo, Japan, 1995.

5. W. I. Kordonski, S. D. Jacobs, D. Golini, E. Fess, D. Strafford, J. Ruckman, and M. Bechtold, "Vertical Wheel Magnetorheological Finishing Machine for Flats, Convex, and Concave Surfaces," pp. 146–149 in *Optical Fabrication and Testing Workshop*, **7**, OSA Technical Digest Series. Edited by the Optical Society of America, Washington, DC, 1996.

6. Zygo Corp., Middlefield, CT 06455.

7. A. Jacobs and T.F. Kilduff, *Engineering Materials Technology*, p. 144, Prentice Hall Inc., Englewood Cliffs, NJ, 1985.

8. J. Lambropoulos, F. Yang, and S. D. Jacobs,"Toward a Mechanical Mechanism for Material Removal in Magnetorheological Finishing," pp. 150–153 in *Optical Fabrication and Testing Workshop*, **7**, OSA Technical Digest Series. Edited by the Optical Society of America, Washington, DC, 1996.

9. D. Golini and W. Czajkowski, "Microgrinding makes ultrasmooth optics fast", *Laser Focus World*, pp. 146-150 (July 1992).

10. OptiPro Systems, Inc., Ontario, NY 14519.

11. Taylor Hobson, Leicester, UK.

Advances in Fusion and Processing of Glass II

SUBSURFACE DAMAGE IN MICROGRINDING OPTICAL GLASSES

John C. Lambropoulos [1], Stephen D. Jacobs [2], Birgit Gillman [3],
Fuqian Yang [3] and Jeff Ruckman [3]
(1) Department of Mechanical Engineering
(2) Laboratory for Laser Energetics and Institute of Optics
(3) Center for Optics Manufacturing
University of Rochester
Rochester, NY 14627

ABSTRACT

Subsurface damage induced by microgrinding of glass is an important characteristic of the resulting surface, since any subsequent manufacturing process must remove the subsurface damage left by the previous process. We address three questions: How can subsurface damage in a brittle material be estimated from the measured surface microroughness? How can subsurface damage among brittle materials be correlated to their near-surface mechanical properties? And, how is the resulting surface quality affected by material removal under loose abrasive microgrinding at fixed nominal pressure (lapping) or by deterministic microgrinding under fixed infeed rate?

INTRODUCTION

In cold processing of optical glasses by microgrinding [1, 2], the resulting brittle material removal rate results in a cracked layer near the glass surface, referred to as subsurface damage (SSD). In addition, there is a corresponding surface microroughness (SR), often found to increase in proportion to SSD, as originally observed by Preston [3]. SSD is a statistical measure and not necessarily equal to the flaw depth that may control mechanical strength of the brittle surface.

Direct measurement of SSD is tedious: The dimple method is often used [4, 5], or wafering methods. Aleinikov [6] showed that SSD induced by lapping of glasses and other brittle ceramics (with hardness changing 30-fold, fracture toughness 6-fold, and Young's modulus 20-fold) was (3.9 ± 0.2) times SR for SiC abrasives (100-150 μm), thus indicating that SSD may be estimated from SR. Aleinikov also found that SSD increased with increasing size of microindentation cracks, see Fig. 1. Thus, microindentation may be used to evaluate propensity to damage in lapping.

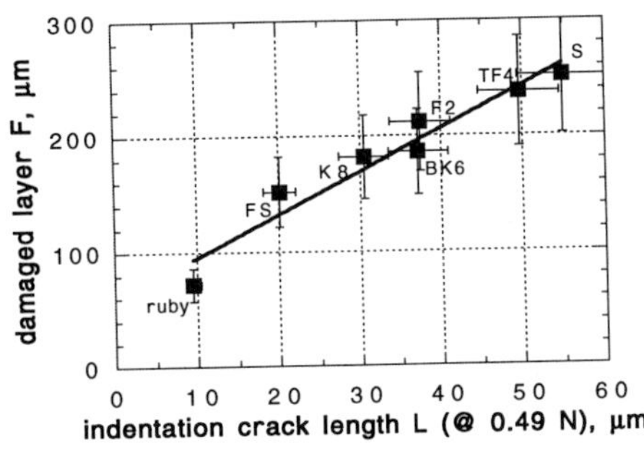

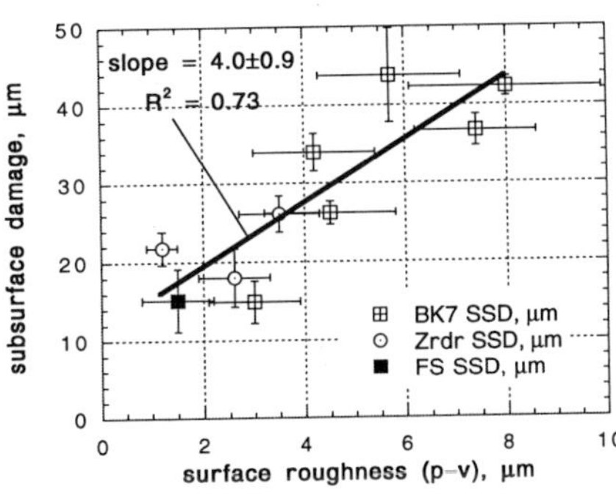

FIG. 1   SSD in lapping vs. indentation crack size (0.49 N) for brittle materials, based on Aleinikov. Russian glass K8 is equivalent to Schott BK7 (Hoya BSC7, Ohara S-BSL7) [7].

FIG. 2   Relation of SSD to SR, as measured in bound diamond abrasive grinding by Edwards and Hed [8].

More recently, Edwards and Hed [8] studied the relation of SSD to SR under bound diamond abrasive conditions (53-65 μm and 180-250 μm in size), and found that for the three glasses they studied (borosilicate crown BK7, zerodur, and fused silica) the average SSD was (6.4±1.3) times the peak-to-valley surface roughness (measured by a profilometer). The factor of 6.4 was arrived at by dividing SSD by SR for each glass. This proportionality factor becomes identical to that of Aleinikov [6] when all three materials tested by Edwards and Hed [8] are treated together, see Fig. 2. Similar observations have been reported for deterministic microgrinding of optical glasses with bound abrasive diamond tools of smaller size (2-4 μm), see Lambropoulos et al. [9]

In addition to correlating SSD with SR, it is possible also to correlate SSD with material mechanical properties for brittle materials. Zhang [10] used metal bond wheels with bound diamond abrasives (40-230 μm in size) to grind structural ceramics under fixed infeed conditions, and reported a subsurface damage depth (consisting of voids induced by the grinding) which correlated with the ductility index $(Kc/H)^2$ of these materials, see Fig. 3. The ductility index [9] is inversely related to the brittleness H/Kc originally introduced by Lawn et al. [11, 12]

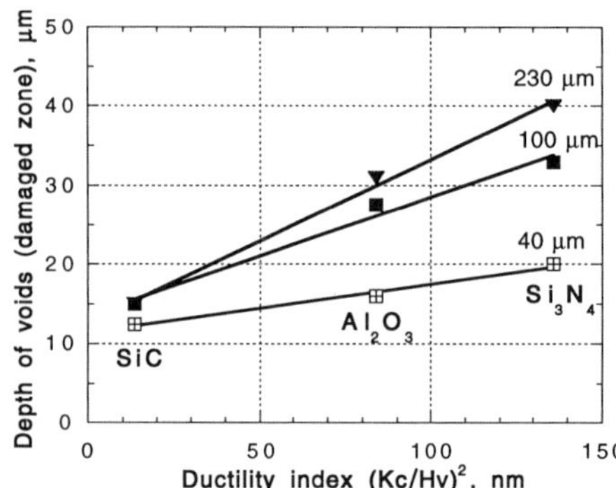

FIG. 3   Subsurface damage vs. mechanical properties of structural ceramics. SSD data from Zhang [10].

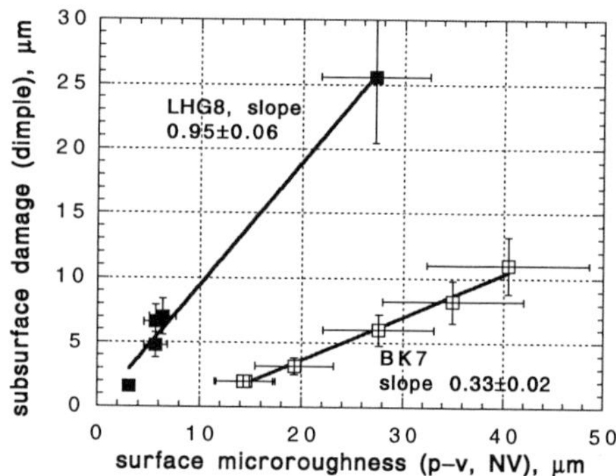

FIG. 4   Correlation of SR (p-v) with SSD for loose abrasive lapping of optical glasses (Al2O3 abrasives).

MICROGRINDING EXPERIMENTS
Lapping: Surface roughness (SR) vs. subsurface damage (SSD)

In all following experiments, surface roughness was measured by a white light interferometer (NewView 100, 0.35x0.26 mm^2, 20x Mirau, 5 measurements per surface), and subsurface damage by the dimple method (typically 3-5 dimples per surface [4, 5]).

The goal of the lapping experiment was to investigate whether surface roughness can provide information about subsurface damage. Loose abrasive lapping experiments were conducted on two glasses, the soft phosphate laser glass LHG8 (63% P_2O_5, 14% BaO, 12% K_2O, 7% Al_2O_3, 4% Nd_2O_3/Nd_2O_5), and the harder borosilicate crown optical glass BK7 (68.9% SiO_2, 10.1% B_2O_3, 8.8% Na_2O, 8.4% K_2O, 2.8% BaO, 1% As_2O_3, % by weight), see Table I.

Five separate LHG8 blocks were lapped on both sides with Al_2O_3 abrasives (median size 30, 9, 5, 3, 1 μm). Measured SSD and SR, after grinding with each abrasive, are shown in Fig. 4.

A similar experiment used BK7 with a wider abrasive size range (median size 40, 30, 20, 9, 5, 3, 1 μm). A single BK7 part was first lapped by 40 μm abrasives. Then the same glass was lapped with 30 μm abrasives, and then with 20, 9, 5, 3, and 1 μm abrasives. SSD and SR were measured at each step. Each lapping step removed between 0.3-1 mm of material, and thus removed all the residual SSD from the previous abrasives used in the sequence. Larger abrasives typically led to higher SSD and higher SR.

The correlations of the subsurface damage to the peak-to-valley surface roughness for lapped LHG8 and BK7 are shown in Fig. 4. For LHG8 the p-v SR is equal to the measured SSD, whereas for BK7 the p-v SR is about 3-5 times the measured SSD. We conclude from these experiments that the p-v SR measured with the white light interferometer provides an upper bound for the SSD measured by the dimple method.

Deterministic microgrinding: Surface roughness (SR) vs. subsurface damage (SSD)

A series of multicomponent optical glasses, as well as fused silica (Corning 7940) were also ground under fixed infeed deterministic microgrinding conditions on the Opticam SM CNC machining platform [13, 14], which can manufacture planar and spherical surfaces, as well as aspheres [9, 13, 14]. Table I below summarizes some of the glass properties.

TABLE I. Thermomechanical properties of optical glasses

Data for density ρ, glass transition temp. T_g, coeff. thermal expansion α, Young's modulus E, and Poisson ratio v are from manufacturers glass catalogs. Hardness H, fracture toughness Kc are from Schulman et al. [15] Knoop hardness is at 1.96 N. The fracture toughness of LaK9 was estimated from that of LaK10.

GLASS	ρ, g/cm^3	T_g, °C	α, 10^{-6} °C^{-1}	E, GPa	v	Hk, GPa	Kc, MPa m$^{1/2}$
LHG8	2.83	485	12.7	50	0.26	2.3	0.43
FS-C7940	2.20	1,090	0.52	73	0.17	5.6	0.75
SF58	5.51	422	9.0	52	0.26	2.7	0.46
SF7	3.80	448	7.9	56	0.23	3.4	0.67
BK7	2.51	559	7.1	81	0.21	5.1	0.82
K7	2.53	513	8.4	69	0.21	4.6	0.95
KzF6	2.54	444	5.5	52	0.21	3.7	1.03
LaK9	3.51	650	6.3	110	0.29	5.7	(0.90)
TaFD5	4.92	670	7.9	126	0.30	7.3	1.54

Three metal bonded diamond abrasive ring tools were sequentially used on each surface (aqueous coolant Loh K-40, relative speed of work and tool of about 30 m/s): 70-80 μm, 10-20 μm, and 2-4 μm at infeed rates of 1 mm/min, 50 μm/min, and 5 μm/min, respectively. Three cuts were done with each tool. After each cut, SR of the optical surface was measured for microgrinding with all three tools, and SSD (three dimples for each cut) for the 2-4 and 10-20 μm tools.

Fig. 5 shows the correlation between the measured p-v and rms SR for the three tools used, with each point representing one of the glasses ground and measured. Fig. 6 shows the correlation of SSD (dimple method) and the p-v SR. It is seen that, as in lapping, the p-v SR may be used as an upper bound for the SSD for the 10-20 and 2-4 μm tools, within the uncertainty in the measurement of SSD and SR.

The effect of glass mechanical properties on SSD is shown in Fig. 7, where we have used the ductility index as the correlating parameter [9]. It is seen that, under fixed infeed grinding conditions, increasing ductility produces higher SSD, as observed in structural ceramics (Fig. 3). Correlations of measured SSD with the critical depth of cut discussed by Bifano et al. [16] or the critical load for fracture initiation discussed by Chiang et al. [17, 18] gave similar trends.

The dependence of SSD on the ductility index is interpreted by a simple model of residual tensile stresses $\sigma \approx \beta \sigma_Y$ (parallel to the surface), where $\beta \approx 0.08$ [19] and σ_Y is glass uniaxial yield stress ($\sigma_Y \approx Hv/2$, see [9]). Thus, crack depth a in presence of such tensile stresses is estimated as

$$K_c = \Omega \left(\beta \; \sigma_Y \right) \sqrt{\pi \; a} \quad \Rightarrow \quad a = \frac{1}{\pi} \left(\frac{K_c}{\Omega \; \beta \; \sigma_Y} \right)^2 \tag{1}$$

$\Omega \approx 1.1$ is a geometric factor accounting for the proximity of the free surface. Typical data for, say, BK7 give a crack depth of 2.1-4.3 μm, i.e. quite comparable to the measured SSD, see Fig. 6.

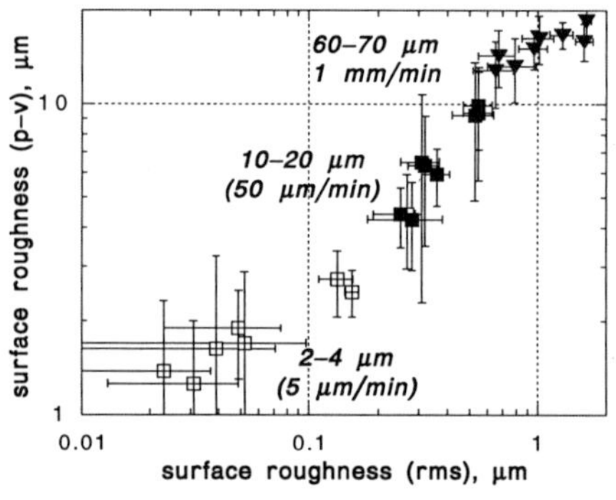

FIG. 5 Correlation of p-v and rms SR under fixed infeed deterministic microgrinding of various optical glasses.

FIG. 6 SSD (dimple method) vs. p-v SR (via NewView 100 white light interferometer) for fixed infeed deterministic microgrinding.

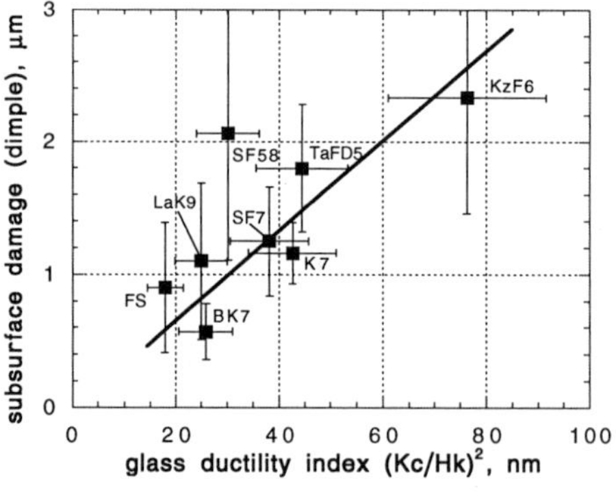

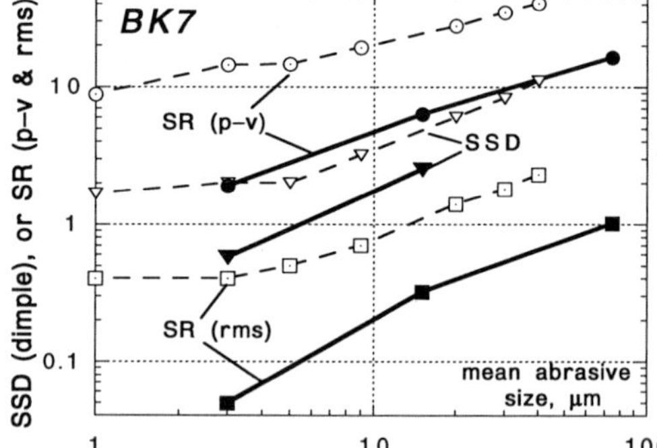

FIG. 7 Dependence of subsurface damage SSD on glass mechanical properties via the ductility index $(Kc/Hk)^2$.

FIG. 8 BK7 surfaces: lapping at fixed pressure (open symbols) vs. deterministic microgrinding at fixed infeed with metal bonded diamond abrasive ring tools (solid symbols).

Advances in Fusion and Processing of Glass II

<u>Comparison of surface quality induced by lapping and deterministic microgrinding</u>

Fig. 8 compares the surface quality of the optical glass BK7 (commonly used in many optical designs) resulting from loose abrasive lapping with Al_2O_3 abrasives (seven sizes spanning 1-40 µm) and from deterministic microgrinding (three sizes spanning 3-75 µm) with bonded diamond abrasives, over a wide range of abrasive sizes. The infeed rates for deterministic microgrinding were 5 µm/min (2-4 µm tool), 50 µm/min (10-20 µm tool), and 1 mm/min (70-80 mm tool). For both lapping and deterministic microgrinding, larger abrasives lead to deeper SSD and higher SR. The lapping results apparently become insensitive to abrasive size for abrasives in the 1-3 µm range.

For a given abrasive size, deterministic microgrinding results in surfaces with lower subsurface damage, and lower surface microroughness (p-v or rms). Such surface features are in addition to any "figure" features extending over the whole aperture of the ground optical surface.

CONCLUSIONS

The quality of a manufactured optical surface can be characterized in a variety of ways, including surface microroughness [9], subsurface damage, surface figure error, residual stresses induced by the grinding process [19, 20], the rate of material removal [21], and the rate of tool wear. In our work we have concentrated on subsurface damage and surface microroughness, and addressed the following questions:

How can subsurface damage in a given brittle material be estimated from the measured surface microroughness? How can subsurface damage among brittle materials be correlated to their near-surface mechanical properties? And, how is the resulting surface quality affected by material removal under loose abrasive microgrinding at fixed nominal pressure (lapping) or by deterministic microgrinding under fixed infeed rate?

We have performed a series of loose abrasive microgrinding (lapping at fixed nominal pressure) and deterministic microgrinding (at fixed infeed) experiments on various optical glasses. We summarize our results as follows:

Peak-to-valley surface microroughness for the optical glasses tested (measured by the white light interferometer, a relatively easy measurement to perform) provides an upper bound to the subsurface damage measured by the more time consuming dimple method;

Subsurface damage in optical glasses under deterministic microgrinding conditions with 2-4 µm bound diamond abrasive tools scales with the glass ductility index $(Kc/H)^2$ in a manner similar to that reported for fixed infeed grinding of structural ceramics [10], and

For a given abrasive size deterministic microgrinding produces lower subsurface damage and lower surface microroughness as compared to lapping.

The issue of residual stresses induced by grinding is also important, and often referred to as the Twyman effect [20]. Although we have not measured residual stresses in this work, our previous work on optical glasses [19] and glass ceramics [22] shows that, for comparable abrasive sizes, deterministic microgrinding induces lower residual stresses than loose abrasive lapping, while maintaining a higher material removal rate and producing a lower surface roughness.

ACKNOWLEDGMENTS

We acknowledge many helpful discussions with and insights from Mr. Don Golini of QED Technologies, Inc. (Rochester, NY) and with Profs Paul Funkenbusch and Stephen Burns of the Mechanical Engineering Department at the University of Rochester.

We also acknowledge surface roughness measurements provided by Mr. Ed Fess from the Center for Optics Manufacturing and microgrinding data by Mr. Bryan Reed from the Department of Mechanical Engineering and Ms. Yuling Hsu from the Center for Optics Manufacturing at the University of Rochester.

REFERENCES

[1]   T. S. Izumitani, *Optical glass*, Chapter 4 (American Institute of Physics, New York, 1986).

[2]   H. H. Karow, *Fabrication methods for precision optics*, Chapter 5 (Wiley, New York, 1993).

[3]   F. W. Preston, The structure of abraded glass surfaces, Trans. Optical Soc. **23**, 141-164 (1921-22).

[4]   A. Lindquist, S. D. Jacobs, and A. Feltz, Surface preparation technique for rapid measurement of sub-surface damage depth, OSA Topical Meeting on Science of Optical Finishing, 10-12 June, Monterey, CA, OSA Technical Digest Series, Vol. **9**, SMC3-1/57 (1989).

[5]   Y. Zhou, P. D. Funkenbusch, D. J. Quesnel, D. Golini, and A. Lindquist, Effect of etching and imaging mode on the measurement of subsurface damage in microground optical glasses, J. Amer. Ceram. Soc. **77**, 3277-3280 (1994).

[6]   F. K. Aleinikov, The effect of certain physical and mechanical properties on the grinding of brittle materials, Sov. Phys. Tech. Phys. **27** (10), 2529-2538 (1957).

[7]   L. B. Glebov & M. N. Tolstoi, Designation of Russian optical glasses, in *CRC Handbook of Laser Science and Technology*, (CRC Press, Boca Raton), 823-826 (1995).

[8]   D. F. Edwards & P. P. Hed, Optical glass fabrication technology. 2: Relationship between surface roughness and subsurface damage, Applied Optics **26**, 4677-4680 (1987).

[9]   J. C. Lambropoulos, T. Fang, P. Funkenbusch, S. Jacobs, M. Cumbo & D. Golini, Surface microroughness of optical glasses under deterministic microgrinding, Applied Optics **35**, 4448-4462 (1996).

[10]   B. Zhang, Surface integrity in grinding of ceramics, Proc. ASPE Spring Topical Mtg. on Precision Grinding of Brittle Materials, pp. 76-81, 3-6 June 1996, Annapolis, Maryland (ASPE, Raleigh, N 1996).

[11]   B. R. Lawn, T. Jensen, & A. Arora, Brittleness as an indentation size effect, J. Mater. Sci. **11**, 5 575 (1975).

[12]   B. R. Lawn and D. B. Marshall, Hardness, toughness, and brittleness, J. Amer. Ceram. Soc. **62**, 347–350 (1979).

[13]   H. H. Pollicove & D. T. Moore, Optics manufacturing technology moves toward automation, Laser Focus World, March (1991).

[14]   H. H. Pollicove & D. T. Moore, Center for Optics Manufacturing Overview, in *Optical Fabrication and Testing Workshop*, Vol. **24**, 1992 OSA Technical Digest Series, 44-47 (1992).

[15]   J. Schulman, T. Fang, and J. Lambropoulos, Brittleness/ductility database for optical glasses, v. Dept. Mechanical Engineering & Center for Optics Manufacturing, Univ. Rochester, 10 Oct.1996.

[16]   T. G. Bifano, T. A. Dow, & R. O. Scattergood, Ductile-regime grinding: A new technology for machining brittle materials, J. Engin. for Industry, Trans. ASME **113**, 184-189 (1991).

[17]   S. S. Chiang, D. B. Marshall, & A. G. Evans, The response of solids to elastic/plastic indentation I. Stresses and residual stresses, J. Appl. Phys. **53**, 298-311 (1982).

[18]   S. S. Chiang, D. B. Marshall, and A. G. Evans, The response of solids to elastic/plastic indentation II. Fracture initiation, J. Appl. Phys. 53, 312-317 (1982).

[19]   J. C. Lambropoulos, Su Xu, Tong Fang, & Don Golini, Twyman effect mechanics in grinding and microgrinding, Appl. Optics **35**, 5704–5713 (1996).

[20]   J. C. Lambropoulos, Su Xu, & Tong Fang, Loose abrasive lapping hardness of optical glasses and its interpretation, Applied Optics **36**, 1501-1516 (1997).

[21]   F. Twyman, 1952, *Prism and lens making*, p. 318 (Hilger & Watts, London, 1952).

[22]   J. C. Lambropoulos, B. E. Gillman, Y. Zhou, S. D. Jacobs, and H. J. Stevens, Glass-ceramic Deterministic microgrinding, lapping, and polishing, to be published in SPIE vol. **31** International Symp. on Optical Science, Engineering, and instrumentation, Session on Opt manufacturing and Testing II (27 July-1 August 1997, San Diego).

CHEMICAL AND STRUCTURAL CHARACTERIZATION OF METAL OXIDE BASED SLURRIES

M. Lauwidjaja, K. Chamma, K. Richardson, and A. Dogariu
Center for Research and Education in Optics and Lasers, (CREOL)
Univesity of Central Florida, Orlando, FL 32816

ABSTRACT

The extent of chemical interaction between workpiece and slurry during the optical polishing process dictates the means by which surface layers in glasses are formed and subsequently removed within the abarasive-water-slurry system. This paper reports results of slurry analyses performed on traditional metal oxide-based abrasive slurries using electrochemical measurements to independently evaluate the effects of materials source, particle size, dilution, electrolyte, and additives on slurry behavior. Electrophoretic mobility measurements and zeta potential determinations were performed for dilute slurry solutions and an enhanced backscattering technique (EBS) was utilized for the structural characterization of particles in dense slurries. Results show that ceria suspensions exhibit strong sensitivity to size, type, and purity which yielded varied IEP data. For alumina slurries, good agreement with prior data was observed, and the acid- and base-stabilized suspensions illustrated a decrease in stability with aging. The EBS measurements of ceria suspensions showed a linear relationship between l^* and concentration, allowing us to probe particle size changes with aging over a broad (2-50% volume solid content) range.

INTRODUCTION

Glass polishing efficiency and rate depend strongly on glass surface-abrasive slurry interaction[1]. This interaction is dictated by the nature and the chemistry of the polishing slurry. To understand and characterize the various forces between the slurry's particles, two different abrasive systems, alumina Al_2O_3 and ceria CeO_2 have been studied. Electrophoretic mobility measurements and zeta potential determinations were performed for dilute slurry solutions and an enhanced backscattering technique (EBS) was utilized for the characterization of particle size and distribution, and particle-particle interactions (agglomeration) in dense slurries. These techniques are complementary in extending our knowledge of particle interactions in these slurries, in that they provide chemical and structural information for slurries spanning a wide range of solid concentration levels.

Zeta potential

Zeta potential (ξ) is most frequently calculated from electro-kinetic-measurements data, and is defined as the average electric potential at the surface of shear near the solid with respect to the bulk fluid potential[2]. Zeta potential is a measure of dispersion stability, and for a suspension, a high zeta potential absolute value implies a stable dispersion. The surface of shear is an imaginary hydrodynamic boundary in the fluid near the solid. The isoelectic point (IEP) or point of zero charge (PZC) is defined as the pH at which there is no net charge (i.e. when the concentrations of positively and negatively charged species at the surface are equal, and hence, neutral). The IEP varies as a function of sample preparation, test technique, and the presence of adsorbed surface contaminants[1]. Such variation has been observed by Hsu[4] in an investigation of ceria raw material source and preparation techniques. In the optical polishing process, ions leached out of the workpiece material and silica by-product removed from the workpiece surface might affect the stability of the slurry. One method to probe a change in stability is by measuring the zeta potential (and IEP) of the slurry before and after polishing process. However, because of the instrument capabilities requiring low solid content samples, multiple dilutions of the slurry (typically 1:100) are required before zeta potential measurements can be done.

Enhanced backscattering (EBS)

The phenomenon of enhanced backscattering (EBS) is a useful alternative to classical scattering or transmission measurements in describing the multiple scattering properties of random media. Until recently, the coherent light propagation through random media has been considered to be somehow degraded, losing its coherence properties. When coherent light is scattered by a random medium such as that found in a suspension of abrasive particles in a slurry, interference effects between the scattered waves which travel through the medium along different paths occur. The random pattern of interference is called *laser speckle*. When the individual scatterers are allowed to move over distances of the order of the wavelength or more, the distribution of intensities in the speckle pattern is averaged out and becomes essentially flat. However, one kind of interference still survives in this average, the interference of the waves emerging from the medium in the directions close to the backward direction and which traveled along the *same* path but in opposite directions. Accordingly there is an enhanced backscattering of intensity. The most exciting feature of this phenomenon is the direct dependence of its angular extension $\Delta\theta$ on the microscopic properties of the scattering medium. For a random medium, the characteristic distance $|\,r_i - r_f\,|$ between two scattering events is usually assimilated to l^*, the transport mean free path of light. On this basis, measurements of the width of enhanced backscattering become a powerful tool for the investigation of the transport mean free path through random media. The shape of the enhanced backscattering peak is described fairly well by a scalar theory that considers that the transport of light inside the scattering medium obeys the classical diffusion equation. In this case the angular profile of the scattered intensity is given by:

$$I(\theta) = \int \{1 + \cos[\,\mathbf{q}(\mathbf{k}_i + \mathbf{k}_f)]\}\, P(\rho)\, d\rho \qquad (1)$$

where $\rho = |\,\mathbf{r}_i - \mathbf{r}_f\,|$ and $P(\rho)$ is the probability that an injected photon will come out at a distance ρ from the injection point. The shape of the enhancement cone depends on the specific form of the $P(\rho)$ and when this evaluated in the frame of classical diffusion, the detailed EBS description becomes:

$$I(\theta) = \frac{3}{10\pi}\left\{1 + \frac{2z_0}{l^*} + \frac{1}{(1 + q_\perp l^*)^2}\left[1 + \frac{1 - \exp(-2q_\perp z_0)}{q_\perp l^*}\right]\right\} \qquad (2)$$

where $q_\perp \approx 2\pi\theta/\lambda$ is the component of $\mathbf{q} = \mathbf{k}_i + \mathbf{k}_f$ along the interface and z_0 is a model-dependent parameter. Based on various scalar or vector theories for the EBS shape, one can evaluate: (i) the transport mean free path l^*, (ii) the average density of the scattering medium, (iii) the average size of the individual scattering centers, by making use, respectively, of: (i) the angular extension of the enhancement cone, (ii) the magnitude of the enhancement, (iii) the polarization-dependent asymmetries of the enhancement cone. This information allows comparison between particle size and density of the medium and the attractive or repulsive forces which contribute to the electrochemical signature provided by zeta potential measurements. These measurements are especially useful in probing dense, highly absorbing media which are opaque in the visible.

EXPERIMENTAL

Zeta Potential Analysis

Numerous CeO_2 and Al_2O_3 slurries were prepared for zeta potential (ZP) measurements using a Brookhaven ZetaPlus[5] , which measures the electrokinetic mobility of particles suspended in a fluid by use of electrophoretic light scattering. The zeta potential, calculated from the electrokinetic mobility using the Smoluchowski equation[6], was measured with the polishing agents suspended in deionized water, except where noted. Aqueous slurries of CeO_2[7] and Al_2O_3[8] were prepared by mixing the powder in deionized water such that the resulting density were 1.14 g/cc^3. The CeO_2 slurries were diluted 1:10 in deionized water while the Al_2O_3 slurries were diluted 1:100 in deionized water, and both suspensions were aged overnight prior to ZP measurement. Commercial base-stabilized Norton Al_2O_3 slurries[9] and acid-stabilized Norton Al_2O_3 slurries[10] were received as suspensions, diluted in deionized water, and aged overnight. The base-stabilized Al_2O_3 was diluted 1:100 while the acid-stabilized suspension was diluted 1:400. Dilution level was determined by the Zeta Plus instrument's ability to measure scattering events and is largely a function of particle size. After aging, small working volumes of the different CeO_2 and Al_2O_3 slurries were prepared with various values of pH. The pH levels were adjusted by addition of HCl or NaOH. The procedure used here differed slightly from that used by Cumbo[3] in his measurements of similar materials. In Cumbo's study, following the incorporation and mixing of HCl or NaOH, and allowing for the particles present to settle, the pH of the supernatant mixture was recorded. In the present study, following the addition of HCl or NaOH, the suspension was shaken vigorously and the pH was measured *prior to* sedimentation. This procedure allows one to determine the zeta potential of the solution with the same concentration of (suspended) solids, as prepared. Following an aging period (2-3 hours) after acid/base addition, the slurries were shaken vigorously and the ZP were measured; this point was defined as time = 0 in studies where changes in ZP were monitored with aging time. For aqueous CeO_2 slurries, base-, and acid-stabilized Al_2O_3 slurries, three "fresh" solutions with "identical" preparation conditions were made for repeatability determination of the instrument. Measured ZP values of the slurries were then plotted as a function of pH, and the corresponding IEP values were obtained by interpolation. For all suspensions, except where noted, the point of the second order polynomial best fit curve where the zeta potential equals zero was defined as the IEP. In repeatability experiments, the tabulated IEP was the mean of the three values. For the solutions of base- and acid- stabilized Al_2O_3 slurries, the ZP values of the slurries, and the corresponding IEP values were plotted as a function of time up to 13 days after solution preparation, to assess suspension stability. In this study, solutions were placed in small sealed vials and aged without agitation. Suspensions were shaken prior to pH and ZP measurements.

Enhanced backscattering.

Enhanced backscattering (EBS) measurements as a function of solid concentration were made on aqueous solutions of CeRite 4250 CeO_2 supplied by Transelco[11]. The average ceria particle size as indicated by the manufacturer is 3.5 μm (Table I, sample (e)). In the measurements, a HeNe laser passes through a beam splitter to the sample contained in a 10 mm thick microcell. For this experiment, the cerium oxide was weighed and mixed in water to give 5,

15, 25, and 40% by volume solutions. The goal of this experiment was to observe how changes in particle density of dense suspensions relate to the shape of the EBS cone and if the linearity observed in prior systems[12] could be seen.

In the experiment, the backscattered light from the sample is reflected by the beam splitter and then focused by a lens onto the plane of a CCD camera. The experimental setup used to capture enhancement cone data for this series is shown in Figure 1.

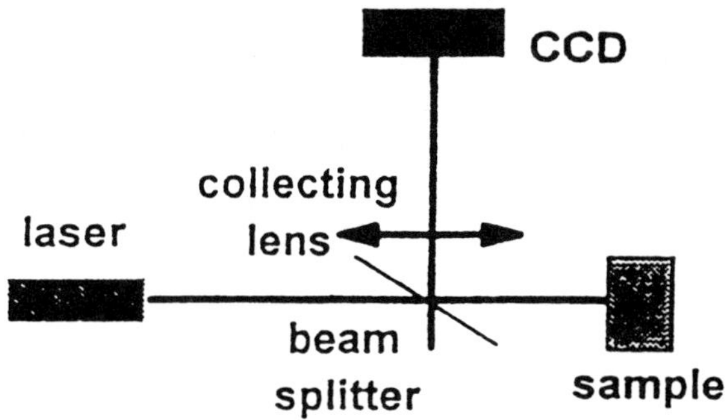

Figure 1: Enhanced backscattering (EBS) experimental set up.

RESULTS AND DISCUSSION

Zeta potential measurement

The physical properties and IEP results of several different CeO_2 slurries are listed in Table I, while the preparation methods for the CeO_2 slurries used in this study are summarized in Table II. The variation in CeO_2 source, purity, particle size, and slurry electrolyte type is reflected in the range of IEP values shown in Table I. It is apparent that the IEP measured for CeO_2 powder (a) was lower compared to the published value obtained by Cook[1] in (d). This might be due to the different type and size of CeO_2 that was used and the preparation method, as a specific source was not cited. The CeO_2 in (a) was used despite the considerably large particle size (for polishing purposes) because of the high (99.9%) purity. To observe the effects of dilution, the concentration of CERAC CeO_2 in (a) was prepared to be 12x more concentrated than that of (b), while keeping the material source and preparation method constant. This dilution caused the IEP of the diluted CeO_2 (b) to decrease compared to that of (a). For CERAC CeO_2 powder (b) and Struers CeO_2 suspension[13] (c), the concentrations were made to be "comparable" to observe the effects of additives in the "less pure" Struers CeO_2 suspension. It can be seen that the additives in the Struers CeO_2 suspension lowered the IEP values considerably. However, because of the proprietary nature of the Struers CeO_2 suspension, the type and amount of the additives are not known. It should also be noted that the IEP value of this suspension (c) was obtained by extrapolation of the second order polynomial best-fit curve because of the limitability of the electrolytic current-handling capability of the instrument. One can also see that although the CeRite CeO_2 powder (in e and f) are of different size, concentration, and purity compared to CERAC CeO_2 (b), and despite the effect of addition of electrolyte in (f), their IEP values are still comparable to the IEP of the dilute CERAC CeO_2 powder (b). The impurities and additives present in the CeRite 4250 CeO_2 (e and f) do not appear to modify their IEP. This may be due to the type and concentration of impurities, and their contribution to the electrostatic interactions which exists between particles.

Table I. Physical properties and IEP results of several different CeO_2 slurries. The order of various CeO_2 listed are based on material sources, concentration, and additives.

ID	source and form	particle size	% purity	IEP measured	IEP published
a.	CERAC powder	44 μm	99.9 %	4.2	
b.	CERAC powder	44 μm	99.9 %	2.8	
c.	Struers suspension	1.0 μm	N/A	0.3[*]	
d.	CeO_2	N/A	N/A		6.8[1]
e.	CeRite 4250	3.5 μm	45-50%	2.75	
f.	CeRite 4250	3.5 μm	45-50%	2.95	
g.	CeRite 480-G	1 μm	N/A		8.8[3]

*IEP value extrapolated from best fit curve, as pH range yields negative values of zeta potential.

Table II. Preparation methods of CeO_2 slurries used in this study.

ID	concentration	solvent	pH adjustments	comments
a.	0.244 M	dI-water	2.5 - 11.5	a. & b. effects of concentration
b.	0.01987 M	dI-water	2 - 11	
c.	~ 0.02 M	dI-water	2 - 11	b. & c. effects of additives
e.	0.00235 M	dI-water	2 - 11	
f.	0.00235 M	0.001 M $NaNO_3$	2 - 11	e. & f. effects of electrolytes

The physical properties and IEP results for several different Al_2O_3 slurries are listed in Table III, while the preparation methods and instruments used are listed in Table IV. It is apparent that depending on the slurry preparation method, instrument used, particle size, type, and additives, the IEP values of the slurries vary between a pH value of 6 to 9. In Figure 2, the zeta potential of fresh acid- and base-stabilized Norton Al_2O_3 suspensions were plotted as a function of pH. In these experiments, for both type of Al_2O_3 slurries, three different solutions prepared under "comparable" conditions were prepared to study the repeatability of the instrument. Figure 2 shows that for fresh solutions, the average IEP of the acid-stabilized Al_2O_3 suspension was 9.4 ± 0.3. For these suspensions, zeta potential measurements at pH levels between 2 and 11 were successful. Similar measurements on fresh base-stabilized suspension (Figure 2) showed an average IEP of 6.6 ± 0.2. The smaller error associated with the average IEP indicated the higher stability of this fluid after initial preparation. Attempts to measure the zeta potential for base-stabilized Al_2O_3 suspension under pH level of 4 and above pH level of 10 failed, due to the instability of these solutions outside the above pH range. For both acid- and base-stabilized suspensions, excellent repeatability was observed in the IEP measurements of the "comparable" solutions.

Figure 3 shows the zeta potential of base-stabilized Norton Al_2O_3 suspension following 13 days of aging (static), as a function of pH. It can be seen in this figure that despite the identical preparation and experimental conditions, since changes in surface charge causing agglomeration are known to occur within the sample volume to various extents, the three solutions showed a tendency to destabilized with time. At time = 13 days, the IEP of the suspension was found to be

Table III. Physical properties and IEP values of aqueous Al_2O_3 slurries.

ID	source and form	particle size	% purity	IEP measured	IEP published
h.	Praxair B powder	0.05 μm	99.98 %	9.1	
i.	Base-stabilized Norton suspension	0.05 μm	N/A	6.6	
j.	Acid-stabilized Norton suspension	0.05 μm	N/A	9.4	
k.	Norton α-Al_2O_3	0.05 μm	N/A		9.3[3]
l.	MAC Al_2O_3 powder	0.3 μm	N/A		6[14]
m.	CS-400 Al_2O_3 powder	0.3 μm	N/A		7.0[15]

Table IV. Preparation methods of Al_2O_3 slurries.

ID	% solids	density (g/cc)	solvent	dilution for exp.	pH adjustments	instrument used to measure ξ potential
h.		1.14	dI-H_2O	1:100	3 - 11	Brookhaven ZetaPlus
i.	23 %	1.22	dI-H_2O	1:100	5 - 10	Brookhaven ZetaPlus
j.	29 %	1.27	dI-H_2O	1:400	2 - 11	Brookhaven ZetaPlus
k.			dI-H_2O	1:10	4, 7, & 10	Brookhaven ZetaPlus
l.			0.01 M BTA			Coulter DELSA 440
m.			dI-H_2O			Micrometrics Mass Transport Analyzer

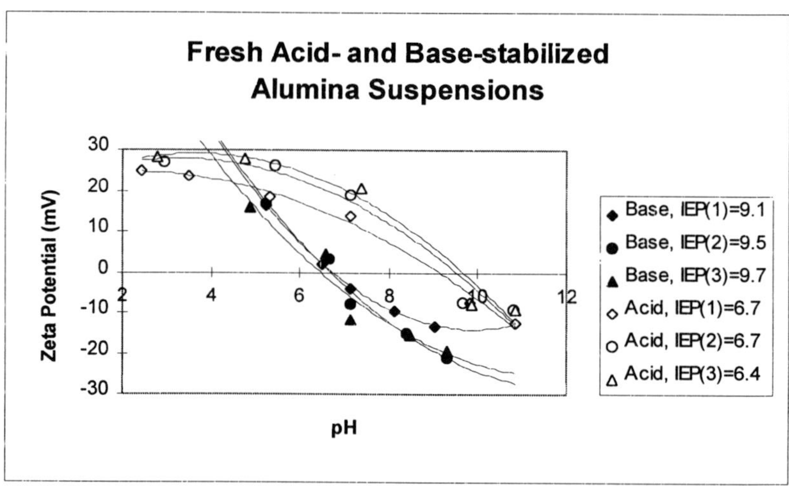

Figure 2: The zeta potential of fresh acid- and base-stabilized Norton Al_2O_3 suspension as a function of pH. Three different solutions were prepared under comparable conditions, and the pH levels were adjusted by addition of HCl or NaOH. The corresponding IEP values were obtained by interpolation using a second order polynomial best-curve fit.

6.7 ± 0.5, while the fresh suspensions' IEP was found to be 6.6 ± 0.2. Within the error of the measurement, the aged suspensions' IEP showed no change, but a slightly larger spread among the three samples occurred. This spread may most likely be due to differences in particle-fluid chemistry within the suspension volume as the fluids are static throughout the experiment. The measured pH of the suspension also changed when measured after 4 days for both types of Al_2O_3 solutions. For the base-stabilized suspension at 4 days, the measured pH of the individual suspensions previously prepared in the low pH ranges (between pH of 5 to 6) increased, and those of the previously high pH (between 8 to 9) decreased, equilibrating at pH values between 6 and 7.5 (Note: for fresh suspensions, the pH values were adjusted between 5 and 10). For the acid-stabilized suspension at 4 days, the measured pH of the individual suspensions previously prepared in the high pH ranges (between 7 and 11) decreased, reaching values between 2.5 and 9.5, and continued to change to values between 2.5 and 8 at 8 days (Note: for fresh solutions, the pH range were between 2 and 11). These changes in pH most probably started within the 4 days period, and are most likely due to the interparticle interactions in the solutions and the amphoteric nature (oxide ability to react with acids and bases) of the Al_2O_3 particles[16].

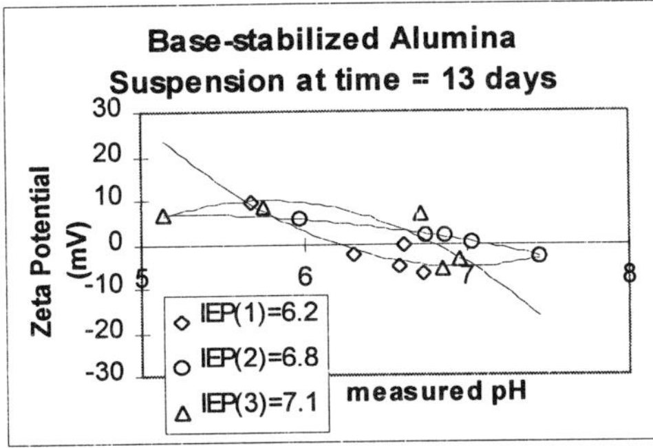

<table>
<tr><td>

Figure 3: The zeta potential of base-stabilized Norton Al_2O_3 suspension aged 13 days as a function of pH.

</td><td>

Figure 4: The IEP of static acid- and base-stabilized Norton Al_2O_3 suspensions as a function of time.

</td></tr>
</table>

Figure 4 shows the IEP of base- and acid-stabilized Norton Al_2O_3 suspensions as a function of time. For the base-stabilized suspension, it was apparent that the IEP values were stable over the entire 13 days of the experiment. However, for the acid-stabilized suspension, the IEP values decreased significantly in the first 4 days of aging, and continued to decrease, approaching the IEP values of the base-stabilized suspension. This behavior suggests the higher stability of the base-stabilized Norton Al_2O_3 suspension compared to the acid-stabilized suspension. Experiments to examine those changes in non-static (stirred) suspensions are on-going.

Enhanced backscattering.

The transport mean free path (l^*) and the absorption coefficient were obtained for the solid particle-containing samples from the values of angular-dependent intensity with the EBS theoretical model developed in the diffusion approximation as described earlier [equation (2)]. These quantities are related to volume fraction, particle size and type[12].

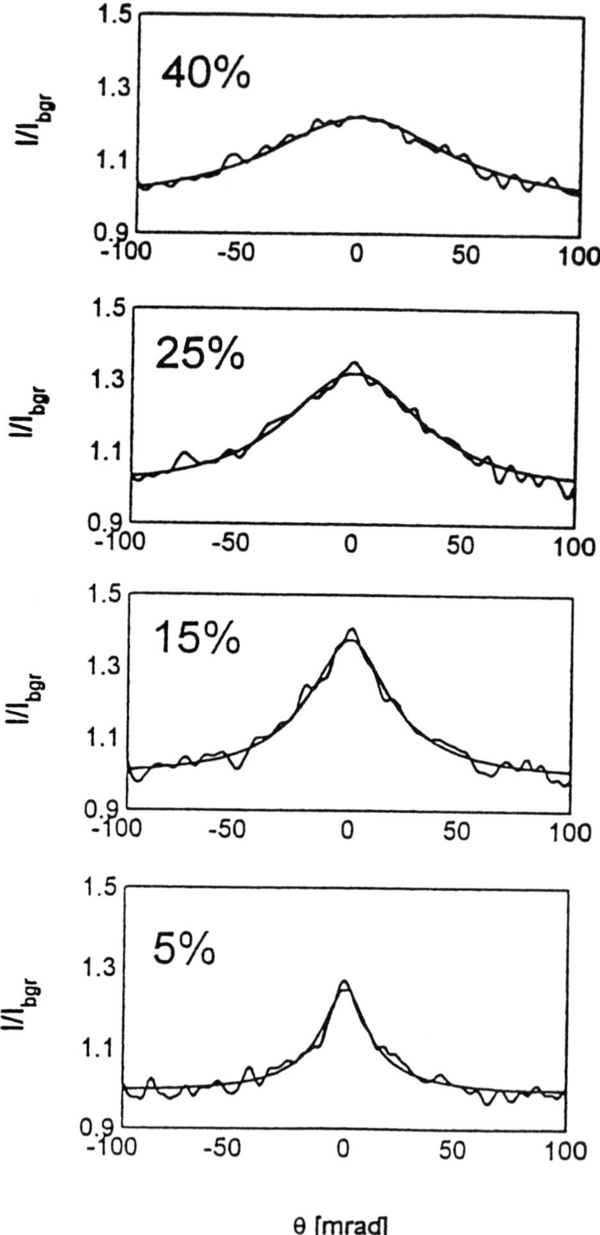

Figure 5: Backscattered intensity as a function of the angular width for CeRite 4250 CeO_2 solutions with increasing concentrations of 5%, 15%, 25% and 40% in volume.

The angular-dependent intensity for the CeRite 4250 CeO_2 solutions of varying concentration is presented in Figure 5. As can be seen, when the dilution level is increased, the average distance between the scattering centers (l^*) increases and, consequently, the EBS cone narrows. Using the above results, the values of the transport mean free path are plotted as a function of concentration using the logarithmic scale, thus obtaining a calibration curve that enables us to determine the average distance between the scattering centers for concentrations ranging from 2 to 50% in volume. One can see from Figure 6 that there is a proportionality between the two quantities, as the concentration increases the average distance between the scattering centers increases, leading to contact between particles, and subsequently aggregation occurs. That distance is comparable to the dimension of the particles for high concentrations, namely around 40% in volume. Knowing the density of particles in such suspension, the same measurement has been used to determine particle size, thus providing a tool to monitor agglomeration in slurry suspensions of other solids[12]. These measurements have proven valuable in the structural analysis of clustering and agglomeration which typically accompanies high solid content suspensions, and allow interpretation of fluid behavior in low solid content measurements (i.e. zeta potential) to be compared to high density results to predict particle interaction processes.

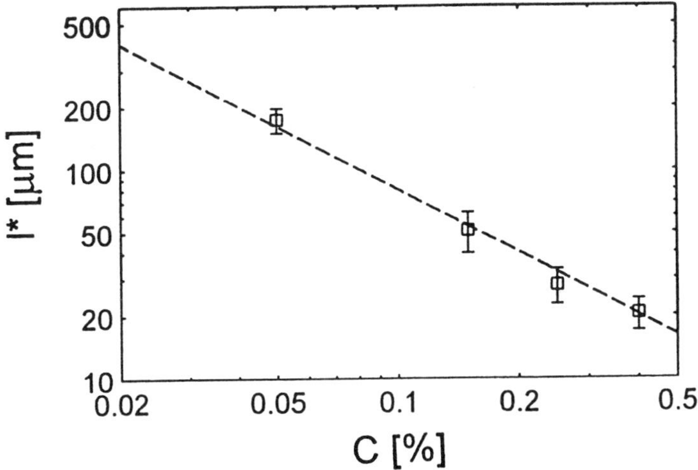

Figure 6: The transport mean free path as a function of volume fraction. Data were obtained by fitting the experimentally recorded EBS with Eq. 2 (see text).

CONCLUSION

Results of our study to evaluate tools for quantifying abrasive particle behavior in polishing slurries yielded several results. Variation of the zeta potential measurements performed on the CeO_2 suspensions in Table I was large, and most IEP values obtained on various CeO_2 suspensions were low compared to the previously published values[1,3], and limited agreement in behavior of "identical" samples was observed. This is believed to be due to variation in material source and particle characteristics, which limit the pH range over which the solutions are stable and yield measurable and/or repeatable zeta potential values. Secondly, the IEP values of different Al_2O_3 slurries were determined from ZP measurements, and showed good agreement with previously published results on similar material types. For base- and acid-stabilized Norton Al_2O_3 suspensions, the repeatability of freshly prepared suspensions were good, however, some degradation was observed with aging. The IEP values of static base-stabilized Norton Al_2O_3 suspensions were stable when aged over the period of 13 days, whereas acid-stabilized suspensions showed a drop in IEP values at 4 days, and continued to decrease, approaching the IEP values of the base-stabilized suspension at longer times.

The above results were complemented by EBS measurements for the determination of volume fraction in dense slurries (2-50% volume solid content). Systematic EBS measurements on ceria suspensions showed a linear relationship between $l*$ and concentration, allowing us to probe particle size changes with aging in dense slurries.

REFERENCES

[1] L. M. Cook,"Chemical Processes in Glass Polishing", *Journal of Non-Crystalline Solids*, **120**, 152-171 (1990).

[2] P. C. Hiemenz, "Principles of Colloid and Surface Chemistry", vol. 4, Marcel Dekker, Inc., New York (1977).

[3] M. J. Cumbo, D. Fairhurst, S. D. Jacobs, and B. E. Puchebner, "Slurry Particle Size Evolution during the Polishing of Optical Glass", *Applied Optics*, **34**, [19] 3743 (1995).

[4] W. P. Hsu, L. Ronnquist, E. Matijivec, "Preparation and Properties of Monodispersed Colloidal Particles of Lanthanide Compounds. 2. Cerium (VI)", *Langmuir*, **4**, 31-37 (1988).

[5] Brookhaven ZetaPlus, Brookhaven Instruments Corporation, Holtsville, N.Y. 11742.

[6] R. J. Hunter, "Zeta Potential in Colloid Science", pp. 59-178, Academic, London (1981).

[7] Cerium oxide, calcined, ~325 mesh, typically 99.9% pure, CERAC Specialty Inorganics, Milwaukee, WI 53233, Item #: C-1064, Lot #: X18099.

[8] Praxair B Al_2O_3 powder, 0.05 µm, 99.98% purity, Cat. #: AL0004, 15% alpha, 85% gamma, Praxair Service Technologies, Inc., Indianapolis, IN 46224.

[9] Nanometer Alumina polishing slurry, product code 9220, 23% solids, pH = 10.03, density = 1.22 g/cc, Norton Company Materials, Worcester, MA 01615-0008.

[10] Nanometer Alumina polishing slurry, pH = 3.88, 29% solids, density = 1.27 g/cc, Norton Company Materials, Worcester, MA 01615-0008.

[11] CeRite 4250 CeO_2 powder, 3.5 µm median diameter, 45-50% purity, Transelco Division of Ferro Corp., PennYan, NY 14527.

[12] A. Dogariu, J. Uozumi, and T. Asakura, "Particle Size Effects on Optical Transport Through Strongly Scattering Media", *Particle & Particle Systems Characterization*, **11**, 250-257 (1994).

[13] Cerium oxide suspension, 1.0 µm, Struers, Cat#: TSCROC, Westlake, OH 44145-1438.

[14] Q. Luo, D. R. Campbell, and S. V. Babu, "Stabilization of Alumina Slurry for Chemical-Mechanical Polishing of Copper.," *Langmuir*, **12** [15] 3563-3566 (1996).

[15] M. Belmonte, R. Moreno, J. S. Moya, and P. Miranzo, "Obtention of Highly Dispersed Platelet-Reinforced Al_2O_3 Composites", *Journal of Materials Science*, **29** [1] 179-183 (1994).

[16] D. F. Shriver, P. Atkins, and C. H. Langford, "Inorganic Chemistry", 2nd ed., W. H. Freeman and Co., New York (1994).

ACKNOWLEDGEMENT

This work was performed with the support of the Center for Optics Manufacturing (COM) at the University of Rochester and the University of Central Florida. We acknowledge the loan of the Brookhaven ZetaPlus instrument from our collegues at COM.